Coplanar Waveguide Circuits, Components, and Systems

Coplanar Waveguide Circuits, Components, and Systems

RAINEE N. SIMONS
NASA Glenn Research Center
Cleveland, Ohio

WILEY-INTERSCIENCE
A JOHN WILEY & SONS, INC., PUBLICATION
NEW YORK · CHICHESTER · WEINHEIM · BRISBANE · SINGAPORE · TORONTO

This book is printed on acid-free paper. ♾

Published simultaneously in Canada.

For ordering and customer service, call 1-800-CALL-WILEY.

Library of Congress Cataloging-in-Publication Data:

Simons, Rainee, 1949–

Coplanar waveguide circuits, components, and systems / Rainee N. Simons.

p. cm. -- (Wiley series in microwave and optical engineering)

ISBN 0-471-16121-7 (cloth : alk. paper)

1. Microwave transmission lines. 2. Wave guides. 3. Antennas (Electronics) I. Title. II. Series.

TK7876.S572 2001
621.381'331--dc21 00-043812

10 9 8 7 6 5 4 3 2 1

To
Joy,
Renita, and
Rona

Contents

Preface

This book is intended to provide a comprehensive coverage of the analysis and applications of coplanar waveguides to microwave circuits and antennas for graduate students in electrical engineering and for practicing engineers.

Coplanar waveguides are a type of planar transmission line used in microwave integrated circuits (MICs) as well as in monolithic microwave integrated circuits (MMICs). The unique feature of this transmission line is that it is uniplanar in construction, which implies that all of the conductors are on the same side of the substrate. This attribute simplifies manufacturing and allows fast and inexpensive characterization using on-wafer techniques.

The first few chapters of the book are devoted to the determination of the propagation parameters of conventional coplanar waveguides and their variants. The remaining chapters are devoted to discontinuities and circuit elements, transitions to other transmission media, directional couplers, hybrids and magic-T, microelectromechanical systems (MEMS) based switches and phase shifters, high-T_c superconducting circuits, tunable devices using ferroelectric materials, photonic bandgap structures, and printed circuit antennas. The author includes several valuable details such as the derivation of the fundamental equations, physical explanations, and numerical examples.

The book is an outgrowth of 15 years of research conducted by the author as a member of the Communications Technology Division (CTD) at the National Aeronautics and Space Administration (NASA), Glenn Research Center (GRC) in Cleveland, Ohio. Over the past few years, interest among engineers in coplanar waveguides has increased tremendously, with some of the concepts being extensively pursued by NASA for future space programs and missions. Numerous articles exist, but there is no collective publication. Thus the decision to publish a book on coplanar waveguides appears to be appropriate.

In the course of writing this book, several persons have assisted the author and offered support. The author first expresses his appreciation to the management of CTD at GRC for providing the environment in which he worked on the book; without their support this book could not have materialized. In particular, he is grateful to Wallace D. Williams, Regis F. Leonard and Charles A. Raquet. The author is further grateful to the engineers and scientists in CTD who shared their time, knowledge, and understanding of this subject. In particular, he would like to thank Samuel A. Alterovitz, Alan N. Downey, Fred Van Keuls, Felix A. Miranda, George E. Ponchak, Maximillian Scardelletti, Joseph D. Warner, Richard R. Kunath, Richard Q. Lee, Hung D. Nguyen, Robert R. Romanofsky, Kurt A. Shalkhauser, and Afroz J. Zaman. In addition the author is grateful to the staff of the clean room and the hybrid/printed circuit fabrication facilities. In particular, he is thankful to William M. Furfaro, Elizabeth A. Mcquaid, Nicholas C. Varaljay, Bruce J. Viergutz and George W. Readus.

The author is grateful to the staff of Publishing Services at GRC for their efficiency in the preparation of the text and illustrations. In particular, he is grateful to Caroline A. Rist, Catherine Gordish, Irene Gorze, and Patricia A. Webb of the co-ordination section, Denise A. Easter and Theresa Young of the manuscript section, and Richard J. Czentorycki, Mary M. Eitel, John L. Jindra, and Nancy C. Mieczkowski of the graphical illustration section. The author is also grateful to the Library at GRC for the help in the literature search.

The author gratefully acknowledges the support and the interactions he has had with Prof. L. P. B. Katehi, Prof. G. M. Rebeiz, Dr. J. R. East, and their students at the University of Michigan, Ann Arbor, for over a decade.

The author thanks Prof. Kai Chang of Texas A&M University, College Station, who suggested and encouraged the writing of this book, and the editorial staff of John Wiley & Sons for the processing of the manuscript.

Finally, the author thanks his wife, Joy, and daughters, Renita and Rona, for their patience during the writing of this book.

RAINEE N. SIMONS
NASA GRC
Cleveland, Ohio

CHAPTER 1

Introduction

A coplanar waveguide (CPW) fabricated on a dielectric substrate was first demonstrated by C. P. Wen [1] in 1969. Since that time, tremendous progress has been made in CPW based microwave integrated circuits (MICs) as well as monolithic microwave integrated circuits (MMICs) [2] to [5].

1.1 ADVANTAGES OF COPLANAR WAVEGUIDE CIRCUITS

1.1.1 Design

A conventional CPW on a dielectric substrate consists of a center strip conductor with semi-infinite ground planes on either side as shown in Figure 1.1. This structure supports a quasi-TEM mode of propagation. The CPW offers several advantages over conventional microstrip line: First, it simplifies fabrication: second, it facilitates easy shunt as well as series surface mounting of active and passive devices [6] to [10]; third, it eliminates the need for wraparound and via holes [6] and [11], and fourth, it reduces radiation loss [6]. Furthermore the characteristic impedance is determined by the ratio of a/b, so size reduction is possible without limit, the only penalty being higher losses [12]. In addition a ground plane exists between any two adjacent lines, hence cross talk effects between adjacent lines are very week [6]. As a result, CPW circuits can be made denser than conventional microstrip circuits. These, as well as several other advantages, make CPW ideally suited for MIC as well as MMIC applications.

1.1.2 Manufacturing

Major advantages gained in manufacturing are, first, CPW lends itself to the use of automatic pick-and-place and bond assembly equipments for surface-mount component placement and interconnection of components, respectively

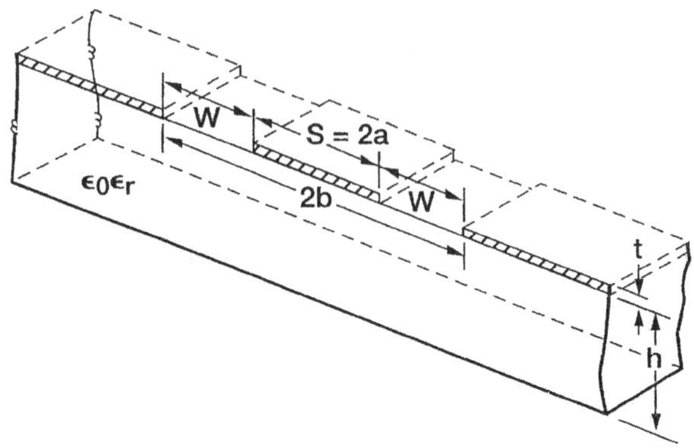

FIGURE 1.1 Schematic of a coplanar waveguide (CPW) on a dielectric substrate of finite thickness.

[6]. Second, CPW allows the use of computer controlled on-wafer measurement techniques for device and circuit characterization up to several tens of GHz [13], [14]. These advantages make CPW based MICs and MMICs cost effective in large volume.

1.1.3 Performance

The quasi-TEM mode of propagation on a CPW has low dispersion and hence offers the potential to construct wide band circuits and components. In CPW amplifier circuits, by eliminating via holes and its associated parasitic source inductance, the gain can be enhanced [15].

1.2 TYPES OF COPLANAR WAVEGUIDES

Coplanar waveguides can be broadly classified as follows:

- Conventional CPW
- Conductor backed CPW
- Micromachined CPW

In a conventional CPW, the ground planes are of semi-infinite extent on either side. However, in a practical circuit the ground planes are made of finite extent. The conductor-backed CPW has an additional ground plane at the bottom surface of the substrate. This lower ground plane not only provides mechanical support to the substrate but also acts as a heat sink for circuits with active devices. A conductor backed CPW is shown in Figure 1.2. The micromachined CPWs are of two types, namely, the microshield line [16] and the CPW suspended by a silicon dioxide membrane above a micromachined groove [17].

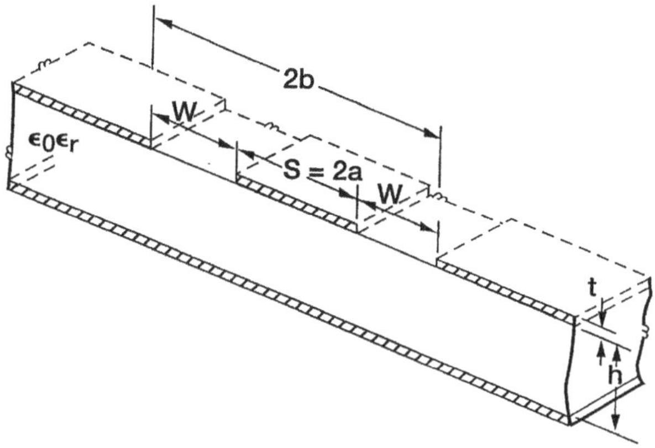

FIGURE 1.2 Schematic of a conductor-backed coplanar waveguide (CBCPW).

These lines are illustrated in Figures 1.3 and 1.4, respectively. The advantages of the microshield line are its extremely wide bandwidth, minimal dispersion and zero dielectric loss. The advantage of the later CPW is that it is compatible with commercial CMOS foundry process and hence, is capable of monolithically integrating CMOS devices and circuits.

1.3 SOFTWARE TOOLS FOR COPLANAR WAVEGUIDE CIRCUIT SIMULATION

Recently accurate models for CPW discontinuities, such as open circuits and short circuits, lumped elements, such as inductors and capacitors, and three- and four-port junctions, such as, tee- and crossjunctions, have become com-

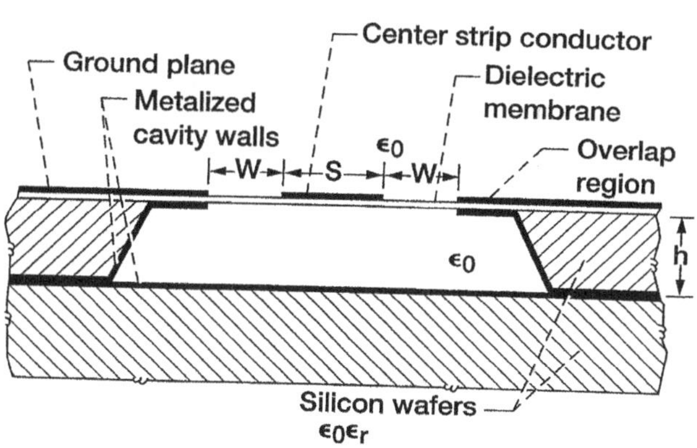

FIGURE 1.3 Cross section of a microshield line. (From Reference [16], © IEEE 1995.)

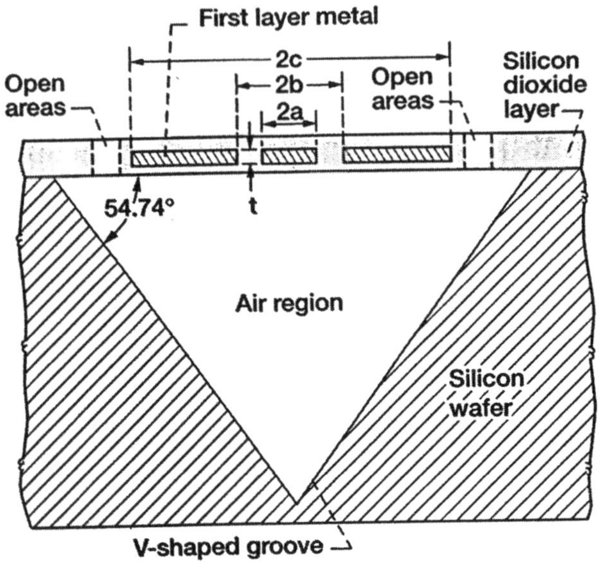

FIGURE 1.4 Cross section of a coplanar waveguide suspended by a silicon dioxide membrane over a micromachined substrate. (From Reference [17], © IEEE 1997.)

mercially available [5], [18] to [21]. In addition electromagnetic simulation software for 2-D and 3-D structures have also become commercially available [21] to [25].

1.4 TYPICAL APPLICATIONS OF COPLANAR WAVEGUIDES

1.4.1 Amplifiers, Active Combiners, Frequency Doublers, Mixers, and Switches

The CPW amplifier circuits include millimeter-wave amplifiers [26], [27], distributed amplifiers [28], [29], cryogenically cooled amplifiers [30], cascode amplifiers [31], transimpedance amplifiers [32], dual gate HEMT amplifiers [33], and low-noise amplifiers [34]. The CPW active combiners and frequency doublers are described in [35] and [36], respectively. The CPW mixer circuits include ultra-small drop in mixers [37], beam lead diode double-balanced mixers [38], harmonic mixers [39], MMIC double-balanced mixers [40], [41] and double-balanced image rejection, MESFET mixers [42]. The CPW PIN diode SPDT switches are described in [43] and [44].

1.4.2 Microelectromechanical Systems (MEMS) Metal Membrane Capacitive Switches

The rapid progress made in the area of semiconductor wafer processing has led to the successful development of MEMS based microwave circuits. In a CPW the conductors are located on the top surface of a substrate which makes it ideally suited for fabricating metal membrane, capacitive, shunt-type switches [45]. CPW MEMS shunt switches with good insertion loss characteristics, reasonable switching voltages, fast switching speed, and excellent linearity have recently been demonstrated [45]. These switches offer, the potential to built new generation of low-loss high-linearity microwave circuits for phased array antennas and communication systems.

1.4.3 Thin Film High-Temperature Superconducting /Ferroelectric Tunable Circuits and Components

Recent advances made in the area of thin film deposition techniques, such as sputtering, laser ablation and chemical vapor deposition, and etching technologies, have resulted in the application of high temperature superconducting (HTS) materials to microwave circuits [46]. The HTS circuits have low microwave surface resistance over a wide range of frequencies. As a result signal propagation takes place along these transmission lines with negligible amount of attenuation. Furthermore the advantage of using CPW is that only one surface of the substrate needs to be coated with HTS material before patterning. Recently HTS low-pass and band-stop CPW filters have been demonstrated in [47] and [48], respectively.

In addition by incorporating ferroelectric materials such as, $SrTiO_3$ with HTS materials such as, $YBa_2Cu_3O_{7-x}$, low-loss, voltage-tunable MMICs with reduced length scales can be constructed [49] and [50]. These MMICs have potential applications in phased array antenna systems and frequency agile communications systems. Recently voltage tunable CPW $YBa_2Cu_3O_{7-x}$/ $SrTiO_3$ phaseshifters, mixers and filters have been demonstrated [50].

1.4.4 Photonic Bandgap Structures

When an electromagnetic wave propagates along a conductor backed CPW considerable amount of energy leakage takes place. The energy that leakes, propagates along the transverse directions away from the line, and excites a parallel plate mode between the CPW top and bottom ground planes. The parasitic parallel plate mode is the leading cause for crosstalk between adjacent circuits. The cross talk can be suppressed by constructing a photonic bandgap lattice on the CPW top ground planes as demonstrated in [51].

1.4.5 Printed Antennas

A radiating element is constructed from a conventional CPW by widening the center strip conductor to form a rectangular or square patch [52]. This patch produces a single-lobe, linearly polarized pattern directed normal to the plane of the conductors. The advantage gained over conventional microstrip patch antenna is lower crosspolarized radiation from the feed [52]. In [53] a conductor backed CPW with a series gap in the center strip conductor is used to couple power to a patch through an aperture in the common ground plane. This design offers the flexibility of inserting semiconductor devices in the series gap of the feed for controlling the coupling.

1.5 ORGANIZATION OF THIS BOOK

This book is organized to serve as a text for a graduate course in MICs and MMICs, as well as a reference volume for scientists and engineers in industry. Chapter 1 gives an overview of the advantages, types, and typical applications of CPW.

Chapters 2 through 5 are devoted to the basic structures such as conventional CPW, conductor backed CPW, CPW with finite-width ground planes, elevated CPW, and CPW suspended inside a conducting enclosure. Analytical expressions to compute, the effective dielectric constant and characteristic impedance of the lines are provided.

Chapter 6 discusses coplanar stripline (CPS) and its variants. Analytical expressions to compute, the effective dielectric constant and the characteristic impedance are provided.

Coupled CPWs have several applications in the design of microwave components such as, directional couplers and filters. In Chapter 7 the even-mode and odd-mode characteristics of both edge coupled as well as broadside coupled CPWs are presented.

When an electromagnetic wave propagates along a CPW it suffers attenuation due to conductor and dielectric losses. In Chapter 8 the attenuation characteristics of conventional, micromachined, and superconducting CPWs are discussed.

Discontinuities such as, open circuits and circuit elements, such as air-bridges, are an integral part of practical CPW circuits. A good understanding of their characteristics is essential for design success. Hence Chapter 9 is devoted to CPW discontinuities.

Transitions between CPW and other transmission media are essential for integrating various components and subsystems into a complete system. Chapter 10 presents transitions between CPW and the following transmission lines: microstrip, slotline, coplanar stripline, balanced stripline, and rectangular waveguide.

Coupling of power from one line to another takes place when the lines are placed in close proximity to each other. In Chapter 11 the design and construction of directional couplers are presented. These couplers can be realized using either edge coupled CPW or broadside coupled CPW. In addition the construction and design of hybrid couplers and magic-Ts are also discussed.

Finally, Chapter 12 presents several emerging applications of CPW. These applications include microelectromechanical systems (MEMS) based switches and phase shifters, high-temperature superconducting circuits, tunable components based on ferroelectric materials, photonic bandgap structures and printed circuit antennas.

REFERENCES

[1] C. P. Wen, "Coplanar Waveguide: A Surface Strip Transmission Line Suitable for Nonreciprocal Gyromagnetic Device Applications," *IEEE Trans. Microwave Theory Tech.*, Vol. 17, No. 12, pp. 1087–1090, Dec. 1969.

[2] J. L. B. Walker, "A Survey of European Activity on Coplanar Waveguide," *1993 IEEE MTT-S Int. Microwave Symp. Dig.*, Vol. 2, pp. 693–696, Atlanta, Georgia, June 14–18, 1993.

[3] A. K. Sharma and T. Itoh (Editors), Special Issue on Modeling and Design of Coplanar Monolithic Microwave and Millimeter-Wave Integrated Circuits, *IEEE Trans. Microwave Theory Tech.*, Vol. 41, No. 9, Sept. 1993.

[4] T. Sporkmann, "The Evolution of Coplanar MMICs over the past 30 Years," *Microwave Journal*, Vol. 41, No. 7, pp. 96–111, July 1998.

[5]. T. Sporkmann, "The Current State of the Art in Coplanar MMICs," *Microwave J.*, Vol. 41, No. 8, pp. 60–74, Aug. 1998.

[6] J. Browne, "Broadband Amps Sport Coplanar Waveguide," *Microwaves RF*, Vol. 26, No. 2, pp. 131–134, Feb. 1987.

[7] Technology Close-Up, *Microwaves RF*, Vol. 27, No. 4, p. 79, April 1988.

[8] J. Browne, "Coplanar Waveguide Supports Integrated Multiplier Systems," *Microwaves RF*, Vol. 28, No. 3, pp.137–138, March 1989.

[9] J. Browne, "Coplanar Circuits Arm Limiting Amp with 100-dB Gain," *Microwaves RF*, Vol. 29, No. 4, pp. 213–220, April 1990.

[10] J. Browne, "Broadband Amp Drops through Noise Floor," *Microwaves RF*, Vol. 31, No. 2, pp. 141–144, Feb. 1992.

[11] J. Browne, "Coplanar MIC Amplifier Bridges 0.5 To 18.0 GHz," *Microwaves RF*, Vol. 26, No. 6, pp. 194–195, June 1987.

[12] R. E. Stegens and D. N. Alliss, "Coplanar Microwave Integrated Circuit for Integrated Subsystems," *Microwave Sys. News Comm. Tech.*, Vol. 17, No. 11, pp. 84–96, Oct. 1987.

[13] E. M. Godshalk and J. Pence, "Low-Cost Wafer Probe Scales 110-GHz Summit," *Microwaves RF*, Vol. 32, No. 3, pp. 162–167, March 1993.

[14] S. M. J. Liu and G. G. Boll, "A New Probe for W-band On-wafer Measurements," *1993 IEEE MTT-S Int. Microwave Symp., Dig.*, Vol. 3, pp. 1335–1338, Atlanta, Georgia, June 14–18, 1993.

[15] R. Majidi-Ahy, M. Riaziat, C. Nishimoto, M. Glenn, S. Silverman, S. Weng, Y. C. Pao, G. Zdasiuk, S. Bandy, and Z. Tan, "5–100 GHz InP CPW MMIC 7-Section Distributed Amplifier," *1990 IEEE Microwave Millimeter-Wave Monolithic Circuits Symp. Dig.*, pp. 31–34, Dallas, Texas, May 7–8, 1990.

[16] T. M. Weller, L. P. B. Katehi, and G. M. Rebeiz, "High Performance Microshield Line Components," *IEEE Trans. Microwave Theory Tech.*, Vol. 43, No. 3, pp. 534–543, March 1995.

[17] V. Milanovic, M. Gaitan, E. D. Bowen, and M. E. Zaghloul, Micromachined Microwave Transmission Lines in CMOS Technology," *IEEE Trans. Microwave Theory Tech.*, Vol. 45, No. 5, pp. 630–635, May 1997.

[18] R. Kulke, T. Sporkmann, D. Kother, I. Wolff, and P. Pogatzki, "Coplanar Elements Support Circuit Designs to 67 GHz, Part 1," *Microwaves RF*, Vol. 33, No. 13, pp. 103–116, Dec. 1994.

[19] R. Kulke, T. Sporkmann, D. Kother, I. Wolff, and P. Pogatzki, "Modeling and Analysis Aid Coplanar Designs, Part 2," *Microwaves RF*, Vol. 34, No. 1, pp. 89–916, Jan. 1995.

[20] R. Kulke, T. Sporkmann, D. Kother, I. Wolff, and P. Pogatzki, "Examine the Applications of Coplanar Circuits, Part 3," *Microwaves RF*, Vol. 34, No. 2, pp. 112–117, Feb. 1995.

[21] Agilent Technologies, Santa Clara, California.

[22] J. C. Rautio, "Free EM Software Analyzes Spiral Inductor on Silicon," *Microwaves RF*, Vol. 38, No. 9, pp. 165–172, Sept. 1999.

[23] Zeland Software, Inc., Fremont, California.

[24] Ansoft Corporation, Pittsburg, Pennsylvania.

[25] Jansen Microwave GmbH, Aachen, Germany.

[26] G. S. Dow, T. N. Ton, and K. Nakano, "Q-Band Coplanar Waveguide Amplifier," *1989 IEEE MTT-S Int. Microwave Symp. Dig.* Vol. 2, pp. 809–812, Long Beach, California, June 13–15, 1989.

[27] K. M. Strohm, J.-F. Luy, F. Schaffler, H. Jorke, H. Kibbel, C. Rheinfelder, R. Doerner, J. Gerdes, F. J. Schmuckle, and W. Heinrich, "Coplanar Ka-Band SiGe-MMIC Amplifier," *Electron. Lett.*, Vol. 31, No. 16, pp. 1353–1354, Aug. 1995.

[28] M. Riaziat, S. Bandy, and G. Zdasiuk, "Coplanar Waveguides for MMICs," *Microwave J.*, Vol. 30, No. 6, pp. 125–131, June 1987.

[29] R. Majidi-Ahy, M. Riaziat, C. Nishimoto, M. Glenn, S. Silverman, S. Weng, Y. C. Pao, G. Zdasiuk, S. Bandy, and Z. Tan, "94 GHz InP MMIC Five-Section Distributed Amplifier," *Electron. Lett.*, Vol. 26, No. 2, pp. 91–92, Jan. 1990.

[30] A. Cappello and J. Pierro, "A 22-24-GHz Cryogenically Cooled GaAs FET Amplifier," *IEEE Trans. Microwave Theory Tech.*, Vol. 32, No. 3, pp. 226–230, March 1984.

[31] R. Majidi-Ahy, C. Nishimoto, M. Riaziat, M. Glenn, S. Silverman, S.-L. Weng, Y.-C. Pao, G. Zdasiuk, S. Bandy, and Z. Tan, "100-GHz High-Gain InP MMIC Cascode Amplifier," *IEEE J. Solid-State Circuits*, Vol. 26, No. 10, pp. 1370–1378, Oct. 1991.

[32] K. W. Kobayashi, L. T. Tran, M. D. Lammert, A. K. Oki, and D. C. Streit, "Transimpedance Bandwidth Performance of an HBT Loss-Compensated Coplanar Waveguide Distributed Amplifier," *Electron. Lett.*, Vol. 32, No. 24, pp. 2287–2288, Nov. 1996.

[33] M. Schefer, H.-P. Meier, B.-U. Klepser, W. Patrick, and W. Bachtold, "Integrated Coplanar MM-Wave Amplifier With Gain Control Using a Dual-Gate InP

HEMT," *IEEE Trans. Microwave Theory Tech.*, Vol. 44, No. 12, pp. 2379–2383, Dec. 1996.

[34] D. Leistner, "Low Noise Amplifier at L- and Ku-Band for Space Applications in Coplanar Technology," *23rd European Microwave Conf. Proc.*, pp. 823–827, Madrid, Spain, Sept. 6–9, 1993.

[35] R. Majidi-Ahy, C. Nishimoto, J. Russell, W. Ou, S. Bandy, G. Zdasiuk, C. Shih, Y. C. Pao, and C. Yuen, "4-40 GHz MMIC Distributed Active Combiner with 3 dB Gain," *Electron. Lett.*, Vol. 28, No. 8, pp. 739–741, April 1992.

[36] M. Riaziat, E. Par, G. Zdasiuk, S. Bandy, and M. Glenn, "Monolithic Millimeter Wave CPW Circuits," *1989 IEEE MTT-S Int. Microwave Symp. Dig.*, Vol. 2, pp. 525–528, Long Beach, CA, June 13–15, 1989.

[37] D. Neuf and S. Spohrer, "Ultrasmall MIC Mixer Designed for ECM Applications," *Microwave Sys. News Comm. Tech.*, Vol. 15, No. 11, pp. 70–80. Oct. 1985.

[38] D. Cahana, "A New, Single Plane, Double-Balanced Mixer," *Appl. Microwave*, Vol. 1, No. 2, pp. 78–83, Aug./Sept. 1989.

[39] J. H. Lepoff, "Mix Harmonics With Dual-Beam-Lead Diodes," *Microwaves RF*, Vol. 23, No. 5, pp. 209–212, May 1984.

[40] J. Eisenberg, J. Panelli, and W. Ou, "A New Planar Double-Double Balanced MMIC Mixer Structure," 1991 *IEEE Microwave Millimeter-Wave Monolithic Circuits Symp. Dig.*, pp. 69–72, Boston, Massachusetts, June 10–11, 1991.

[41] J. A. Eisenberg, J. S. Panelli, and W. Ou, "Slotline and Coplanar Waveguide Team to Realize a Novel MMIC Double Balanced Mixer," *Microwave J.*, Vol. 35, No. 9, pp. 123–131, Sept. 1992.

[42] D. Neuf and S. Spohrer, "Double Balanced, Coplanar, Image Rejection Mixer Uses Monolithic MESFET Quad," *1991 IEEE MTT-S Int. Microwave Symp. Dig.*, Vol. 2, pp. 843–846, Boston, Massachusetts, June 10–14, 1991.

[43] R. W. Waugh and R. M. Waugh, "SPDT Switch Serves PCN Applications," *Microwaves RF*, Vol. 33, No. 1, pp. 111–118, Jan. 1994.

[44] G. E. Ponchak and R. N. Simons, "Channelized Coplanar Waveguide PIN-Diode Switches," *19th European Microwave Conf. Proc.*, pp. 489–494, London, England, Sept. 4–7, 1989.

[45] C. L. Goldsmith, Z. Yao, S. Eshelman and D. Denniston, "Performance of Low-Loss RF MEMS Capacitive Switches," *IEEE Microwave Guided Wave Lett.*, Vol. 8, No. 8, pp. 269–271, Aug. 1998.

[46] M. Nisenoff and W. J. Meyers (Editors), Special Issue on the Microwave and Millimeter Wave Applications of High Temperature Superconductivity, *IEEE Trans. Microwave Theory Tech.*, Vol. 44, No. 7, Part II, July 1996.

[47] W. Chew, L. J. Bajuk, T. W. Cooley, M. C. Foote, B. D. Hunt, D. L. Rascoe, and A. L. Riley, "High-T_c Superconductor Coplanar Waveguide Filter," *IEEE Electron. Device Lett.*, Vol. 12, No. 5, pp. 197–199, May 1991.

[48] S. Wallage, J. L. Tauritz, G. H. Tan, P. Hadley, and J. E. Mooij, "High T_c Superconducting CPW Bandstop Filters for Radio Astronomy Front Ends," *IEEE Trans. Appl. Superconductivity*, Vol. 7, No. 2, pp. 3489–3491, June 1997.

[49] D. C. DeGroot, J. A. Beall, R. B. Marks, and D. A. Rudman, "Microwave Properties of Voltage-Tunable $YBa_2Cu_3O_{7-x}/SrTiO_3$ Coplanar Waveguide

Transmission Lines," *IEEE Trans. Appl. Superconductivity*, Vol. 5, No. 2, pp. 2272–2275, June 1995.

[50] A. T. Findikoglu, Q. X. Jia, and D. W. Reagor, "Superconductor/Nonlinear-Dielectric Bilayers for Tunable and Adaptive Microwave Devices," *IEEE Trans. Appl. Superconductivity*, Vol. 7, No. 2, pp. 2925–2928, June 1997.

[51] F.-R. Yang, K.-P. Ma, Y. Qian, and T. Itoh, "A Uniplanar Compact Photonic-Bandgap (UC-PBG) Structure and Its Applications for Microwave Circuits," *IEEE Trans. Microwave Theory Tech.*, Vol. 47, No. 8, pp. 1509–1514, Aug. 1999.

[52] J. W. Greiser, "Coplanar Stripline Antenna," *Microwave J.*, Vol. 19, No. 10, pp. 47–49, Oct. 1976.

[53] R. Q. Lee and R. N. Simons, "Coplanar Waveguide Aperture-Coupled Microstrip Patch Antenna," *IEEE Microwave Guided Wave Lett.*, Vol. 2, No. 4, pp. 138–139, April 1992.

CHAPTER 2

Conventional Coplanar Waveguide

2.1 INTRODUCTION

The coplanar waveguide (CPW) proposed by C. P. Wen in 1969 consisted of a dielectric substrate with conductors on the top surface [1]. The conductors formed a center strip separated by a narrow gap from two ground planes on either side. The dimensions of the center strip, the gap, the thickness and permittivity of the dielectric substrate determined the effective dielectric constant (ε_{eff}), characteristic impedance (Z_0) and the attenuation (α) of the line. This basic structure has become known as the conventional CPW.

In Section 2.2 closed form expressions for ε_{eff} and Z_0 for CPW variants are presented. These expressions are derived using conformal mapping techniques. The conformal mapping technique assumes a quasi-static TEM mode of propagation along the line. Section 2.3 briefly explains iterative techniques to determine quasi-static ε_{eff} and Z_0. The iterative methods considered are the relaxation method and the hybrid method. Section 2.4 presents a detailed analysis of CPW using the spectral domain method. In this method the frequency dependence ignored in the conformal mapping technique is taken into consideration. This section is supported by Appendixes 2A and 2B which present the steps involved in deriving the dyadic Green's function and the time average power flow.

Sections 2.5 and 2.6 present an empirical formula for dispersion and synthesis formulas for dispersion and characteristic impedance respectively. Section 2.7 presents the characteristics of CPW with elevated or buried center strip conductor. Using these CPW structures very high Z_0 can be achieved. Section 2.8 presents the characteristics of CPW with ground plane or center strip conductor underpasses. Using these CPW structures very low Z_0 can be achieved. Section 2.9 presents the field components of conventional CPW.

Section 2.10 presents closed form expressions for ε_{eff} and Z_0 for CPW on cylindrical surfaces. Finally, Section 2.11 presents the effect of metal thickness on ε_{eff} and Z_0 of conventional CPW.

2.2 CONVENTIONAL COPLANAR WAVEGUIDE ON A MULTILAYER DIELECTRIC SUBSTRATE

2.2.1 Analytical Expressions Based on Quasi-static Conformal Mapping Techniques to Determine Effective Dielectric Constant and Characteristic Impedance

The cross-sectional view of a two coplanar waveguide (CPW) structures on multilayer dielectric substrates are shown in Figures 2.1(*a*) and (*b*). These two CPW structures are designated as sandwiched CPW and CPW on a double-layer substrate respectively. In these figures the CPW center strip conductor width S is equal to $2a$ and the distance of separation between the two semi-infinite ground planes in $2b$. Consequently the slot width W is equal to $b - a$. The two dielectric substrate thicknesses are designated as h_1, h_2 and as h_2, $h_1 - h_2$ in the case of sandwiched CPW and CPW on a double-layer substrate, respectively. The corresponding relative permittivities are designated as ε_{r1} and ε_{r2}, respectively. Two metal covers that act as a shield are placed at a distance of h_3 and h_4 from the CPW conductors. The thickness of the CPW conductors is t.

In the analysis that follows, the CPW conductors and the dielectric substrates are assumed to have perfect conductivity and relative permittivity, respectively. Hence the structure is considered to be loss less. Further the dielectric substrate materials are considered to be isotropic.

In this section expressions for determining ε_{eff} and Z_0 using conformal mapping techniques are presented. The assumptions made are that the conductor thickness t is zero and magnetic walls are present along all the dielectric boundaries including the CPW slots. The CPW is then divided into several partial regions and the electric field is assumed to exist only in that partial region. In this manner the capacitance of each partial region is determined separately. The total capacitance is then the sum of the partial capacitances [2]. Expressions for the partial capacitances of the sandwiched CPW will be derived first and later extended to the case of CPW on a double-layer dielectric.

The total capacitance C_{CPW} of the sandwiched CPW is the sum of the partial capacitances C_1, C_2, and C_{air} of the three partial regions shown in Figures 2.2(*a*) to (*c*). That is,

$$C_{CPW} = C_1 + C_2 + C_{air}. \tag{2.1}$$

In this equation C_1 and C_2 are the partial capacitance of the CPW with only the lower and the upper dielectric layers, respectively. Futher C_{air} is the partial capacitance of the CPW in the absence of all the dielectric layers.

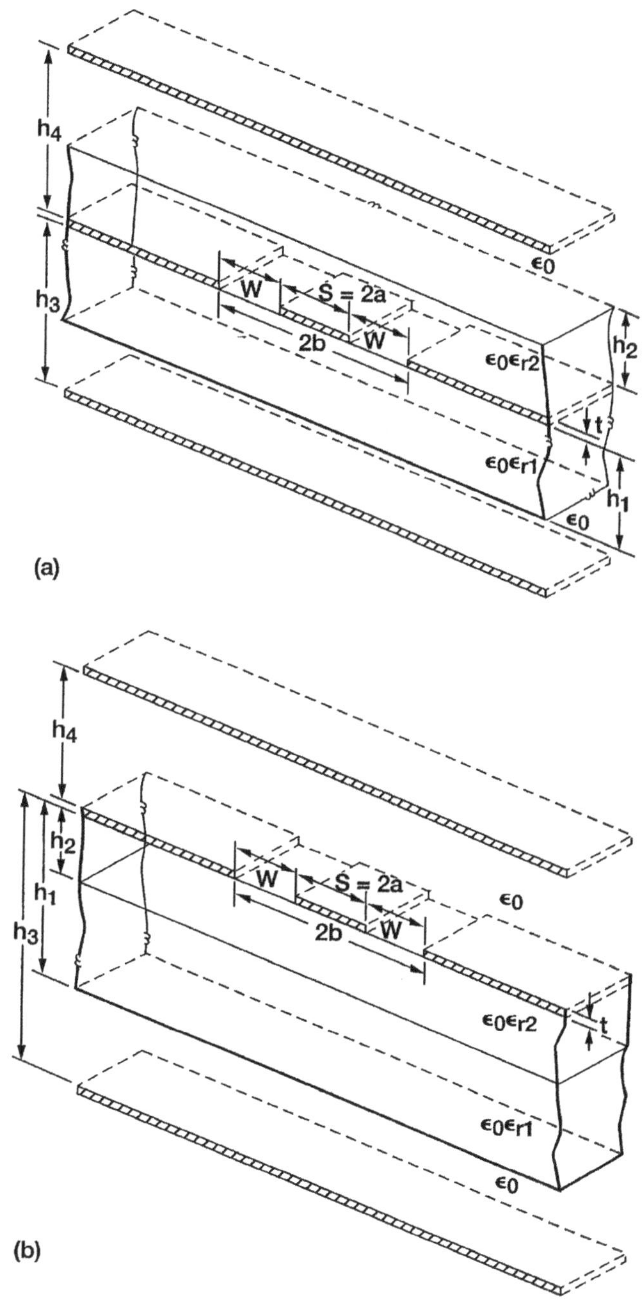

FIGURE 2.1 Schematic of a CPW with top and bottom metal cover: (*a*) Sandwiched between two dielectric substrates; (*b*) on a double-layer dielectric substrate.

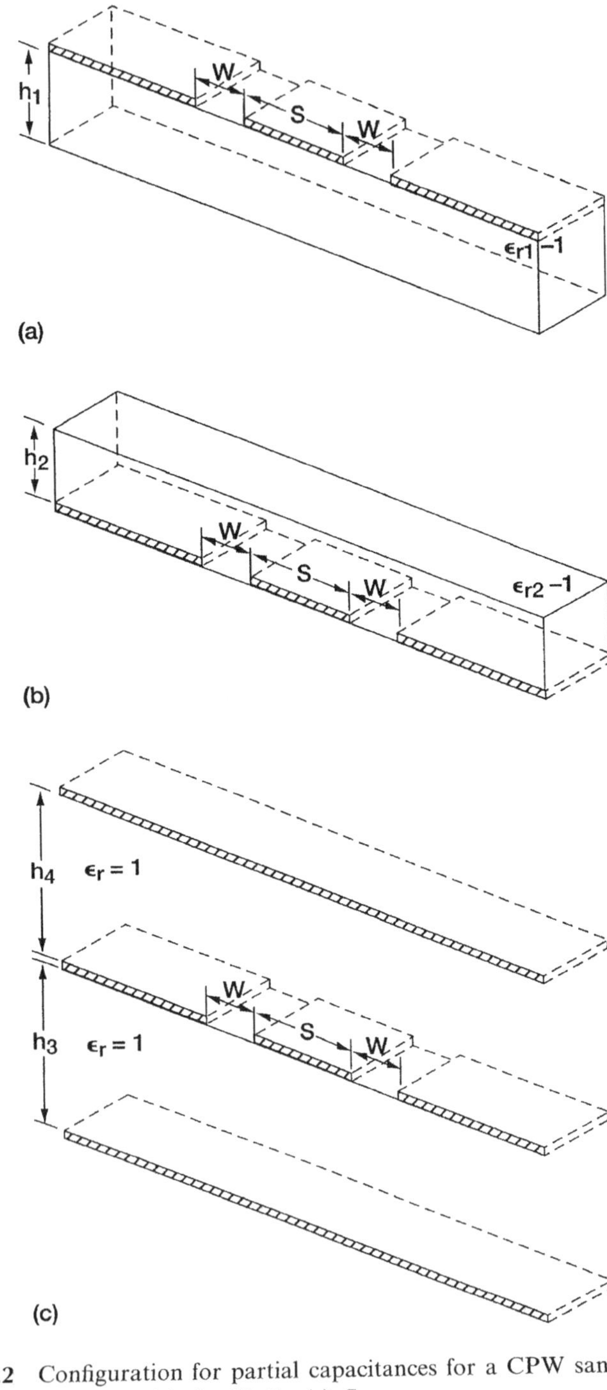

FIGURE 2.2 Configuration for partial capacitances for a CPW sandwiched between two dielectric substrates: (*a*) C_1; (*b*) C_2; (*c*) C_{air}.

Calculation of C_1 The capacitance C_1 of the lower partial dielectric region is given by [3]

$$C_1 = 2\varepsilon_0(\varepsilon_{r1} - 1)\,\frac{K(k_1)}{K(k_1')}, \tag{2.2}$$

where the modulus of the complete elliptic integrals $K(k_1)$ and $K(k_1')$ are [3]

$$k_1 = \frac{\sinh(\pi S/4h_1)}{\sinh\{[\pi(S + 2W)]/4h_1\}} \tag{2.3}$$

$$k_1' = \sqrt{1 - k_1^2}. \tag{2.4}$$

Calculation of C_2 The capacitance C_2 of the upper partial dielectric region is given by [3]

$$C_2 = 2\varepsilon_0(\varepsilon_{r2} - 1)\,\frac{K(k_2)}{K(k_2')}, \tag{2.5}$$

where

$$k_2 = \frac{\sinh(\pi S/4h_2)}{\sinh\{[\pi(S + 2W)]/4h_2\}}, \tag{2.6}$$

$$k_2' = \sqrt{1 - k_2^2}. \tag{2.7}$$

Calculation of C_{air} The capacitance C_{air} is given by [3]

$$C_{air} = 2\varepsilon_0\,\frac{K(k_3)}{K(k_3')} + 2\varepsilon_0\,\frac{K(k_4)}{K(k_4')}, \tag{2.8}$$

where

$$k_3 = \frac{\tanh(\pi S/4h_3)}{\tanh\{[\pi(S + 2W)]/4h_3\}}, \tag{2.9}$$

$$k_4 = \frac{\tanh(\pi S/4h_4)}{\tanh\{[\pi(S + 2W)]/4h_4\}}, \tag{2.10}$$

$$k_3' = \sqrt{1 - k_3^2}, \tag{2.11}$$

$$k_4' = \sqrt{1 - k_4^2}. \tag{2.12}$$

Calculation of Effective Dielectric Constant (ε_{eff}), Phase Velocity (v_{ph}), and Characteristic Impedance (Z_0) Substituting Eqs. (2.2), (2.5), and (2.8) into

Eq. (2.1) gives

$$
\begin{aligned}
C_{\mathrm{CPW}} &= 2\varepsilon_0(\varepsilon_{r1}-1)\frac{K(k_1)}{K(k_1')} + 2\varepsilon_0(\varepsilon_{r2}-1)\frac{K(k_2)}{K(k_2')} + 2\varepsilon_0\left[\frac{k(k_3)}{K(k_3')}+\frac{K(k_4)}{K(k_4')}\right] \\
&= 2\varepsilon_0\left[\frac{K(k_3)}{K(k_3')}+\frac{K(k_4)}{K(k_4')}\right]\left\{1+(\varepsilon_{r1}-1)\frac{K(k_1)}{K(k_1')}\left[\frac{k(k_3)}{K(k_3')}+\frac{K(k_4)}{K(k_4')}\right]^{-1}\right. \\
&\qquad \left.+(\varepsilon_{r2}-1)\frac{K(k_2)}{K(k_2')}\left[\frac{k(k_3)}{K(k_3')}+\frac{K(k_4)}{K(k_4')}\right]^{-1}\right\}.
\end{aligned}
\tag{2.13}
$$

Under quasi-static approximation $\varepsilon_{\mathrm{eff}}$ is defined as [3]

$$\varepsilon_{\mathrm{eff}} = \frac{C_{\mathrm{CPW}}}{C_{\mathrm{air}}}. \tag{2.14}$$

Substituting Eqs. (2.8) and (2.13) into (2.14) gives

$$\varepsilon_{\mathrm{eff}} = 1 + q_1(\varepsilon_{r1}-1) + q_2(\varepsilon_{r2}-1) \tag{2.15}$$

where the terms q_1 and q_2 are called the partial filling factors, and they are equal to

$$q_1 = \frac{K(k_1)}{K(k_1')}\left[\frac{K(k_3)}{K(k_3')}+\frac{K(k_4)}{K(k_4')}\right]^{-1}, \tag{2.16}$$

$$q_2 = \frac{K(k_2)}{K(k_2')}\left[\frac{K(k_3)}{K(k_3')}+\frac{K(k_4)}{K(k_4')}\right]^{-1}. \tag{2.17}$$

Further v_{ph} and Z_0 are defined as [3]

$$v_{\mathrm{ph}} = \frac{c}{\sqrt{\varepsilon_{\mathrm{eff}}}}, \tag{2.18}$$

$$Z_0 = \frac{1}{C_{\mathrm{CPW}}v_{\mathrm{ph}}} \tag{2.19}$$

where c is the velocity of light in free space. Equations (2.8), (2.14), (2.18), and (2.19) give

$$Z_0 = \frac{1}{cC_{\mathrm{air}}\sqrt{\varepsilon_{\mathrm{eff}}}}, \tag{2.20a}$$

$$= \frac{60\pi}{\sqrt{\varepsilon_{\mathrm{eff}}}}\left[\frac{K(k_3)}{K(k_3')}+\frac{K(k_4)}{K(k_4')}\right]^{-1}. \tag{2.20b}$$

For the CPW on a double-layer dielectric substrate shown in Figure 2.1(*b*), the partial capacitances are determined from the structures illustrated in Figure 2.3(*a*) to (*c*). Since these structures resemble those in the previous example, Eqs. (2.2), (2.3), (2.8), to (2.10) are still valid. However, the only change is in the equation for the partial capacitance C_2, which is as follows [3]:

$$C_2 = 2\varepsilon_0(\varepsilon_{r2} - \varepsilon_{r1})\frac{K(k_2)}{K(k'_2)}. \tag{2.21}$$

Equation (2.6) for the modulus of the elliptical integral is still valid. In a manner similar to the previous case by combining the above equations an expression for ε_{eff} is obtained which is as follows [3]:

$$\varepsilon_{eff} = 1 + q_1(\varepsilon_{r1} - 1) + q_2(\varepsilon_{r2} - \varepsilon_{r1}). \tag{2.22}$$

Equations (2.16) and (2.17) for the partial filling factors q_1 and q_2 are valid in this case also. Lastly Eq. (2.20) holds good for the characteristic impedance. In the sections that follow several limiting cases will be discussed and expressions for ε_{eff} and Z_0 presented.

2.2.2 Conventional Coplanar Waveguide on an Infinitely Thick Dielectric Substrate

This structure is schematically illustrated in Figure 2.4. In order for the equations derived earlier to be applicable, we have to set $h_1 = \infty$, $\varepsilon_{r2} = 1$ and $h_3 = h_4 = \infty$. When $h_1 = \infty$, Eqs. (2.3) and (2.4) reduce to

$$k_1 = k_0 = \frac{S}{S + 2W}, \tag{2.23}$$

$$k'_1 = k'_0. \tag{2.24}$$

Hence Eq. (2.2) for C_1 becomes

$$C_1 = 2\varepsilon_0(\varepsilon_{r1} - 1)\frac{K(k_0)}{K(k'_0)}. \tag{2.25}$$

When ε_{r2} is set equal to 1 in Eq. (2.5), C_2 becomes zero, that is,

$$C_2 = 0. \tag{2.26}$$

Lastly, when $h_3 = h_4 = \infty$, Eqs. (2.9) and (2.10) become

$$k_3 = k_4 = k_0 = \frac{S}{S + 2W} \tag{2.27}$$

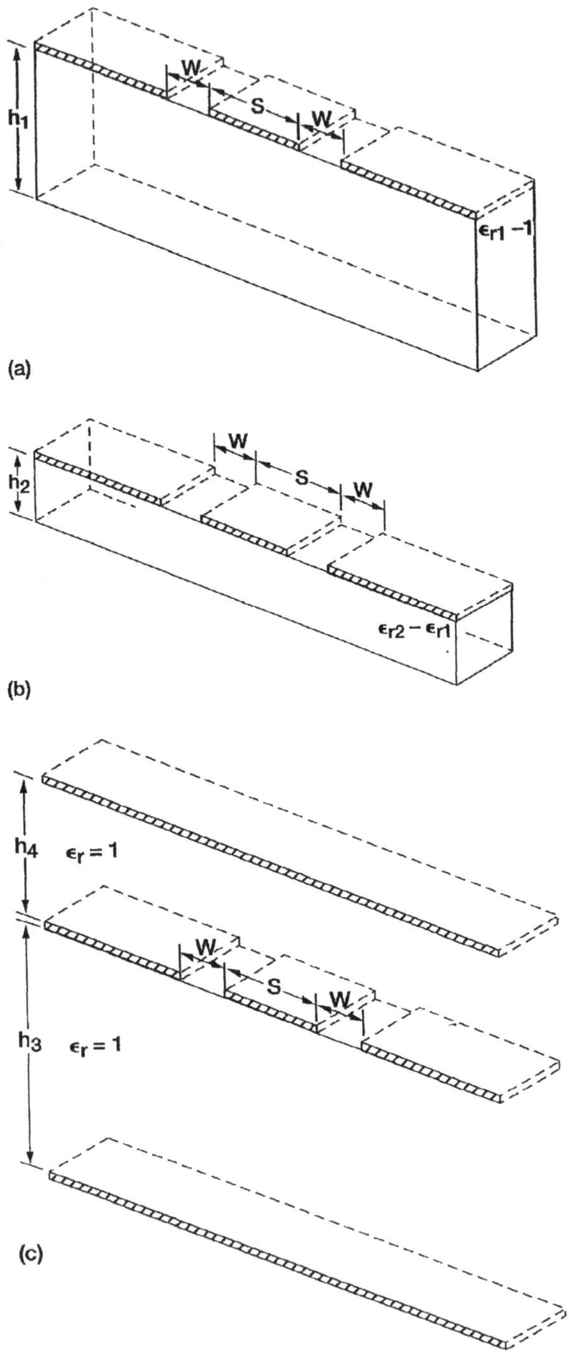

FIGURE 2.3 Configuration for partial capacitances for a CPW on a double layer dielectric substrate: (*a*) C_1, (*b*) C_2, (*c*) C_{air}.

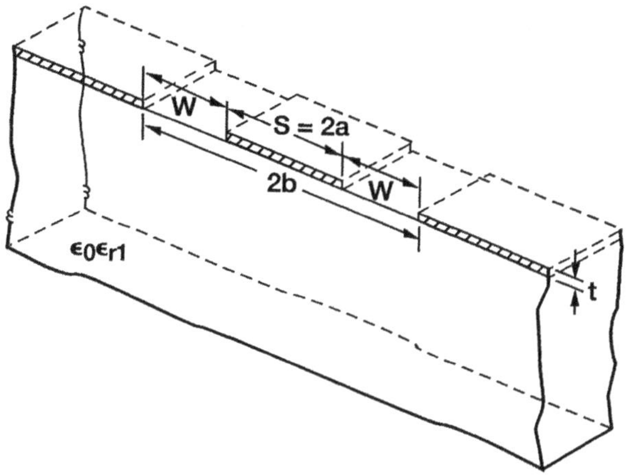

FIGURE 2.4 Schematic of a CPW on an infinitely thick dielectric substrate.

and hence Eq. (2.8) for C_{air} simplifies to

$$C_{\text{air}} = 4\varepsilon_0 \frac{K(k_0)}{K(k_0')}. \tag{2.28}$$

Substituting Eqs. (2.25), (2.26), and (2.28) into Eq. (2.1) gives

$$C_{\text{CPW}} = 2\varepsilon_0(\varepsilon_{r1} + 1)\frac{K(k_0)}{K(k_0')}. \tag{2.29}$$

Substituting Eqs. (2.28) and (2.29) into (2.14) gives

$$\varepsilon_{\text{eff}} = \frac{1 + \varepsilon_{r1}}{2}. \tag{2.30}$$

Substituting Eq. (2.30) into (2.18) gives

$$v_{\text{ph}} = \frac{c}{\sqrt{(1 + \varepsilon_{r1})/2}}, \tag{2.31}$$

and Eqs. (2.19), (2.29), and (2.31) gives

$$Z_0 = \frac{30\pi}{\sqrt{(\varepsilon_{r1} + 1)/2}} \frac{K(k_0')}{K(k_0)}. \tag{2.32}$$

The expression for ε_{eff} and Z_0 are identical to those given by Wen [1].

2.2.3 Conventional Coplanar Waveguide on a Dielectric Substrate of Finite Thickness

Consider the structure schematically illustrated in Figure 2.5. In this case $\varepsilon_{r2} = 1$ and $h_3 = h_4 = \infty$. Hence Eq. (2.2) gives

$$C_1 = 2\varepsilon_0(\varepsilon_{r1} - 1)\frac{K(k_1)}{K(k_1')}, \tag{2.33}$$

where k_1 and k_1' are given by Eqs. (2.3) and (2.4), respectively. From Eqs. (2.5), when $\varepsilon_{r2} = 1$, we have

$$C_2 = 0. \tag{2.34}$$

From Eqs. (2.8) to (2.10), when $h_3 = h_4 = \infty$ we have

$$k_3 = k_4 = k_0 = \frac{S}{S + 2W} \qquad \text{(2.35)(a)}$$

and

$$C_{air} = 4\varepsilon_0 \frac{K(k_0)}{K(k_0')}. \qquad \text{(2.35)(b)}$$

Substituting Eqs. (2.33) to (2.35) into Eq. (2.1) gives

$$C_{CPW} = 2\varepsilon_0(\varepsilon_{r1} - 1)\frac{K(k_1)}{K(k_1')} + 4\varepsilon_0 \frac{K(k_0)}{K(k_0')} \tag{2.36}$$

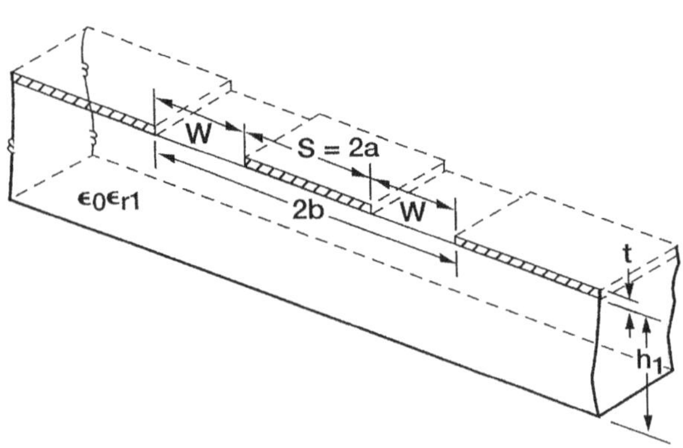

FIGURE 2.5 Schematic of a CPW on a dielectric substrate of finite thickness.

which yields from Eq. (2.14),

$$\varepsilon_{\text{eff}} = \frac{C_{\text{CPW}}}{C_{\text{air}}} = 1 + \frac{(\varepsilon_{r1} - 1)}{2} \frac{K(k_1)}{K(k_1')} \frac{K(k_0')}{K(k_0)}. \tag{2.37}$$

Lastly, from Eq. (2.20),

$$Z_0 = \frac{1}{cC_{\text{air}}\sqrt{\varepsilon_{\text{eff}}}} = \frac{30\pi}{\sqrt{\varepsilon_{\text{eff}}}} \frac{K(k_0')}{K(k_0)} \tag{2.38}$$

The expression above of ε_{eff} and Z_0 is identical to those given by [4], [5]. The computed ε_{eff} and Z_0 are presented in Figures 2.6(*a*) and (*b*), respectively, for a CPW on an alumina substrate.

It is worth mentioning here that in a related study [6] it has been shown that when the substrate thickness is less than twice the slot width, the deviation from the results of infinite dielectric (Section 2.2.2) is about 10 to 15 percent. Hence the propagation characteristics of lines with high Z_0, and therefore large slot widths on thin substrates, should be determined using Eqs. (2.37) and (2.38).

In [7] the Z_0 computed using Eq. (2.38) and by the spectral domain method are compared. The spectral domain computations are carried out at 1 GHz to avoid the effects of frequency dispersions. Table 2.1 presents this comparison. The results demonstrate that the accuracy of the conformal mapping results is better than 1 percent for a wide range of physical dimensions and available dielectric materials.

Although the conformal mapping expressions are rigorously valid at zero frequency, they can be used for the design of GaAs MMICs at millimeter wave frequencies [7]. The upper frequency limit is determined by comparing the computed quasi-static ε_{eff} and Z_0 with the spectral domain values and observing the frequency at which the two sets of results deviate more than a few percent. This comparison is presented later in Section 2.6.6.

2.2.4 Conventional Coplanar Waveguide on a Finite Thickness Dielectric Substrate and with a Top Metal Cover

An upper metal shielding is inevitably present in microwave monolithic integrated circuits (MMICs) and also in hybrid circuits when flip-chip active elements are inserted [8]. This structure is schematically illustrated in Figure 2.7. In this case, $\varepsilon_{r2} = 1$ and $h_3 = \infty$. Proceeding in a manner similar to the previous cases, ε_{eff} and Z_0 are given by

$$\varepsilon_{\text{eff}} = 1 + q_3(\varepsilon_{r1} - 1), \tag{2.39}$$

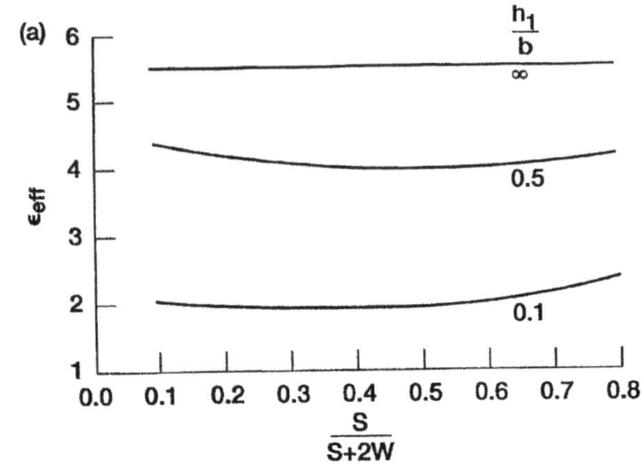

FIGURE 2.6 Computed characteristics of CPW as a function of $S/(S+2W)$, with the normalized substrate h_1/b as a parameter, $\varepsilon_{r1} = 10$: (*a*) Effective dielectric constant; (*b*) characteristic impedance. (From Reference [4], copyright © IEE.)

where

$$q_3 = \frac{K(k_1)/K(k'_1)}{[K(k_0)/K(k'_0)] + [K(k_4)/K(k'_4)]}, \tag{2.40}$$

where k_0, k_1, and k_4 are given by Eqs. (2.23), (2.3), and (2.10), respectively:

$$Z_0 = \frac{60\pi}{\sqrt{\varepsilon_{eff}}} \cdot \frac{1}{[K(k_0)/K(k'_0)] + [K(k_4)/K(k'_4)]}. \tag{2.41}$$

There equations are identical to those in [5].

TABLE 2.1 Comparison of $Z_0(\Omega)$ of Conventional CPW with Spectral Domain Technique at 1 GHz, h_1 = 200 μm, $t = 0$

		Conformal mapping			Spectral Domain		
a/b	b (μm)	$\varepsilon_{r1}=20$	$\varepsilon_{r1}=12.9$	$\varepsilon_{r1}=2.25$	$\varepsilon_{r1}=20$	$\varepsilon_{r1}=12.9$	$\varepsilon_{r1}=2.25$
	50	54.49	67.95	140.75	55.76	68.28	141.40
	170	57.52	70.29	142.86	57.52	70.27	142.97
0.2	230	59.00	72.13	144.83	59.02	71.95	143.95
	350	62.89	76.44	147.85	62.60	75.93	146.30
	50	42.04	51.47	106.57	42.22	51.69	106.99
	170	43.88	53.60	108.47	43.86	53.56	108.32
0.4	230	45.28	55.21	109.82	45.20	55.05	109.34
	350	48.46	58.82	112.62	48.24	58.44	111.41
	50	33.32	40.80	84.45	33.48	40.99	84.83
0.6	170	34.87	42.59	86.04	34.86	42.56	85.84
	230	35.99	43.87	87.12	35.93	43.76	86.68
	350	38.41	46.63	89.24	38.26	46.36	88.38
	50	25.68	31.45	65.09	25.86	31.66	65.51
0.8	170	26.81	32.75	66.03	26.80	32.71	66.03
	230	27.56	33.61	66.59	27.53	33.54	66.59

Source: From Reference [7], © 1992 IEEE.

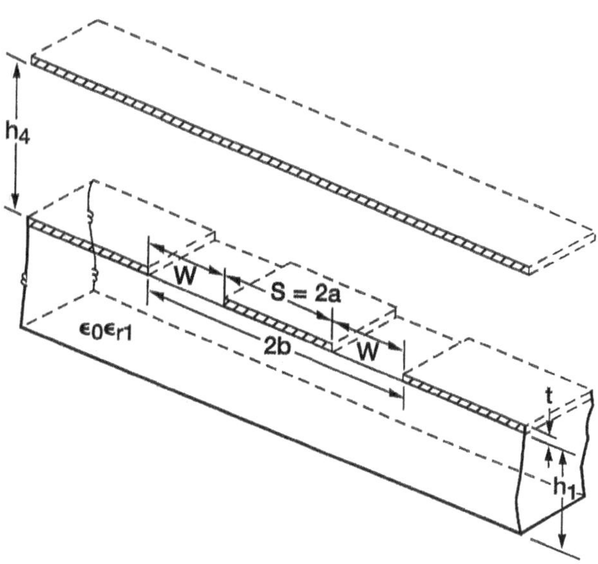

FIGURE 2.7 Schematic of a CPW on a dielectric substrate of finite thickness and with a top metal cover.

2.2.5 Conventional Coplanar Waveguide Sandwiched between Two Dielectric Substrates

This structure is schematically illustrated in Figure 2.8. In this case $h_3 = h_4 = \infty$, hence Eqs. (2.9) and (2.10) simplify to

$$k_3 = k_4 = k_0 = \frac{S}{S + 2W}, \tag{2.42}$$

and therefore

$$K(k_3) = K(k_4) = K(k_0).$$

Substituting for $K(k_3)$ and $K(k_4)$ into Eqs. (2.16) and (2.17) results in

$$q_1 = \frac{1}{2} \frac{k(k_1)}{K(k_1')} \frac{K(k_0')}{K(k_0)}, \tag{2.43}$$

$$q_2 = \frac{1}{2} \frac{K(k_2)}{K(k_2')} \frac{K(k_0')}{K(k_0)}. \tag{2.44}$$

The C_{air} from Eq. (2.8) simplifies to

$$C_{air} = 4\varepsilon_0 \frac{K(k_0)}{K(k_0')}. \tag{2.45}$$

Last, on substituting the expression above into Eqs. (2.15) and (2.20), we obtain

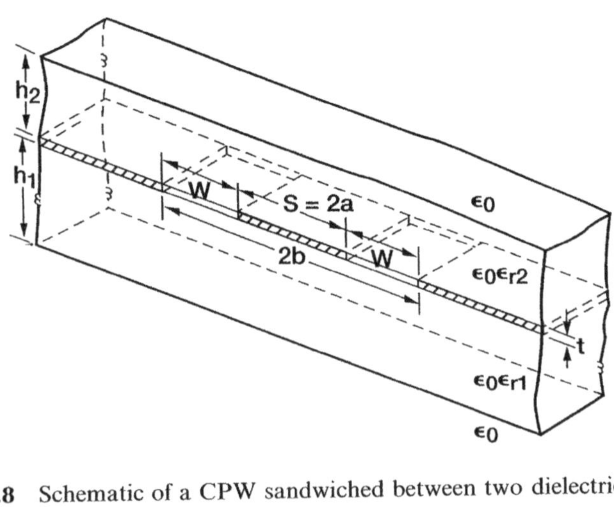

FIGURE 2.8 Schematic of a CPW sandwiched between two dielectric substrates.

$$\varepsilon_{\text{eff}} = 1 + q_1(\varepsilon_{r1} - 1) + q_2(\varepsilon_{r2} - 1), \tag{2.46}$$

$$Z_0 = \frac{30\pi}{\sqrt{\varepsilon_{\text{eff}}}} \frac{K(k_0')}{K(k_0)}. \tag{2.47}$$

2.2.6 Conventional Coplanar Waveguide on a Double-Layer Dielectric Substrate

There are three variants to this structure and the first variant is shown in Figure 2.9(*a*). This variant follows from Figure 2.1 by letting h_3 and h_4 equal to infinity. Under these boundary conditions, Eqs. (2.42) to (2.44) derived earlier are also valid here. An expression for ε_{eff} is obtained by substituting Eqs. (2.43) and (2.44) into Eq. (2.22), and this is as follows:

$$\varepsilon_{\text{eff}} = 1 + \frac{(\varepsilon_{r1} - 1)}{2} \frac{K(k_1)}{K(k_1')} \frac{K(k_0')}{K(k_0)} + \frac{(\varepsilon_{r2} - \varepsilon_{r1})}{2} \frac{K(k_2)}{K(k_2')} \frac{K(k_0')}{K(k_0)}. \tag{2.48}$$

This equation for ε_{eff} when substituted into Eq. (2.47) results in the Z_0 of the structure. In [9] both ε_{eff} and Z_0 are computed as a function of h_2 in the range of 5 to 50 μm for a CPW with $S = 50$ μm, $W = 30$ μm, $\varepsilon_{r1} = 13$, $\varepsilon_{r2} = 4$, and the thickness of the lower substrate, $(h_1 - h_2)$ equal to 500 μm. The computed results are shown to be in good agreement with those obtained using a full wave analysis.

The second variant is shown in Fig. 2.9(*b*). This structure has an infinitely thick lower dielectric substrate; that is, h_1 is equal to infinity. When $h_1 = \infty$, the capacitance C_1 is given by Eq. (2.25). Equation (2.21) for the capacitance C_2 is also valid here. Further for h_3 and h_4 equal to infinity, the capacitance C_{air} is given by Eq. (2.28). Knowing C_1, C_2, and C_{air} allows one to derive an expression of ε_{eff} as follows:

$$\varepsilon_{\text{eff}} = \left(\frac{\varepsilon_{r1} + 1}{2}\right) + \frac{(\varepsilon_{r2} - \varepsilon_{r1})}{2} \frac{K(k_2)}{K(k_2')} \frac{K(k_0')}{K(k_0)}. \tag{2.49}$$

The Z_0 from Eq. (2.20a) is

$$Z_0 = \frac{30\pi}{\sqrt{\varepsilon_{\text{eff}}}} \frac{K(k_0')}{K(k_0)}. \tag{2.50}$$

In [7] the Z_0 computed using Eq. (2.50) and by the spectral domain method are compared for a wide range of parameters. The spectral domain computations are carried out at 1 GHz to avoid the effect of frequency dispersion. Table 2.2 presents this comparison. The comparison shows that the accuracy of the conformal mapping results in better than 1 percent for wide range of physical dimensions and available dielectric materials.

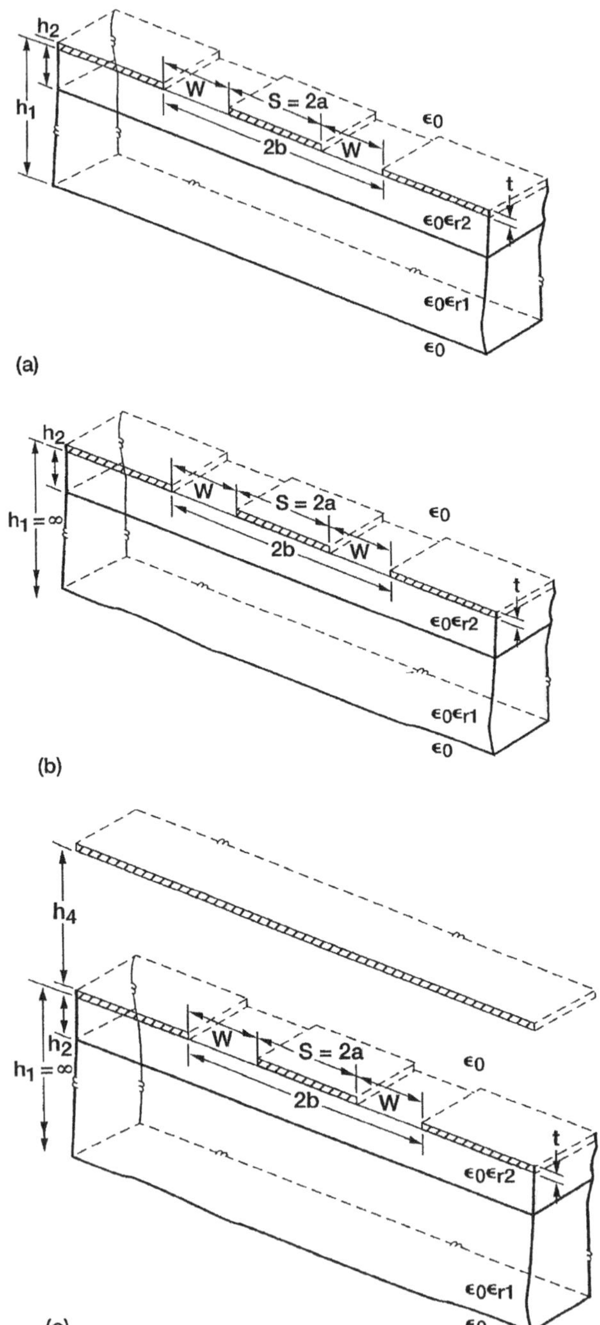

FIGURE 2.9 Variants of a CPW on a double-layer dielectric substrate: (*a*) Open CPW on finite thickness substrates; (*b*) open CPW on infinitely thick support substrate: (*c*) covered CPW on infinitely thick support substrate.

TABLE 2.2 Comparison of $Z_0(\Omega)$ of CPW on a Double-Layer Dielectric Substrate with Spectral Domain Technique at 1 GHz, $h_1 = \infty$, $h_2 = 200\,\mu m$, $t = 0$

		Conformal Mapping			Spectral Domain		
$W\,(\mu m)$	$S\,(\mu m)$	$\varepsilon_{r2}=20$, $\varepsilon_{r1}=10$	$\varepsilon_{r2}=12.9$, $\varepsilon_{r1}=3.78$	$\varepsilon_{r2}=12.9$, $\varepsilon_{r1}=10$	$\varepsilon_{r2}=20$, $\varepsilon_{r1}=10$	$\varepsilon_{r2}=12.9$, $\varepsilon_{r1}=3.78$	$\varepsilon_{r2}=12.9$, $\varepsilon_{r1}=10$
	20	45.51	55.96	55.91	45.85	56.37	56.33
	60	33.24	40.90	40.79	33.38	41.08	40.98
20	120	27.68	34.11	33.90	27.12	34.16	33.96
	200	24.52	30.30	29.98	24.50	30.28	29.92
	800	19.01	23.73	22.60	18.83	23.72	22.47
	20	61.82	76.09	75.84	62.01	76.32	76.09
	60	45.81	56.46	56.07	45.83	56.50	56.14
60	120	37.73	46.62	46.01	37.67	46.56	46.00
	200	32.95	40.88	39.96	32.81	40.72	39.89
	800	24.41	30.99	28.69	24.05	30.56	28.43
	20	70.56	86.97	86.36	70.60	87.03	86.48
	60	53.26	65.80	64.98	53.18	65.71	64.97
100	120	44.09	54.65	53.92	43.90	54.44	53.42
	200	38.47	47.91	46.39	38.19	47.60	46.22
	800	27.98	35.76	32.61	27.48	35.16	32.25
	20	83.77	103.81	101.75	83.24	103.18	101.48
	60	65.28	81.22	78.85	64.80	80.66	78.94
200	120	54.87	68.62	65.81	54.31	67.95	65.44
	200	48.95	60.58	57.26	47.50	59.80	56.22
	800	34.42	44.60	39.47	33.66	43.65	38.95

Source: From Reference [7], © 1992 IEEE.

To demonstrate that the quasi-static TEM mode has very small dispersion, the ε_{eff} and Z_0 are computed using the spectral domain technique [10] and presented in Table 2.3 [7]. Table 2.3 shows that in the case of a CPW with slot width and strip width of 200 and 120 μm, respectively, on a 200 μm thick GaAs substrate ($\varepsilon_r = 12.9$) backed by a thick quartz substrate ($\varepsilon_r = 3.78$), the deviation in ε_{eff} and Z_0 at 45 GHz from its value at 1 GHz is as small as 2.08 and 2.19 percent, respectively. This frequency limit extends to 65 GHz with deviation in ε_{eff} and Z_0 of 2.23 and -0.29 percent, respectively, when the slot width is reduced from 200 to 20 μm.

To study the effect of the lower substrate on the CPW propagation characteristics, the Z_0 is computed as a function of the strip width. Figure 2.10 presents the computed characteristics [7]. In this figure the slot width is held constant, and the relative permittivity of the lower substrate is used as a parameter. It is observed that the Z_0 is less sensitive to changes in the permittivity of the supporting dielectric for relatively small values of the ratio

TABLE 2.3 Dispersion in Z_0 and ε_{eff} of CPW on a Double-Layer Dielectric Substrate Calculated Using Spectral Domain Technique for $\varepsilon_{r1}=3.78$, $h_1=\infty$, $\varepsilon_{r2}=12.9$, $h_2=200\,\mu m$, $t=0$

	$W = 200\,\mu m$ and $S = 120\,\mu m$				$W = 20\,\mu m$ and $S = 120\,\mu m$			
f (GHz)	Z_0 (Ω)	ΔZ_0 (percent)	ε_{eff}	$\Delta\varepsilon_{eff}$ (percent)	Z_0 (Ω)	ΔZ_0 (percent)	ε_{eff}	$\Delta\varepsilon_{eff}$ (percent)
1	67.97	—	6.2932	—	34.16	—	6.8431	—
5	68.09	0.18	6.3014	0.05	34.17	0.03	6.8448	0.02
10	68.32	0.51	3.3210	0.17	34.19	0.09	6.8490	0.09
15	68.57	0.88	6.3481	0.35	34.20	0.12	6.8552	0.18
20	68.81	1.24	6.3816	0.56	34.22	0.18	6.8630	0.29
25	69.03	1.56	6.4208	0.80	34.23	0.20	6.8724	0.43
30	69.22	1.84	6.4651	1.08	34.24	0.23	6.8833	0.59
35	69.36	2.05	6.5141	1.39	34.24	0.23	6.8955	0.77
40	69.44	2.16	6.5675	1.73	34.23	0.20	6.9091	0.96
45	69.46	2.19	6.6248	2.08	34.22	0.18	6.9239	1.18
50	—	—	—	—	32.20	0.12	6.9400	1.42
55	—	—	—	—	34.16	0.00	6.9573	1.67
60	—	—	—	—	34.12	−0.12	6.9758	1.94
65	—	—	—	—	34.06	−0.29	6.9955	2.23

Source: From Reference [7], © 1992 IEEE.

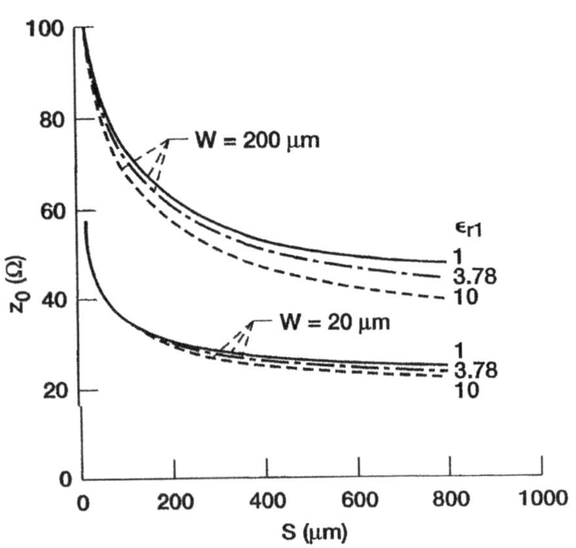

FIGURE 2.10 Characteristic impedance as a function of the strip width with the slot width and the relative permittivity of the infinitely thick support substrate as parameters, $h_1 = \infty$, $\varepsilon_{r2} = 12.9$, $h_2 = 200\,\mu m$, $t = 0$. (From Reference [7], © IEEE 1992.)

S/h_2 and W/h_2. However, Z_0 is more sensitive for large values of S/h_2 and or W/h_2.

The third variant is shown in Fig. 2.9(c). In this case the structure has a top metal cover. The capacitances C_1 and C_2 are given by Eqs. (2.25) and (2.21), respectively. The capacitance C_{air} is given by

$$C_{air} = 2\varepsilon_0 \left[\frac{K(k_0)}{K(k_0')} + \frac{K(k_4)}{K(k_4')}\right]. \tag{2.51}$$

Knowing C_1, C_2, and C_{air} an expression for ε_{eff} is derived and is as follows:

$$\varepsilon_{eff} = 1 + q_1(\varepsilon_{r1} - 1) + q_2(\varepsilon_{r2} - \varepsilon_{r1}). \tag{2.52}$$

The partial filling factors q_1 and q_2 are

$$q_1 = \frac{K(k_0)}{K(k_0')}\left[\frac{K(k_0)}{K(k_0')} + \frac{K(k_4)}{K(k_4')}\right]^{-1}, \tag{2.53}$$

$$q_2 = \frac{K(k_2)}{K(k_2')}\left[\frac{K(k_0)}{K(k_0')} + \frac{K(k_4)}{K(k_4')}\right]^{-1}. \tag{2.54}$$

The Z_0 is

$$Z_0 = \frac{60\pi}{\sqrt{\varepsilon_{eff}}}\left[\frac{K(k_0)}{K(k_0')} + \frac{K(k_4)}{K(k_4')}\right]. \tag{2.55}$$

The height of the shield above a certain height has negligible effect on the propagation characteristics and hence its presence can be ignored. This height is called the critical height h_{4c}. A reasonable theoretical definition for a theoretical critical shield height to ground plane separation h_{4c}/b is the height h_{4c} above which the absolute difference between the Z_0 of the CPW in Figs. 2.9(*b*) and (*c*) is less than 0.1 percent. Figure 2.11 shows the ratio h_{4c}/b as a function of the strip width with the slot width as a parameter. As a concluding remark it is worth mentioning that in [11] the conformal mapping technique has been extended to a CPW on a three-layer dielectric substrate.

2.2.7 Experimental Validation

Effective Dielectric Constant Equation (2.37) for ε_{eff} of a conventional CPW on a finite thickness substrate (Fig. 2.5) has been validated in [2] by comparing with the time domain reflectometer (TDR) measurements [12] made on CPW fabricated on 0.65 mm thick alumina substrate ($\varepsilon_r = 9.8$). This comparison made at 4 GHz shows that when W/h_1 is approximately 1.0 (thick substrate),

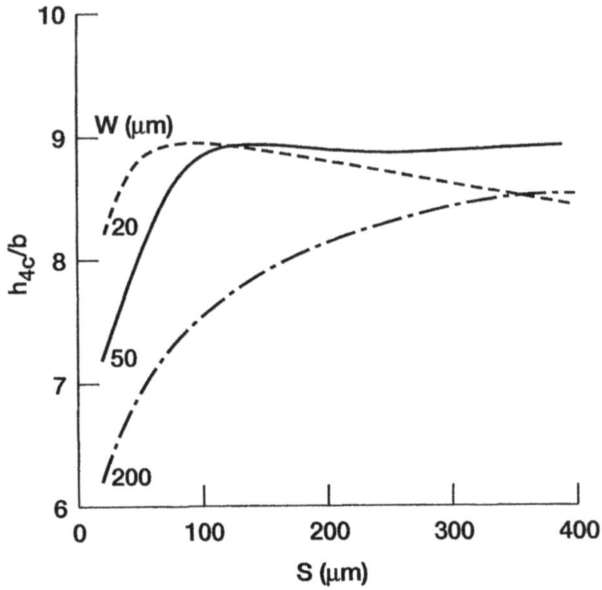

FIGURE 2.11 Critical shield height to ground plane separation ratio h_{4c}/b as a function of the strip width with the slot width as a parameter, $\varepsilon_{r1} = 3.78$, $h_1 = \infty$, $\varepsilon_{r2} = 12.9$, $h_2 = 200\,\mu$m, and $t = 0$. (From Reference [7], © IEEE 1992.)

the error is on the order of 0.4 percent and that when W/h_1 is approximately 3.6 (thin substrate), the error is on the order of 4.5 percent [2].

Characteristic Impedance In order to validate Eq. (2.38), the computed Z_0 is compared in Figure 2.12 with the experimental values obtained from [13], [14] and [15]. The Z_0 is experimentally determined using a TDR [13], [15] or from scattering parameters (S-parameters) [14] measurements. It should be pointed out that Eq. (2.38) assumes that the ground planes extend to infinity and the conductor thickness is zero, but the CPW circuits in [13], [14] and [15] have finite size ground planes in the range of 0.05 in. to 0.138 in. and the conductor thickness is in the range of 1.8 to 15 μm. Further, as discussed earlier, the accuracy of the computed results depends on the ratios W/h_1 and $S/(S + 2W)$. For a CPW on a thick substrate with narrow slots, W/h_1 is less than or equal to 0.5 and $S/(S + 2W)$ is greater or equal to 0.5. The difference between the measured and the computed Z_0 is small and is less than or equal to 3.3 percent. For a CPW on a thick substrate with fairly wide slots, W/h_1 is less than or equal to 0.5 and $S/(S + 2W)$ is in the range of 0.4 to 0.5. The difference in Z_0 increases to about 5.8 percent. Last, for a CPW on a thin substrate with very wide slots, W/h_1 is in the range of 0.5 to 2.0 and $S(S + 2W)$ is less than or equal to 0.4; the difference in Z_0 ranges from a few percent to as much as 11 percent.

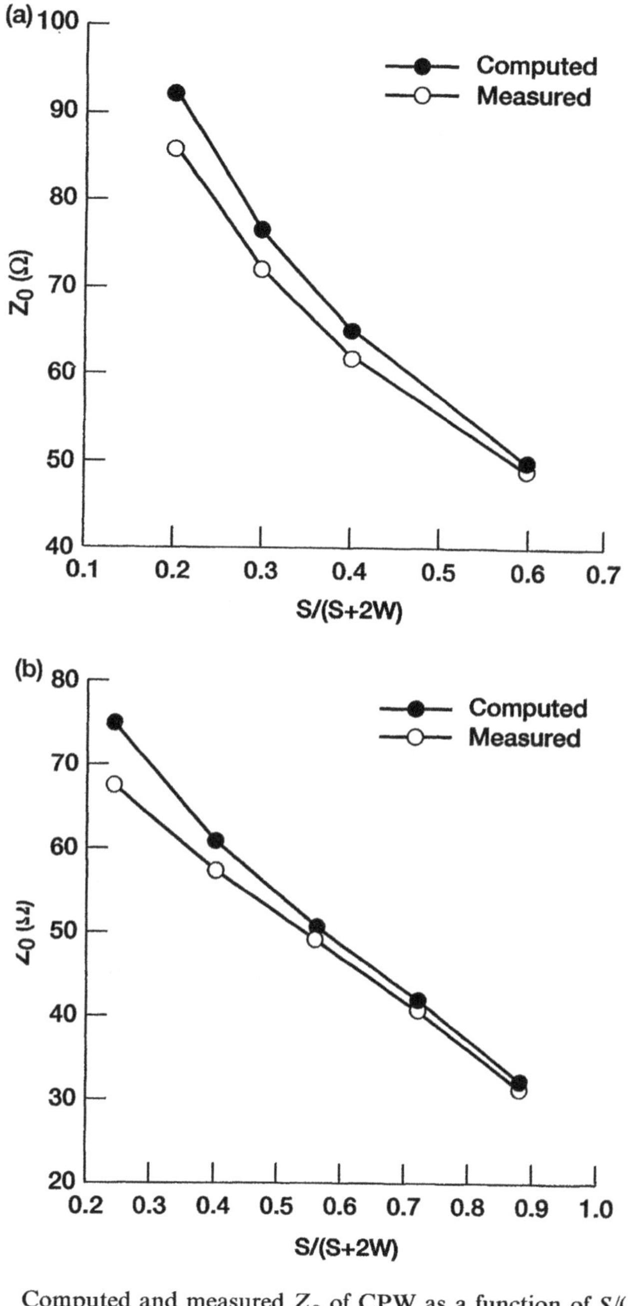

FIGURE 2.12 Computed and measured Z_0 of CPW as a function of $S/(S + 2W)$: (*a*) $\varepsilon_r = 9.2$, $h_1 = 0.05$ in., $S = 0.05$ in.; (*b*) $\varepsilon_r = 9.7$, $h_1 = 0.64$ mm, $S + 2W = 1.0$ mm; (*c*) $\varepsilon_r = 9.6$, $h_1 = 0.025$ in., $S = 0.02$ in.

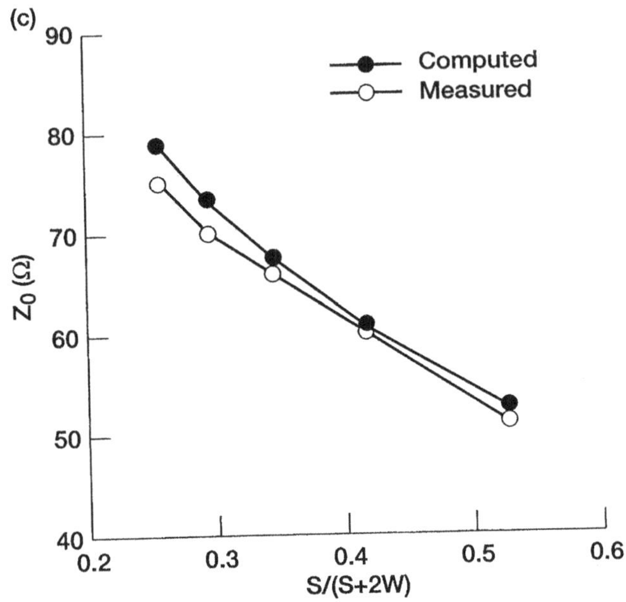

FIGURE 2.12 Continued.

2.3 QUASI-STATIC TEM ITERATIVE TECHNIQUES TO DETERMINE ε_{eff} AND Z_0

Section 2.2 discussed the application of conformal mapping method to CPW and its variants. The conformal mapping method resulted in closed form equations to calculate ε_{eff} and Z_0. The purpose of this section is to acquaint the reader with other quasi-static TEM methods, such as the relaxation method [16] and the hybrid method [17]. These methods are iterative in nature and hence is different from the conformal mapping method. They can also take into consideration nonuniform geometry and or irregular boundaries.

2.3.1 Relaxation Method

In this method the two-dimensional Laplace equation is solved along the cross section of the CPW to obtain ε_{eff} and Z_0. This is done by expanding the Laplace equation to form a simultaneous difference equation. In the numerical model the CPW is assumed to be housed inside a metal enclosure that is held at zero potential. Further the CPW conductor thickness, which is much smaller than the substrate thickness, is assumed to be zero. The numerical model is validated by comparing the computed Z_0 with the experimentally measured values. The experimental CPW lines are fabricated on a dielectric substrate of relative permittivity 9.4 and thickness 0.61 mm. A TDR is used to measure the

Z_0. The maximum difference between the measured and modeled Z_0 values is 3 percent [16].

2.3.2 Hybrid Method

The hybrid method combines the Galerkin finite-element method and the conformal-mapping technique. In this method, first, Wen's [1] mapping function is adopted, which transforms the original infinite domain into a finite image domain. This mapping also overcomes the field singularity difficulty around the strip edges. The problem is then solved in this image domain by the Galerkin finite-element method.

The computed ε_{eff} for conventional CPW agrees well with the conformal mapping results only for large a/b ratios [17]. The computed Z_0 is almost the same as that provided by the conformal mapping method [17]. An interesting feature of this method is that it can provide the magnitude of the field components in the cross-section of the structure.

2.4 FREQUENCY-DEPENDENT TECHNIQUES FOR DISPERSION AND CHARACTERISTIC IMPEDANCE

2.4.1 Spectral Domain Method

In this section the spectral domain analysis [18, 19] is presented initially for a single slot line and later extended to the case of a coupled slot line. The schematic of a single slot line on a dielectric substrate is shown in Figure 2.13(*a*). This structure has three regions that are defined as follows:

Region 1 for $h_1 < y$
Region 2 for $0 \leqslant y \leqslant h_1$
Region 3 for $y < 0$

The structure supports a hybrid mode which can be considered as a superposition of TE and TM modes.

In a planar waveguide, the fields associated with the TM and TE modes may be derived from the scalar electric potential function $\phi(x, y)$ and from the scalar magnetic potential function $\psi(x, y)$ respectively. These functions are related to the electric vector potential function $\vec{\phi}(x, y)$ and to the magnetic vector potential function $\vec{\psi}(x, y)$ as follows:

$$\vec{\phi}(x, y) = \phi(x, y)e^{\gamma z}\hat{k}, \tag{2.56a}$$

$$\vec{\psi}(x, y) = \psi(x, y)e^{\gamma z}\hat{k}, \tag{2.56b}$$

where $\hat{k}$ denotes the unit vector in the z-direction and γ is the propagation

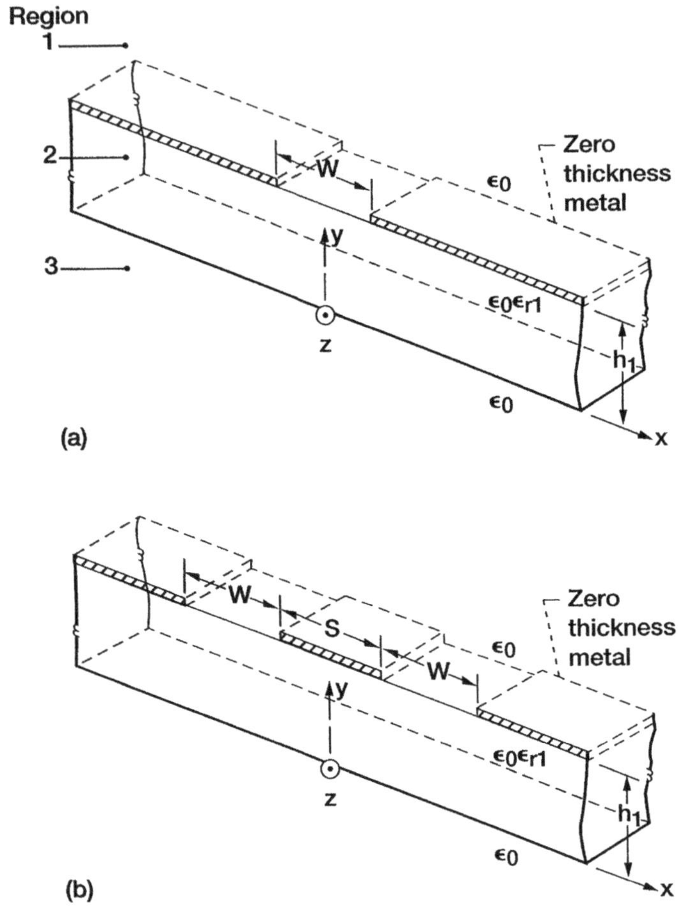

FIGURE 2.13 Schematic of (*a*) slot line; (*b*) coupled slot line.

constant along the longitudinal z-direction. In the loss less case γ is equal to

$$\gamma = j\beta. \tag{2.56c}$$

The scalar potential functions are solutions to the scalar Helmholtz equations in the three spatial regions and are as follows:

$$[\nabla_{xy}^2 + (\omega^2\mu_i\varepsilon_i + \gamma^2)]\phi_i(x, y) = 0, \tag{2.57a}$$

$$[\nabla_{xy}^2 + (\omega^2\mu_i\varepsilon_i + \gamma^2)]\psi_i(x, y) = 0, \tag{2.57b}$$

where ∇_{xy}^2 denotes the two-dimensional Laplacian operator in the transverse x- and y-directions, ω is the angular frequency, ε_i and μ_i are the permittivity and permeability of the three spatial regions, respectively, and $i = 1, 2, 3$ defines the three spatial regions. Further let

$$\omega^2 \mu_i \varepsilon_i + \gamma^2 = k_i^2 - \beta^2 = k_{ci}^2, \qquad i = 1, 2, 3. \tag{2.57c}$$

Using Maxwell's equations, it is possible to show that in a source-free medium, the fields associated with the TM mode can be derived from an electric vector potential function $\vec{\phi}(x, y)$ as follows:

$$\left.\begin{aligned} \vec{E} &= \nabla \times \nabla \times \vec{\phi}(x, y) \\ \vec{H} &= j\omega\varepsilon \nabla \times \vec{\phi}(x, y) \end{aligned}\right\}. \tag{2.58a}$$

On substituting Eq. (2.56a), we have

$$\left.\begin{aligned} \vec{E}_{ti} &= \gamma \nabla_{ti} \phi(x, y) \exp(\gamma z) \\ \vec{H}_{ti} &= -j\omega\varepsilon_i \hat{k} \times \nabla_{ti} \phi(x, y) \exp(\gamma z) \\ E_{zi} &= k_{ci}^2 \phi_i(x, y) \exp(\gamma z) \end{aligned}\right\}, \tag{2.58b}$$

where

$$\nabla_t = \frac{\partial}{\partial x} \hat{\imath} + \frac{\partial}{\partial y} \hat{\jmath} \tag{2.58c}$$

and $\hat{\imath}$ and $\hat{\jmath}$ are unit vectors in the x- and y-directions.

Similarly the fields associated with the TE modes can be derived from a magnetic vector potential function $\vec{\psi}(x, y)$ as follows:

$$\left.\begin{aligned} \vec{E} &= -j\omega\mu \nabla \times \vec{\psi}(x, y) \\ \vec{H} &= \nabla \times \nabla \times \vec{\psi}(x, y) \end{aligned}\right\}. \tag{2.59a}$$

On substitution of Eq. (2.56b), we have

$$\left.\begin{aligned} \vec{E}_{ti} &= j\omega\mu \hat{k} \times \nabla_{ti} \psi(x, y) \exp(\gamma z) \\ \vec{H}_{ti} &= \gamma \nabla_{ti} \psi_i(x, y) \exp(\gamma z) \\ H_{zi} &= k_{ci}^2 \psi_i(x, y) \exp(\gamma z) \end{aligned}\right\}. \tag{2.59b}$$

Thus from Eqs. (2.58b) and (2.59b), we obtain

$$E_{xi} = \left(\gamma \frac{\partial \phi_i(x, y)}{\partial x} - j\omega\mu_i \frac{\partial \psi_i(x, y)}{\partial y}\right) \exp(\gamma z), \tag{2.60a}$$

$$E_{yi} = \left(\gamma \frac{\partial \phi_i(x, y)}{\partial y} + j\omega\mu_i \frac{\partial \psi_i(x, y)}{\partial x}\right) \exp(\gamma z), \tag{2.60b}$$

$$H_{xi} = \left(\gamma \frac{\partial\psi_i(x, y)}{\partial x} + j\omega\varepsilon_i \frac{\partial\phi_i(x, y)}{\partial y}\right)\exp(\gamma z), \tag{2.60c}$$

$$H_{yi} = \left(\gamma \frac{\partial\psi_i(x, y)}{\partial y} - j\omega\varepsilon_i \frac{\partial\phi_i(x, y)}{\partial x}\right)\exp(\gamma z), \tag{2.60d}$$

where $i = 1, 2, 3$ defines the three regions.

Equations (2.57a) and (2.57b) are second-order partial differential equations. A simple way to solve these equations is to transform them by an integral transform into an ordinary differential equations and then solve them analytically. The appropriate boundary conditions for Eqs. (2.57a) and (2.57b) are, first, $\phi_i(x, y)$ as well as $\psi_i(x, y)$ approach zero as x and or y approach $\pm\infty$. Second, the time average power flow is finite; that is, $\int_{-\infty}^{+\infty} |\phi_i(x, y)|^2\, dx$ or $\int_{-\infty}^{+\infty} |\psi_i(x, y)|^2\, dx$ is bounded. Hence $\phi_i(x, y)$ and $\psi_i(x, y)$ are continuous, absolutely integrable, and square integrable. These are also the necessary conditions for a Fourier transform to exist. Therefore the scalar potential functions can be transformed into the Fourier domain by a Fourier transformation defined as

$$\mathscr{F}\{\phi(x, y)\} = \tilde{\phi}(\alpha, y) = \int_{-\infty}^{+\infty} \phi(x, y)e^{j\alpha x}\, dx \tag{2.61}$$

where the transform is expressed by a tilde above the symbol. Applying the Fourier transformation above to Eqs. (2.57a) and (2.57b) and making use of the identities

$$\mathscr{F}\left(\frac{\partial\phi_i(x, y)}{\partial x}\right) = -j\alpha\tilde{\phi}_i(\alpha, y) \tag{2.62}$$

and

$$\mathscr{F}\left(\frac{\partial^2\phi_i(x, y)}{\partial x^2}\right) = -\alpha^2\tilde{\phi}_i(\alpha, y), \tag{2.63}$$

the following ordinary differential equations are obtained

$$\frac{d^2}{dy^2}\tilde{\phi}_i(\alpha, y) = \gamma_i^2\tilde{\phi}_i(\alpha, y), \tag{2.64}$$

$$\frac{d^2}{dy^2}\tilde{\psi}_i(\alpha, y) = \gamma_i^2\tilde{\psi}_i(\alpha, y), \tag{2.65}$$

where

$$\gamma_i^2 = \alpha^2 - \gamma^2 - \omega^2\mu_i\varepsilon_i = \alpha^2 + \beta^2 - k_i^2. \tag{2.66}$$

The solutions to Eqs. (2.64) and (2.65) are given by [19]

$$\tilde{\phi}_1(\alpha, y) = A^e(\alpha)\exp[-\gamma_1(y - h_1)], \tag{2.67a}$$

$$\tilde{\phi}_2(\alpha, y) = B^e(\alpha)\sinh \gamma_2 y + C^e(\alpha)\cosh \gamma_2 y, \tag{2.67b}$$

$$\tilde{\phi}_3(\alpha, y) = D^e(\alpha)\exp(\gamma_3 y), \tag{2.67c}$$

and

$$\tilde{\psi}_1(\alpha, y) = A^h(\alpha)\exp[-\gamma_1(y - h_1)], \tag{2.68a}$$

$$\tilde{\psi}_2(\alpha, y) = B^h(\alpha)\sinh \gamma_2 y + C^h(\alpha)\cosh \gamma_2 y, \tag{2.68b}$$

$$\tilde{\psi}_3(\alpha, y) = D^h(\alpha)\exp(\gamma_3 y), \tag{2.68c}$$

For the slot line under consideration $k_1 = k_3 = k_0$ and $k_0 < \beta < k_2$. Hence, for small values of the transform variable α, that is, $|\alpha| < \sqrt{k_2^2 - \beta^2}$, γ_2^2 can be less than zero. In that case the hyperbolic functions in Eqs. (2.67) and (2.68) are replaced by trigonometric functions. The eight unknown coefficients $A^e(\alpha)$ to $D^h(\alpha)$ are related to the horizontal electric- and magnetic-field components at the interfaces $y = 0$ and $y = D$ by the continuity conditions, to the surface current density on the metal and the electric field in the slot at $y = h_1$. Mathematically this is represented as follows [19]:

At $y = 0$,

$$E_{z2}(x, 0, z) = E_{z3}(x, 0, z), \tag{2.69a}$$

$$E_{x2}(x, 0, z) = E_{x3}(x, 0, z), \tag{2.69b}$$

$$H_{z2}(x, 0, z) = H_{z3}(x, 0, z), \tag{2.69c}$$

$$H_{x2}(x, 0, z) = H_{x3}(x, 0, z). \tag{2.69d}$$

At $y = h_1$,

$$E_{z1}(x, h_1, z) = E_{z2}(x, h_1, z), \tag{2.70a}$$

$$E_{x1}(x, h_1, z) = E_{x2}(x, h_1, z), \tag{2.70b}$$

$$E_{z1}(x, h_1, z) = \begin{cases} E_z(x)e^{\gamma z}, & |x| < W/2 \\ 0, & \text{elsewhere} \end{cases}, \tag{2.70c}$$

$$E_{x1}(x, h_1, z) = \begin{cases} E_x(x)e^{\gamma z}, & |x| < W/2 \\ 0, & \text{elsewhere} \end{cases}, \tag{2.70d}$$

$$H_{z1}(x, h_1, z) - H_{z2}(x, h_1, z) = \begin{cases} j_x(x)e^{\gamma z}, & |x| > W/2 \\ 0, & \text{elsewhere} \end{cases}, \tag{2.70e}$$

$$H_{x1}(x, h_1, z) - H_{x2}(x, h_1, z) = \begin{cases} j_z(x)e^{\gamma z}, & |x| > W/2 \\ 0, & \text{elsewhere} \end{cases}. \tag{2.70f}$$

Where $E_x(x)$ and $E_z(x)$ are the unknown electric fields across the slot at $y = h_1$ and $j_x(x)$ and $j_z(x)$ are the unknown current density functions on the conductors at $y = h_1$.

If we denote the Fourier transform of the x- and z-directional electric field and current density components by [19]

$$\tilde{E}_x(\alpha) = \mathscr{F}\{E_x(x)\}, \tag{2.71a}$$

$$\tilde{E}_z(\alpha) = \mathscr{F}\{E_z(x)\}, \tag{2.71b}$$

$$\tilde{J}_x(\alpha) = \mathscr{F}\{j_x(x)\}, \tag{2.71c}$$

$$\tilde{J}_z(\alpha) = \mathscr{F}\{j_z(x)\}. \tag{2.71d}$$

We obtain a set of coupled equations as derived in Appendix 2A. The set is expressed as [19]

$$\begin{bmatrix} M_1(\alpha, \beta) & M_2(\alpha, \beta) \\ M_3(\alpha, \beta) & M_4(\alpha, \beta) \end{bmatrix} \begin{bmatrix} \tilde{J}_x(\alpha) \\ \tilde{J}_z(\alpha) \end{bmatrix} = \begin{bmatrix} \tilde{E}_x(\alpha) \\ \tilde{E}_z(\alpha) \end{bmatrix}, \tag{2.72}$$

where the elements of the M-matrix are the Fourier transforms of dyadic Green's function components. By inverting the M-matrix, a new matrix N and a second set of coupled equations [19] are obtained

$$\begin{bmatrix} N_1(\alpha, \beta) & N_2(\alpha, \beta) \\ N_3(\alpha, \beta) & N_4(\alpha, \beta) \end{bmatrix} \begin{bmatrix} \tilde{E}_x(\alpha) \\ \tilde{E}_z(\alpha) \end{bmatrix} = \begin{bmatrix} \tilde{J}_x(\alpha) \\ \tilde{J}_z(\alpha) \end{bmatrix}. \tag{2.73}$$

The coupled Eqs. (2.73) are further simplified if a inner product is defined on the space of complex functions of the real variable α over the domain $-\infty < \alpha < +\infty$ as

$$\langle f(\alpha), g(\alpha) \rangle = \int_{-\infty}^{+\infty} f(\alpha) g^*(\alpha) d\alpha, \tag{2.74}$$

where the asterisk above the symbol denotes complex conjugate. Taking the inner product with the Fourier transform of the electric-field components yields

$$\langle N_1(\alpha, \beta)\tilde{E}_x(\alpha), \tilde{E}_x(\alpha)\rangle + \langle N_2(\alpha, \beta)\tilde{E}_z(\alpha), \tilde{E}_x(\alpha)\rangle = \langle \tilde{J}_x(\alpha), \tilde{E}_x(\alpha)\rangle, \tag{2.75a}$$

$$\langle N_3(\alpha, \beta)\tilde{E}_x(\alpha), \tilde{E}_z(\alpha)\rangle + \langle N_4(\alpha, \beta)\tilde{E}_z(\alpha), \tilde{E}_z(\alpha)\rangle = \langle \tilde{J}_z(\alpha), \tilde{E}_z(\alpha)\rangle, \tag{2.75b}$$

The right-hand sides of Eq. (2.75) are zero as is evident from Parseval's theorem. For example,

$$\langle \tilde{J}_x(\alpha), \tilde{E}_x(\alpha)\rangle = \int_{-\infty}^{+\infty} \tilde{J}_x(\alpha)\tilde{E}_x(\alpha)d\alpha = 2\pi \int_{-\infty}^{+\infty} j_x(x)E_x(x)dx = 0$$

because $j_x(x) = 0$ for $|x| < W/2$ and $E_x(x) = 0$ for $|x| > W/2$ from the boundary conditions (2.70). Hence the integrand $j_x(x)\ E_x(x) = 0$ for any value of x. Up to this point, the formulation of the problem is exact, since no approximations have been made.

The next step is to solve Eqs. (2.75a) and (2.75b) simultaneously by means of Galerkin's method. This is accomplished by first expanding $E_x(x)$ and $E_z(x)$ in a set of complete basis functions as follows [19]:

$$E_x(x) = \sum_{n=1}^{\infty} a_n e_{xn}(x), \tag{2.76a}$$

$$E_z(x) = \sum_{n=1}^{\infty} b_n e_{zn}(x). \tag{2.76b}$$

If we restrict the analysis to a one-term approximation, $\tilde{E}_x(\alpha) \approx \mathscr{F}\{e_x(x)\}$ and $\tilde{E}_z(\alpha) \approx \mathscr{F}\{e_z(x)\}$. Then Eq. (2.75) becomes

$$\begin{aligned} a\int_{-\infty}^{+\infty} N_1(\alpha, \beta)|\tilde{E}_x(\alpha)|^2 d\alpha + b\int_{-\infty}^{+\infty} N_2(\alpha, \beta)\tilde{E}_z(\alpha)(\tilde{E}_z(\alpha))^*\, d\alpha = 0,\\ a\int_{-\infty}^{+\infty} N_3(\alpha, \beta)\tilde{E}_x(\alpha)(\tilde{E}_z(\alpha))^* d\alpha + b\int_{-\infty}^{+\infty} N_4(\alpha, \beta)|\tilde{E}_z(\alpha)|^2 d\alpha = 0. \end{aligned} \tag{2.77}$$

The dispersion characteristics of the slot line is determined by varying β such that the determinant of the coefficient matrix of (2.77) is zero for a given set of physical parameters at a desired frequency of operation.

A choice of the basis function which approximates the field closely is [19]

$$\begin{aligned} e_x(x) &= \begin{cases} \dfrac{1}{\sqrt{(W/2)^2 - x^2}}, & |x| \leqslant \dfrac{W}{2} \\ 0, & \text{elsewhere} \end{cases},\\ e_z(x) &= \begin{cases} x\sqrt{(W/2)^2 - x^2}, & |x| \leqslant W/2 \\ 0, & \text{elsewhere} \end{cases}. \end{aligned} \tag{2.78}$$

The Fourier transform of Eqs. (2.78) is

$$\tilde{E}_x(\alpha) = \pi B_0\left(\frac{\alpha W}{2}\right),$$
$$\tilde{E}_z(\alpha) = j\frac{\sqrt{2}}{4}\pi\frac{W}{\alpha}B_2\left(\frac{\alpha W}{2}\right), \tag{2.79}$$

where B_0 and B_2 are the Bessel functions of the first kind of order zero and two, respectively.

Characteristic Impedance The characteristic impedance Z_0 for a slot line is defined as [19]

$$Z_0 = \frac{V_0^2}{2P_{\text{avg}}}, \tag{2.80}$$

where V_0 is the slot voltage and P_{avg} is the time average power flow. To reduce the algebraic complexity, the z-directed electric field component given in Eq. (2.78) is neglected. This is referred to as the first-order approximation. The slot voltage V_0 is expressed as

$$V_0 = \int_{-W/2}^{+W/2} E_x(x)dx. \tag{2.81}$$

By performing an analytical integration for E_x given by Eq. (2.78), V_0 is equal to π. P_{avg} is defined as [19]

$$P_{\text{avg}} = \frac{1}{2}\text{Re}\left\{\iint_S \vec{E}_t \times (\vec{H}_t)^* \cdot \vec{a}_z dxdy\right\}. \tag{2.82}$$

The subscript t indicates transverse field components and S is the cross-sectional area. Assuming that a wave propagating in the $-z$-direction, the equation above reduces to

$$P_{\text{avg}} = \frac{1}{2}\text{Re}\left\{\int_{-\infty}^{+\infty}\int_{-\infty}^{+\infty}(E_y H_x^* - E_x H_y^*)dxdy\right\}. \tag{2.83}$$

For a wave propagating in the $-z$-direction, the z dependence is of the form $e^{\gamma z}$, where $\gamma = j\beta$ for a loss less case. The E_x, E_y, H_x, and H_y components can be expressed in terms of the scalar potential functions given in Eqs. (2.60a) to (2.60d):

$$E_y H_x^* = \beta^2 \frac{\partial \phi_i(x, y)}{\partial y} \frac{\partial \psi_i(x, y)}{\partial x} + \omega \varepsilon_i \beta \left(\frac{\partial \phi_i(x, y)}{\partial y} \right)^2 + \omega \mu_i \beta \left(\frac{\partial \psi_i(x, y)}{\partial x} \right)^2 + \omega^2 \mu_i \varepsilon_i \frac{\partial \phi_i(x, y)}{\partial y} \frac{\partial \psi_i}{\partial z}, \tag{2.84}$$

$$E_x H_y^* = \beta^2 \frac{\partial \phi_i(x, y)}{\partial x} \frac{\partial \psi_i(x, y)}{\partial y} - \omega \varepsilon_i \beta \left(\frac{\partial \phi_i(x, y)}{\partial x} \right)^2 - \omega \mu_i \beta \left(\frac{\partial \psi_i(x, y)}{\partial x} \right)^2 + \omega^2 \mu_i \varepsilon_i \frac{\partial \phi_i(x, y)}{\partial x} \frac{\partial \psi_i(x, y)}{\partial y}, \tag{2.85}$$

Equation (2.82) with the substitution above takes on the following form:

$$P_{\text{avg}} = \frac{1}{2} \operatorname{Re} \int_{-\infty}^{+\infty} \int_{-\infty}^{+\infty} \left\{ \omega \mu_i \beta \left[\left(\frac{\partial \psi_i(x, y)}{\partial x} \right)^2 + \left(\frac{\partial \psi_i(x, y)}{\partial y} \right)^2 \right] + \omega \varepsilon_i \beta \left[\left(\frac{\partial \phi_i(x, y)}{\partial x} \right)^2 + \left(\frac{\partial \phi_i(x, y)}{\partial y} \right)^2 \right] + (\beta^2 + k_i^2) \left(\frac{\partial \phi_i(x, y)}{\partial y} \frac{\partial \psi_i(x, y)}{\partial x} - \frac{\partial \phi_i(x, y)}{\partial x} \frac{\partial \psi_i(x, y)}{\partial y} \right) \right\} dxdy, \tag{2.86}$$

where $k_i^2 = \omega^2 \mu_i \varepsilon_i$. The limits of integration are infinite since the slot line is an open structure.

The next step is to transform Eq. (2.85) into the spectral domain. Since the limits of integration extends to infinity, the transformation is made possible through the use of Parseval's relation, which is given as

$$\int_{-\infty}^{+\infty} f^*(x) g(x) dx = \int_{-\infty}^{+\infty} \mathscr{F}^*[f] \mathscr{F}[g] dt. \tag{2.87}$$

To facilitate the use of Parseval's relation, the first term in Eq. (2.86) is expressed as the product of two complex quantities as follows:

$$\left(\frac{\partial \psi_i(x, y)}{\partial x} \right)^2 + \left(\frac{\partial \psi_i(x, y)}{\partial y} \right)^2 = \left(\frac{\partial \psi_i(x, y)}{\partial x} + j \frac{\partial \psi_i(x, y)}{\partial y} \right) \times \left(\frac{\partial \psi_i(x, y)}{\partial x} - j \frac{\partial \psi_i(x, y)}{\partial y} \right). \tag{2.88}$$

A similar relation holds good for the second term in Eq. (2.86) and making use of the identity

$$\mathscr{F}[f'(x)] = -j\alpha \mathscr{F}[f], \tag{2.89}$$

the time average power flow is

$$
\begin{aligned}
P_{\text{avg}} = \frac{1}{4\pi} \operatorname{Re} \int_{-\infty}^{+\infty} \int_{-\infty}^{+\infty} \Bigg[& -\alpha^2 \beta\omega\varepsilon_i |\tilde{\phi}_i(\alpha, y)|^2 - \beta\omega\mu_i \left| \frac{\partial \tilde{\psi}_i(\alpha, y)}{\partial y} \right|^2 \\
& - \beta\omega\varepsilon_i \left| \frac{\partial \tilde{\phi}_i(\alpha, y)}{\partial y} \right|^2 - \alpha^2 \beta\omega\mu_i |\tilde{\psi}_i(\alpha, y)|^2 \\
& - j\alpha\beta^2 \left(\tilde{\phi}_i(\alpha, y) \frac{\partial \tilde{\psi}_i(\alpha, y)^*}{\partial y} + \frac{\partial \tilde{\phi}_i(\alpha, y)}{\partial y} \tilde{\psi}_i(\alpha, y)^* \right) \\
& + j\alpha k_1^2 \left(\tilde{\phi}_i(\alpha, y)^* \frac{\partial \tilde{\psi}_i(\alpha, y)}{\partial y} + \frac{\partial \tilde{\phi}_i(\alpha, y)^*}{\partial y} \tilde{\psi}_i(\alpha, y) \right) \Bigg] \, d\alpha \, dy. \qquad (2.90)
\end{aligned}
$$

Since $\tilde{\phi}_i(\alpha, y)$ and $\tilde{\psi}_i(\alpha, y)$ take different forms in the three spatial regions, as in Eqs. (2.67) and (2.68), the equation above has to be evaluated for each region separately. In addition the double integral reduces to a single integral by performing the integration with respect to y analytically. In Appendix 2B the integrands for the three regions are separately derived.

Coupled Slot Lines The structure shown in Figure 2.13(b) is considered as coupled slot lines. The two natural modes supported by this structure are the even and odd modes. These modes are defined as

$$
\begin{aligned}
E_x(x) &= E_x(-x) \qquad \text{for the even mode,} \\
E_x(x) &= -E_x(-x) \qquad \text{for the odd mode.}
\end{aligned} \qquad (2.91)
$$

The dyadic Green's function given in Eq. (2.73) is a function of α, β, ε_{r1}, h_1, and the operating frequency, but it is independent of the slot configuration. The dimensions of the slot enter into the computations only through the basis function. Thus it is only necessary to modify Eq. (2.78) to suit the coupled slot line geometry and field distribution.

A choice of the basis function that approximates the fields of the coupled slot line closely is

$$
e_x(x) = \begin{cases} \dfrac{1}{\sqrt{(W/2)^2 - \{x - [(S+W)/2]\}^2}}, & \dfrac{S}{2} < x < \dfrac{S}{2} + W \\[2ex] \dfrac{\pm 1}{\sqrt{(W/2)^2 - \{x + [(S+W)/2]^2}}, & -\dfrac{S}{2} - W < x < -\dfrac{S}{2} \\[2ex] 0, & \text{elsewhere.} \end{cases} \qquad (2.92)
$$

The + sign for the even mode and − sign for the odd mode. Figure 2.14 shows graphically the assumed field distribution for coupled slot lines. In comparing

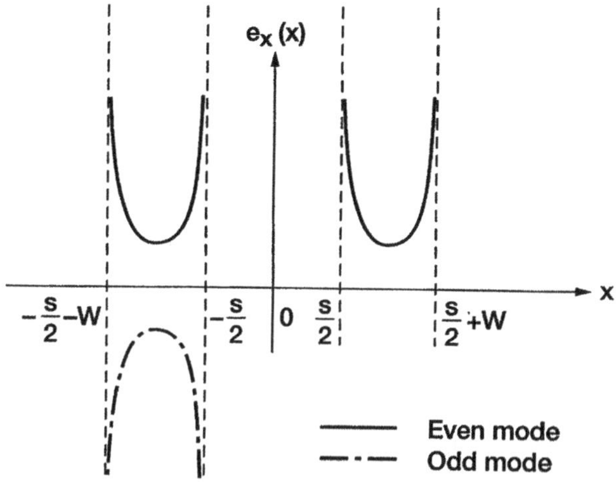

FIGURE 2.14 Assumed field distribution for coupled slots.

Eq. (2.92) with Eq. (2.78), we notice that the former is a shifted version of the latter. Hence, to obtain the Fourier transform of Eq. (2.92), the shifting property of the Fourier transform is used. That is, if

$$\mathscr{F}\{\phi(x)\} = \tilde{\phi}(\alpha) = \int_{-\infty}^{+\infty} \phi(x)e^{j\alpha x}dx, \quad \text{then} \quad \mathscr{F}\{\phi(x - x_0)\} = e^{j\alpha x_0}\tilde{\phi}(\alpha). \tag{2.93}$$

Hence the Fourier transform of the electric field in Eq. (2.92) for the even mode is

$$\tilde{E}_{xe}(\alpha) = (e^{j\alpha(S+W)/2} + e^{-j\alpha(S+W)/2})\tilde{E}_x(\alpha) = 2\cos\left(\alpha\,\frac{S+W}{2}\right)\tilde{E}_x(\alpha) \tag{2.94}$$

and for the odd mode is

$$\tilde{E}_{xo}(\alpha) = (e^{j\alpha(S+W)/2} - e^{-j\alpha(S+W)/2})\tilde{E}_x(\alpha) = +2j\sin\left(\alpha\,\frac{S+W}{2}\right)\tilde{E}_x(\alpha) \tag{2.95}$$

where $\tilde{E}_x(x)$ is given by Eq. (2.78). A similar transformation holds good for the z-directed electric field $e_z(x)$. The expression for the characteristic impedance is changed to [19]

$$Z_{0,o,e} = \frac{V_0^2}{P_{avg,o,e}}, \tag{2.96}$$

since the total time average power surrounding the transmission lines is now due to two lines.

Computed Characteristics for Coupled Slot Lines The computed even-mode and odd-mode wavelength ratio (λ_g/λ), where λ_g and λ are the guide wavelength and free space wavelength, respectively, is shown in Figure 2.15(*a*) [19]. The corresponding characteristic impedance is shown in Figure 2.15(*b*) [19].

The validity of the foregoing results is checked in the limit where the separation S between the slots tends to be very small. In this limit the even-mode characteristic impedance $Z_{0,e}$ approaches one-half of Z_0, where Z_0 is the characteristic impedance of a single slot line with twice the width of the slot in the coupled structure. The dispersion characteristics for both the structures, however, is the same.

From the above characteristics some interesting observations can be made. First, the characteristic impedance of the CPW is one-half of the odd-mode characteristic impedance $Z_{0,o}$ for equal slot width. Second, for large S/h_1 as frequency increases, $Z_{0,e}$ and $Z_{0,o}$ converge to Z_0. Where Z_0 is the characteristic impedance of a single slot with no coupling. Third, for a fixed h_1/λ, the ratio λ_g/λ for the even-mode first increases and then decreases as S/h_1 increases from a small to a large value. This is because for small separation, the metal strip between the slots has negligible effect on the wave, and therefore the wave propagates as if it were on a slot of width $2(W/h_1) + S/h_1$. As the separation increases the slot width effectively increases and hence λ_g/λ increases. As S/h_1 continues to increases, the waves on the two slot lines start to decouple and eventually propagate as if on two independent slot lines of width W/h_1. The ratio λ_g/λ therefore decreases.

As a concluding remark, it may be useful to indicate other techniques that have been used to analyze CPW structures. In [20] and [21] the CPW was analyzed by modeling the structure as a capacitive iris in a rectangular waveguide. In [22] the conformal mapping technique and the variational reaction theory were combined and solved using the finite element method (FEM). Finally, in [23] the CPW was analyzed using the finite difference time domain technique (FDTD).

2.4.2 Experimental Validation

The computed ε_{eff} using the spectral-domain technique for CPW on alumina and GaAs substrates are compared with the measured values over the frequency range of 1 to 25 GHz [24]. The experimental values for ε_{eff} are determined using an automatic network analyzer with time-domain option. Figure 2.16 presents the above comparison.

The computed ε_{eff} valves for CPW on Duroid and Cuflon substrates based on the technique described in [20] are compared with the experimentally measured valves over the frequency range of 2 to 18 GHz [25]. The experimental valves are determined from the measured resonance frequencies and the physical length of a pair of series-gap coupled straight resonators fabricated on these substrates. Figure 2.17(*a*) and (*b*) presents this comparison.

The computed Z_0 using the spectral-domain technique for CPW on GaAs substrate is compared with the experimentally measured values [26]. The CPW

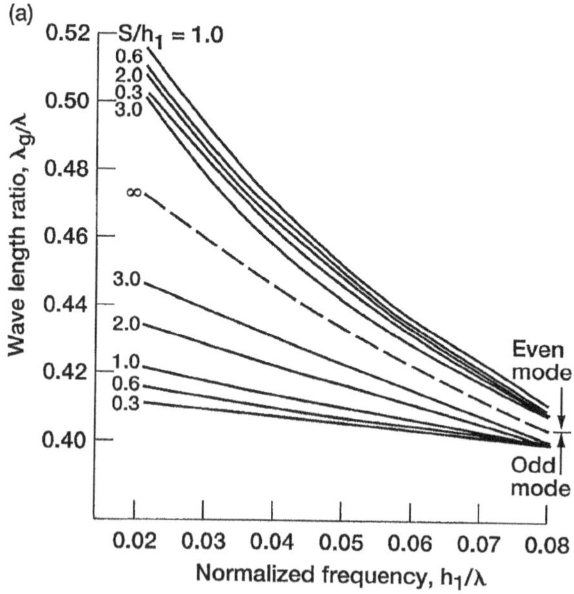

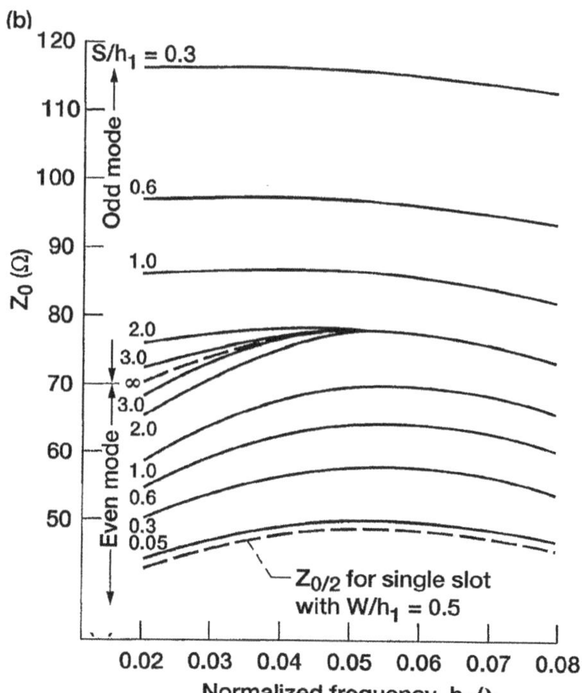

FIGURE 2.15 Computed even-mode and odd-mode characteristics as a function of frequency with the separation as a parameter, $W/h_1 = 0.25$, $\varepsilon_{r1} = 11$: (*a*) Wavelength ratio; (*b*) characteristic impedance. (From Reference [19], © IEEE 1975.)

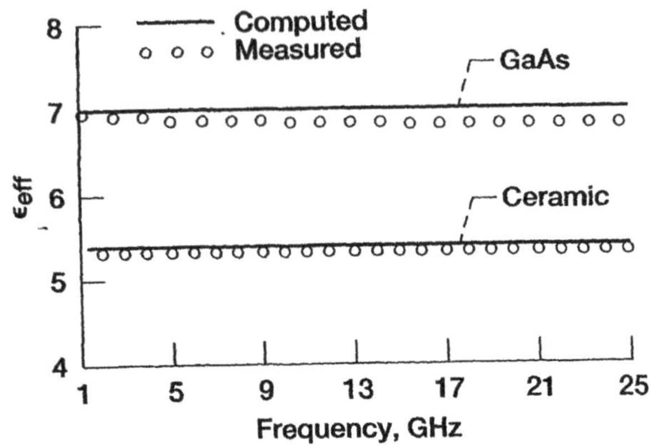

FIGURE 2.16 Measured and computed effective dielectric constant as a function of the frequency: GaAs: $\varepsilon_r = 12.9$, $S = 0.075$ mm, $W = 0.05$ mm, $h_1 = 0.4$ mm, $t = 0.002$ mm; Ceramic: $\varepsilon_r = 9.8$, $S = 0.2$ mm, $W = 0.01$ mm, $h_1 = 0.635$ mm, $t = 0.003$ mm. (From Reference [24], with permission from Microwave Exhibitions and Publishers.)

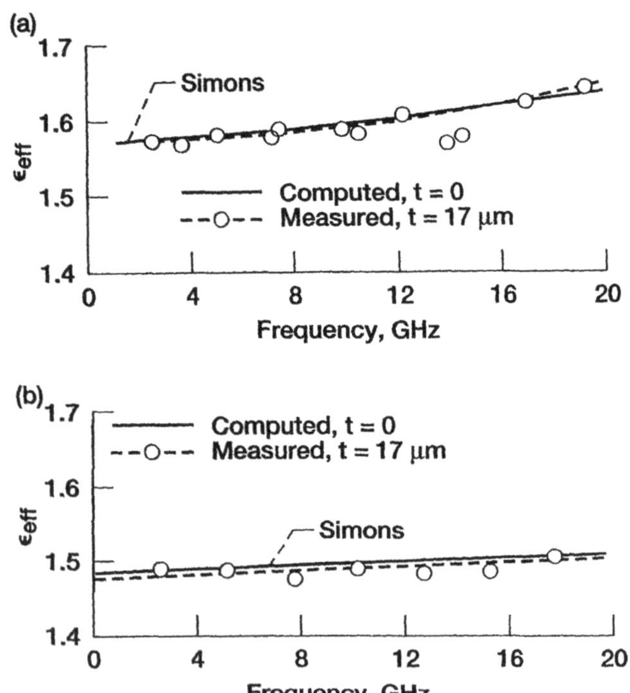

FIGURE 2.17 Measured and computed effective dielectric constant as a function of the frequency: (*a*) $\varepsilon_r = 2.2$, $h_1 = 0.125$ in., $W = 0.01$ in., $S/(S + 2W) = 0.9$; (*b*) $\varepsilon_r = 2.1$, $h_1 = 0.062$ in., $W = 0.01$ in. $S/(S + 2W) = 0.86$.

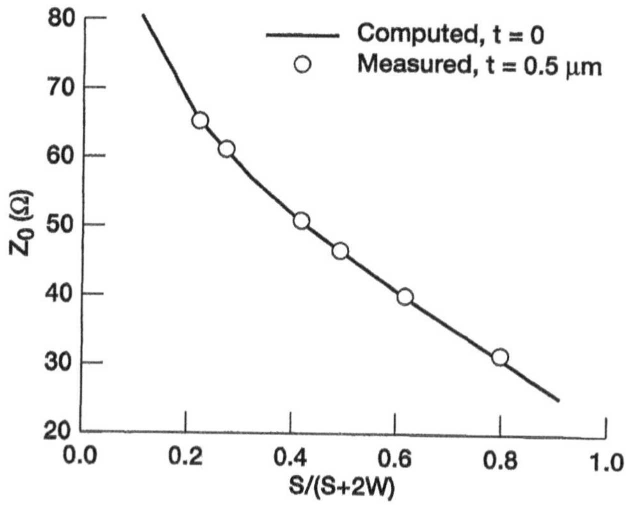

FIGURE 2.18 Computed and measured characteristic impedance as a function of $S/(S+2W)$: $\varepsilon_r = 12.9$, $h_1 = 0.5\,\text{mm}$, $S = 75\,\mu\text{m}$. (From Reference [26], © IEEE 1991.)

impedance is determined from the measured two-port S-parameters of through lines. The S-parameters are obtained using on-wafer RF probes and a calibrated automatic network analyzer. Figure 2.18 presents this comparison.

2.5 EMPIRICAL FORMULA TO DETERMINE DISPERSION BASED ON SPECTRAL DOMAIN RESULTS

An empirical formula to compute ε_{eff} of CPW shown in Figure 2.5 has been obtained. This is done by curve fitting the dispersion data obtained using the spectral domain method. The expression is valid into the terahertz regime and given below [27], [28]:

$$\sqrt{\varepsilon_{\text{eff}}(f)} = \sqrt{\varepsilon_q} + \frac{(\sqrt{\varepsilon_{r1}} - \sqrt{\varepsilon_q})}{(1 + aF^{-b})}, \tag{2.97}$$

where

f = frequency

$F = f/f_{\text{TE}}$, the normalized frequency

$f_{\text{TE}} = c/(4h_1\sqrt{\varepsilon_{r1} - 1})$, the cutoff frequency for the lowest-order TE mode

ε_q = effective permittivity at the quasi-static limit

$b \approx 1.8$, is a constant independent of the dimensions

The factor a is computed from the expression [28]

$$\log(a) \approx u \log\left(\frac{S}{W}\right) + v \tag{2.98}$$

where u and v depend on the substrate thickness h_1 as follows [28]:

$$\left.\begin{aligned} u &\approx 0.54 - 0.64q + 0.015q^2 \\ v &\approx 0.43 - 0.86q + 0.540q^2 \end{aligned}\right\} \tag{2.99}$$

and $q = \log(S/h_1)$.

The preceding formula for $\varepsilon_{\mathrm{eff}}$ is accurate to within 5 percent for the following range of parameters [28]:

$$\left.\begin{aligned} 0.1 < \frac{S}{W} < 5 \\ 0.1 < \frac{S}{h_1} < 5 \\ 1.5 < \varepsilon_{r1} < 50 \\ 0 < \frac{f}{f_{\mathrm{TE}}} < 10 \end{aligned}\right\}. \tag{2.100}$$

A similar set of equations was presented in [29].

2.5.1 Comparison of Coplanar Waveguide Dispersion with Microstrip

Equation (2.97) is used to compare the dispersion characteristics of a $50\,\Omega$ CPW ($S = 85\,\mu$m, $W = 50\,\mu$m) with a $50\,\Omega$ microstrip ($W = 73\,\mu$m) on an identical GaAs substrate ($\varepsilon_{r1} = 13$, $h_1 = 100\,\mu$m). The cutoff frequency f_{TE} for this substrate is about 216.5 GHz. Figure 2.19 presents the $\sqrt{\varepsilon_{\mathrm{eff}}}$ for the two transmission lines as a function of the normalized frequency [28]. The figure shows that the quasi-static valve of the effective permittivity ε_q is lower for the CPW compared to the microstrip. This is expected because the CPW has greater fringe fields. At infinite frequencies the effective permittivity, $\varepsilon_{\mathrm{eff}}$ approaches ε_{r1} in both cases. This implies that ultrashort pulses having bandwidths greater than 700 GHz ($\log\,(f/f_{\mathrm{TE}}) \geqslant 0.5$) will suffer greater dispersion in the CPW. On the other hand, since the $\varepsilon_{\mathrm{eff}}$ increases with frequency more gradually for CPW, longer pulses with narrow bandwidth will suffer lower dispersion.

In general, the dispersion for both CPW and microstrip can be reduced by reducing the substrate thickness. For CPW, dispersion can be slightly reduced by decreasing the strip and slot dimensions, but the $\varepsilon_{\mathrm{eff}}$ at low frequencies remains lower than the corresponding microstrip and thus it is intrinsically more dispersive for short pulses.

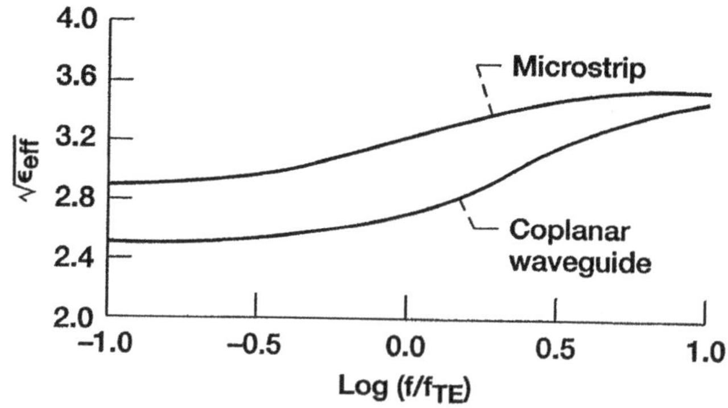

FIGURE 2.19 Computed effective dielectric constant as a function of normalized frequency. (From Reference [28], © IEEE 1986.)

2.6 SYNTHESIS FORMULAS TO DETERMINE ε_{eff} AND Z_0 BASED ON QUASI-STATIC EQUATIONS

Closed form formulas to computer ε_{eff} and Z_0 of CPW shown in Figure 2.5 are available [30] to [32]. These formulas are obtained by function approximation and by curve-fitting the quasi-static equations. The synthesis formulas are valid over a very wide range of relative substrate permittivity ($1.5 \leqslant \varepsilon_{r1} \leqslant 120$). Consequently these formulas are broad enough for most CPW circuit design, which includes printed antennas on low-permittivity substrates as well as monolithic microwave integrated circuits on high-permittivity semiconductor substrates.

The synthesis formulas for the strip width S or the slot width W are expressed in terms of ε_{r1}, h_1, Z_0 and either the slot width W or the strip width S, respectively. The CPW is schematically illustrated in Figure 2.5. The expression for the strip width will be first presented.

When

$$\frac{W}{h_1} \leqslant \frac{10}{[3(1+\ln \varepsilon_{r1})]} \quad \text{and} \quad \frac{S}{h_1} \leqslant \frac{80}{[3(1+\ln \varepsilon_{r1})]},$$

the strip width is given by

$$S = W \times G(\varepsilon_{r1}, h_1, Z_0, W), \tag{2.101}$$

with

$$G = 0.25 \exp\left(\frac{30\pi^2}{Z_0 \varepsilon_{eff}^{0.5}}\right) + \exp\left(-\frac{30\pi^2}{Z_0 \varepsilon_{eff}^{0.5}}\right) - 1 \qquad \text{for } Z_0 < \frac{60\sqrt{2}\pi}{(\varepsilon_{r1}+1)^{0.5}}, \tag{2.102}$$

$$G = \left[0.125\exp\left(\frac{Z_0\varepsilon_{\text{eff}}^{0.5}}{60}\right) - 0.5\right]^{-1} \qquad \text{for } Z_0 \geqslant \frac{60\sqrt{2}\pi}{(\varepsilon_{r1}+1)^{0.5}}. \tag{2.103}$$

In these expressions, ε_{eff} is given by

$$\varepsilon_{\text{eff}}(\varepsilon_{r1}, h_1, Z_0, W) = T_1[1 + (\varepsilon_{r1} - 1)^{1.5} Z_0 T_2] \qquad \text{for } \varepsilon_{r1} < 6.0, \tag{2.104}$$

$$\varepsilon_{\text{eff}}(\varepsilon_{r1}, h_1, Z_0, W) = (\varepsilon_{r1} + 1) T_3 T_4 \qquad \text{for } \varepsilon_{r1} \geqslant 6.0, \tag{2.105}$$

with

$$T_1 = \operatorname{sech}\left\{\left[\frac{\varepsilon_r}{\varepsilon_r + 1}\right]^5 \left[\frac{47.56}{Z_0}\right]^2 \times \exp\left[\left(1 + 0.0064 Z_0 \frac{W}{h_1}\right) \times \ln\left(0.6 + \frac{W}{h_1}\right)\right]\right\}, \tag{2.106}$$

$$T_2 = \left(\frac{1}{837.5}\right) \ln\left[2 \times \frac{(1+g)}{(1-g)}\right] \qquad \text{for } 0.841 \leqslant g < 1, \tag{2.107}$$

$$T_2 = \left(84.85 \ln\left\{\frac{2 \times [1 + (1 - g^4)^{0.25}]}{[1 - (1 - g^4)^{0.25}]}\right\}\right)^{-1} \qquad \text{for } 0 < g \leqslant 0.841, \tag{2.108}$$

$$g = \left(\frac{\{\exp[\pi(1+p)W/2h_1] - \exp(\pi W/2h_1)\}}{\{\exp[\pi(2+p)W/2h_1] - 1\}}\right)^{0.5}, \tag{2.109}$$

$$p = 0.25\exp\left[\frac{30\sqrt{2}\pi^2}{Z_0(\varepsilon_{r1}+1)^{0.5}}\right] + \exp\left[-\frac{30\sqrt{2}\pi^2}{Z_0(\varepsilon_{r1}+1)^{0.5}}\right] - 1$$
$$\text{for } Z_0 < \frac{60\sqrt{2}\pi}{(\varepsilon_{r1}+1)^{0.5}}, \tag{2.110}$$

$$p = \left\{0.125\exp\left[\frac{Z_0(\varepsilon_{r1}+1)^{0.5}}{60\sqrt{2}}\right] - 0.5\right\}^{-1} \qquad \text{for } Z_0 \geqslant \frac{60\sqrt{2}\pi}{(\varepsilon_{r1}+1)^{0.5}}, \tag{2.111}$$

$$T_3 = \operatorname{sech}\left\{\frac{\varepsilon_{r1}^6}{(\varepsilon_r+1)^7}\left[\frac{150.4}{Z_0}\right]^2 \times \exp\left[\left(1 + 0.0008\varepsilon_{r1} Z_0 \frac{W}{h_1}\right)\right.\right.$$
$$\left.\left. \times \ln\left(0.3 + \frac{W}{h_1}\right)\right]\right\}, \tag{2.112}$$

$$T_4 = 0.5 + [0.02125 - 0.345Q - 0.0005(0.25 + Q)\varepsilon_{r1}] \frac{QW}{h_1}$$
$$- \left\{1 + \exp\left[3.5 - 1.55 \ln\left(\frac{W}{h_1}\right)\right]\right\}^{-1}, \tag{2.113}$$

$$Q = \left\{1 - 0.5 + 0.25 \times \exp\left[\frac{30\sqrt{2}\pi^2}{Z_0(\varepsilon_{r1}+1)^{0.5}}\right]^{-1}\right\}^2 \qquad \text{for } Z_0 < \frac{60\sqrt{2}\pi}{(\varepsilon_{r1}+1)^{0.5}}, \tag{2.114}$$

$$Q = 4\exp\left\{-\left(\frac{Z_0}{120}\right) \times [2(\varepsilon_{r1}+1)]^{0.5}\right\} \qquad \text{for } Z_0 \geqslant \frac{60\sqrt{2}\pi}{(\varepsilon_{r1}+1)^{0.5}}. \tag{2.115}$$

In an analogous manner the slot width can be expressed.
When

$$\frac{S}{h_1} \leqslant \frac{80}{[3(1+\ln\varepsilon_{r1})]} \quad \text{and} \quad \frac{W}{h_1} \leqslant \frac{10}{[3(1+\ln\varepsilon_{r1})]}$$

the slot width W is given by

$$W = \frac{S}{G(\varepsilon_{r1}, h_1, Z_0, S)}, \tag{2.116}$$

with

$$G = 0.25\exp\left[\frac{30\pi^2}{Z_0\varepsilon_{\text{eff}}^{0.5}}\right] + \exp\left[-\frac{30\pi^2}{Z_0\varepsilon_{\text{eff}}^{0.5}}\right] - 1 \qquad \text{for } Z_0 < \frac{60\sqrt{2}\pi}{(\varepsilon_{r1}+1)^{0.5}}, \tag{2.117}$$

$$G = \left[0.125\exp\left[\frac{Z_0\varepsilon_{\text{eff}}^{0.5}}{60}\right] - 0.5\right]^{-1} \qquad \text{for } Z_0 \geqslant \frac{60\sqrt{2}\pi}{(\varepsilon_{r1}+1)^{0.5}}, \tag{2.118}$$

where ε_{eff} is equal to

$$\varepsilon_{\text{eff}}(\varepsilon_{r1}, h_1, Z_0, S) = (1 + T_A) \times [1 + (\varepsilon_{r1} - 1)^{1.5} Z_0 T_B],$$

with

$$T_A = \tanh\left\{\frac{\varepsilon_{r1}^7}{(\varepsilon_{r1}+1)^8}\left[\frac{60}{Z_0}\right]^2 \times \exp\left[\left(1 + \frac{0.002 Z_0 S}{Q h_1}\right) \times \ln\left(\frac{S}{Q h_1}\right)\right]\right\} \qquad \text{for } \varepsilon_{r1} < 6, \tag{2.119}$$

$$T_A = \tanh\left\{\frac{\varepsilon_{r1}^5}{(\varepsilon_{r1}+1)^6}\left[\frac{60}{Z_0}\right]^2 \times \exp\left[\left(1 + \frac{0.002\varepsilon_{r1} Z_0 S}{Q h_1}\right) \times \ln\left(\frac{S}{Q h_1}\right)\right]\right\} \qquad \text{for } \varepsilon_{r1} \geqslant 6, \tag{2.120}$$

$$T_B = \left(\frac{1}{837.5}\right) \times \ln\left[\frac{2(1+g)}{1-g}\right] \qquad \text{for } 0.841 \leqslant g < 1, \tag{2.121}$$

$$T_B = \left(84.85 \ln\left\{2 \times \left[\frac{1 + (1 - g^4)^{0.25}}{1 - (1 - g^4)^{0.25}}\right]\right\}\right)^{-1} \quad \text{for } 0 < g \leqslant 0.841, \tag{2.122}$$

$$g = \left\{\frac{\sinh(\pi S/4h_1)}{\sinh\{[\pi(1 + 2/Q)S]/4h_1\}}\right\}^{0.5}, \tag{2.123}$$

$$Q = 0.25 \exp\left[\frac{30\sqrt{2}\pi^2}{Z_0(\varepsilon_{r1} + 1)^{0.5}}\right] + \exp\left[-\frac{30\sqrt{2}\pi^2}{Z_0(\varepsilon_{r1} + 1)^{0.5}}\right] - 1 \quad \text{for } Z_0 < \frac{60\sqrt{2}\pi}{(\varepsilon_{r1} + 1)^{0.5}}, \tag{2.124}$$

$$Q = \left\{0.125 \times \exp\left[\frac{Z_0(\varepsilon_{r1} + 1)^{0.5}}{60\sqrt{2}}\right] - 0.5\right\}^{-1} \quad \text{for } Z_0 \geqslant \frac{60\sqrt{2}\pi}{(\varepsilon_{r1} + 1)^{0.5}}. \tag{2.125}$$

The accuracy of Eq. (2.101) is 1 percent for $\varepsilon_{r1} > 6.0$ and better than 3 percent for the entire ε_{r1} range, while the accuracy of Eq. (2.116) is 1 percent for $\varepsilon_{r1} > 6.0$ and better than 2 percent for the entire ε_{r1} range.

2.7 COPLANAR WAVEGUIDE WITH ELEVATED OR BURIED CENTER STRIP CONDUCTOR

The conventional CPW on a GaAs substrate has a Z_0 approximately in the range of 30 to 80 Ω. In applications, such as, reduced-size hybrid couplers, broadband bias network, subpicosecond sampling circuits, etc., there is a need for CPW with Z_0 greater than 80 Ω. A straight forward approach to achieve high Z_0 for a given ground-to-ground separation $(S + 2W)$ is by reducing the center strip conductor width S. The disadvantage of this approach is the steep increase in attenuation due to ohmic losses. In addition the Z_0 becomes very sensitive to the strip width-to-slot width ratio and any small deviation causes significant change in Z_0. A solution to this problem is to either elevate or bury the center strip conductor as shown in Figure 2.20 [33] to [38]. In Figure 2.20(*a*) and (*b*) the center strip conductor is either elevated and supported by or buried beneath several dielectric layers respectively. In Figure 2.20(*c*), the center strip conductor is fabricated as a sequence of air-bridges and hence is supported by posts at regular intervals. In this case, for the purpose of numerical simulation, the center strip conductor is considered to be elevated and supported by an air dielectric. The propagation characteristics of the lines above will be discussed next. The technique employed to compute the Z_0 of CPW with elevated center strip conductor on dielectric layers has not been revealed in the original reference. The characteristics for CPW with elevated center strip conductor on air dielectric were obtained using the Sonnet Software.

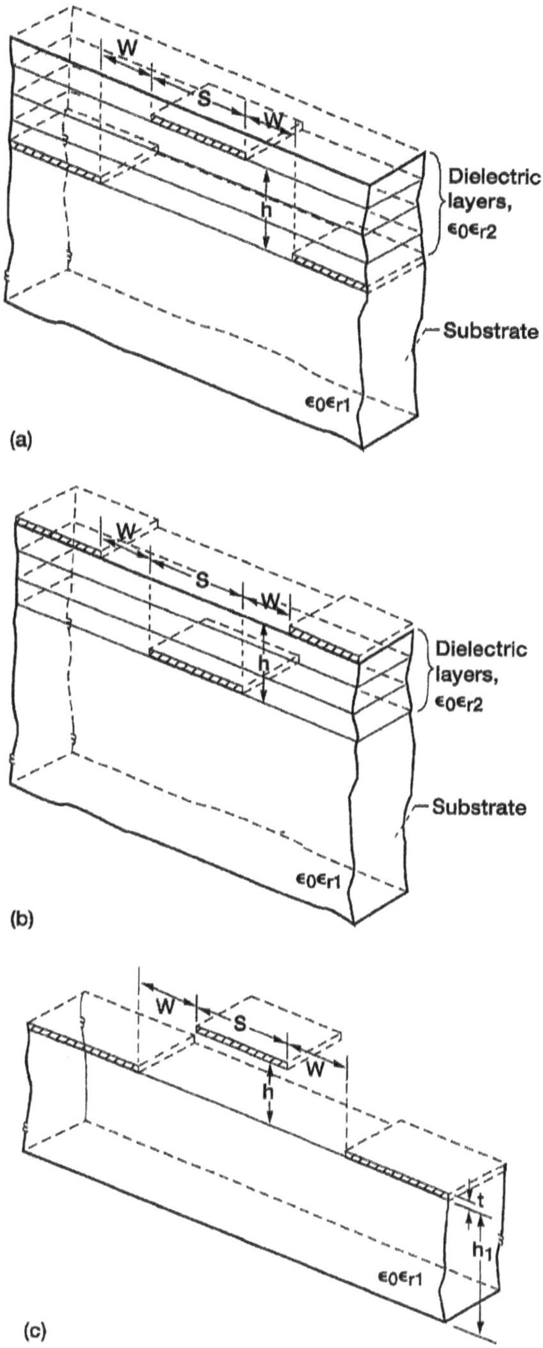

FIGURE 2.20 Coplanar Waveguide with (*a*) elevated center strip conductor on dielectric layers, (*b*) buried center strip conductor, and (*c*) elevated center strip conductor on air dielectric.

2.7.1 CPW with Elevated Center Strip Conductor Supported on Dielectric Layers

This structure in shown in Fig. 20(a). The substrate is GaAs ($\varepsilon_r = 12.9$), and the dielectric layers are polyimide ($\varepsilon_r = 3.3$). In a practical structure each polyimide layer is 2.5 μm thick and is formed using a spin-coating method to ensure uniform thickness [34], [35]. For this structure the computed Z_0 as a function of the ground-to-ground separation with the thickness h of the polyimide layer as a parameter is shown in Figure 2.21. The strip width S is held fixed at 9 μm. The Z_0 for a conventional CPW with h equal to zero is also shown on the same figure for comparison. As an example, for an elevated CPW with Z_0 equal to 100 Ω, the ground-to-ground separation in 40 μm for h equals 10 μm, while a conventional CPW requires a ground-to-ground separation in excess of 120 μm [35]. Thus in MMICs, for a desired ground-to-ground separation, high Z_0 can be achieved by using an elevated CPW that does not require the center strip width to be excessively narrow.

2.7.2 CPW with Elevated Center Strip Conductor Supported on Posts

This structure is shown in Figure 2.20(*c*). Normal air-bridge fabrication process is used to elevate the center strip conductor to a height of 2 to 3 μm above the GaAs substrate. Table 2.4 compares the computed ε_{eff}, Z_0 and attenuation of conventional CPW with the elevated CPW for identical strip and slot widths

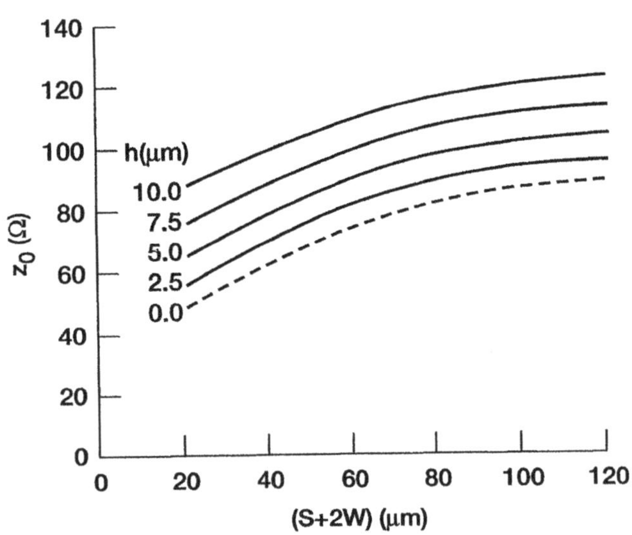

FIGURE 2.21 Computed characteristic impedance as a function of the ground-to-ground separation for elevated CPW with weight of the dielectric layer as a parameter: $\varepsilon_{r1} = 12.9$, $\varepsilon_{r2} = 3.3$, S = 9 μm. (From Reference [35], © IEEE 1992.)

TABLE 2.4 Comparison of Computed ε_{eff}, Z_0 and Attenuation of Conventional CPW with the Elevated CPW for Identical S and W

	Conventional CPW (Computed)	Elevated CPW Computed	Elevated CPW Measured
ε_{eff}	6.4	2.62	2.62
Z_0 (Ω)	60	101	97
α (dB/mm)	0.25	0.13	0.1

Note: $S=12\,\mu$m, $W=18\,\mu$m, $h=2.6\,\mu$m, $t=2.9\,\mu$m, $h_1=500\,\mu$m, $\varepsilon_{r1}=12.9$, $\sigma_{gold}=30$ S/μm, $\tan\delta_e=3\times10^{-4}$, $F=20$ GHz.

[38]. The results show that by elevating the center strip the Z_0 increases from 60 to 100 Ω, which is about 66 percent, while the attenuation decreases from 0.25 to 0.13 dB/mm, which is about 50 percent. The table also shows that the computed and measured propagation parameters of the elevated CPW are in good agreement. Table 2.5 compares the strip width and attenuation of conventional CPW with elevated CPW for identical Z_0 and ground-to-ground separation. Clearly, the strip width of the conventional CPW is significantly smaller than that of the elevated CPW. A narrower strip width is consequently responsible for the higher attnuation. Figure 2.22 shows the computed ε_{eff}, Z_0 and attenuation as a function of the ground-to-ground separation for elevated CPW with air dielectric. The strip width and the height h are held fixed at 10 and 2.5 μm, respectively. The range of Z_0 that can be realized is from about 100 to 150 Ω.

As a concluding remark it is interesting to note that the capacitance per unit length and the inductance per unit of the elevated CPW with air dielectric is nearly four times smaller and nearly equal to that of conventional CPW respectively [36]. Hence this structure is well suited for picosecond sampling circuits [37].

TABLE 2.5 Comparison of Strip Width and Attenuation of Conventional CPW with Elevated CPW for Identical Z_0 and $S+2W$

	Conventional CPW	Elevated CPW
S (μm)	1.5	12
α (dB/mm)	0.5	0.1

Note: $Z_0=97\,\Omega$, $S+2W=48\,\mu$m, $h=2.6\,\mu$m, $t=2.9\,\mu$m, $h_1=500\,\mu$m, $\varepsilon_{r1}=12.9$, $\sigma_{gold}=30$ S/μm, $\tan\delta_e=3\times10^{-4}$, $F=20$ GHz.

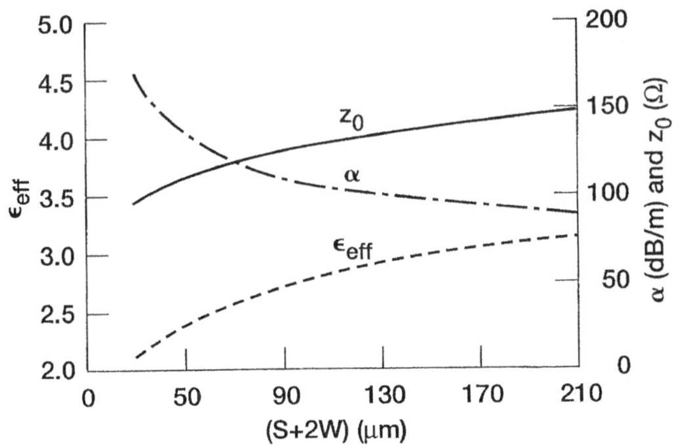

FIGURE 2.22 Computed effective dielectric constant, characteristic impedance, and attenuation as a function of ground-to-ground separation for elevated CPW with air dielectric: $h_1 = \infty$, $\varepsilon_{r1} = 12.9$, $h = 2.5\,\mu\text{m}$, $t = 2.5\,\mu\text{m}$, $S = 10\,\mu\text{m}$, $F = 20$ GHz, $\varepsilon_{r2} = 1.0$. (From Reference [38], © IEEE 1996.)

2.8 COPLANAR WAVEGUIDE WITH GROUND PLANE OR CENTER STRIP CONDUCTOR UNDERPASSES

In a conventional CPW low Z_0 is achieved by fabricating a very narrow slot width and a very wide center strip conductor. The disadvantages of this approach are, first, that the current density at the slot edges is high, which increases conductor losses; second, the wide strip conductor has potential to couple power from the dominant CPW mode to spurious unwanted propagation modes. Therefore conventional CPW is not recommended for lines with Z_0 less than about 30 Ω.

Using multilayer CPW technology, those problems can be addressed. In the new geometry either the ground planes can be extended underneath the center strip conductor, as shown in Figure 2.23(*a*), or the center strip conductor can be extended under the ground planes, as shown in Figure 2.23(*b*) [39, 40]. Because the conductors in the new configurations are overlapping, an almost arbitrarily low Z_0 can be realized without significantly increasing conductor losses. Table 2.6 summarizes the characteristics of the two types of CPW.

2.9 COPLANAR WAVEGUIDE FIELD COMPONENTS

In the design of coplanar waveguide circuits and devices, such as ferrite isolators and circulators, electro-optic modulators, and traveling wave antennas, knowledge of the field components is essential. In this section, the field

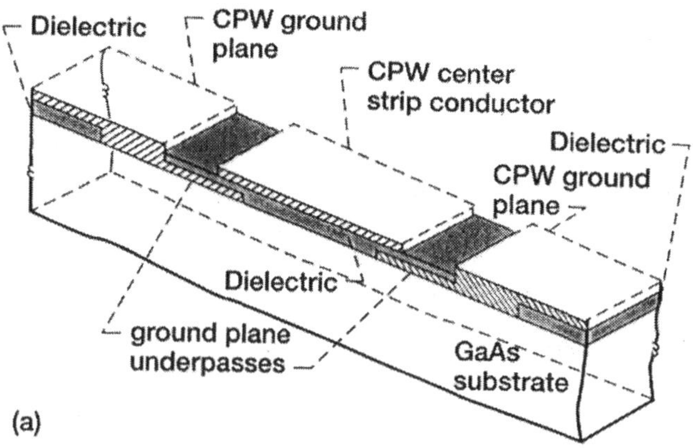

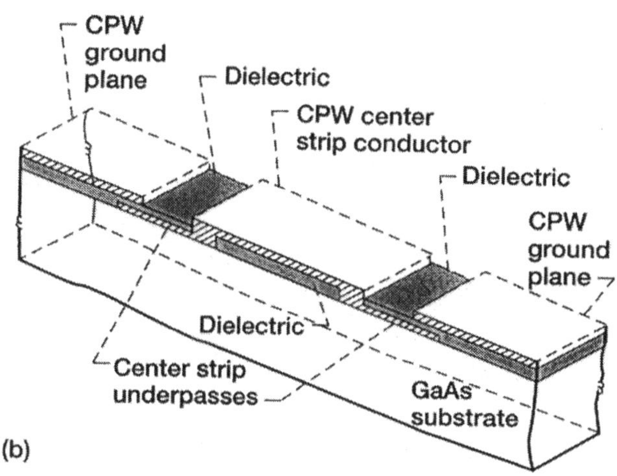

FIGURE 2.23 Low impedance coplanar waveguide: (*a*) With ground plane underpasses; (*b*) with center strip conductor underpasses.

TABLE 2.6 Coplanar Waveguide with Underpass Conductors

	Ground Plane Underpasses			Center Strip Conductor Underpasses
Center strip conductor, μm	20	40	120	40
Slot width, μm	10	10	10	10
Overlap, μm	5	10	20	10
Z_0, Ω	20	15	7.5	10
Loss (dB/mm) at 10 GHz	—	—	—	0.2

Note: Substrate is semi-insulating GaAs of thickness 200 μm and $\varepsilon_r = 12.85$.

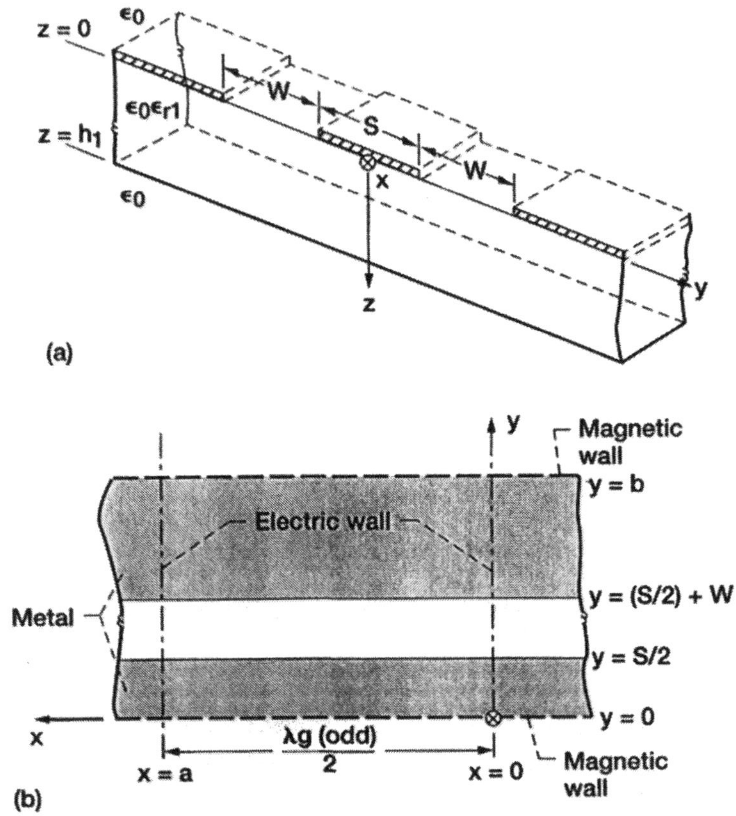

FIGURE 2.24 (*a*) Schematic of the coplanar waveguide; (*b*) Development of waveguide model for coplanar waveguide: magnetic walls at $y = 0$ and $y = b$ and electric walls at $x = 0$ and $x = a$.

components in the air region above the slots, in the substrate and in the air region below the substrate for a CPW are presented [41].

In order to simplify the notation, the coordinate system shown for the CPW in Figure 2.13(*b*) is modified as in Figure 2.24(*a*). For the case of odd excitation, a magnetic wall is placed at the $y = 0$ plane; it then suffices to restrict the analysis to the right half of the structure. The coplanar waveguide problem is reduced to a rectangular waveguide problem by inserting electric wall in the planes perpendicular to the slot at $x = 0$ and $x = a = \lambda_g/2$ (λ_g is the slot mode wavelength) and magnetic wall at $y = b$; this is illustrated in Figure 2.24(*b*) [20].

On the air side of the slot ($z \leqslant 0$), the E_y and E_z components of the electric field and H_x, H_y, and H_z components of the magnetic field exist. From Maxwell's equations it follows that the E_x component of the electric field on the air side of the slot is zero. On the substrate side of the slot E_x, E_y, and E_z

component of the electric field and H_x, H_y and H_z component of the magnetic field exist. The E_y component of the electric field and the H_x component of the magnetic field are determined as explained in [20], while the rest of the electric field and the magnetic field components are determined by the application of Maxwell's equations.

The rectangular coordinates x, y, z, the slot width W, substrate thickness h_1, the center strip conductor width S, and relative permittivity of the substrate material ε_{r1} are indicated in Figure 24(a). A factor $\exp[j(\omega t - 2\pi x/\lambda g)]$ is assumed for each field component, implying wave propagation in the $+x$-direction only; V_0 is the voltage directly across the slot

$$V_0 = \int_{S/2}^{(S/2)+W} E_y dy. \tag{2.126}$$

Air Side of the Slot $z \leqslant 0$

$$E_y = \frac{2V_0}{b} \sum_{n>0}^{\infty} \left[\frac{\sin n\pi/2}{n\pi\delta/2} \sin\frac{n\pi\delta}{2}\right] \sin\frac{n\pi y}{b} e^{-\gamma_n|z|}, \tag{2.127}$$

$$E_z = -\frac{2V_0}{b} \sum_{n>0}^{\infty} \frac{1}{F_n} \left[\frac{\sin n\pi\delta/2}{n\pi\delta/2} \sin\frac{n\pi\delta}{2}\right] \cos\frac{n\pi y}{b} e^{-\gamma_n|z|}, \tag{2.128}$$

$$H_x = -j\frac{2V_0}{\eta b}\left(\frac{\lambda}{\lambda_g}\right)^2 \frac{2b}{\lambda} \sum_{n>0}^{\infty} \frac{1-(\lambda_g/\lambda)^2}{nF_n} \cdot \left[\frac{\sin n\pi\delta/2}{n\pi\delta/2} \sin\frac{n\pi\delta}{2}\right] \sin\frac{n\pi y}{b} e^{-\gamma_n|z|}, \tag{2.129}$$

$$H_y = \frac{2V_0}{\eta b} \frac{\lambda}{\lambda_g} \sum_{n>0}^{\infty} \frac{1}{F_n} \left[\frac{\sin n\pi\delta/2}{n\pi\delta/2} \sin\frac{n\pi\delta}{2}\right] \cos\frac{n\pi y}{b} e^{-\gamma_n|z|}, \tag{2.130}$$

$$H_z = \frac{2V_0}{\eta b} \frac{\lambda}{\lambda_g} \sum_{n>0}^{\infty} \left[\frac{\sin n\pi\delta/2}{n\pi\delta/2} \sin\frac{n\pi\delta}{2}\right] \cos\frac{n\pi y}{b} e^{-\gamma_n|z|}. \tag{2.131}$$

Substrate Side of the Slot $0 \leqslant z \leqslant h_1$

$$E_x = j\frac{2V_0}{\lambda_g} \sum_{n>0}^{\infty} \frac{2}{n[1+(2b/n\lambda_g)^2} \left[\frac{\sin n\pi\delta/2}{n\pi\delta/2} \sin\frac{n\pi\delta}{2}\right] \cdot \left\{\cos\frac{n\pi y}{b} [\coth q_n - \tanh r_n] \sinh \gamma_{n1} z\right\}, \tag{2.132}$$

$$E_y = \frac{2V_0}{b} \sum_{n>0}^{\infty} \left[\frac{\sin n\pi\delta/2}{n\pi\delta/2} \sin\frac{n\pi\delta}{2}\right] \sin\frac{n\pi y}{b} \cdot \left\{\cosh \gamma_{n1} z - \left[\frac{\tanh r_n + (2b/n\lambda_g)^2 \coth q_n}{1+(2b/\lambda_g)^2}\right] \sinh \gamma_{n1} z\right\}, \tag{2.133}$$

$$E_z = -\frac{2V_0}{b}\sum_{n>0}^{\infty}\frac{1}{F_{n1}}\left[\frac{\sin n\pi\delta/2}{n\pi\delta/2}\sin\frac{n\pi\bar{\delta}}{2}\right]\cos\frac{n\pi y}{b}\cdot\{\sinh\gamma_{n1}z - \tanh r_n\cosh\gamma_{n1}z\}, \tag{2.134}$$

$$H_x = j\frac{2V_0}{b\eta}\left(\frac{\lambda}{\lambda_g}\right)^2\frac{2b}{\lambda}\sum_{n>0}^{\infty}\frac{1}{nF_{n1}}\cdot\left[\frac{\sin n\pi\delta/2}{n\pi\delta/2}\sin\frac{n\pi\bar{\delta}}{2}\right]\sin\frac{n\pi y}{b}$$
$$\cdot\left\{\left[\frac{F_{n1}^2\coth q_n - \varepsilon_r(\lambda_g/\lambda)\tanh r_n}{1+(2b/n\lambda_g)^2}\right]\cosh\gamma_{n1}z - [1-\varepsilon_r(\lambda_g/\lambda)^2]\sinh\gamma_{n1}z\right\}, \tag{2.135}$$

$$H_y = -\frac{2V_0}{\eta b}\frac{\lambda}{\lambda_g}\sum_{n>0}^{\infty}\frac{1}{F_{n1}}\left[\frac{\sin n\pi\delta/2}{n\pi\delta/2}\sin\frac{n\pi\bar{\delta}}{2}\right]\cos\frac{n\pi y}{b}$$
$$\cdot\left\{\left[\frac{F_{n1}^2\coth q_n + \varepsilon_r(2b/n\lambda)^2\tanh r_n}{1+(2b/n\lambda_g)^2}\right]\cdot\cosh\gamma_{n1}z - \sinh\gamma_{n1}z\right\}, \tag{2.136}$$

$$H_z = \frac{2V_0}{\eta b}\frac{\lambda}{\lambda_g}\sum_{n>0}^{\infty}\left[\frac{\sin n\pi\delta/2}{n\pi\delta/2}\sin\frac{n\pi\bar{\delta}}{2}\right]\sin\frac{n\pi y}{b}\cdot[\cosh\gamma_{n1}z - \coth q_n\sinh\gamma_{n1}z]. \tag{2.137}$$

Substrate Side of the Slot $z \geqslant h_1$

The expressions for the field components on the substrate side of the slot $z \geqslant h_1$ are derived from Eqs. (2.132) to (2.137) by replacing $\gamma_{n1}z$ by $\gamma_{n1}h_1$. Further the equations are multiplied by the factor $\exp[-\gamma_n(z-h_1)]$, indicating that the fields decay exponentially. Symbols not defined above are $\eta = 376.7\,\Omega$, $\delta = W/b$, $\bar{\delta} = (S+W)/b$, and

$$F_n = \frac{b\gamma_n}{n\pi} = \sqrt{1+\left(\frac{2bv}{n\lambda}\right)^2}, \tag{2.138}$$

$$F_{n1} = \frac{b\gamma_{n1}}{n\pi} = \sqrt{1-\left(\frac{2bu}{n\lambda}\right)^2}, \tag{2.139}$$

$$v = \sqrt{\left(\frac{\lambda}{\lambda_g}\right)^2 - 1}, \quad u = \sqrt{\varepsilon_{r1} - \left(\frac{\lambda}{\lambda_g}\right)^2}, \tag{2.140}$$

$$r_n = \gamma_{n1}h_1 + \tanh^{-1}\left(\frac{F_{n1}}{\varepsilon_{r1}F_n}\right), \tag{2.141}$$

$$q_n = \gamma_{n1}h_1 + \coth^{-1}\left(\frac{F_n}{F_{n1}}\right). \tag{2.142}$$

λ and λ_g are the free space wavelength and guide wavelength, respectively.

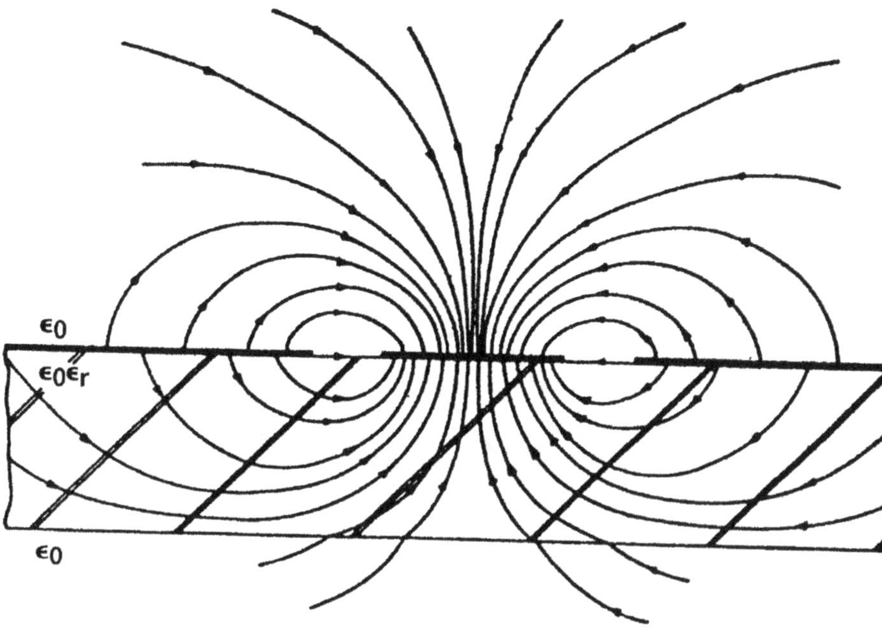

FIGURE 2.25 Computed electric-field distribution in the cross section ($x = 0$ plane).

Figures 2.25 and 2.26 illustrate the computed electric field and magnetic field, respectively, in the cross section of the coplanar waveguide. The ε_{r1} is 16, $h_1/\lambda = 0.07$, $S/h_1 = 1$, $W/h_1 = 0.4$, the frequency is equal to 3 GHz, and $b \to \infty$ Since the expressions involve summing an infinite series, the following criterion for terminating the series at n_t is adopted: $n_t = n_0/(1 + z/z_1)$, where n_0 and z_1 are constants. In the above case $n_0 = 1000$ and $z_1 = 0.005$ in. are found suitable.

It is observed that the electric-field lines extend across the slot while the magnetic-field lines are perpendicular to the air-dielectric interface in the slot. The electric and magnetic field in the right half of the structure are in a direction opposite to the electric and magnetic field in the left half of the structure. Furthermore part of the magnetic-field lines encircle the center conducting strip separating the two slots. Hence it should be possible to realize CPW circulators whose function is dominated by the transverse magnetic field component. The longitudinal view in Figure 2.27 shows that in the air regions the magnetic-field lines curve and return to the slot at half-wavelength intervals. Consequently a wave propagating along the structure has an elliptically polarized magnetic field. Hence it should be possible to successfully exploit the elliptically polarized magnetic field in the design of CPW resonance isolators and differential phase shifters.

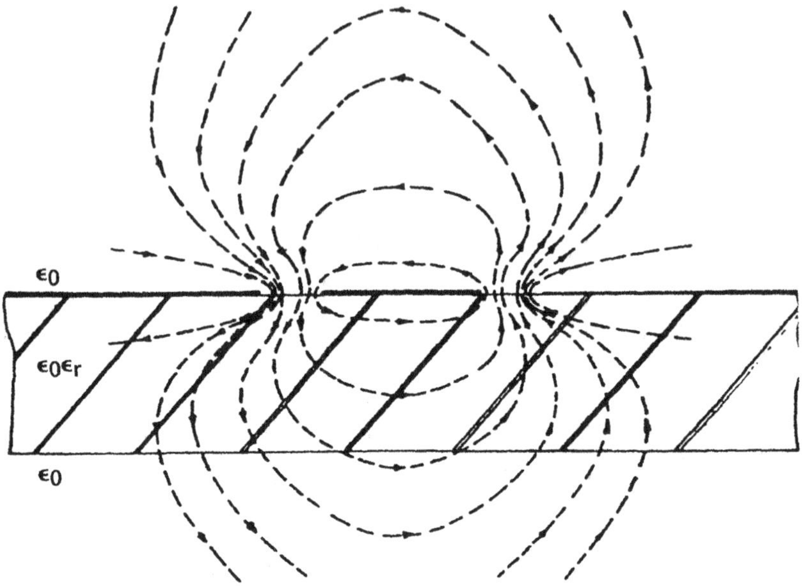

FIGURE 2.26 Computed magnetic-field distribution in the cross section ($x = 0$ plane).

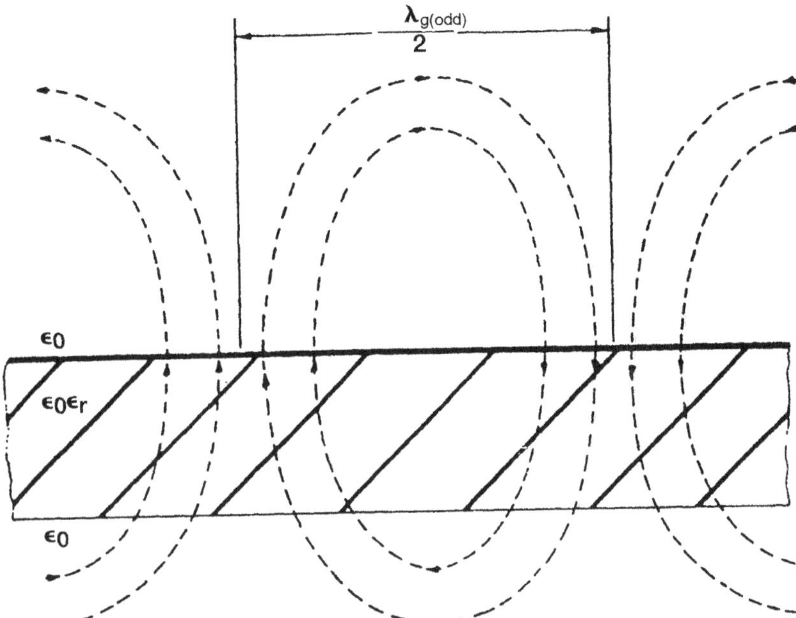

FIGURE 2.27 Computed magnetic field in the longitudinal section ($y = (S + W)/2$ plane) through the slot.

2.10 COPLANAR WAVEGUIDE ON A CYLINDRICAL SURFACE

2.10.1 Analytical Expressions Based on Quasi-static Conformal Mapping Technique

A CPW on the outside and the inside surfaces of a cylindrical dielectric tube is shown in Figures 2.28(*a*) and (*b*). In the figure the CPW center strip conductor width and the slot width are designated as S and W, respectively. The inner and outer radius and the relative permittivity of the dielectric tube are designated as a, b, and ε_r, respectively. The CPW ground-to-ground separation is designated as d. The thickness t_m of the metallizations is assumed to be negligible. The dielectric material that constitutes the tube is assumed to be loss less, and the conductivity of the metal coating is assumed to be perfect.

The analysis of the CPW on the outside surface [42] is first presented. The sequence of conformal mapping to transform the structure of Figure 2.28(*a*) into a parallel plate capacitor is shown in Figure 2.29. The first step is to transform the CPW on the outside surface of a cylinder into a planar CPW with finite ground planes as shown in Figure 2.29(*a*) through the mapping [42]

$$z = -j\ln\left(\frac{b}{r}\right) + \phi. \tag{2.143}$$

The second step is to calculate the capacitance of the structure with the dielectric replaced by free space. To facilitate this, the first quadrant of Figure 2.29(*a*) is transformed into the upper half of the t-plane as shown in Figure 2.29(*b*) through the mapping [42]

$$t = z^2 \tag{2.144}$$

and then into a parallel plate region as shown in Figure 2.29(*c*) through the mapping [42]

$$w = \int_{to}^{t} \frac{dt}{\sqrt{t(t - t_1)(t - t_2)(t - t_3)}}. \tag{2.145}$$

The capacitance of the structure considering all four quadrants is given by [42]

$$C_a = 4\varepsilon_0 \frac{K(k_1)}{K(\sqrt{1 - k_1^2})}, \tag{2.146}$$

where

$$k_1 = \frac{S}{S + 2W}\sqrt{\frac{1 - (S + 2W)^2/4b^2\pi^2}{1 - S^2/4b^2\pi^2}} \tag{2.147}$$

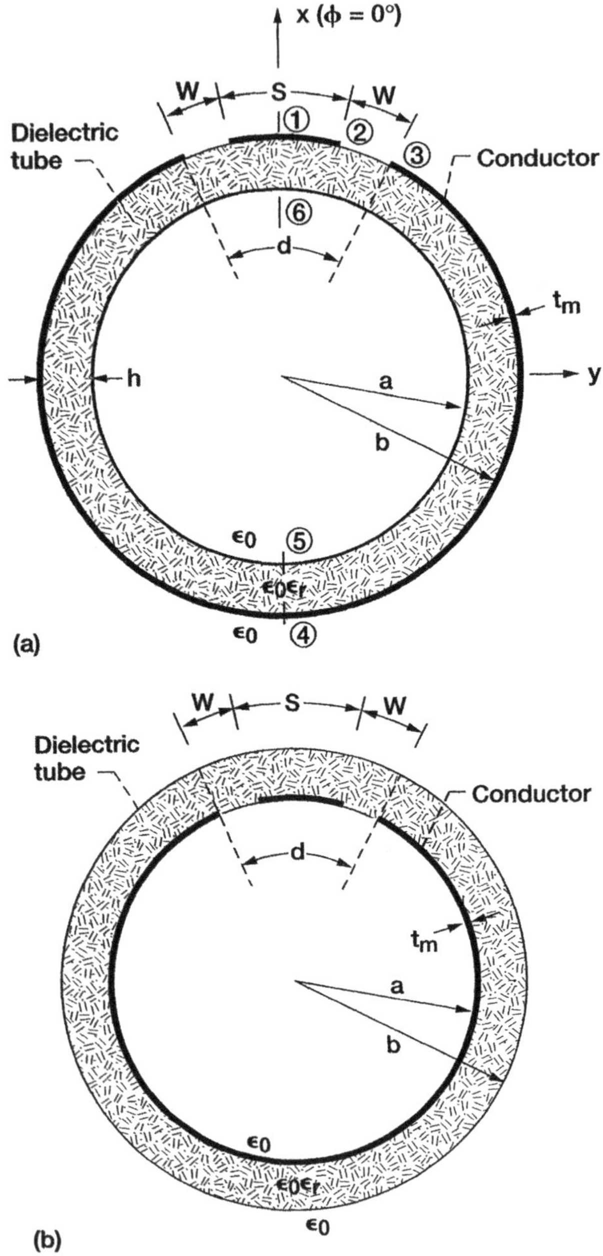

FIGURE 2.28 The geometry of a cylindrical coplanar waveguide: (*a*) On the outside surface, (*b*) on the inside surface.

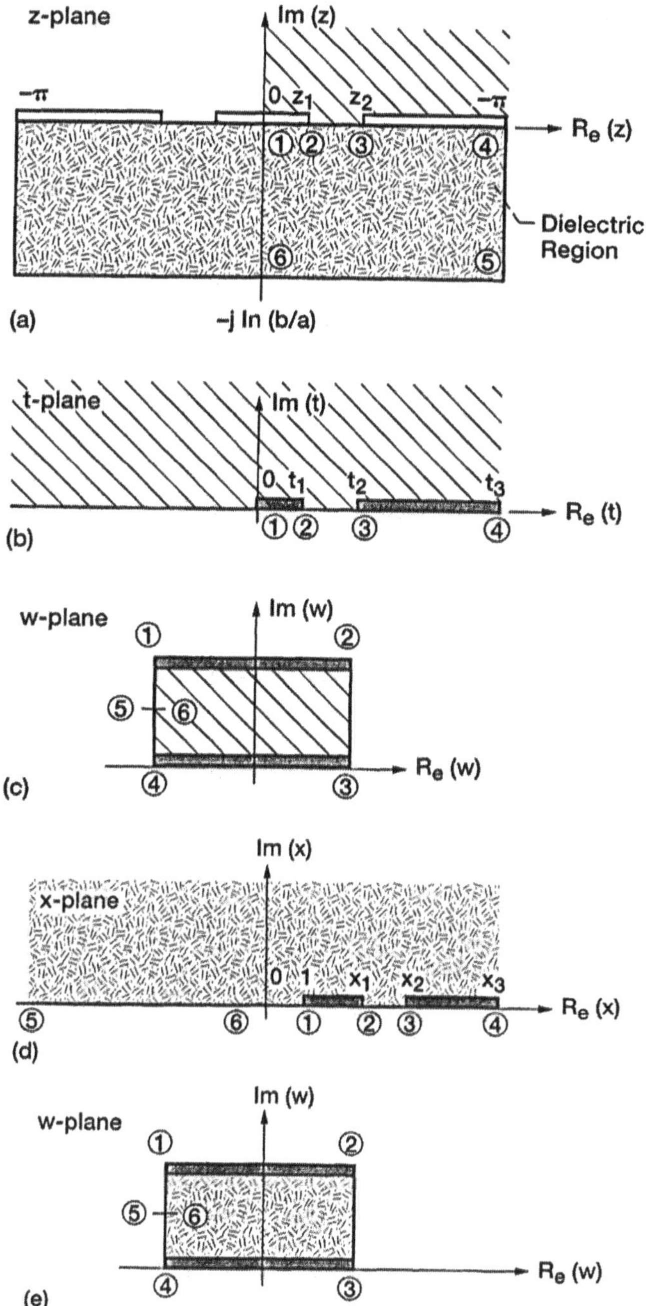

FIGURE 2.29 Conformal mapping steps for a CPW on the outside surface of a cylinder: (*a*) Intermediate transformation into a planar structure; (*b*) intermediate transformation for the air region; (*c*) mapping into a parallel plate capacitor for the air region, $\mathrm{Re}(z) > 0$; (*d*) intermediate transformation for the dielectric region, $\mathrm{Re}(z) > 0$; (*e*) mapping into parallel plate capacitor for the dielectric region, $\mathrm{Re}(z) > 0$.

and K is the complete elliptic integral of the first kind. The third step is to calculate the capacitance of the structure with the dielectric substrate replaced by a substrate of relative permittivity $\varepsilon_r - 1$. To facilitate this computation, the dielectric-air interfaces are replaced by magnetic walls. Then, through the mapping [42]

$$x = \cosh^2\left[\frac{\pi z}{2\ln(b/a)}\right] \tag{2.148}$$

and

$$w = \int_{x_0}^{x} \frac{dx}{\sqrt{(x-1)(x-x_1)(x-x_2)(x-x_3)}}, \tag{2.149}$$

the right half of dielectric region in Figure 2.29(a) is first transformed into the upper half of the x-plane and then transformed into a parallel plate region as shown in Figure 2.29(d) and (e). The capacitance of the structure considering both the left and the right half is [42]

$$C_\mathrm{d} = 2\varepsilon_0(\varepsilon_r - 1)\frac{K(k_2)}{K(\sqrt{1-k_2^2})}, \tag{2.150}$$

where

$$k_2 = \frac{\sinh(AS)}{\sinh(A(S+2W))}\sqrt{\frac{1-\sinh^2(A(S+2W))/\sinh^2(2Ab\pi)}{1-\sinh^2(AS)/\sinh^2(2Ab\pi)}}, \tag{2.151}$$

$$A = \frac{\pi}{4b\ln(b/a)}. \tag{2.152}$$

The total capacitance C per unit length of the original structure is the sum of the capacitances C_a and C_d. That is,

$$C = C_\mathrm{a} + C_\mathrm{d} \tag{2.153}$$

The effective dielectric constant is given by [42]

$$\begin{aligned}\varepsilon_\mathrm{eff} &= \frac{C}{C_\mathrm{a}} \\ &= 1 + \frac{C_\mathrm{d}}{C_\mathrm{a}} \\ &= 1 + \frac{(\varepsilon_r - 1)}{2}\cdot\frac{K(\sqrt{1-k_1^2})}{K(k_1)}\cdot\frac{K(k_2)}{K(\sqrt{1-k_2^2})}.\end{aligned} \tag{2.154}$$

The characteristic impedance is given by [42]

$$
\begin{aligned}
Z_0 &= \frac{\sqrt{\varepsilon_{\text{eff}}}}{Cc} \\
&= \frac{120\pi\varepsilon_0}{C_{\text{a}}\sqrt{\varepsilon_{\text{eff}}}} \\
&= \frac{30\pi}{\sqrt{\varepsilon_{\text{eff}}}} \frac{K(\sqrt{1-k_1^2})}{K(k_1)},
\end{aligned}
\tag{2.155}
$$

where c is the velocity of light in free space. The effective dielectric constant and characteristic impedance for a CPW on the inside surface of a cylindrical dielectric tube has the same form as Eqs. (2.154) and (2.155), except that a and b are interchanged [42]. Finally, it may be mentioned that the foregoing structures have been analyzed using full-wave techniques and the results can be found in [43] and [44].

2.10.2 Computed Effective Dielectric Constant and Characteristic Impedance

The quasi-static ε_{eff} and Z_0 are computed via Eqs. (2.154) and (2.155), respectively. The computed ε_{eff} and Z_0 as a function of S/d for a CPW on the outside surface of a cylinder are presented in Figure 2.30(a) and (b) [42]. The characteristics are illustrated for two different ground plane separations, namely $d = h$ and $d = 4h$. In Figure 2.30(a) and (b) the computed ε_{eff} and Z_0 for a planar CPW [5] are included for comparison. The curvature effect is significant for large CPW, that is, when $d = 4h$. It is also observed that in decreasing the curvature, that is, increasing the ratio a/b, the ε_{eff} decreases. In the limit where the curvature becomes very small, that is, where the ratio a/b approaches unity, the ε_{eff} approaches that of the planar structure. The Z_0, however, is much less sensitive to the curvature, specially for small CPW, that is, when $d = h$. These results contrast with the case of a CPW on the inside surface for which the ε_{eff} increases as the curvature decreases.

2.11 EFFECT OF METALLIZATION THICKNESS ON COPLANAR WAVEGUIDE CHARACTERISTICS

A conventional coplanar waveguide is shown in Figure 2.5. For this structure the analysis presented in earlier sections ignored the thickness of the metal conductors. However, in the analysis presented in [45], [46] and [47] conductor thickness is taken into account. In these references the analysis is carried out using quasi-static method, frequency dependent network analytical method

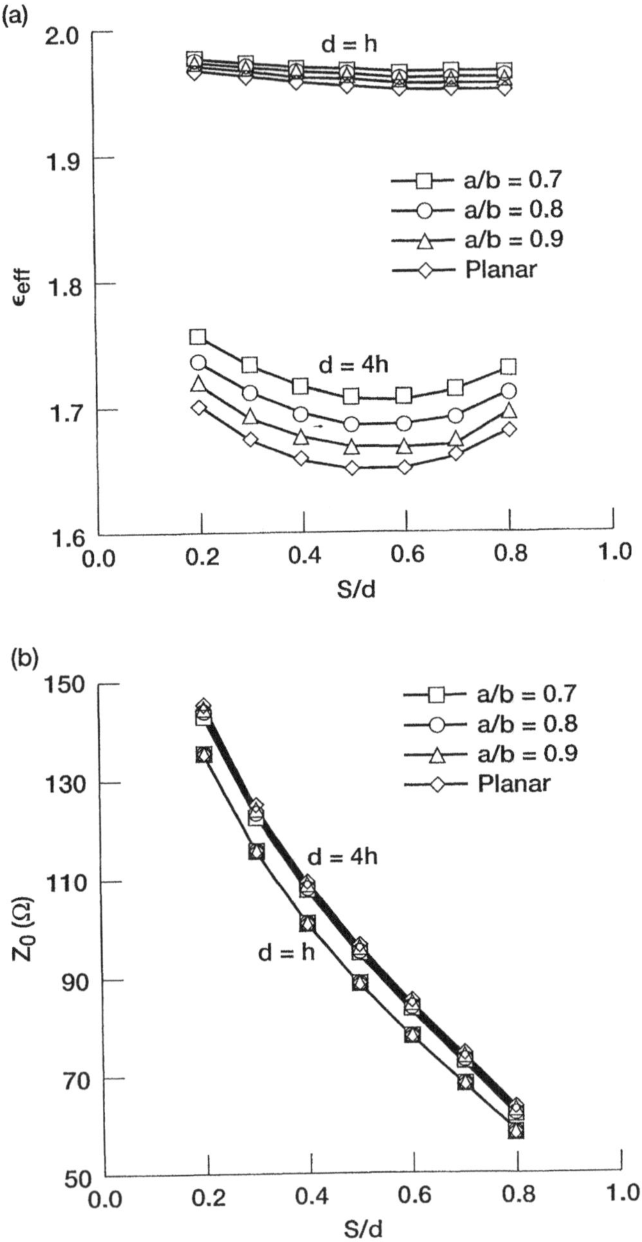

FIGURE 2.30 Computed characteristics for a coplanar waveguide on the outside surface of a dielectric tube, $\varepsilon_r = 3.0$, $h = 1.524\,\text{mm}$: (*a*) Effective dielectric constant; (*b*) characteristic impedance. (From Reference [42], © IEEE 1997.)

of electromagnetic fields, and spectral domain method, respectively. The computed Z_0 using the spectral domain method [26] with metal thickness t and separation between ground planes $S + 2W$ as parameters is presented in Figure 2.31(*a*) and (*b*). Figure 2.31(*a*) shows that for fixed $S + 2W$ and aspect ratio $S/(S + 2W)$, the Z_0 decreases as t increases. Figure 2.31(*b*) shows that for fixed t and $S/(S + 2W)$, the Z_0 increases as $S + 2W$ increases. In [45] it has been shown that the effect of t on Z_0 is smaller for substrates of higher dielectric constant.

The computed ε_{eff} is shown in Figure 2.31(*c*) for fixed $S + 2W = 25\ \mu m$. The ε_{eff} deviates appreciably as t increases. Notice that increasing $S + 2W$ from $25\ \mu m$ to $100\ \mu m$ has a small effect on ε_{eff}. In [45] it has been shown that the effect of t on ε_{eff} is larger for substrates of higher dielectric constant.

APPENDIX 2A: SPECTRAL DOMAIN DYADIC GREEN'S FUNCTION COMPONENTS

Equations for the tangential field components E_{zi}, H_{zi}, E_{xi}, and H_{xi} from Eqs. (2.58b), (2.59b), (2.60a), and (2.60c) are substituted into the set of continuity Eqs. (2.69a) to (2.69d) at the interface $y = 0$. The resulting equations are transformed into the Fourier domain by taking the transform with respect to x. This process yields the following equations at the interface $y = 0$:

$$k_{c2}^2 \tilde{\phi}_2(\alpha, 0) = k_{c0}^2 \tilde{\phi}_3(\alpha, 0), \tag{2A.1a}$$

$$-j\alpha\gamma\tilde{\phi}_2(\alpha, 0) - j\omega\mu_0 \frac{\partial}{\partial y} \tilde{\psi}_2(\alpha, 0) = -j\alpha\gamma\tilde{\phi}_3(\alpha, 0) - j\omega\mu_0 \frac{\partial}{\partial y} \tilde{\psi}_3(\alpha, 0), \tag{2A.1b}$$

$$k_{c2}^2 \tilde{\psi}_2(\alpha, 0) = k_{c0}^2 \tilde{\psi}_3(\alpha, 0), \tag{2A.1c}$$

$$-j\alpha\gamma\tilde{\psi}_2(\alpha, 0) + j\omega\varepsilon_2 \frac{\partial}{\partial y} \tilde{\phi}_2(\alpha, 0) = -j\alpha\gamma\tilde{\psi}_3(\alpha, 0) + j\omega\varepsilon_0 \frac{\partial}{\partial y} \tilde{\phi}_3(\alpha, 0), \tag{2A.1d}$$

where the transform is expressed by a tilde above the symbol.

Next the equations for the tangential fields are substituted into the set of continuity Eqs. (2.70a) to (2.70f) at the interface $y = h_1$. Once again, upon taking the Fourier transform, the following equations are obtained:

$$k_{c0}^2 \tilde{\phi}_1(\alpha, h_1) = k_{c2}^2 \tilde{\phi}_2(\alpha, h_1), \tag{2A.2a}$$

$$-j\alpha\gamma\tilde{\phi}_1(\alpha, h_1) - j\omega\mu_0 \frac{\partial}{\partial y} \tilde{\psi}_1(\alpha, h_1) = -j\alpha\gamma\tilde{\phi}_2(\alpha, h_1) - j\omega\mu_0 \frac{\partial}{\partial y} \tilde{\psi}_2(\alpha, h_1), \tag{2A.2b}$$

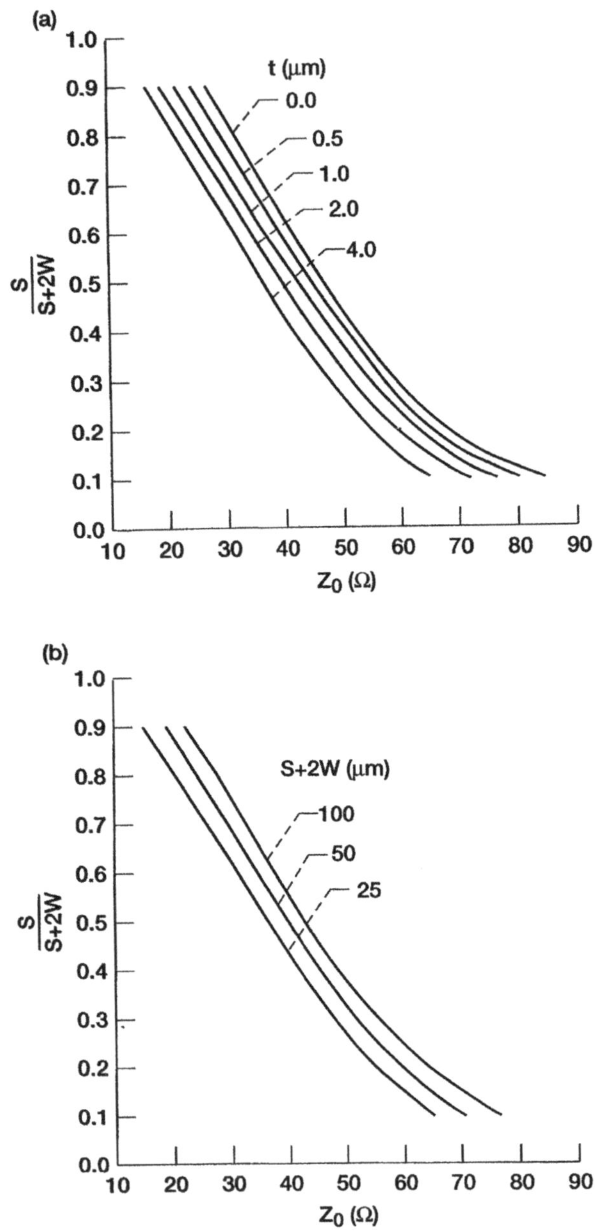

FIGURE 2.31 Computed characteristics of conventional coplanar waveguide, $\varepsilon_{r1} = 12.9$, $h_1 = 500\,\mu\text{m}$: (*a*) The aspect ratio as a function of Z_0 with metallization thickness as parameter, $S + 2W = 25\,\mu\text{m}$; (*b*) the aspect ratio as a function of Z_0 with $S + 2W$ as parameter, $t = 4\,\mu\text{m}$; (*c*) the aspect ratio as a function normalized metallization thickness with ε_{eff} as parameter, $S + 2W = 25\,\mu\text{m}$. (From Reference [26], © IEEE 1991.)

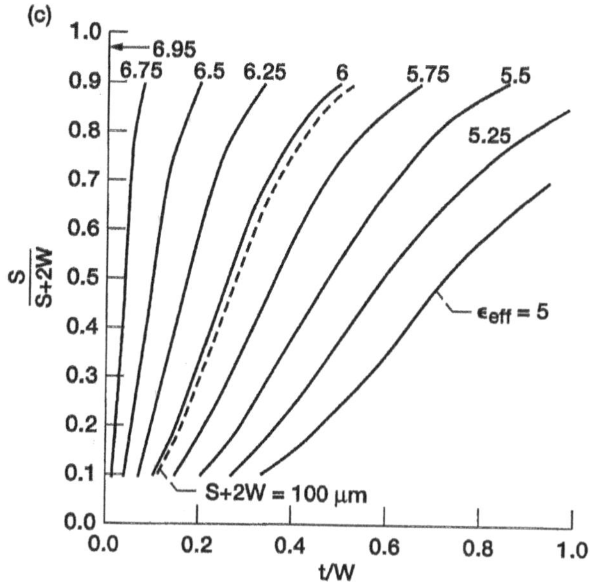

FIGURE 2.31 Continued.

$$k_{c0}^2\tilde{\phi}_1(\alpha, h_1) = \tilde{E}_z(\alpha), \tag{2A.2c}$$

$$-j\alpha\gamma\tilde{\phi}_1(\alpha, h_1) - j\omega\mu_0 \frac{\partial}{\partial y}\tilde{\psi}_1(\alpha, h_1) = \tilde{E}_x(\alpha), \tag{2A.2d}$$

$$k_{c0}^2\tilde{\psi}_1(\alpha, h_1) - k_{c2}^2\tilde{\psi}_2(\alpha, h_1) = \tilde{J}_x(\alpha), \tag{2A.2e}$$

$$-j\alpha\gamma\tilde{\psi}_1(\alpha, h_1) + j\omega\varepsilon_0 \frac{\partial}{\partial y}\tilde{\phi}_1(\alpha, h_1) + j\alpha\gamma\tilde{\psi}_2(\alpha, h_1) - j\omega\varepsilon_2 \frac{\partial}{\partial y}\tilde{\phi}_2(\alpha, h_1)$$
$$= \tilde{J}_z(\alpha, h_1), \tag{2A.2f}$$

where

$$\mu_1 = \mu_2 = \mu_3 = \mu_0,$$
$$\varepsilon_1 = \varepsilon_3 = \varepsilon_0,$$

and

$$\varepsilon_2 = \varepsilon_{r1} \text{ so that}$$
$$k_{c1} = k_{c3} = k_{c0}.$$

The two Helmholtz Eqs. (2.57a) and (2.57b) whose solutions are Eqs. (2.67) and (2.68) are now substituted into Eqs. (2A.1) and (2A.2). The coefficients A^e through D^h are functions of α as indicated within parenthesis. However, to keep the notations simple, within the parentheses α with the coefficients is omitted in the mathematical steps that follow:

$$k_{c2}^2 C^e = k_{c0}^2 D^e, \tag{2A.3a}$$

$$-j\alpha\gamma C^e - j\omega\mu_0\gamma_2 B^h = -j\alpha\gamma D^e - j\omega\mu_0\gamma_3 D^h, \tag{2A.3b}$$

$$k_{c2}^2 C^h = k_{c0}^2 D^h, \tag{2A.3c}$$

$$-j\alpha\gamma C^h + j\omega\varepsilon_2\gamma_2 B^e = -j\alpha\gamma D^h + j\omega\varepsilon_0 D^e, \tag{2A.3d}$$

$$k_{c0}^2 A^e = k_{c2}^2[B^e \sinh\gamma_2 h_1 + C^e \cosh\gamma_2 h_1], \tag{2A.4a}$$

$$\begin{aligned}-j\alpha\gamma A^e + j\omega\mu_0\gamma_1 A^h = &-j\alpha\gamma[B^e \sinh\gamma_2 h_1 + C^e \cosh\gamma_2 h_1]\\ &-j\omega\mu_0\gamma_2[B^h \cosh\gamma_2 h_1 + C^h \sinh\gamma_2 h_1],\end{aligned} \tag{2A.4b}$$

$$k_{c0}^2 A^e = \tilde{E}_z(\alpha), \tag{2A.4c}$$

$$-j\alpha\gamma A^e + j\omega\mu_0\gamma_1 A^h = \tilde{E}_x(\alpha), \tag{2A.4d}$$

$$k_{c0}^2 A^h - k_{c2}^2[B^h \sinh\gamma_2 h_1 + C^h \cosh\gamma_2 h_1] = \tilde{J}_x(\alpha), \tag{2A.4e}$$

$$\begin{aligned}&-j\alpha\gamma A^h - j\omega\varepsilon_0\gamma_1 A^e + j\alpha\gamma[B^h \sinh\gamma_2 h_1 + C^h \cosh\gamma_2 h_1] - j\omega\varepsilon_2\gamma_2\\ &\qquad \times [B^e \cosh\gamma_2 h_1 + C^e \sinh\gamma_2 h_1] = \tilde{J}_z(\alpha).\end{aligned} \tag{2A.4f}$$

In the mathematical steps that follow, the unknown coefficients A^e through A^h are replaced to obtain a direct relation between the current density and the electric field components.

From Eqs. (2A.3a) and (2A.3c),

$$D^e = \left(\frac{k_{c2}}{k_{c0}}\right)^2 C^e, \tag{2A.5}$$

$$D^h = \left(\frac{k_{c2}}{k_{c0}}\right)^2 C^h. \tag{2A.6}$$

Substituting D^e and D^h into Eqs. (2A.3b) and (2A.3d), B^h and B^e are expressed as

$$B^h = \frac{\alpha\gamma}{\omega\mu_0\gamma_2}\left[\left(\frac{k_{c2}}{k_{c0}}\right)^2 - 1\right] C^e + \frac{\gamma_3}{\gamma_2}\left(\frac{k_{c2}}{k_{c0}}\right)^2 C^h, \tag{2A.7}$$

$$B^e = -\frac{\alpha\gamma}{\omega\mu_0\gamma_2}\left[\left(\frac{k_{c2}}{k_{c0}}\right)^2 - 1\right]\frac{\mu_0}{\varepsilon_2} C^h + \frac{\gamma_3}{\gamma_2}\left(\frac{k_{c2}}{k_{c0}}\right)^2 \frac{\varepsilon_0}{\varepsilon_2} C^e. \tag{2A.8}$$

Let us define two new variables F_1 and F_2 such that

$$F_1 = \frac{\alpha\gamma}{\omega\mu_0\gamma_2}\left[\left(\frac{k_{c2}}{k_{c0}}\right)^2 - 1\right], \tag{2A.9}$$

$$F_2 = \frac{\gamma_3}{\gamma_2}\left(\frac{k_{c2}}{k_{co}}\right)^2. \tag{2A.10}$$

With these substitutions the equations for B^h and B^e simplify to

$$B^h = F_1 C^e + F_2 C^h, \tag{2A.11}$$

$$B^e = \frac{\varepsilon_0}{\varepsilon_2} F_2 C^e - \frac{\mu_0}{\varepsilon_2} F_1 C^h. \tag{2A.12}$$

Substitute Eq. (2A.12) into Eq. (2A.4a) to obtain

$$A^e = \left(\frac{k_{c2}}{k_{co}}\right)^2 \left(\frac{\varepsilon_0}{\varepsilon_2} F_2 \sinh \gamma_2 h_1 + \cosh \gamma_2 h_1\right) C^e - \left\{\left(\frac{k_{c2}}{k_{co}}\right)^2 \frac{\mu_0}{\varepsilon_2} F_1 \sinh \gamma_2 h_1\right\} C^h. \tag{2A.13}$$

Define two new variables G_1 and G_2 as follows:

$$G_1 = \left(\frac{k_{c2}}{k_{co}}\right)^2 \left(\frac{\varepsilon_0}{\varepsilon_2} F_2 \sinh \gamma_2 h_1 + \cosh \gamma_2 h_1\right), \tag{2A.14}$$

$$G_2 = -\left(\frac{k_{c2}}{k_{co}}\right)^2 \frac{\mu_0}{\varepsilon_2} F_1 \sinh \gamma_2 h_1. \tag{2A.15}$$

With these substitutions

$$A^e = G_1 C^e + G_2 C^h. \tag{2A.16}$$

Substitute Eqs. (2A.11), (2A.12), and (2A.16) into Eq. (2A.4b) to obtain

$$\begin{aligned} A^h = &\left[\frac{\alpha\gamma}{\omega\mu_0\gamma_1}\left(G_1 - \frac{\varepsilon_0}{\varepsilon_2} F_2 \sinh \gamma_2 h_1 - \cosh \gamma_2 h_1\right) - \frac{\gamma_2}{\gamma_1} F_1 \cosh \gamma_2 h_1\right] C^e \\ &+ \left[\frac{\alpha\gamma}{\omega\mu_0\gamma_1}\left(G_2 + \frac{\mu_0}{\varepsilon_2} F_1 \sinh \gamma_2 h_1\right) - \frac{\gamma_2}{\gamma_1}(F_2 \cosh \gamma_2 h_1 + \sinh \gamma_2 h_1)\right] C^h. \end{aligned} \tag{2A.17}$$

Define two new variables G_3 and G_4 as follows:

$$G_3 = \frac{\alpha\gamma}{\omega\mu_0\gamma_1}\left(G_1 - \frac{\varepsilon_0}{\varepsilon_2} F_2 \sinh \gamma_2 h_1 - \cosh \gamma_2 h_1\right) - \frac{\gamma_2}{\gamma_1} F_1 \cosh \gamma_2 h_1, \tag{2A.18}$$

$$G_4 = \frac{\alpha\gamma}{\omega\mu_0\gamma_1}\left(G_2 + \frac{\mu_0}{\varepsilon_2} F_1 \sinh \gamma_2 h_1\right) - \frac{\gamma_2}{\gamma_1}(F_2 \cosh \gamma_2 h_1 + \sinh \gamma_2 h_1). \tag{2A.19}$$

Then

$$A^{\rm h} = G_3 C^{\rm e} + G_4 C^{\rm h}. \tag{2A.20}$$

Equations (2A.16) and (2A.20) are rewritten in matrix form as

$$\begin{bmatrix} G_1 & G_2 \\ G_3 & G_4 \end{bmatrix} \begin{bmatrix} C^{\rm e} \\ C^{\rm h} \end{bmatrix} \begin{bmatrix} A^{\rm e} \\ A^{\rm h} \end{bmatrix}, \tag{2A.21a}$$

that is,

$$[G][C] = [A]. \tag{2A.21b}$$

Inverting the $[G]$ and solving for $[C]$ yields

$$[C] = [G]^{-1}[A]. \tag{2A.22a}$$

That is,

$$C^{\rm e} = J_1 A^{\rm e} + J_2 A^{\rm h}, \tag{2A.22b}$$

$$C^{\rm h} = J_3 A^{\rm e} + J_4 A^{\rm h}, \tag{2A.22c}$$

where

$$J_1 = G_4/\det G$$
$$J_2 = -G_2/\det G$$
$$J_3 = -G_3/\det G$$
$$J_4 = G_1/\det G$$

with $\det G = G_1 G_4 - G_2 G_3$.

Substituting Eqs. (2A.22b) and (2A.22c) into Eqs. (2A.11) and (2A.12), we have

$$B^{\rm e} = \left(\frac{\varepsilon_0}{\varepsilon_2} F_2 J_1 - \frac{\mu_0}{\varepsilon_2} F_1 J_3\right) A^{\rm e} + \left(\frac{\varepsilon_0}{\varepsilon_2} F_2 J_2 - \frac{\mu_0}{\varepsilon_2} F_1 J_4\right) A^{\rm h}, \tag{2A.23}$$

$$B^{\rm h} = (F_1 J_1 + F_2 J_3) A^{\rm e} + (F_1 J_2 + F_2 J_4) A^{\rm h}. \tag{2A.24}$$

Define the four new variables K_1, K_2, K_3, and K_4 as follows:

$$K_1 = \frac{\varepsilon_0}{\varepsilon_2} F_2 J_1 - \frac{\mu_0}{\varepsilon_2} F_1 J_3, \tag{2A.25a}$$

$$K_2 = \frac{\varepsilon_0}{\varepsilon_2} F_2 J_2 - \frac{\mu_0}{\varepsilon_2} F_1 J_4, \tag{2A.25b}$$

$$K_3 = F_1 J_1 + F_2 J_3, \tag{2A.25c}$$

$$K_4 = F_1 J_2 + F_2 J_4. \tag{2A.25d}$$

Then

$$B^{\mathrm{e}} = K_1 A^{\mathrm{e}} + K_2 A^{\mathrm{h}}, \tag{2A.26a}$$

$$B^{\mathrm{h}} = K_3 A^{\mathrm{e}} + K_4 A^{\mathrm{h}}. \tag{2A.26b}$$

Substitute Eqs. (2A.22b), (2A.22c), (2A.26a), and (2A.26b) into Eqs. (2A.4e) and (2A.4f) to obtain

$$-k_{c2}^2(K_3 \sinh \gamma_2 h_1 + J_3 \cosh \gamma_2 h_1)A^{\mathrm{e}} + k_{c0}^2\left[1 - \frac{k_{c2}^2}{k_{c0}^2}(K_4 \sinh \gamma_2 h_1 + J_4 \cosh \gamma_2 h_1)\right]A^{\mathrm{h}} = \tilde{J}_x(\alpha), \tag{2A.27a}$$

$$\begin{aligned} j\omega\varepsilon_0\gamma_1 &\left\{\frac{\alpha\gamma}{\omega\varepsilon_0\gamma_1}(K_3 \sinh \gamma_2 h_1 + J_3 \cosh \gamma_2 h_1) - \frac{\varepsilon_2}{\varepsilon_0}\frac{\gamma_2}{\gamma_1}(K_1 \cosh \gamma_2 h_1 + J_1 \sinh \gamma_2 h_1) - 1\right\}A^{\mathrm{e}} \\ &+ j\omega\varepsilon_2\gamma_2\left\{\frac{\alpha\gamma}{\omega\varepsilon_2\gamma_2}(K_4 \sinh \gamma_2 h_1 + J_4 \cosh \gamma_2 h_1 - 1) - (K_2 \cosh \gamma_2 h_1 + J_2 \sinh \gamma_2 h_1)\right\}A^{\mathrm{h}} = \tilde{J}_z(\alpha) \end{aligned} \tag{2A.27b}$$

Define four new variables L_1, L_2, L_3, and L_4 as follows:

$$L_1 = -(K_3 \sinh \gamma_2 h_1 + J_3 \cosh \gamma_2 h_1), \tag{2A.28a}$$

$$L_2 = 1 - \left(\frac{k_{c2}}{k_{c0}}\right)^2 (K_4 \sinh \gamma_2 h_1 + J_4 \cosh \gamma_2 h_1), \tag{2A.28b}$$

$$L_3 = \frac{\alpha\gamma}{\omega\varepsilon_0\gamma_1}(K_3 \sinh \gamma_2 h_1 + J_3 \cosh \gamma_2 h_1) - \frac{\varepsilon_2}{\varepsilon_0}\frac{\gamma_2}{\gamma_1}(K_1 \cosh \gamma_2 h_1 + J_1 \sinh \gamma_2 h_1) - 1, \tag{2A.28c}$$

$$L_4 = \frac{\alpha\gamma}{\omega\varepsilon_2\gamma_2}(K_4 \sinh \gamma_2 h_1 + J_4 \cosh \gamma_2 h_1 - 1) - (K_2 \cosh \gamma_2 h_1 + J_2 \sinh \gamma_2 h_1). \tag{2A.28d}$$

Then

$$k_{c2}^2 L_1 A^e + k_{c0}^2 L_2 A^h = \tilde{J}_x(\alpha), \tag{2A.29a}$$

$$j\omega\varepsilon_0\gamma_1 L_3 A^e + j\omega\varepsilon_2\gamma_2 L_4 A^h = \tilde{J}_z(\alpha). \tag{2A.29b}$$

Equations (2A.29a) and (2A.29b) are rewritten in matrix form as

$$\begin{bmatrix} k_{c2}^2 & k_{c0}^2 L_2 \\ j\omega\varepsilon_0\gamma_1 L_3 & j\omega\varepsilon_2\gamma_2 L_4 \end{bmatrix} \begin{bmatrix} A^e \\ A^h \end{bmatrix} = \begin{bmatrix} \tilde{J}_x(\alpha) \\ \tilde{J}_z(\alpha) \end{bmatrix} \tag{2A.29c}$$

that is,

$$[L][A] = [J]. \tag{2A.29d}$$

Inverting the $[L]$ and solving for $[A]$ yields A^e and A^h in terms of the current density components

$$[A] = [L]^{-1}[J]. \tag{2A.30a}$$

That is,

$$A^e = \frac{j\omega\varepsilon_2\gamma_2 L_4}{\det L}\tilde{J}_x(\alpha) - \frac{K_{c0}^2 L_2}{\det L}\tilde{J}_z(\alpha), \tag{2A.30b}$$

$$A^h = -\frac{j\omega\varepsilon_0\gamma_1 L_3}{\det L}\tilde{J}_x(\alpha) + \frac{K_{c2}^2 L_1}{\det L}\tilde{J}_z(\alpha), \tag{2A.30c}$$

where

$$\det L = j\omega\varepsilon_2\gamma_2 k_{c0}^2 \left[\left(\frac{k_{c2}}{k_{c0}}\right)^2 L_1 L_4 - \frac{\varepsilon_0}{\varepsilon_2}\frac{\gamma_1}{\gamma_2} L_2 L_3\right]. \tag{2A.30d}$$

When A^e and A^h are substituted in Eqs. (2A.4c) and (2A.4d), all the unknown coefficients are eliminated, and the electric-field components are expressed in terms of the current density components:

$$\left(\frac{\alpha\gamma\omega\varepsilon_2\gamma_2 L_4}{\det L} + \frac{\omega^2\mu_0\varepsilon_0\gamma_1^2 L_3}{\det L}\right)\tilde{J}_x(\alpha) + \left(\frac{j\alpha\gamma k_{c0}^2 L_2}{\det L} + \frac{j\omega\mu_0\gamma_1 k_{c1}^2 L_1}{\det L}\right)\tilde{J}_z(\alpha) = \tilde{E}_x(\alpha), \tag{2A.31}$$

$$\left(\frac{k_{c0}^2 j\omega\varepsilon_2\gamma_2 L_4}{\det L}\right)\tilde{J}_x(\alpha) - \left(\frac{k_{c0}^4 L_2}{\det L}\right)\tilde{J}_z(\alpha) = \tilde{E}_z(\alpha). \tag{2A.32}$$

Define five new variables M_1, M_2, M_3, M_4, and M_5 as follows:

$$M_1 = \left(-\frac{j\alpha\gamma}{k_{c0}^2} L_4 - \frac{j\omega\mu_0\varepsilon_0\gamma_1^2}{k_{c0}^2\varepsilon_2\gamma_2} L_3\right)\frac{1}{M_5}, \tag{2A.33}$$

$$M_2 = \left(\frac{\alpha\gamma}{\omega\varepsilon_2\gamma_2} L_2 + \frac{\mu_0\gamma_1}{\varepsilon_2\gamma_2}\left(\frac{k_{c1}}{k_{c0}}\right)^2 L_1\right)\frac{1}{M_5}, \tag{2A.34}$$

$$M_3 = \frac{L_4}{M_5}, \tag{2A.35}$$

$$M_4 = \frac{jk_{c0}^2}{\omega\varepsilon_2\gamma_2}\frac{L_2}{M_5}, \tag{2A.36}$$

$$M_5 = \left(\frac{k_{c2}}{k_{c0}}\right)^2 L_1 L_4 - \frac{\varepsilon_0}{\varepsilon_2}\frac{\gamma_1}{\gamma_2} L_2 L_3. \tag{2A.37}$$

Then

$$\begin{bmatrix} M_1 & M_2 \\ M_3 & M_4 \end{bmatrix}\begin{bmatrix} J_x \\ J_z \end{bmatrix} = \begin{bmatrix} E_x \\ E_z \end{bmatrix}. \tag{2A.38}$$

APPENDIX 2B: TIME AVERAGE POWER FLOW IN THE THREE SPATIAL REGIONS

The expressions for the time average power flow in the three spatial regions indicated in Figure 2.13(a) are derived from Eq. (2.90). The first step is to substitute the solutions to the Helmholtz's equations, namely, Eqs. (2.67) and (2.68) into Eq. (2.90). There are eight terms under the double integral sign. The double integral is reduced to a single integral by performing the integration with respect to y analytically. Analogous to Appendix 2A, within parentheses α associated with the coefficients A^e through D^h is omitted. Further $\mu_1 = \mu_2 = \mu_3 = \mu_0$, $\varepsilon_1 = \varepsilon_3 = \varepsilon_0$, and $\varepsilon_2 = \varepsilon_{r1}$. Hence from Eqs. (2.57c) and (2.66), $\omega^2\mu_i\varepsilon_i = k_i^2$ $(i = 1, 2, 3)$, $k_1 = k_3 = k_0$, and $\gamma_1 = \gamma_3$. The tilde and the asterisk denote transform and complex conjugate, respectively.

Region 1, $h_1 \leqslant y \leqslant \infty$ The integration with respect to y is shown below for each of the eight terms of Eq. (2.90):

$$\frac{1}{4\pi}\,\text{Re}\int_{-\infty}^{+\infty}\int_{h_1}^{+\infty} -\alpha^2\beta\omega\varepsilon_1\,|\tilde{\phi}_1|^2\,dy\,d\alpha$$

$$= -\frac{1}{4\pi}\text{Re}\int_{-\infty}^{+\infty}\int_{h_1}^{+\infty} \alpha^2\beta\omega\varepsilon_0|A^e\exp[-\gamma_1(y-h_1)]|^2\,dy\,d\alpha,$$

$$I_{11} = -\frac{1}{4\pi}\,\mathrm{Re}\int_{-\infty}^{+\infty}\frac{\alpha^2}{2\gamma_1}\,\beta\omega\varepsilon_0|A^{\mathrm{e}}|^2\,d\alpha; \tag{2B.1}$$

$$\frac{1}{4\pi}\,\mathrm{Re}\int_{-\infty}^{+\infty}\int_{h_1}^{+\infty} -\beta\omega\mu_1\left|\frac{\partial\tilde{\psi}_1}{\partial y}\right|^2 dy d\alpha$$

$$= -\frac{1}{4\pi}\mathrm{Re}\int_{-\infty}^{+\infty}\int_{h_1}^{+\infty}\beta\omega\mu_0|A^{\mathrm{h}}\{\exp-\gamma_1(y-h_1)\}(-\gamma_1)|^2\,dy d\alpha,$$

$$I_{12} = -\frac{1}{4\pi}\,\mathrm{Re}\int_{-\infty}^{+\infty}\frac{\beta\omega\mu_0\gamma_1^2}{2\gamma_1}\,|A^{\mathrm{h}}|^2\,d\alpha; \tag{2B.2}$$

$$\frac{1}{4\pi}\,\mathrm{Re}\int_{-\infty}^{+\infty}\int_{h_1}^{+\infty} -\beta\omega\varepsilon_1\left|\frac{\partial\tilde{\phi}_1}{\partial y}\right|^2 dy d\alpha$$

$$= -\frac{1}{4\pi}\mathrm{Re}\int_{-\infty}^{+\infty}\int_{h_1}^{+\infty}\beta\omega\varepsilon_0|A^{\mathrm{e}}\{\exp-\gamma_1(y-h_1)\}(-\gamma_1)|^2\,dy d\alpha,$$

$$I_{13} = -\frac{1}{4\pi}\,\mathrm{Re}\int_{-\infty}^{+\infty}\frac{\beta\omega\varepsilon_0\gamma_1^2}{2\gamma_1}\,|A^{\mathrm{e}}|^2\,d\alpha; \tag{2B.3}$$

$$\frac{1}{4\pi}\,\mathrm{Re}\int_{-\infty}^{+\infty}\int_{h_1}^{+\infty} -\alpha^2\beta\omega\mu_1\,|\tilde{\psi}_1|^2\,dy d\alpha$$

$$= -\frac{1}{4\pi}\mathrm{Re}\int_{-\infty}^{+\infty}\int_{h_1}^{+\infty}\alpha^2\beta\omega\mu_0|A^{\mathrm{h}}\exp[-\gamma_1(y-h_1)]|^2\,dy d\alpha,$$

$$I_{14} = -\frac{1}{4\pi}\,\mathrm{Re}\int_{-\infty}^{+\infty}\frac{\alpha^2\beta\omega\mu_0}{2\gamma_1}\,|A^{\mathrm{h}}|^2\,d\alpha; \tag{2B.4}$$

$$\frac{1}{4\pi}\,\mathrm{Re}\int_{-\infty}^{+\infty}\int_{h_1}^{+\infty} -j\alpha\beta^2\tilde{\phi}_1\frac{\partial\tilde{\psi}_1^*}{\partial y}\,dy d\alpha$$

$$= -\frac{1}{4\pi}\mathrm{Re}\int_{-\infty}^{+\infty}\int_{h_1}^{+\infty} j\alpha\beta^2 A^{\mathrm{e}}\exp[-\gamma_1(y-h_1)]$$

$$\times[A^{\mathrm{h}}\{\exp-\gamma_1(y-h_1)\}(-\gamma_1)]^*dy d\alpha,$$

$$I_{15} = \frac{1}{4\pi}\,\mathrm{Re}\int_{-\infty}^{+\infty}\frac{j\alpha\beta^2}{2}\,A^{\mathrm{e}}[A^{\mathrm{h}}]^*\,d\alpha; \tag{2B.5}$$

$$\frac{1}{4\pi}\,\mathrm{Re}\int_{-\infty}^{+\infty}\int_{h_1}^{+\infty} -j\alpha\beta^2\,\frac{\partial\tilde{\phi}_1}{\partial y}\,\tilde{\psi}_1^*\,dy d\alpha$$

$$= -\frac{1}{4\pi}\mathrm{Re}\int_{-\infty}^{+\infty}\int_{h_1}^{+\infty} j\alpha\beta^2 A^{\mathrm{e}}\{\exp-\gamma_1(y-h_1)\}(-\gamma_1)$$

$$\times[A^{\mathrm{h}}\{\exp[-\gamma_1(y-h_1)]]^*dy d\alpha,$$

$$I_{16} = \frac{1}{4\pi}\,\mathrm{Re}\int_{-\infty}^{+\infty}\frac{j\alpha\beta^2}{2}\,A^{\mathrm{e}}[A^{\mathrm{h}}]^*\,d\alpha; \tag{2B.6}$$

$$\frac{1}{4\pi}\,\mathrm{Re}\int_{-\infty}^{+\infty}\int_{h_1}^{+\infty} j\alpha k_1^2\tilde{\phi}_1^*\,\frac{\partial\tilde{\psi}_1}{\partial y}\,dy\,d\alpha$$

$$= \frac{1}{4\pi}\mathrm{Re}\int_{-\infty}^{+\infty}\int_{h_1}^{+\infty} j\alpha k_0^2[A^{\mathrm{e}}\exp[-\gamma_1(y-h_1)]]^*$$

$$\times[A^{\mathrm{h}}\{\exp-\gamma_1(y-h_1)\}(-\gamma_1)]\,dy\,d\alpha,$$

$$I_{17} = -\frac{1}{4\pi}\,\mathrm{Re}\int_{-\infty}^{+\infty}\frac{j\alpha k_0^2}{2}\,[A^{\mathrm{e}}]^*A^{\mathrm{h}}\,d\alpha; \qquad (2B.7)$$

$$\frac{1}{4\pi}\,\mathrm{Re}\int_{-\infty}^{+\infty}\int_{h_1}^{+\infty} j\alpha k_1^2\,\frac{\partial\tilde{\phi}_1^*}{\partial y}\,\tilde{\psi}_1\,dy\,d\alpha$$

$$= \frac{1}{4\pi}\int_{-\infty}^{+\infty}\int_{h_1}^{+\infty} j\alpha k_0^2[A^{\mathrm{e}}\{\exp-\gamma_1(y-h_1)\}(-\gamma_1)]^*$$

$$\times[A^{\mathrm{h}}\exp[-\gamma_1(y-h_1)]]\,dy\,d\alpha,$$

$$I_{18} = -\frac{1}{4\pi}\,\mathrm{Re}\int_{-\infty}^{+\infty}\frac{j\alpha k_0^2}{2}\,[A^{\mathrm{e}}]^*A^{\mathrm{h}}\,d\alpha. \qquad (2B.8)$$

Substituting Eqs. (2B.1) to (2B.8) into Eq. (2.90), the following expression for P_{avg} in region 1 is obtained:

$$P_{\mathrm{avg}} = -\frac{1}{8\pi}\,\mathrm{Re}\int_{-\infty}^{+\infty}\Bigg[\beta\omega\,\frac{\alpha^2+\gamma_1^2}{2\gamma_1}\,(\varepsilon_0|A^{\mathrm{e}}|^2+\mu_0|A^{\mathrm{h}}|^2)$$

$$+j2\alpha(k_0^2[A^{\mathrm{e}}]^*A^{\mathrm{h}} - \beta^2A^{\mathrm{e}}[A^{\mathrm{h}}]^*)\Bigg]\,d\alpha. \qquad (2B.9)$$

A negative sign in front of the integral sign is because a wave propagating in the $-z$-direction is assumed.

Region 2, $0 \leqslant y \leqslant h_1$ We first consider the case where $\gamma_2^2 > 0$ and γ_2 is real. The integration with respect to y is shown below for each of the eight terms of Eq. (2.90):

$$\frac{1}{4\pi}\,\mathrm{Re}\int_{-\infty}^{+\infty}\int_{0}^{h_1} -\alpha^2\beta\omega\varepsilon_2|\tilde{\phi}_2|^2\,dy\,d\alpha$$

$$= \frac{1}{4\pi}\,\mathrm{Re}\int_{-\infty}^{+\infty}\int_{0}^{h_1} -\alpha^2\beta\omega\varepsilon_2(B^{\mathrm{e}}\sinh\gamma_2 y + C^{\mathrm{e}}\cosh\gamma_2 y)$$

$$\times(B^{\mathrm{e}*}\sinh\gamma_2 y + C^{\mathrm{e}*}\cosh\gamma_2 y)\,dy\,d\alpha,$$

$$I_{21} = \frac{1}{4\pi} \operatorname{Re} \int_{-\infty}^{+\infty} -\alpha^2 \beta\omega\varepsilon_2 \left\{ (|B^e|^2 + |C^e|^2) \frac{\sinh 2\gamma_2 h_1}{4\gamma_2} + \left(\frac{B^e C^{e*} + C^e B^{e*}}{2} \right) \frac{(\cosh 2\gamma_2 h_1 - 1)}{2\gamma_2} + (|C^e|^2 - |B^e|^2) \frac{h_1}{2} \right\} d\alpha; \tag{2B.10}$$

$$\frac{1}{4\pi} \operatorname{Re} \int_{-\infty}^{+\infty} \int_0^{h_1} -\alpha^2 \beta\omega\mu_2 |\tilde{\psi}_2|^2 dy d\alpha$$

$$= \frac{1}{4\pi} \operatorname{Re} \int_{-\infty}^{+\infty} \int_0^{h_1} -\alpha^2 \beta\omega\mu_0 (B^h \sinh \gamma_2 y + C^h \cosh \gamma_2 y) \times (B^{h*} \sinh \gamma_2 y + C^{h*} \cosh \gamma_2 y) dy d\alpha;$$

$$I_{22} = \frac{1}{4\pi} \operatorname{Re} \int_{-\infty}^{+\infty} -\alpha^2 \beta\omega\mu_0 \left\{ (|B^h|^2 + |C^h|^2) \frac{\sinh 2\gamma_2 h_1}{4\gamma_2} + \left(\frac{B^h C^{h*} + C^h B^{h*}}{2} \right) \frac{(\cosh 2\gamma_2 h_1 - 1)}{2\gamma_2} + (|C^h|^2 - |B^h|^2) \frac{h_1}{2} \right\} d\alpha; \tag{2B.11}$$

$$\frac{1}{4\pi} \operatorname{Re} \int_{-\infty}^{+\infty} \int_0^{h_1} -\beta\omega\mu_2 \left| \frac{\partial \tilde{\psi}_2}{\partial y} \right|^2 dy d\alpha$$

$$= \frac{1}{4\pi} \operatorname{Re} \int_{-\infty}^{+\infty} \int_0^{h_1} -\beta\omega\mu_0 \gamma_2^2 (B^h \cosh \gamma_2 y + C^h \sinh \gamma_2 y) \times (B^{h*} \cosh \gamma_2 y + C^{h*} \sinh \gamma_2 y) dy d\alpha,$$

$$I_{23} = \frac{1}{4\pi} \operatorname{Re} \int_{-\infty}^{+\infty} -\beta\omega\mu_0 \gamma_2^2 \left\{ (|B^h|^2 + |C^h|^2) \frac{\sinh 2\gamma_2 h_1}{4\gamma_2} + \left(\frac{B^h C^{h*} + C^h B^{h*}}{2} \right) \frac{(\cosh 2\gamma_2 h_1 - 1)}{2\gamma_2} + (|B^h|^2 - |C^h|^2) \frac{h_1}{2} \right\} d\alpha; \tag{2B.12}$$

$$\frac{1}{4\pi} \operatorname{Re} \int_{-\infty}^{+\infty} \int_0^{h_1} -\beta\omega\varepsilon_2 \left| \frac{\partial \tilde{\phi}_2}{\partial y} \right|^2 dy d\alpha$$

$$= \frac{1}{4\pi} \operatorname{Re} \int_{-\infty}^{+\infty} \int_0^{h_1} -\beta\omega\varepsilon_2 \gamma_2^2 (B^e \cosh \gamma_2 y + C^e \sinh \gamma_2 y) \times (B^{e*} \cosh \gamma_2 y + C^{e*} \sinh \gamma_2 y) dy d\alpha,$$

$$I_{24} = \frac{1}{4\pi} \operatorname{Re} \int_{-\infty}^{+\infty} - \beta\omega\varepsilon_2\gamma_2^2 \left\{ (|B^e|^2 + |C^e|^2) \frac{\sinh 2\gamma_2 h_1}{4\gamma_2} \right.$$

$$\left. + \left(\frac{B^e C^{e*} + C^e B^{e*}}{2} \right) \frac{(\cosh 2\gamma_2 h_1 - 1)}{2\gamma_2} + (|B^e|^2 - |C^e|^2) \frac{h_1}{2} \right\} d\alpha. \tag{2B.13}$$

Equations (2B.10) through (2B.13) which correspond to the first four terms of Eq. (2.90), when combined together, result in the following equation:

$$I_{21} + I_{22} + I_{23} + I_{24}$$

$$= -\frac{1}{4\pi} \operatorname{Re} \int_{-\infty}^{+\infty} \omega\beta \left\{ \sinh 2\gamma_2 h_1 \left[\frac{\alpha^2 + \gamma_2^2}{4\gamma_2} (\varepsilon_2(|B^e|^2 + |C^e|^2) + \mu_0(|B^h|^2 + |C^h|^2)) \right] \right.$$

$$+ \frac{D(\alpha^2 - \gamma_2^2)}{2} [\varepsilon_2(|C^e|^2 - |B^e|^2) + \mu_0(|C^h|^2 - |B^h|^2)] + (\cosh 2\gamma_2 h_1 - 1)$$

$$\left. \times \left[\frac{\alpha^2 + \gamma_2^2}{2\gamma_2} \left(\frac{\varepsilon_2(B^e C^{e*} + C^e B^{e*})}{2} + \frac{\mu_0(B^h C^{h*} + C^h B^{h*})}{2} \right) \right] \right\}. \tag{2B.14}$$

Continuing the integration process, the fifth through eighth terms of Eq. (2.90) are as follows:

$$\frac{1}{4\pi} \operatorname{Re} \int_{-\infty}^{+\infty} \int_0^{h_1} -j\alpha\beta^2 \tilde{\phi} \frac{\partial \tilde{\psi}^*}{\partial y} dy d\alpha$$

$$= \frac{1}{4\pi} \operatorname{Re} \int_{-\infty}^{+\infty} \int_0^{h_1} - j\alpha\beta^2\gamma_2(B^e \sinh \gamma_2 y + C^e \cosh \gamma_2 y)$$

$$\times (B^{h*} \cosh \gamma_2 y + C^{h*} \sinh \gamma_2 y) d\alpha dy,$$

$$I_{25} = -\frac{1}{4\pi} \operatorname{Re} \int_{-\infty}^{+\infty} j\alpha\beta^2 \left\{ \frac{\sinh 2\gamma_2 h_1}{4} (B^e C^{h*} + C^e B^{h*}) \right.$$

$$\left. + \frac{(\cosh 2\gamma_2 h_1 - 1)}{2} \frac{(B^e B^{h*} + C^e C^{h*})}{2} + (C^e B^{h*} - B^e C^{h*}) \frac{\gamma_2 h_1}{2} \right\} d\alpha; \tag{2B.15}$$

$$\frac{1}{4\pi} \operatorname{Re} \int_{-\infty}^{+\infty} \int_0^{h_1} -j\alpha\beta^2 \frac{\partial \tilde{\phi}}{\partial y} \tilde{\psi}^* dy d\alpha$$

$$= \frac{1}{4\pi} \operatorname{Re} \int_{-\infty}^{+\infty} \int_0^{h_1} - j\alpha\beta^2\gamma_2(B^e \cosh \gamma_2 y + C^e \sinh \gamma_2 y)$$

$$\times (B^{h*} \sinh \gamma_2 y + C^{h*} \cosh \gamma_2 y) d\alpha dy,$$

$$I_{26} = -\frac{1}{4\pi}\,\mathrm{Re}\int_{-\infty}^{+\infty} j\alpha\beta^2 \left\{\frac{\sinh 2\gamma_2 h_1}{4}\,(B^{\mathrm{e}}C^{\mathrm{h}*} + C^{\mathrm{e}}B^{\mathrm{h}*}) + \frac{(\cosh 2\gamma_2 h_1 - 1)}{2}\,\frac{(B^{\mathrm{e}}B^{\mathrm{h}*} + C^{\mathrm{e}}C^{\mathrm{h}*})}{2} + (B^{\mathrm{e}}C^{\mathrm{h}*} - C^{\mathrm{e}}B^{\mathrm{h}*})\,\frac{\gamma_2 h_1}{2}\right\} d\alpha; \tag{2B.16}$$

$$\frac{1}{4\pi}\,\mathrm{Re}\int_{-\infty}^{+\infty}\int_{0}^{h_1} j\alpha k^2 \left(\phi^* \frac{\partial\tilde{\psi}}{\partial y} + \frac{\partial\tilde{\phi}^*}{\partial y}\,\tilde{\psi}\right) dy\, d\alpha$$

$$= \frac{1}{4\pi}\,\mathrm{Re}\int_{-\infty}^{+\infty}\int_{0}^{h_1} j\alpha k^2 \gamma_2 \times [(B^{\mathrm{e}*}\sinh\gamma_2 y + C^{\mathrm{e}*}\cosh\gamma_2 y)$$
$$\times (B^{\mathrm{h}}\cosh\gamma_2 y + C^{\mathrm{h}}\sinh\gamma_2 y) + (B^{\mathrm{e}*}\cosh\gamma_2 y + C^{\mathrm{e}*}\sinh\gamma_2 y)$$
$$\times (B^{\mathrm{h}}\sinh\gamma_2 y + C^{\mathrm{h}}\cosh\gamma_2 y)]\,d\alpha\, dy,$$

$$I_{27} + I_{28} = \frac{1}{4\pi}\,\mathrm{Re}\int_{-\infty}^{+\infty} j\alpha k^2 \left[(B^{\mathrm{e}*}B^{\mathrm{h}} + C^{\mathrm{e}*}C^{\mathrm{h}})\,\frac{(\cosh 2\gamma_2 h_1 - 1)}{2} + (B^{\mathrm{e}*}C^{\mathrm{h}} + C^{\mathrm{e}*}B^{\mathrm{h}})\,\frac{\sinh 2\gamma_2 h_1}{2}\right] d\alpha. \tag{2B.17}$$

Combining Eqs. (2B.15) through (2B.17) results in the following:

$$I_{25} + I_{26} + I_{27} + I_{28}$$
$$= -\frac{1}{4\pi}\,\mathrm{Re}\int_{-\infty}^{+\infty} j\alpha \left\{\frac{\sinh 2\gamma_2 h_1}{2}\,[\beta^2(B^{\mathrm{e}}C^{\mathrm{h}*} + C^{\mathrm{e}}B^{\mathrm{h}*}) - k^2(B^{\mathrm{e}*}C^{\mathrm{h}} + C^{\mathrm{e}*}B^{\mathrm{h}})] \times \frac{(\cosh 2\gamma_2 h_1 - 1)}{2}\,[\beta^2(B^{\mathrm{e}}B^{\mathrm{h}*} + C^{\mathrm{e}}C^{\mathrm{h}*}) - k^2(B^{\mathrm{e}*}C^{\mathrm{h}} + C^{\mathrm{e}*}B^{\mathrm{h}})]\right\} d\alpha. \tag{2B.18}$$

Last, combining Eqs. (2B.14) and (2B.18), the expression for P_{avg} in region 2 is obtained:

$$P_{\mathrm{avg}} = -\frac{1}{4\pi}\,\mathrm{Re}\int_{-\infty}^{+\infty} \left\{\frac{\omega\beta}{2}\left\{\sinh 2\gamma_2 h_1 \left[\frac{\alpha^2 + \gamma_2^2}{2\gamma_2}\,(\varepsilon_2(|B^{\mathrm{e}}|^2 + |C^{\mathrm{e}}|^2) + \mu_0(|B^{\mathrm{h}}|^2 + |C^{\mathrm{h}}|^2))\right]\right.\right.$$
$$+ h_1(\alpha^2 - \gamma_2^2)[\varepsilon_2(|C^{\mathrm{e}}|^2 - |B^{\mathrm{e}}|^2) + \mu_0(|C^{\mathrm{h}}|^2 - |B^{\mathrm{h}}|^2)\Big]$$
$$\left.\left. + (\cosh 2\gamma_2 h_1 - 1)\left[\frac{\alpha^2 + \gamma_2^2}{2\gamma_2}\,(\varepsilon_2(B^{\mathrm{e}}C^{\mathrm{e}} + C^{\mathrm{e}}B^{\mathrm{e}*}) + \mu_0(B^{\mathrm{h}}C^{\mathrm{h}*} + c^{\mathrm{h}}B^{\mathrm{h}*}))\right]\right\}\right.$$

$$+\frac{j\alpha}{2}\{\sinh 2\gamma_2 h_1[\beta^2(B^{e}C^{h*} + C^{e}B^{h*}) - k^2(B^{e*}C^{h} + C^{e*}B^{h})]$$

$$+ (\cosh 2\gamma_2 h_1 - 1)[\beta^2(B^{e}B^{h*} + C^{e}C^{h*}) - k^2(B^{e*}C^{h} + C^{e*}B^{h})]\}\Bigg\} d\alpha \tag{2B.19}$$

A negative sign in front of the integral sign is because a wave propagating in the $-z$-direction is assumed.

Region 3, $-\infty \leqslant y \leqslant 0$ Proceeding in a manner analogous to region 1, the P_{avg} is given by

$$P_{avg} = -\frac{1}{8\pi}\,\mathrm{Re}\int_{-\infty}^{+\infty}\Bigg[\omega\beta\,\frac{\alpha^2+\gamma_1^2}{\gamma_1}\,(\varepsilon_0|D^{e}(\alpha)|^2 + \mu_0|D^{h}(\alpha)|^2)$$

$$+ j2\alpha(\beta^2 D^{e}(\alpha)[D^{h}(\alpha)]^* - k_0^2[D^{e}(\alpha)]^* D^{h}(\alpha))\Bigg] d\alpha. \tag{2B.20}$$

REFERENCES

[1] C. P. Wen, "Coplanar Waveguide: A Surface Strip Transmission Line Suitable for Nonreciprocal Gyromagnetic Device Applications," *IEEE Trans. Microwave Theory Tech.*, Vol. 17, No. 12, pp. 1087–1090, Dec. 1969.

[2] C. Veyres and V. F. Hanna, "Extension of the Application of Conformal Mapping Techniques to Coplanar Lines with Finite Dimensions," *Int. J. Electron.*, Vol. 48, No. 1, pp. 47–56, Jan. 1980.

[3] S. Gevorgian, L. J. P. Linner and E. L. Kollberg, "CAD Models for Shielded Multilayered CPW," *IEEE Trans. Microwave Theory Tech.*, Vol. 43, No. 4, pp. 772–779, April 1995.

[4] G. Ghione and C. Naldi, "Analytical Formulas for Coplanar Lines in Hybrid and Monolithic MICs," *Electron. Lett.*, Vol. 20, No. 4, pp. 179–181, Feb. 1984.

[5] G. Ghione and C. U. Naldi, "Coplanar Waveguides for MMIC Applications: Effect of Upper Shielding, Conductor Backing, Finite-Extent Ground Planes, and Line-to-Line Coupling," *IEEE Trans. Microwave Theory Tech.*, Vol. 35, No. 3, pp. 260–267, March 1987.

[6] M. E. Davis, E. W. Williams, and A. C. Celestini, "Finite-Boundary Corrections to the Coplanar Waveguide Analysis," *IEEE Trans. Microwave Theory Tech.*, Vol. 21, No. 9, pp. 594–596, Sept. 1973.

[7] S. S. Bedair and I. Wolff, "Fast, Accurate and Simple Approximate Analysis Formulas for Calculating the Parameters of Supported Coplanar Waveguides for (M)MIC's," *IEEE Trans. Microwave Theory Tech.*, Vol. 40, No. 1, pp. 41–48, Jan. 1992.

[8] G. Ghione, C. Naldi, and R. Zich, "Coplanar Lines in GaAs Hybrid Structures: Influence of Flip-Chip Insertion," *Proc. MELCON'83*, Vol. 1, sec. B2, No. 6, 1983.

[9] S. S. Gevorgian, "Basic Characteristics of Two Layered Substrate Coplanar Waveguides," *Electron. Lett.*, Vol. 30, No. 5, pp. 1236–1237, July 1994.

[10] R. H. Jansen, "Hybrid Mode Analysis of End Effects of Planar Microwave and Millimeterwave Transmission Line," *Proc. IEE*, Vol. 128, Pt. H, No. 2, pp. 77–86, April 1981.

[11] S. S. Gevorgian, P. Linner, and E. Kollberg, "Analytical Models for Shielded and Multilayered CPW," *24th European Microwave Conf. Proc.*, Vol. 1, pp. 263–267, Cannes, France, Sept. 5–8, 1994.

[12] E. Mueller, "Measurement of the Effective Relative Permittivity of Unshielded Coplanar Waveguides," *Electron. Lett.*, Vol. 13, No. 24, pp. 729–730, Nov. 1977.

[13] P. A. J. Dupuis and C. K. Campbell, "Characteristic Impedance of Surface-Strip Coplanar Waveguides," *Electron Lett.*, Vol. 9, No. 16, pp. 354–355, Aug. 1973.

[14] J.-P. Becker and D. Jager, "Electrical Properties of Coplanar Transmission Lines on Lossless and Lossy Substrates," *Electron. Lett.*, Vol. 15, No. 3, pp. 88–90, Feb. 1979.

[15] A. A. R. Riad, S. M. Riad, M. Ahmad, F. W. Stephenson, and R. A. Ecker, "Thick-Film Coplanar Strip and Slot Transmission Lines for Microwave and Wideband Integrated Circuits," *Int. Microelectronics Symp. Dig.*, pp. 18–21, Reno, Nevada, Nov. 15–17, 1982.

[16] T. Hatsuda, "Computation of the Characteristics of Coplanar-Type Strip Lines by the Relaxation Method," *IEEE Trans. Microwave Theory Tech.*, Vol. 20, No. 6, pp. 413–416, June 1972.

[17] C.-N. Chang, Y.-C. Wong, and C. H Chen, "Hybrid Quasistatic Analysis for Multilayer Coplanar Lines," *IEE Proc.*, Part-H, Vol. 138, No. 4, pp. 307–312, Aug. 1991.

[18] T. Itoh and R. Mittra, "Dispersion Characteristics of Slot Lines," *Electron. Lett.*, Vol. 7, No. 13, pp. 364–365, July 1971.

[19] J. B. Knorr and K.-D. Kuchler, "Analysis of Coupled Slots and Coplanar Strips on Dielectric Substrate," *IEEE Trans. Microwave Theory Tech.*, Vol. 23, No. 7, pp. 541–548, July 1975.

[20] R. N. Simons, "Suspended Coupled Slotline Using Double Layer Dielectric," *IEEE Trans. Microwave Theory Tech.*, Vol. 29, No. 2, pp. 162–165, Feb. 1981.

[21] R. N. Simons, "Propagation Characteristics of Some Novel Coplanar Waveguide Transmission Lines on GaAs at MM-Wave Frequencies," 1986 Conf. on Millimeter Wave/Microwave Measurements and Standards for Miniaturized Systems, Redstone Arsenal, Alabama, Nov. 6–7, 1986 (Also NASA Tech. Memo No. 89839).

[22] C.-N. Chang, Y.-C. Wong, and C. H. Chen, "Full-Wave Analysis of Coplanar Waveguides by Variational Conformal Mapping Technique," *IEEE Trans. Microwave Theory Tech.*, Vol. 38, No. 9, pp. 1339–1344, Sept. 1990.

[23] G.-C. Liang, Y.-W. Liu, and K. K. Mei, "Full-Wave Analysis of Coplanar Waveguide and Slotline Using the Time-Domain Finite-Difference Method," *IEEE Trans. Microwave Theory Tech.*, Vol. 37, No. 12, pp. 1949–1957, Dec. 1989.

[24] G. Kibuuka, R. Bertenburg, M. Naghed, and I. Wolff, "Coplanar Lumped Elements and their Application in Filters on Ceramic and Gallium Arsenide Substrates," *19th European Microwave Conf. Proc.*, pp. 656–661, London, England, Sept. 4–7, 1989.

[25] R. N. Simons and G. E. Ponchak, "Modeling of Some Coplanar Waveguide Discontinuities," *IEEE Trans. Microwave Theory Tech.*, Vol. 36, No. 12, pp. 1796–1803, Dec. 1988.

[26] W. H. Haydl, T. Kitazawa, J. Braunstein, R. Bosch, and M. Schlechtweg, "Millimeterwave Coplanar Transmission Lines on Gallium Arsenide, Indium Phosphide and Quartz with Finite Metalization Thickness," *1991 IEEE MTT-S Int. Microwave Symp. Dig.*, Vol. 2, pp. 691–694, Boston, MA, June 10–14, 1991.

[27] E. Yamashita, K. Atsuki, and T. Ueda, "An approximate dispersion formula of microstrip lines for computer-aided design of microwave integrated circuits," *IEEE Trans. Microwave Theory Tech.*, Vol. 27, No. 12, pp. 1036–1038, Dec. 1979.

[28] G. Hasnain, A. Dienes, and J. R. Whinnery, "Dispersion of Picosecond Pulses in Coplanar Transmission Lines," *IEEE Trans. Microwave Theory Tech.*, Vol. 34, No. 6, pp. 738–741, June 1986.

[29] S. S. Gevorgian, T. Martinsson, A. Deleniv, E. Kollberg, and I. Vendik, "Simple and Accurate Dispersion Expression for the Effective Dielectric Constant of Coplanar Waveguides," *IEE Proc. Microwave Antennas Propag.*, Vol. 144, No. 2, pp. 145–148, April 1997.

[30] T. Q Deng, M. S. Leong, and P. S. Kooi, "Accurate and Simple Closed-Form Formulas for Coplanar Waveguide Synthesis," *Electron. Lett.*, Vol. 31, No. 23, pp. 2017–2019, Nov. 1995.

[31] T. Deng, "CAD Model for Coplanar Waveguide Synthesis," *IEEE Trans. Microwave Theory Tech.*, Vol. 44, No. 10, pp. 1733–1738, Oct. 1996.

[32] T. Q. Deng, M. S. Leong, P. S. Kooi, and T. S. Yeo, "Synthesis Formulas Simplify Coplanar-Waveguide Design," *Microwaves RF*, Vol. 36, No. 3, pp. 84–98, March 1997.

[33] D. P. McGinnis and J. B. Beyer, "A Broad-Band Microwave Superconducting Thin-Film Transformer," *IEEE Microwave Theory Tech.*, Vol. 36, No. 11, pp. 1521–1525, Nov. 1988.

[34] T. Tokumitsu, T. Hiraoka, H. Nakamoto, and T. Takenaka, "Multilayer MMIC Using a 3 μm × 3-Layer Dielectric Film Structure," *1990 IEEE MTT-S Int. Microwave Symp. Dig.*, Vol. 2, Dallas, TX, pp. 831–834, May 8–10, 1990.

[35] H. Kamitsuna, "A Very Small, Low-Loss MMIC Rat-Race Hybrid Using Elevated Coplanar Waveguides," *IEEE Microwave Guided Wave Lett.*, Vol. 2, No. 8, pp. 337–339, Aug. 1992.

[36] S. Hofschen and I. Wolff, "Simulation of an Elevated Coplanar Waveguide Using 2-D FDTD," *IEEE Microwave Guided Wave Lett.*, Vol. 6, No. 1, pp. 28–30, Jan. 1996.

[37] U. Bhattacharya, S. T. Allen, and M. J. W. Rodwell, "DC-725 GHz Sampling Circuits and Subpicosecond Nonlinear Transmission Lines Using Elevated Coplanar Waveguide," *IEEE Microwave Guided Wave Lett.*, Vol. 5, No. 2, pp. 50–52, Feb. 1995.

[38] F. Schnieder, R. Doerner, and W. Heinrich, "High-Impedance Coplanar Waveguides with Low Attenuation," *IEEE Microwave and Guided Wave Letters*, Vol. 6, No. 3, pp. 117–119, March 1996.

[39] M. Gillick and I. D. Robertson, "Ultra low Impedance CPW Transmission Lines of Multilayer MMIC's," *1993 IEEE Microwave Millimeter-Wave Monolithic Circuits Symp. Dig.*, pp. 127–130, Atlanta, GA, June 14–15, 1993.

[40] M. Gillick, I. D. Robertson, and J. S. Joshi, "A 12–36 GHz MMIC 3 dB Coplanar Waveguide Directional Coupler," *22nd European Microwave Conf. Proc.*, pp. 724–728, Espoo, Finland, Aug. 24–27, 1992.

[41] R. N. Simons and R. K. Arora, "Coupled Slot Line Field Components," *IEEE Trans. Microwave Theory Tech.*, Vol. 30, No. 7, pp. 1094–1099, July 1982.

[42] H.-C. Su and K.-L. Wong, "Quasistatic Solutions of Cylindrical Coplanar Waveguides," *Microwave Optical Technology Lett.*, Vol. 14, No. 6, pp. 347–351, April 1997.

[43] H.-C. Su and K.-L. Wong, "Dispersion Characteristics of Cylindrical Coplanar Waveguides," *IEEE Trans. Microwave Theory Tech.*, Vol. 44, No. 11, pp. 2120–2122, Nov. 1996.

[44] H.-C. Su and K.-L. Wong, "Full-Wave Analysis of the Effective Relative Permittivity of a Coplanar Waveguide Printed Inside a Cylindrical Substrate," *Microwave Optical Tech. Lett.*, Vol. 12, No. 2, pp. 94–97, June 1996.

[45] T. Kitazawa and Y.Hayashi, "Quasistatic Characteristics of a Coplanar Waveguide with Thick Metal Coating," *IEE Proc.*, Vol. 133, Pt.H, No. 1, pp. 18–20, Feb. 1986.

[46] T. Kitazawa, Y. Hayashi, and M. Suzuki, "A Coplanar Waveguide with Thick Metal-Coating," *IEEE Trans. Microwave Theory Tech.*, Vol. 24, No. 9, pp. 604–608, Sept. 1976.

[47] T. Kitazawa and T. Itoh, "Asymmetrical Coplanar Waveguide with Finite Metallization Thickness Containing Anisotropic Media," *1990 IEEE MTT-S Int. Microwave Symp. Dig.*, Vol. 2, Dallas, TX, pp. 673–676, May 8–10, 1990.

CHAPTER 3

Conductor-Backed Coplanar Waveguide

3.1 INTRODUCTION

This chapter presents the characteristics of a coplanar waveguide with a lower ground plane. The lower ground plane provides mechanical strength for a thin and fragile wafer, such as GaAs, and acts as a heat sink for circuits with active devices. This configuration of the CPW is known as the conductor-backed coplanar waveguide (CBCPW). There are several other variants of this basic structure, and these are presented in [1].

The chapter commences with expressions based on quasi-static analysis for the effective dielectric constant (ε_{eff}) and characteristic impedance (Z_0) of CBCPW with and without a top metal cover. The computed ε_{eff} and Z_0 as a function of the center conductor strip width S and slot width W are presented. In addition the frequency dependant guide wavelength ratio λ_g/λ_0, where λ_g and λ_0 are the guide wavelength and free space wavelength, respectively, and Z_0 is also presented for comparison. In Section 3.3, the effect of conducting lateral walls on the dominant mode of the CBCPW is discussed and the closed form equation to calculate the Z_0 is presented. The effect of lateral electric and magnetic walls on the higher order mode propagation constant is discussed in Section 3.4. Two cases are considered separately, first, perfect conductors and lossless dielectric, second, conductors with finite thickness, finite conductivity, and lossless or lossy dielectric.

In Section 3.5 a new variant of the CBCPW, namely the channelized coplanar waveguide (CCPW), is presented. In a CCPW, by judicious choice of the lateral wall separation, the dielectric filled rectangular waveguide mode is suppressed. Further, since the ground planes are electrically connected by the side walls to the conductor backing, the excitation of spurious parallel plate modes is eliminated.

In Section 3.6 realization of the lateral walls in a practical circuit by the use of metal-filled vias is explained. In the last section analytical closed form expressions describing the quasi-TEM field patterns in both the air and the dielectric regions of the CBCPW are presented. Experimental results to validate the theoretical results are presented in almost all of the sections.

The CBCPW in this chapter is assumed to by symmetric and fabricated on an isotropic homogeneous dielectric substrate of arbitrary thickness h and relative permittivity ε_r. Further, if not specified, the substrate metallization and the metal cover are assumed to have perfect conductivity.

3.2 CONDUCTOR-BACKED COPLANAR WAVEGUIDE ON A DIELECTRIC SUBSTRATE OF FINITE THICKNESS

3.2.1 Analytical Expressions Based on Quasi-static TEM Conformal Mapping Technique to Determine Effective Dielectric Constant and Characteristic Impedance

A CBCPW with ground planes infinite in the lateral direction is shown in Figure 3.1. A quasi-static TEM mode is assumed to propagate on this structure. Based on this approximation, Wen's [2] analysis for conventional CPW is extended in [3] to [7] to the CBCPW. The analysis provides simple analytical expressions for ε_{eff} and Z_0 as a function of the geometry. These expressions are given below:

$$\varepsilon_{\text{eff}} = \frac{1 + \varepsilon_r \dfrac{K(k')}{K(k)} \dfrac{K(k_3)}{K(k_3')}}{1 + \dfrac{K(k')}{K(k)} \dfrac{K(k_3)}{K(k_3')}}, \tag{3.1}$$

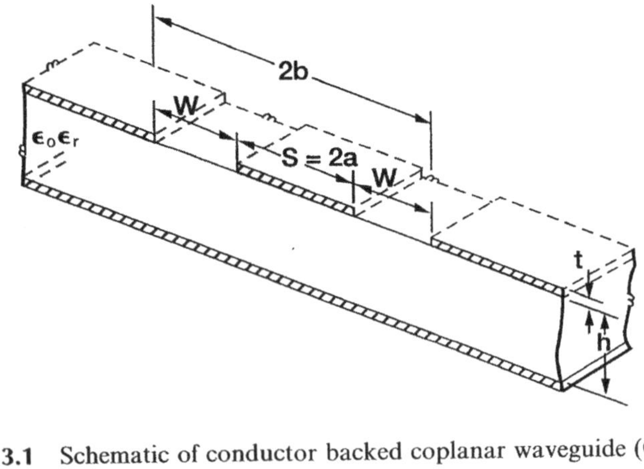

FIGURE 3.1 Schematic of conductor backed coplanar waveguide (CBCPW).

$$Z_0 = \frac{60\pi}{\sqrt{\varepsilon_{\text{eff}}}} \frac{1}{\dfrac{K(k)}{K(k')} + \dfrac{K(k_3)}{K(k_3')}}, \tag{3.2}$$

where

$$k = a/b$$
$$k_3 = \tanh(\pi a/2h)/\tanh(\pi b/2h)$$
$$k' = \sqrt{1.0 - k^2}$$
$$k_3' = \sqrt{1.0 - k_3^2}$$

and $K(k)$ is the complete elliptic integral of the first kind. Figure 3.2(*a*) and (*b*) presents the computed ε_{eff} and Z_0, respectively, for the CBCPW.

3.2.2 Experimental Validation

The ε_{eff} and characteristic impedance are not directly measured, instead, they are calculated from the measured complex *S*-parameters of the through lines. This procedure is preferred because the *S*-parameters can be very accurately measured with the computer controlled network analyzer and a RF probe station. To obtain expressions that relate the ε_{eff} and characteristic impedance to the measured *S*-parameters [8], consider the uniform through line of length L shown in Figure 3.3(*a*). The complex propagation constant and characteristic impedance of the through line are γ and Z, respectively. When signals of equal magnitude and phase are applied to the line at port 1 and port 2, the line is said to be excited in the even mode. For this excitation a magnetic wall (open circuit) is placed at the center, between port 1 and port 2, as shown in Figure 3.3(*b*). The total signal S_e leaving port 1 is therefore

$$s_e = S_{11} + S_{12}. \tag{3.3}$$

The signal S_e also can be viewed as the reflected signal by the magnetic wall and can be expressed as

$$S_e = \frac{z \coth \gamma l - 1}{z \coth \gamma l + 1}, \tag{3.4}$$

where $\gamma = \alpha + j\beta$,

$$l = \frac{L}{2},$$

$$z = \frac{Z}{Z_0}.$$

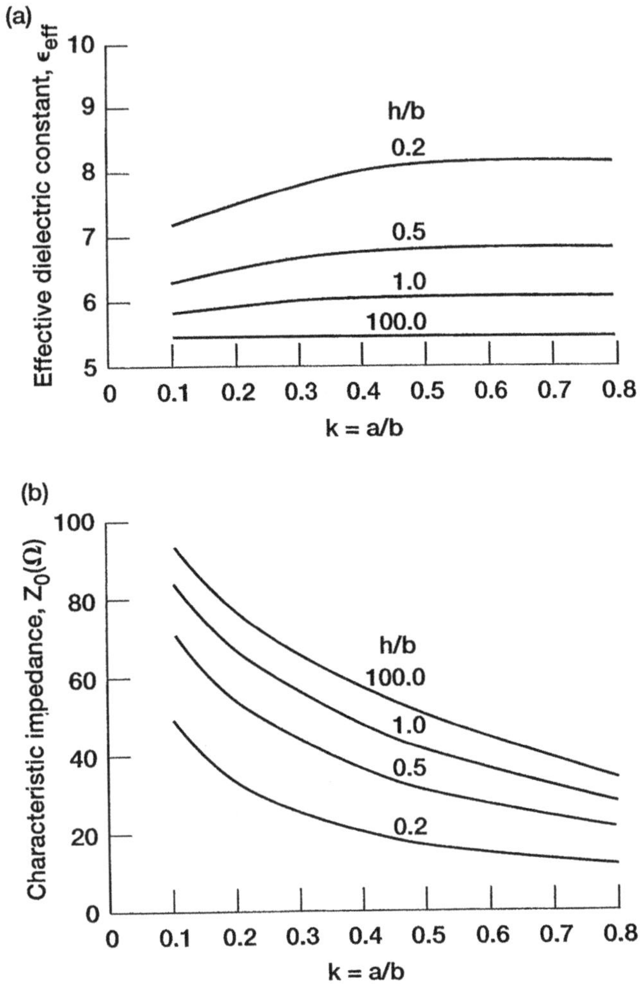

FIGURE 3.2 Computed characteristics of CBCPW as a function of the shape ratio $k = a/b$, with the normalized substrate thickness h/b as a parameters, $\varepsilon_r = 10$: (*a*) Effective dielectric constant, ε_{eff}; (*b*) characteristic impedance, Z_0. (From Reference [4], copyright © IEE.)

In the equations above α and β are the attenuation constant and the propagation constant of the through line, respectively, and Z_0 is the characteristic impedance of the input and output lines respectively. Equation (3.4) can be rewritten as

$$z \coth \gamma l = \frac{1 + S_e}{1 - S_e}. \tag{3.5}$$

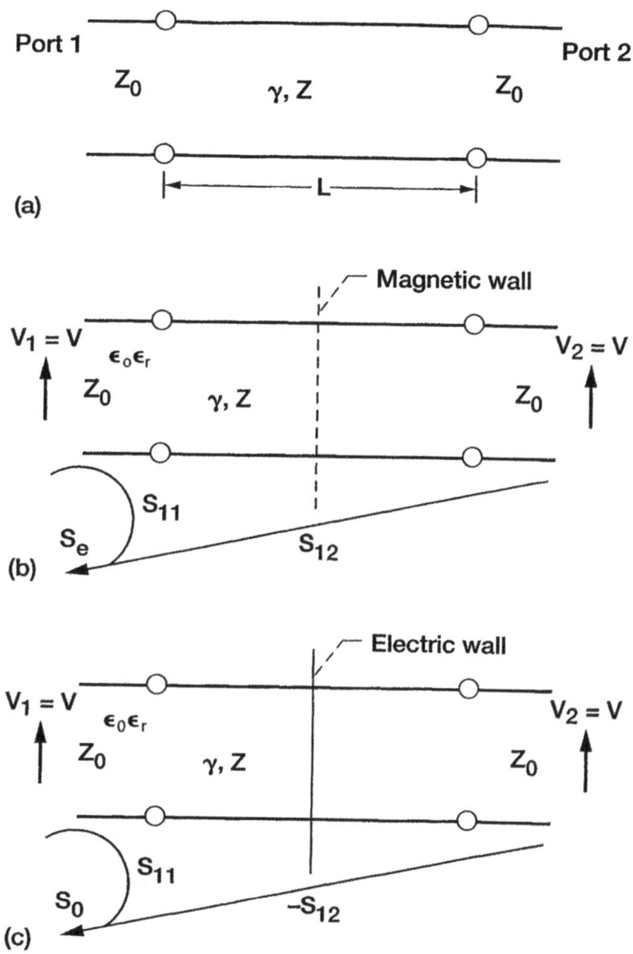

FIGURE 3.3 Uniform transmission line: (*a*) Configuration; (*b*) even excitation; (*c*) odd excitation.

In a similar manner, assuming an odd-mode excitation, an electric wall (short circuit) is placed at the center as shown in Figure 3.3(*c*). The corresponding equations are

$$S_o = S_{11} - S_{12}. \tag{3.6}$$

and

$$z \tanh \gamma l = \frac{1 + S_o}{1 - S_o}. \tag{3.7}$$

TABLE 3.1 Dimensions and Characteristic Impedance of CBCPW through Lines ($h = 100\,\mu m$, $\varepsilon_r = 12.9$)

S (μm)	W (μm)	$Z(\Omega)$ ($t = 0$)	$Z(\Omega)$ ($t = 1.5\,\mu$m)	$Z(\Omega)$ (measured)
51	50	50	47.8	48
27	20	50	45.3	46
14	10	50	41.9	44

Source: From Reference [8], with permission from *Microwave Journal.*

Equations (3.5) and (3.7), when taken together, yield the complex propagation constant and characteristic impedance of the line [8]:

$$\tanh^2 \gamma l = \frac{\left(\dfrac{1 + S_0}{1 - S_0}\right)}{\left(\dfrac{1 + S_e}{1 - S_e}\right)} \tag{3.8}$$

and

$$z^2 = \left(\frac{1 + S_e}{1 - S_e}\right)\left(\frac{1 + S_0}{1 - S_0}\right) \tag{3.9}$$

To carry out the experiments, CBCPW through lines are fabricated on a 100 μm thick GaAs substrate ($\varepsilon_r = 12.9$) [8]. To keep the attenuation due to conduction loss small, the metalization thickness is chosen to be greater than the skin depth. Hence the topside metal forming the center strip conductor and the coplanar ground planes are evaporated gold of 1.5 μm thickness. The bottom side metal is made thicker as in a practical circuit, where it serves as a good heat sink also. Thus the bottom side metal is gold plated to a thickness greater than 3 μm. Further, to ensure that the through lines supports only a quasi-static TEM mode, the topside coplanar ground planes are connected to the bottom ground plane by filled metal vias [8]. The metal vias short circuit the electric fields of the parasitic parallel-plate mode and thus suppress its propagation. Alternatively, the CPW through lines can be placed in a narrow metal channel, which serves the same purpose as the vias and is demonstrated in [9] and [10].

The dimensions of the through lines are summarized in Table 3.1. The theoretical ε_{eff} determined from Eq. (3.1) and ε_{eff} calculated from Eq. (3.8) using the measured S-parameters are found to be in good agreement up to about 20 GHz [8]. However, at very low frequencies, typically below 1 GHz, the experimental ε_{eff} is higher than the theoretical ε_{eff}. Further this effect is more pronounced as the strip conductor width is reduced. This phenomenon can be explained by the skin effect due to finite conductivity and finite thickness of the gold metal [8].

The theoretical characteristic impedance obtained from Eq. (3.2), which neglects the conductor thickness t, is observed to be higher than the characteristic impedance obtained from Eq. (3.9) using the measured S-parameters. In all practical circuits the conductor thickness is finite. Hence an effective center strip and slot widths are obtained by applying the following correction [8]:

$$\Delta = 1.25t \, \frac{1 + \ln(2h/t)}{\pi}. \tag{3.10}$$

The theoretical characteristics impedance after this correction is observed to be lower than the measured values. The correction factor over estimates the thickness effect. The theoretical characteristic impedance with and without the correction are also summarized in Table 3.1. In general, the metal thickness has more effect on the lines with narrower slot width.

3.2.3 Analytical Expressions for CBCPW ε_{eff} and Z_0 in the Presence of a Top Metal Cover

A CBCPW with a top metal cover is shown in Figure 3.4. The ε_{eff} and Z_0 for this structure are given by [5]

$$\varepsilon_{\text{eff}} = 1 + q_2(\varepsilon_r - 1), \tag{3.11}$$

where

$$\left.\begin{aligned}
q_2 &= \frac{\dfrac{K(k_3)}{K(k_3')}}{\dfrac{K(k_3)}{K(k_3')} + \dfrac{K(k_4)}{K(k_4')}} \\
k_3 &= \frac{\tanh(\pi a/2h)}{\tanh(\pi b/2h)} \\
k_4 &= \frac{\tanh(\pi a/2h_1)}{\tanh(\pi b/2h_1)} \\
k_i' &= \sqrt{1.0 - k_i^2}, \quad i = 3 \text{ or } 4
\end{aligned}\right\} \tag{3.12}$$

and

$$Z_0 = \frac{60\pi}{\sqrt{\varepsilon_{\text{eff}}}} \frac{1}{\dfrac{K(k_3)}{K(k_3')} + \dfrac{K(k_4)}{K(k_4')}}, \tag{3.13}$$

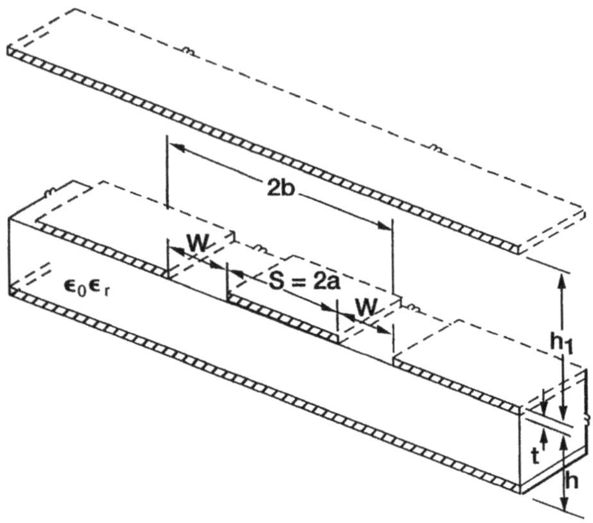

FIGURE 3.4 Schematic of conductor backed coplanar waveguide with top metal cover.

where $K(k)$ is the complete elliptical integral of the first kind. If the cover height h_1 tends to infinity, then k_4 tends to a/b, and the expressions above reduce to those in Section 3.2.1.

It is interesting to note that there are four limiting cases [6] for the CBCPW with a top metal cover. The first case as shown in Figure 3.5(a) is when the height of the cover h_1 is comparable to the substrate thickness h and the slot width W is less than the critical slot width W_C. The critical slot width is defined as a slot width W_C beyond which the expressions for the microstrip line [11] and [12] can replace those of the CPW within a required accuracy [6]. This case is the normal CBCPW with a top metal cover.

The second case is when the height of the top metal cover h_1 is much greater than the substrate thickness h but the slot width W is still less than W_C. On this structure shown in Figure 3.5(b) propagation is similar to a CBCPW.

The third case is when the height of the cover h_1 is comparable to the substrate thickness h but the slot width W is much larger than W_C. Here propagation takes place as on a covered microstrip line [11] as shown in Figure 3.5(c).

The last case is when the height of the cover is much greater than the substrate thickness h and the slot width W is also much greater than W_C. On this structure propagation is as on a open microstrip line [12] as shown in Figure 3.5(d).

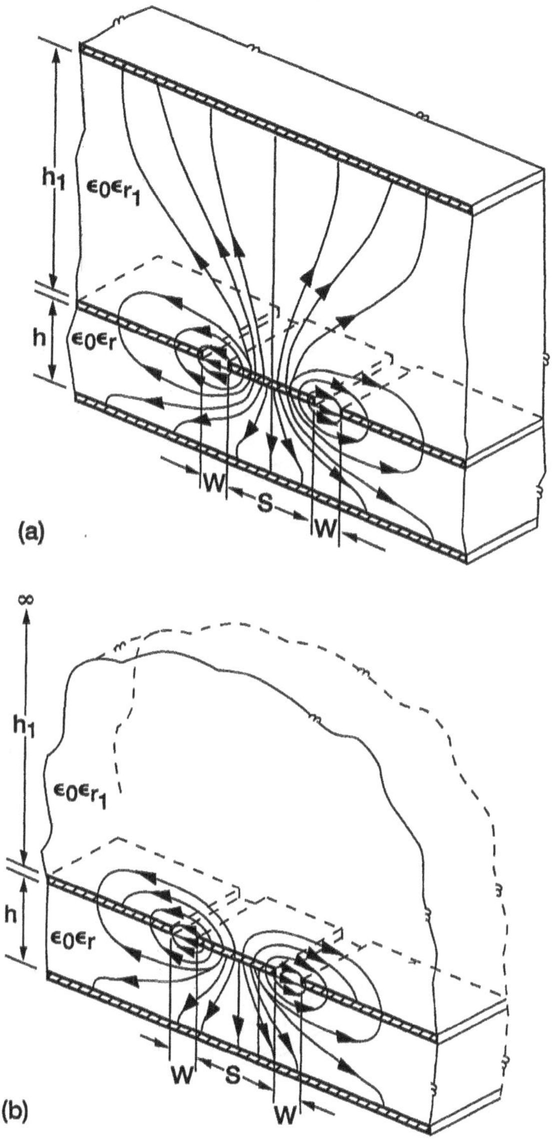

FIGURE 3.5 Limiting cases of CBCPW with top cover as the height of the cover and or the slot width varies: (*a*) Normal CBCPW with a top cover; (*b*) CBCPW; (*c*) microstrip line with top cover; (*d*) microstrip line.

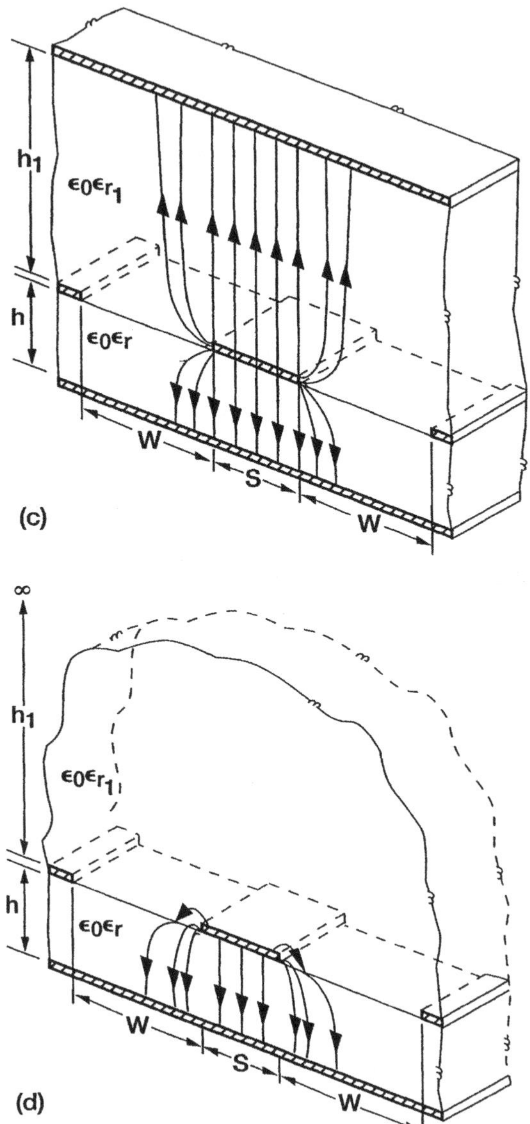

FIGURE 3.5 Continued

3.2.4 Dispersion and Characteristic Impedance from Full-Wave Analysis

The CBCPW structure shown in Figure 3.1 without the top metal cover has also been analyzed assuming infinite lateral extent using spectral domain technique [13]. Figure 3.6(*a*) and (*b*) shows the computed guide wavelength

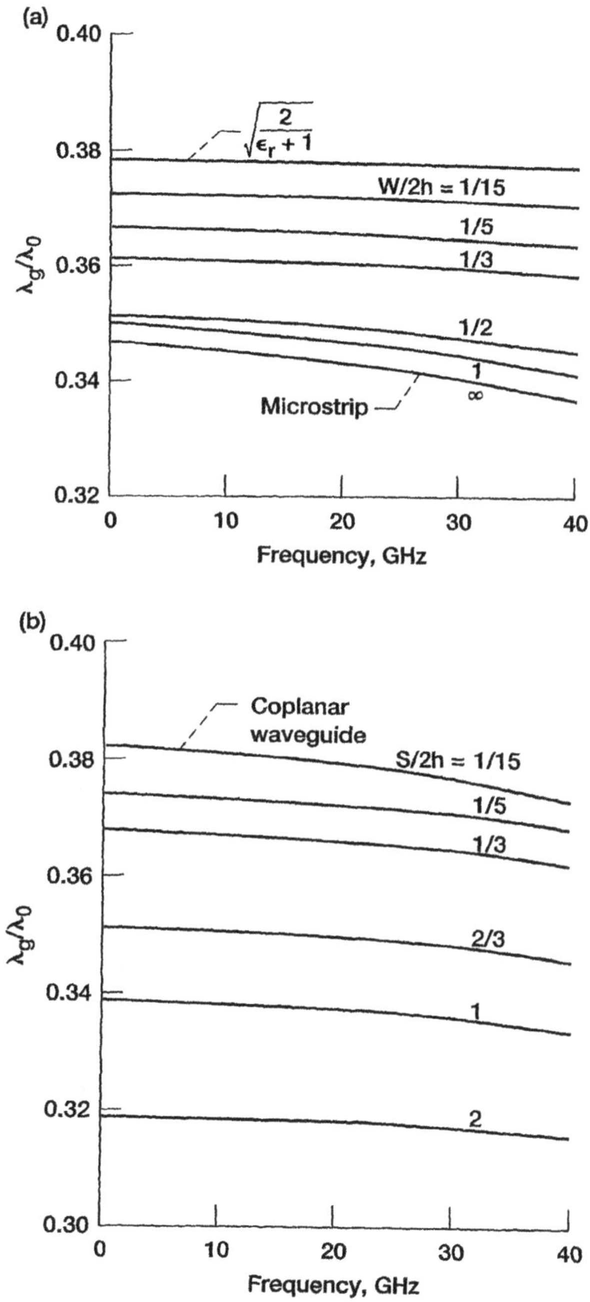

FIGURE 3.6 Computed normalized guide wavelength ratio λ_g/λ_0 as a function of the frequency: (*a*) Normalized slot width $W/2h$ as a parameter, $\varepsilon_r = 13$, $h = 150\mu$m, $S/2h = 1/3$; (*b*) normalized strip width $S/2h$ as a parameter, $\varepsilon_r = 13$, $S/2 = 100\,\mu$m, $W/S = 1/2$. (From Reference [13], copyright © IEE.)

ratio (λ_g/λ_0), where λ_g and λ_0 are the guide wavelength and free space wavelength, respectively, as a function of frequency with the slot width W and substrate thickness h as parameters. Since the CBCPW structure is a combination of both microstrip line and coplanar waveguide, the properties tend to approach that of a microstrip line or that of a coplanar waveguide depending on the dimensions. For example, for a fixed substrate thickness h as W increases, the ratio λ_g/λ_0 approaches that of a microstrip line. On the other hand, for a fixed slot width W and as the substrate thickness h increases, the ratio λ_g/λ_0 approaches that of a conventional coplanar waveguide (CPW). These properties are illustrated in Fig. 3.6(*a*) and (*b*).

The characteristic impedance is defined as [13]

$$Z_0 = \frac{Z_0(\text{air})}{(\lambda_0/\lambda_g)} \tag{3.14}$$

Where $Z_0(\text{air})$ is the characteristic impedance when $\varepsilon_r = 1$. The computed Z_0 is shown in Figure 3.7.

3.3 EFFECT OF CONDUCTING LATERAL WALLS ON THE DOMINANT MODE PROPAGATION CHARACTERISTICS OF CBCPW AND CLOSED FORM EQUATIONS FOR Z_0

The cross section of a CBCPW with conducting lateral walls is shown in Figure 3.8. Depending on the distance of separation, the walls interact with the fields guided by the CBCPW structure and modify its propagation characteristics. This structure has been analyzed using quasi-static TEM approximation [14], and the ε_{eff} and Z_0 are obtained by numerically solving Laplace's equation.

In carrying out the computations the distance g of the side wall from the outer edge of the slot is set equal to $S/2 + W$. With g equal to $S/2 + W$ and the ratio $S/(S + 2W) \leqslant 0.8$, the computed Z_0 is within 1 percent of its value with g very large. This, ensures that the side walls have minimum effect on the propagation. The accuracy of this numerical method is determined by comparing the results in the microstrip limit, that is, when $S/(S + 2W)$ approaches zero, and also in the coplanar limit, that is, when h/S approaches infinity with other results. For example, when $\varepsilon_r = 9.9$ and 12.9, the computed Z_0 in the microstrip limit agrees with [15] and [13] to within 1 and 7 percent, respectively. In the CPW limit as $h/S \rightarrow \infty$, the results asymptotically approach [2].

Based on the solution above a closed form empirical expression is presented for Z_0 [14], [16] for the CPW geometry shown in Figure 3.8 as follows:

$$Z_0 = \left[\frac{5q}{1 + 5q} \cdot \frac{1}{Z_m} + \frac{1}{1 + q} \cdot \frac{1}{Z_c}\right]^{-1}, \tag{3.15}$$

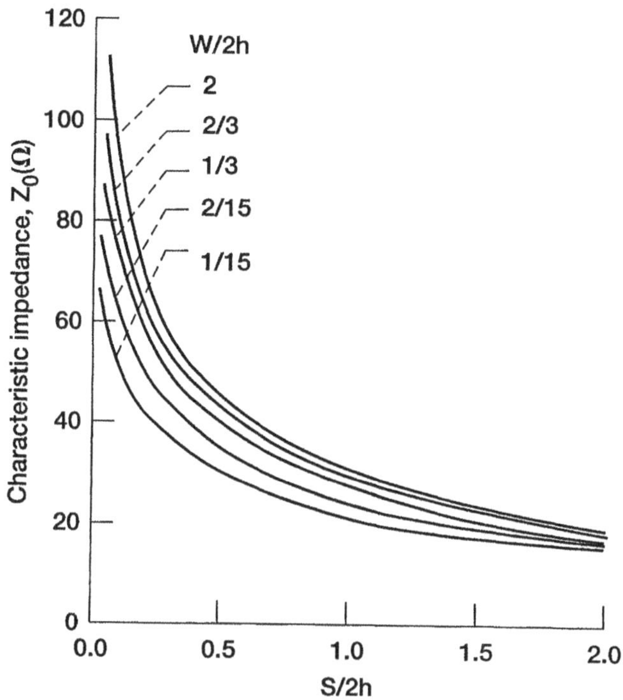

FIGURE 3.7 Computed characteristic impedance Z_0 as a function of the normalized strip width $S/2h$ with the normalized slot width $W/2h$ as a parameter, $\varepsilon_r = 13$, $h = 150\,\mu\text{m}$. (From Reference [13], copyright © IEE.)

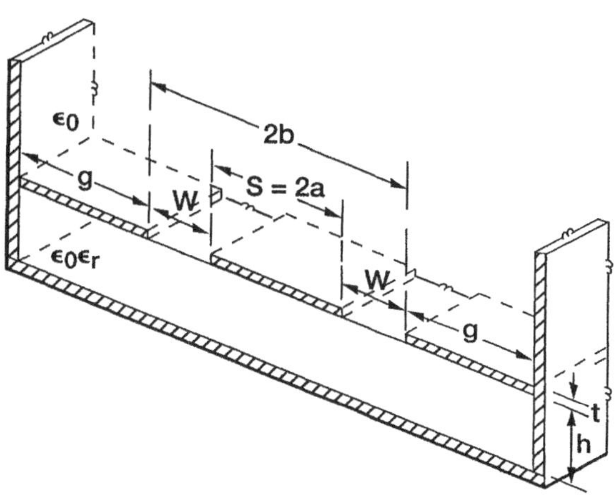

FIGURE 3.8 Schematic of CBCPW with conducting lateral walls.

where

$$q = \frac{S}{h}\left(\frac{S+2W}{S} - 1\right)\left\{3.6 - 2\exp\left[-\frac{(\varepsilon_r + 1)}{4}\right]\right\}. \tag{3.16}$$

This function reproduces the microstrip impedance Z_m exactly as q goes to infinity, and similarly for the CPW impedance Z_c as q goes to zero.

Approximate expressions for Z_m and Z_c are given below [17].

For $S/h \leqslant 1$

$$Z_m = \frac{\eta}{2\pi\sqrt{\varepsilon_{effm}}} \ln\left\{\frac{8h}{S} + 0.25\frac{S}{h}\right\}. \tag{3.17}$$

For $S/h \geqslant 1$,

$$Z_m = \frac{\eta}{\sqrt{\varepsilon_{effm}}}\left\{\frac{S}{h} + 1.393 + 0.667\ln\left(\frac{S}{h} + 1.444\right)\right\}^{-1}, \tag{3.18}$$

where $\eta = 120\pi\ \Omega$ and

$$\varepsilon_{effm} = \frac{\varepsilon_r + 1}{2} + \frac{\varepsilon_r - 1}{2}\left(1 + \frac{10h}{S}\right)^{-1/2}. \tag{3.19}$$

The maximum error in Z_m and ε_{effm} relative to [15], [18] is less than 2 percent. The Z_c is given by [17], [19]

$$Z_c = \frac{30\pi}{\sqrt{\varepsilon_{effc}}}\frac{K(k')}{K(k)}, \tag{3.20}$$

where

$$k = \frac{S}{S+2W} \quad \text{and} \quad k' = \sqrt{1-k^2},$$

$$\varepsilon_{effc} = \frac{\varepsilon_r + 1}{2}\left\{\tan\left[0.775\ln\left(\frac{h}{W}\right) + 1.75\right]\right.$$
$$\left. + \frac{kW}{h}[0.04 - 0.7k + 0.01(1 - 0.1\varepsilon_r)(0.25 + k)]\right\}. \tag{3.21}$$

The maximum error in ε_{effc} is 1.5 percent for $\varepsilon_r \geqslant 9$, $h/W \geqslant 1$, and $0 \leqslant k \leqslant 0.7$ when compared to [20]. With the foregoing substitutions, the expression in Eq. (3.15) for Z_0 is accurate to within 2.2 percent for $0.25 \leqslant h/S \leqslant 3$, $0.1 \leqslant S/(S + 2W) \leqslant 0.8$ and $\varepsilon_r \leqslant 30$.

3.3.1 Experimental Validation

Experimental validation is carried out by measuring the time domain response (TDR) of a multisection CBCPW structure [14] shown in Figure 3.9(*a*). In this structure the aspect ratio of the CBCPW, namely $S/(S + 2W)$, is made to vary from a low to a high value while the characteristic impedance Z_0 is held constant at 50 Ω. Thus the structure consists of alternate low and high aspect ratio uniform CBCPW sections cascaded by tapered CBCPW section. The taper profile is normally chosen according to a graph of $(S + 2W)/S$ as a function of S/h for Z_0 equal to 50 Ω. The numerical analysis above is based on

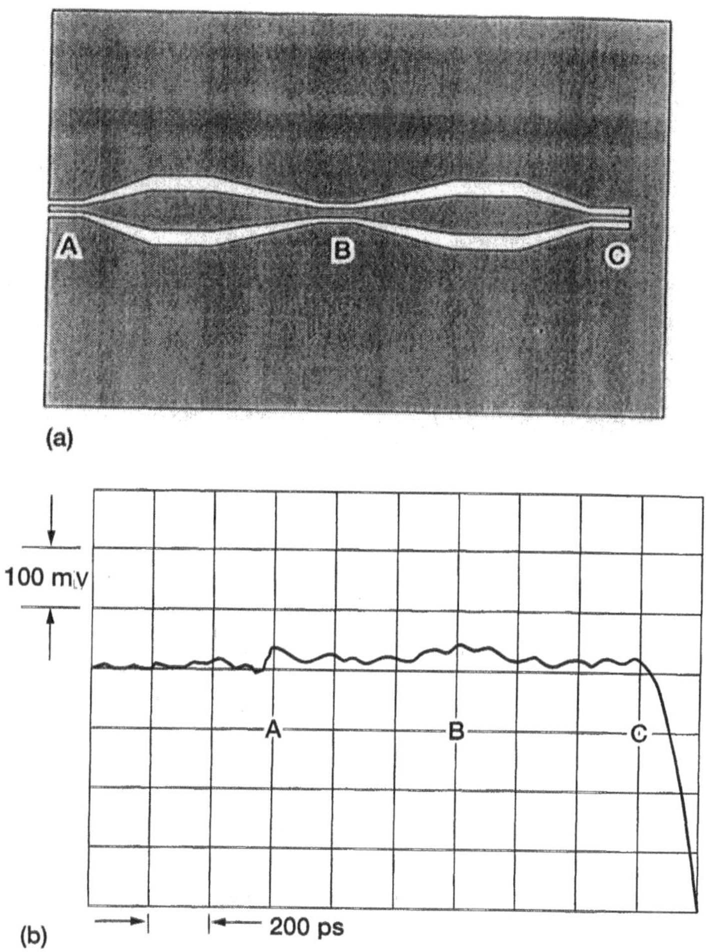

FIGURE 3.9 Tapered CBCPW test structure on a dielectric substrate, $\varepsilon_r = 4.7$, $h = 0.125$ in.: (*a*) Schematic of the conductor pattern; (*b*) measured TDR response, horizontal scale, 200 ps. (From Reference [14], © 1983 IEEE.)

a two-dimensional solution of Laplace's equation, and therefore it does not give the effect of the rate of tapering on Z_0. Hence in the structure above, a simple linear taper with an angle of 20° is chosen. The overall length of the structure is about 5 in.

The reflection coefficient is measured by a TDR and the response is shown in Figure 3.9(*b*). The pulse used in the measurement has a rise time of 25 ps. The measured response shows 50 Ω nature of the entire structure [14]. Points *A* to *C* in Figure 3.9(*b*) refer to the locations on the structure. The worst-case deviation from 50 Ω occurs due to over etch where the line width is a minimum, that is, point *B*. The response drops to zero beyond point *C*, since the structure is terminated in a short circuit.

3.4 EFFECT OF LATERAL WALLS ON THE HIGHER-ORDER MODE PROPAGATION ON CBCPW

3.4.1 Perfect Conductors and Lossless Dielectric

Figure 3.10 shows a CBCPW with either electric or magnetic lateral walls. This structure is modeled as a thick capacitive iris in a parallel plate waveguide backed by a dielectric substrate and short circuited at one end in [22]. The thick iris is also considered as a parallel plate waveguide. This model is similar to that initially proposed for a slot line in [23]. For the purpose of analysis, the whole structure is divided into three regions, namely the air region, the iris region, and the dielectric substrate region. In each region the electromagnetic fields are expressed as a series expansion in terms of TE and TM modes (with

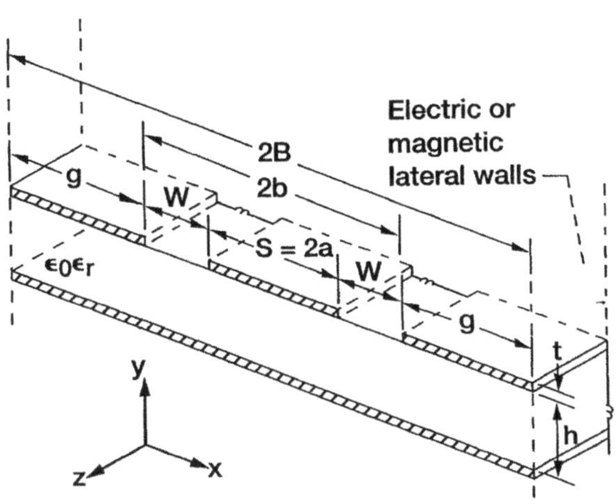

FIGURE 3.10 Analytical model for CBCPW with electric or magnetic lateral walls.

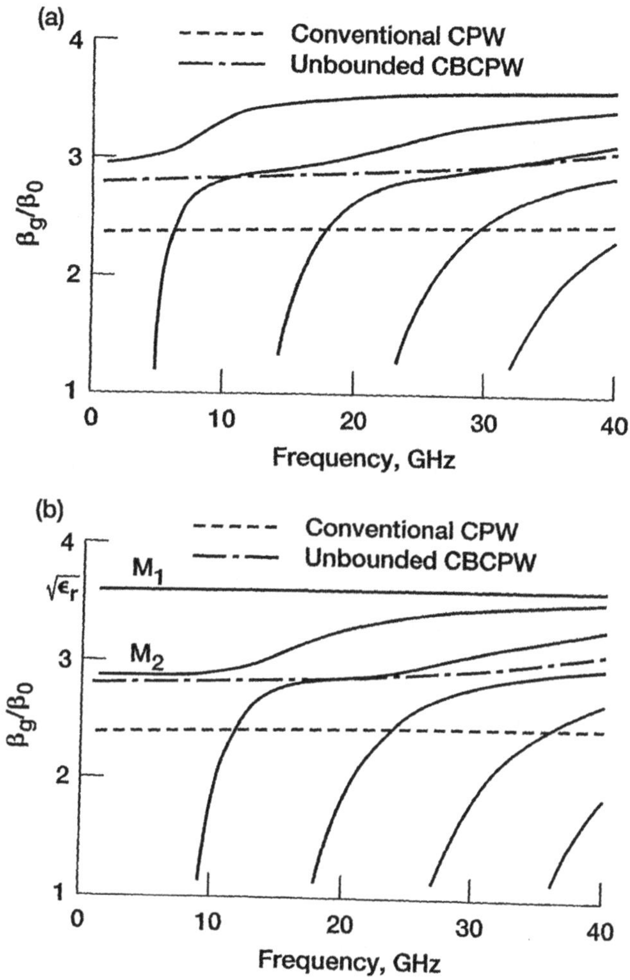

FIGURE 3.11 Computed normalized propagation constant β_g/β_0 as a function of frequency for CBCPW, $\varepsilon_r = 13$, $h = 150\,\mu\text{m}$, $S = 100\,\mu\text{m}$, $W = 200\,\mu\text{m}$, $t = 1\,\mu\text{m}$, $B = 5\,\text{mm}$: (*a*) Bounded by lateral electric wall; (*b*) bounded by lateral magnetic wall. (From Reference [22], © 1983 IEEE.)

respect to *y*). After applying the appropriate boundary conditions, the system of equations are solved to obtain the propagation constant along the *z*-direction.

The higher-order modes in a CBCPW with lateral electric or conducting walls is first discussed. Figure 3.11(*a*) shows the computed normalized propagation constant β_g/β_0 for the dominant as well as the higher-order modes for the electric wall case. Where β_g and β_0 are the propagation constant in the guide and free space, respectively. The figure also includes the quasi-TEM mode propagation constant for a conventional CPW as well as for an unbounded CBCPW. The computed results show that the dominant mode for the CBCPW with lateral electric walls is a quasi-TEM mode with zero cutoff frequency. The presence of this mode is also verified using spectral domain method independently in [24]. This mode has a propagation constant higher than that of the conventional CPW and the unbounded CBCPW. The cutoff frequency of the higher-order modes depends on the distance between the lateral electric walls. This fact is also verified independently in [24].

Figure 3.11(*b*) shows the computed propagation constant for the CBCPW with lateral magnetic walls. The computed results show that on this structure two quasi-TEM modes with zero cutoff frequency can propagate. These modes correspond to the two electrostatic fields that are obtained by putting at equal or different potentials the center strip conductor and the top ground planes. These two modes are designated as M_1 and M_2 in Figure 3.11(*b*). It could be easily visualized that mode M_1 with higher propagation constant and consequently higher $\varepsilon_{\mathrm{eff}}$ has a field distribution the same as a parallel plate waveguide perturbed by two slots on the top wall. The $\varepsilon_{\mathrm{eff}}$ is equal to ε_r at all frequencies. The mode designated M_2 and the higher-order modes are almost identical to those in Figure 3.11(*a*). Finally, the computations show that if the substrate thickness is increased from150 to 500 μm with all other parameters remaining constant, the quasi-TEM mode propagation constant of the conventional CPW and the unbounded CBCPW approach each other but the cutoff frequencies of the higher-order modes are unaffected.

3.4.2 Conductors with Finite Thickness, Finite Conductivity, and Lossless or Lossy Dielectric

A CBCPW with top strip conductor and top ground planes with finite thickness and finite conductivity is analyzed using two different methods in [25] to [27]. The two methods are the mode-matching method [25] and the method of lines [26], [27], respectively. In the mode-matching methods, in order to realistically model a practical structure, the lateral walls and the bottom ground plane are assumed to have finite conductivity and perfect conductivity, respectively. Further in [25] a metal cover with perfect conductivity is placed above the structure. Since the guided energy is confined to the slots, it is assumed that the top metal cover, which is at a sufficiently large distance, will not perturb significantly the guiding characteristics of the structure. In the method of lines, the lateral walls and the bottom ground plane are assumed to have perfect conductivity and finite conductivity, respectively. Figure 3.12(*a*) and (*b*) shows the cross-sectional geometry of the CBCPW considered in the analysis above.

Consider the CBCPW shown in Figure 3.12(*a*). The computed normalized guide wavelength ratio λ_0/λ_g and normalized leakage constant α/k_0, where $k_0 = 2\pi/\lambda_0$, for the dominant and the first higher-order mode using the mode-matching method for two different lateral wall separation are shown in

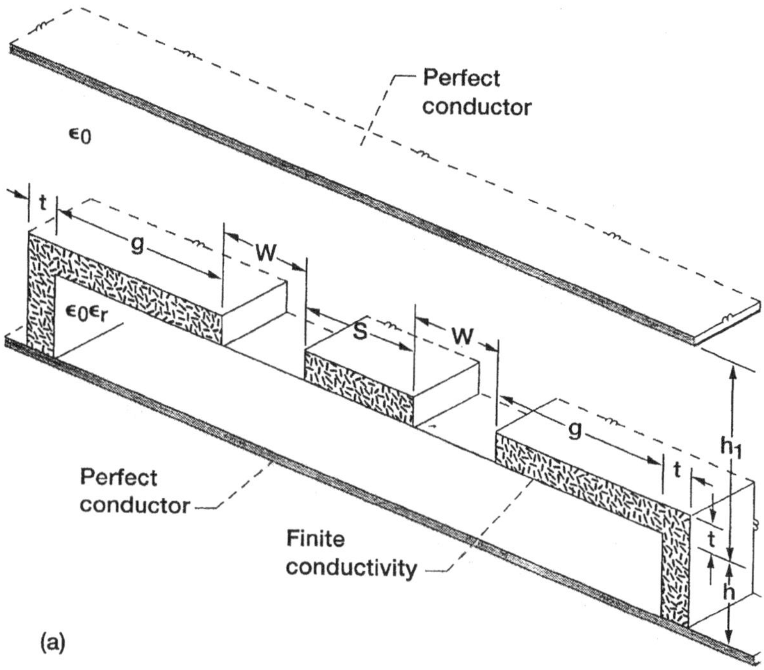

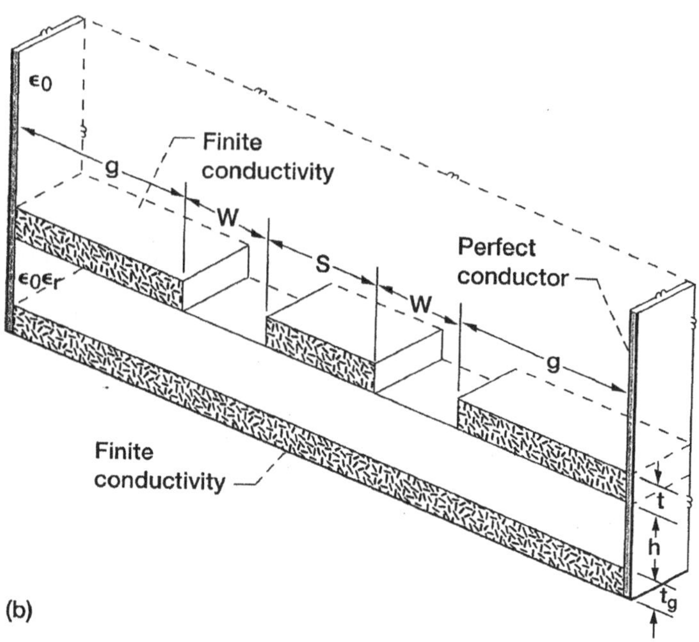

FIGURE 3.12 Cross-sectional geometry of CBCPW considered in the (*a*) mode-matching method and (*b*) method of lines.

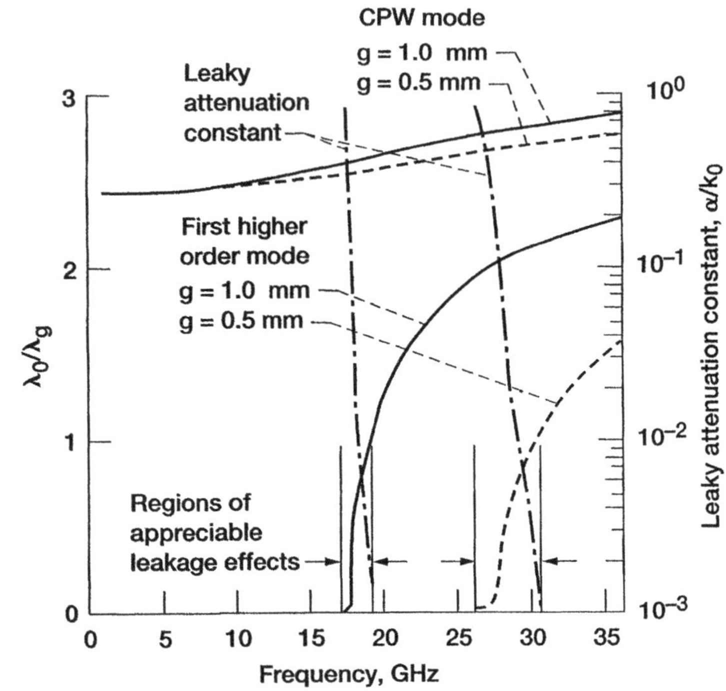

FIGURE 3.13 Computed normalized guide wavelength ratio λ_g/λ_0 and normalized leaky attenuation constant α/k_0 of first higher-order mode as a function of frequency for CBCPW bounded by lateral conducting walls: $\varepsilon_r = 10.2$, $h = 0.635\,\text{mm}$, $t = 17\,\mu\text{m}$, $\sigma = 5.8 \times 10^7$ S/m, $W = 0.508\,\text{mm}$, $S = 0.508\,\text{mm}$. (From Reference [25], copyright © IEE.)

Figure 3.13. The two later wall separations correspond to $g = 0.5$ mm and $g = 1.0\,\text{mm}$, respectively. In the region in which the higher-order mode propagates with λ_0/λ_g and leaky constant α/k_0 both less than unity, an appreciable leakage effect is observed. The leaky mode radiates power into space at an angle to the transmission line. The leaky higher-order mode is cutoff when its normalized leaky attenuation constant is greater than unity. Below the appreciable leakage region, discontinuities in a CBCPW circuit cannot convert the CPW mode to the leaky higher-order mode and thus no power leakage occurs. Above this region, the higher-order mode becomes bounded and mode conversion will take place. In Figure 3.13, the single-mode frequency range is 17 and 26 GHz for the two lateral wall separations, respectively. The corresponding appreciable leakage regions are 17 to 19 and 26 to 30.5 GHz, respectively. Likewise consider the CBCPW shown in Figure 3.12(*b*). The parameters for this structure are $\varepsilon_r = 9.6$, $h = 0.254\,\text{mm}$, $W = 0.15\,\text{mm}$, $S = 0.3\,\text{mm}$, $g = 1.0\,\text{mm}$, $t = 3\,\mu\text{m}$, and t_g is equal to one skin depth δ. The method of lines [26] predicts the cutoff frequencies for the first two higher-order modes to lie between 23 and 25 GHz. In the next section experimental validation is provided for the computed results.

3.4.3 Experimental Validation

To experimentally validate the numerical results of the mode matching method [25], a 50 Ω CBCPW through line is characterized by measuring the scattering parameters. The through line geometry and parameters are presented in Figures 3.12(*a*) and 3.13, respectively. The lateral wall separation for the experiment is such that $g = 1.0$ mm. The transmission coefficient S_{21} is measured over the frequency range of dc to 26.5 GHz. The measurement shows that the through line is resonance free up to 17 GHz [25]. This is in agreement with the numerical results discussed in the previous section. Beyond 17 up to 26.5 GHz, the measured S_{21} indicate the presence of several resonances. The first two resonances occur at about 17.7 and 18.4 GHz [25]. These resonances occur in the frequency range pertaining to appreciable leakage in accordance with the analysis. Further these resonances are attributed to the first higher-order mode.

To experimentally validate the numerical results of the method of lines [26], a CBCPW through line is characterized by measuring the reflection coefficient S_{11} over the frequency range of dc to 40.0 GHz. The CBCPW geometry is presented in Figure 3.12(*b*), and the parameters are indicated in the previous section. The measured S_{11} is below -10.0 dB up to about 24 GHz indicating good impedance match and single-mode propagation. The S_{11} increases beyond this frequency, and has a peak at about 25 GHz. This is possible if the power normally carried by the fundamental mode is now converted into a higher-order mode, which causes a high reflection or return loss [26]. This observation is in agreement with the numerical result discussed in the previous section.

3.5 CHANNELIZED COPLANAR WAVEGUIDE

This section presents a new variant of the CBCPW. The new structure has low height side conducting walls, which together with the conductor backing constitute a metal channel and hence is appropriately termed as channelized coplanar waveguide (CCPW) [9], [10]. The structure is shown in Figure 3.14.

The channel width 2B is chosen such that the dielectric filled rectangular waveguide mode is cutoff. As a consequence the ground planes are of finite extent with width g typically less than 1.5S. Further, since the ground planes are electrically connected by the side walls to the conductor backing, the excitation of spurious parallel plate modes are eliminated. Figure 3.15 shows the measured and computed ε_{eff} for a CCPW. The computed ε_{eff} is based on the model reported in [21]. The agreement between the measured and modeled ε_{eff} for a CCPW on a Duroid substrate ($\varepsilon_r = 2.2$) is excellent.

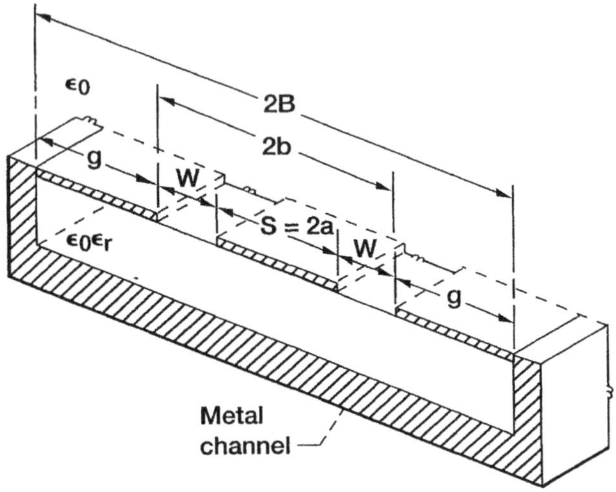

FIGURE 3.14 Schematic of channelized coplanar waveguide (CCPW).

In [30] the computed λ_g/λ_0 and Z_0 for two variants of the structure above, namely the suspended CPW and inverted CPW, respectively, are represented. In these structures a very small air gap is introduced between the substrate and the lower ground plane. In [31] the characteristics of a suspended CCPW is described.

3.6 REALIZATION OF LATERAL WALLS IN PRACTICAL CIRCUITS

The previous sections showed that if closely spaced conducting lateral walls are introduced, the cutoff frequency of the higher-order modes can be raised. Thereby increasing the useful bandwidth of the circuit. However, introducing

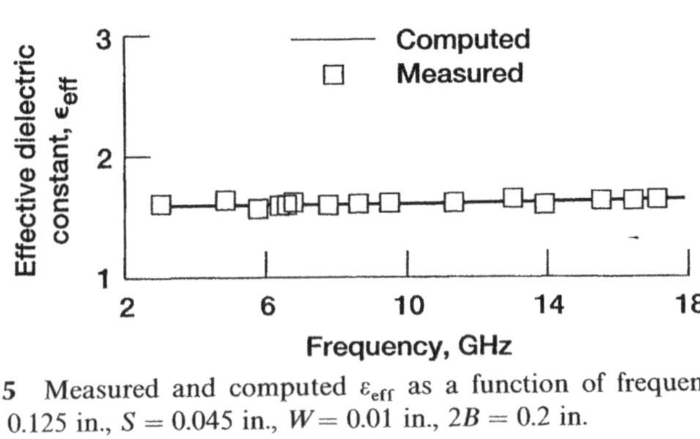

FIGURE 3.15 Measured and computed ε_{eff} as a function of frequency for CCPW: $\varepsilon_r = 2.2$, $h = 0.125$ in., $S = 0.045$ in., $W = 0.01$ in., $2B = 0.2$ in.

these walls is not a trivial task because of small feature size of the circuits. A practical solution to this problem is to tie the top ground plane to the bottom conductor by metal-filled via holes [8], [28], [29]. To demonstrate the efficacy of this solution, a CBCPW on a 0.01 in. thick alumina substrate ($\varepsilon_r = 9.6$), with $S = 0.2$ mm and $W = 0.15$ mm, is fabricated. the spacing between the via holes in this circuit is 0.7 mm and the via hole row is at a distance of 0.2 mm from the edge of the slot [28]. This configuration of the CBCPW is interpreted as a rectangular waveguide with side walls 0.9 mm apart, which allows a fundamental waveguide mode to start propagating at about 55 GHz. Broadband measurements made on this through line using a coaxial test fixture and a calibrated network analyzer shows that the scattering parameters (S_{11}, S_{21}) deteriorates at approximately this frequency [28].

REFERENCES

[1] J. L. B. Walker, "A Survey of European Activity on Coplanar Waveguide," in *1993 IEEE MTT-S Int. Microwave Symp. Dig.*, Vol. 2, Atlanta, GA, pp. 693–696, June 14–18, 1993.

[2] C. P. Wen, "Coplanar Waveguide: A Surface Strip Transmission Line Suitable for Nonreciprocal Gyromagnetic Device Applications," *IEEE Trans. Microwave Theory Tech.*, Vol. 17, No. 12, pp. 1087–1090, Dec. 1969.

[3] C. Veyres and V. Fouad Hanna, "Extension of the Application of Conformal Mapping Techniques to Coplanar Lines with Finite Dimensions," *Int. J. Electron.*, Vol. 48, No. 1, pp. 47–56, Jan. 1980.

[4] G. Ghione and C. U. Naldi, "Parameters of Coplanar Waveguides with Lower Ground Plane," *Electron. Lett.*, Vol. 19, No. 18, pp. 734–735, Sept. 1983.

[5] G. Ghione and C. U. Naldi, "Coplanar Waveguides for MMIC Applications: Effect of Upper Shielding, Conductor Backing, Finite-Extent Ground Planes, and Line-to-Line Coupling," *IEEE Trans. Microwave Theory Tech.*, Vol. 35, No. 3, pp. 260–267, March 1987.

[6] S. S. Bedair and I. Wolff, "Fast and Accurate Analytic Formulas for Calculating the Parameters of a General Broadside—Coupled Coplanar Waveguide for (M)MIC Applications," *IEEE Trans. Microwave Theory Tech.*, Vol. 37, No. 5, pp. 843–850, May 1989.

[7] Y.-C. Wang and J. A. Okoro, "Impedance Calculations for Modified Coplanar Waveguides," *Int. J. Electron.*, Vol. 68, No. 5, pp. 861–875, May 1990.

[8] Y.-C. Shih, "Broadband Characterization of Conductor-Backed Coplanar Waveguide Using Accurate On-Wafer Measurement Techniques," *Microwave J.*, Vol. 34, No. 4, pp. 95–105, April 1991.

[9] R. N. Simons, G. E. Ponchak, K. S. Martzaklis, and R. R. Romanofsky, "Channelized Coplanar Waveguide: Discontinuities, Junctions, and Propagation Characteristics," in 1989 *IEEE MTT-S Int. Microwave Symp. Dig.*, Vol. 3, Long Beach, CA, pp. 915–918, June 13–15, 1989.

[10] R. N. Simons, G. E. Ponchak, K. S. Martzaklis, and R. R. Romanofsky, "Experimental Investigations on Channelized Coplanar Waveguide," NASA TM–102494, April, 1990.

[11] S. S. Bedair and M. I. Sobhy, "Accurate Formulas for the Computer-Aided Design of Shielded Microstrip Circuits," *Proc. Inst. Elec. Eng.*, Vol. 127, pt. H, pp. 305–308, 1980.

[12] E. Hammerstad and O. Jensen, "Accurate Models for Microstrip Computer-Aided Design," in *IEEE MTT-S Int. Microwave Symp. Dig.*, Washington, DC, pp. 407–409, May 28–29, 1980.

[13] Y.-C. Shih and T. Itoh, "Analysis of Conductor-Backed Coplanar Waveguide," *Electron. Lett.*, Vol. 18, No. 12, pp. 538–540, June 1982.

[14] D. A. Rowe and B. Y. Lao, "Numerical Analysis of Shielded Coplanar Waveguides," *IEEE Trans. Microwave Theory Tech.*, Vol. 31, No. 11, pp. 911–915, Nov. 1983.

[15] H. A. Wheeler, "Transmission Line Properties of Parallel Strips Separated by a Dielectric Sheet," *IEEE Trans. Microwave Theory Tech.*, Vol. 13, No. 3, pp. 172–185, March 1965.

[16] D. Neuf and S. Spohrer, "Ultrasmall MIC Mixer Designed for ECM Applications," *Microwave System News and Communication Technology*, Vol. 15, No. 10, pp. 70–80, Oct. 1985.

[17] K. C. Gupta, R. Garg, and R. Chadha, *Computer-Aided Design of Microwave Circuits*, Dedham, MA: Artech House, 1981, pp. 61 and 69.

[18] M. V. Schneider, "Microstrip Lines for Microwave Integrated Circuits," *Bell Systems Tech. J.*, Vol. 48, pp. 1421–1444, 1969.

[19] G. Ghione and C. Naldi, "Analytical Formulas for Coplanar Lines in Hybrid and Monolithic MICs," *Electron. Lett.*, Vol. 20, No. 4, pp. 179–181, Feb. 1984.

[20] M. E. Davis, E. W. Williams and A. C. Celestini, "Finite-Boundary Corrections to the Coplanar Waveguide Analysis," *IEEE Trans. Microwave Theory Tech.*, Vol. 21, No. 9, pp. 594–596, Sept. 1973.

[21] R. N. Simons, "Suspended Coupled Slotline Using Double Layer Dielectric," *IEEE Trans. Microwave Theory Tech.*, Vol. 29, No. 2, pp. 162–165, Feb. 1981.

[22] G. Leuzzi, A. Silbermann, and R. Sorrentino, "Mode Propagation in Laterally Bounded Conductor-Backed Coplanar Waveguides," in *1983 IEEE MTT-S Int. Microwave Symp. Dig.*, Boston, Massachusetts, pp. 393–395, May 31–June 3, 1983.

[23] S. B. Cohn, "Slot Line on a Dielectric Substrate," *IEEE Trans. Microwave Theory Tech.*, Vol. 17, No. 10, pp. 768–778, Oct. 1969.

[24] M. A. Magerko, L. Fan, and K. Chang, "A Discussion on the Coupling Effects in Conductor-Backed Coplanar Waveguide MIC's with Lateral Sidewalls," in *1993 IEEE MTT-S Int. Microwave Symp. Dig.*, Vol. 2, Atlanta, GA, pp. 947–950, June 14–18, 1993.

[25] C.-C. Tien, C.-K. C. Tzuang, and J. Monroe, "Effect of Lateral Walls on the Propagation Characteristics of Finite Width Conductor-Backed Coplanar Waveguides," *Electron. Lett.*, Vol. 29, No. 15, pp. 1357–1358, July 1993.

[26] K. Wu and R. Vahldieck, "Field Distribution and Dispersion Characteristics of Fundamental and Higher-Order Modes in Miniature Hybrid MIC (MHMIC)

Considering Finite Conductor Thickness and Conductivity," in *1991 IEEE MTT-S Int. Microwave Symp. Dig.*, Vol. 3, Boston, MA, pp. 995–998, June 10–14, 1991.

[27] K. Wu and R. Vahldieck, "Rigorous Analysis of the Characteristics Impedance in Conductor-Backed Miniature Coplanar Waveguides Considering Multiple Layers of Lossy and Finite Thickness Metal," in *1992 IEEE MTT-S Int. Microwave Symp. Dig.*, Vol. 2, Albuquerque, NM, pp. 987–990, June 1–5, 1992.

[28] M. Yu, R. Vahldieck, and J. Huang, "Comparing Coax Launcher and Wafer probe Excitation for 10 mil Conductor Backed CPW with Via Holes and Airbridges," in *1993 IEEE MTT-S Int. Microwave Symp. Dig.*, Vol. 2, Atlanta, GA, pp. 705–708, June 14–18, 1993.

[29] M.-J. Tsai, C. Chen, N. G. Alexopoulos, and T.-S. Horng, "Multiple Arbitrary Shape Via-Hole and Air-Bridge Transitions in Multilayered Structures," *IEEE Trans. Microwave Theory Tech.*, Vol. 44, No. 12, pp. 2504–2511, Dec. 1996.

[30] R. N. Simons, "Propagation Characteristics of Some Novel Coplanar Waveguide Transmission Lines on GaAs at MM-Wave Frequencies," *1986 Conf. on Millimeter Wave/Microwave Measurements and Standards for Miniaturized Systems*, Redstone Arsenal, AL, Nov. 6–7, 1986 (Also NASA TM-89839).

[31] K. Wu, Y. Xu, and R. G. Bosisio, "Theoretical and Experimental Analysis of Channelized Coplanar Waveguides (CCPW) for Wideband Applications of Integrated Microwave and Millimeter-Wave Circuits," *IEEE Trans. Microwave Theory Tech.*, Vol. 42, No. 9, pp. 1651–1659, Sept. 1994.

CHAPTER 4

Coplanar Waveguide with Finite-Width Ground Planes

4.1 INTRODUCTION

In this chapter the characteristics of a conventional coplanar waveguide (CPW) and a conductor-backed coplanar waveguide (CBCPW) with finite-width top ground planes are presented. The reason for making the top ground planes of finite extent is because this more closely models a practical CPW circuit. In addition, by taking a substrate with conductor backing, a CPW circuit mounted either on a metal base as in a package or in a test fixture for performance characterization is accurately modeled.

Section 4.2 commences with analytical expressions to determine the effective dielectric constant ε_{eff} and characteristic impedance Z_0 of a CPW with finite-width ground planes (FW-CPW) supported on a multilayer dielectric substrate. There expressions are derived using quasi-static TEM conformal mapping techniques. It is also shown that these expressions simplify when only one or two dielectric substrates are present as in the case of a conventional CPW or a sandwiched CPW. Computed ε_{eff} and Z_0 for the simplified structure are presented. The section concludes with a quantitative discussion on the dispersion and Z_0 obtained using full-wave analysis.

Section 4.3 commences with a qualitative discussion on the modes that are supported by a conductor-backed CPW with top ground planes of finite width (FW-CBCPW). However, in this chapter the discussion is limited to the dominant CPW mode and the microstrip-like mode (MSL). In addition the lower ground plane is considered either to be finite or infinite in extent. The section concludes with a quantitative discussion on the dispersion characteristics of the above two modes obtained using full-wave analysis.

Section 4.4 presents three models to predict the resonance observed in the characteristics of FW-CBCPW 50 Ω through lines. The accuracy of the predicted values are demonstrated by comparing with experimentally measured values.

4.2 CONVENTIONAL COPLANAR WAVEGUIDE WITH FINITE-WIDTH GROUND PLANES ON A DIELECTRIC SUBSTRATE OF FINITE THICKNESS

4.2.1 Analytical Expressions Based on Quasi-static TEM Conformal Mapping Techniques to Determine Effective Dielectric Constant and Characteristic Impedance

Initial studies on coplanar waveguide (CPW) with finite-width ground planes were limited to dielectric substrates of infinite thickness [1]. In [1] simple expression to calculate the normalized phase velocity ratio v_{ph}/c', where c' is the velocity of light in free space, and characteristic impedance Z_0 are presented. However, in a practical circuit, the thickness of the substrate is always finite. Hence subsequent studies considered a coplanar waveguide with finite width ground planes on (a) a dielectric substrate of finite thickness [2] to [4], (b) sandwiched between two dielectric substrates [5] and (c) sandwiched between multiple dielectric substrates [6].

A FW-CPW sandwiched between multiple dielectric substrates is shown in Figure 4.1. On this structure a quasi-static TEM mode is assumed to propagate. Using this approximation, the ε_{eff}, phase velocity v_{ph}, and Z_0 are given by [3]

$$\varepsilon_{eff} = \frac{C}{C_0}, \tag{4.1}$$

$$v_{ph} = \frac{c'}{\sqrt{\varepsilon_{eff}}}, \tag{4.2}$$

$$Z_0 = \frac{1}{C v_{ph}}, \tag{4.3}$$

where c' is the speed of light in free space, C is the capacitance per unit length of the line, and C_0 is the capacitance per unit length of the line in the absence of dielectric substrates. Thus, to determine ε_{eff} and Z_0, one needs only to find the capacitance C and C_0. To find these capacitances, we assume that the boundaries of the dielectric layers are along the electric-field lines. In which case magnetic walls can be placed along the dielectric boundaries without disturbing the fields and the capacitance of the line can be divided into partial capacitances [3] and [6]. By this assumption the capacitance of the line shown in Figure 4.1 can be written as the superposition of six partial capacitances as follows [6]:

$$C = C_0 + C_1 + C_2 + C_3 + C_4 + C_5. \tag{4.4}$$

The configurations of there capacitances are shown in Figure 4.2(a) to (f).

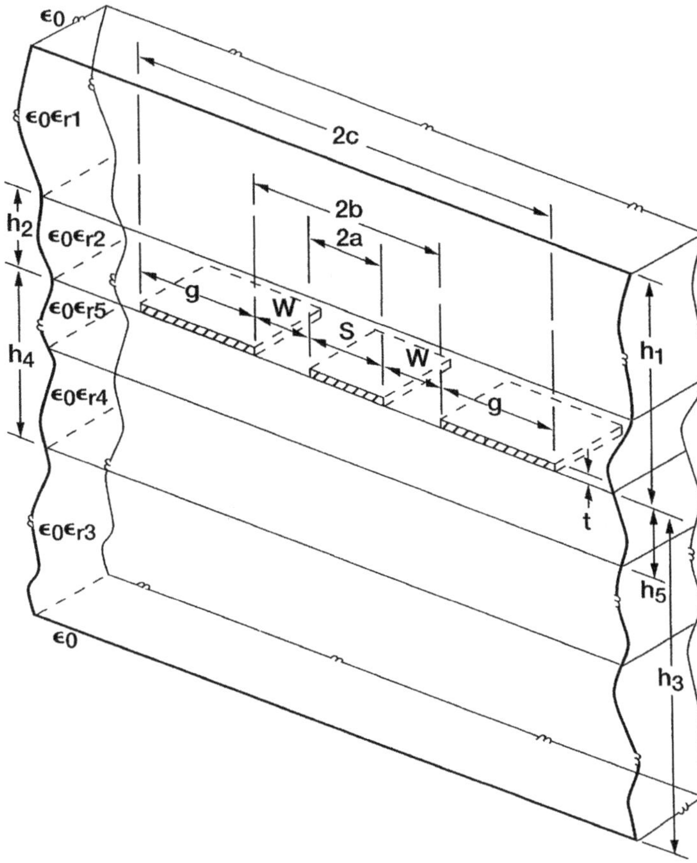

FIGURE 4.1 Schematic of a FW-CPW on a multilayer dielectric substrate.

Calculation of C_0 As seen from Figure 4.2(*a*), C_0 is the capacitance of the line in the absence of all dielectrics. This capacitance is given by [3]

$$C_0 = 4\varepsilon_0 \frac{K(k')}{K(k)}, \tag{4.5}$$

where K is the complete elliptical integral of the first kind. The arguments k and k' are dependent on the geometry of the line and are given by

$$k = \frac{c}{b}\sqrt{\frac{b^2 - a^2}{c^2 - a^2}}, \tag{4.6}$$

$$k' = \sqrt{1 - k^2} = \frac{a}{b}\sqrt{\frac{c^2 - b^2}{c^2 - a^2}}. \tag{4.7}$$

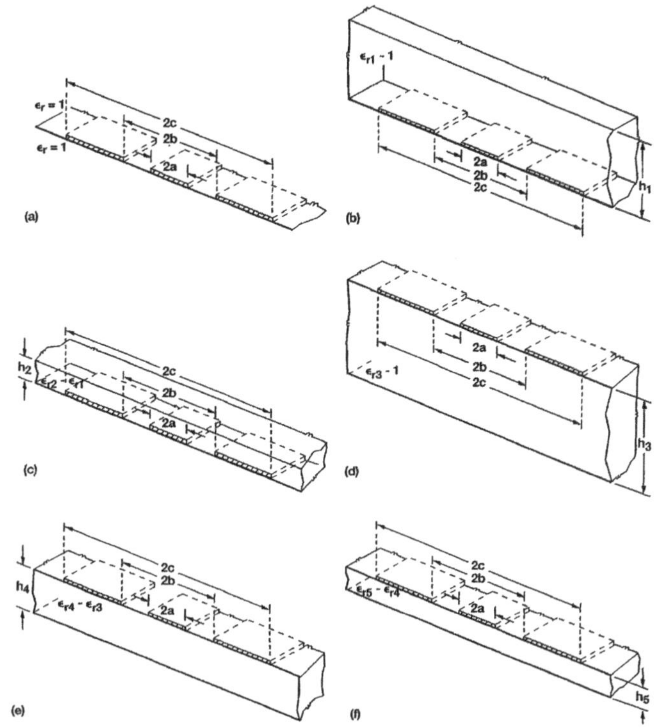

FIGURE 4.2 Configurations for finding the capacitances: (*a*) C_0; (*b*) C_1; (*c*) C_2; (*d*) C_3, (*e*) C_4; (f) C_5.

Calculation of C_1 The configuration of C_1 is shown in Figure 4.2(*b*). The assumption here is that the electric field exists only in the dielectric layer of thickness h_1 and relative dielectric constant $\varepsilon_{r1} - 1$. The capacitance C_1 is given by [3]

$$C_1 = 2\varepsilon_0(\varepsilon_{r1} - 1)\,\frac{k(k_1')}{K(k_1)}, \tag{4.8}$$

where

$$k_1 = \frac{\sinh(\Pi c/2h_1)}{\sinh(\Pi b/2h_1)}\sqrt{\frac{\sinh^2(\Pi b/2h_1) - \sinh^2(\Pi a/2h_1)}{\sinh^2(\Pi c/2h_1) - \sinh^2(\Pi a/2h_1)}}, \tag{4.9}$$

$$k_1' = \sqrt{1 - k_1^2} = \frac{\sinh(\Pi a/2h_1)}{\sinh(\Pi b/2h_1)}\sqrt{\frac{\sinh^2(\Pi c/2h_1) - \sinh^2(\Pi b/2h_1)}{\sinh^2(\Pi c/2h_1) - \sinh^2(\Pi a/2h_1)}}. \tag{4.10}$$

Calculation of C_2, C_3, C_4, and C_5 The configurations of C_2, C_3, C_4, and C_5 are shown in Figure 4.2(*c*) to (*f*). The assumption here is that the electric field exists only in the dielectric layers of thickness h_2, h_3, h_4, and h_5 with relative dielectric constants of $\varepsilon_{r2} - \varepsilon_{r1}$, $\varepsilon_{r3} - 1$, $\varepsilon_{r4} - \varepsilon_{r3}$, and $\varepsilon_{r5} - \varepsilon_{r4}$, respectively. The capacitances are given by [3]

$$C_2 = 2\varepsilon_0(\varepsilon_{r2} - \varepsilon_{r1})\frac{K(k'_2)}{K(k_2)}, \tag{4.11}$$

$$C_3 = 2\varepsilon_0(\varepsilon_{r3} - 1)\frac{K(k'_3)}{K(k_3)}, \tag{4.12}$$

$$C_4 = 2\varepsilon_0(\varepsilon_{r4} - \varepsilon_{r3})\frac{K(k'_4)}{K(k_4)}, \tag{4.13}$$

$$C_5 = 2\varepsilon_0(\varepsilon_{r5} - \varepsilon_{r4})\frac{K(k'_5)}{K(k_5)}, \tag{4.14}$$

where

$$k_i = \frac{\sinh(\Pi c/2h_i)}{\sinh(\Pi b/2h_i)}\sqrt{\frac{\sinh^2(\Pi b/2h_i) - \sinh^2(\Pi a/2h_i)}{\sinh^2(\Pi c/2h_i) - \sinh^2(\Pi a/2h_i)}}, \tag{4.15}$$

$$k'_i = \sqrt{1 - k_i^2} = \frac{\sinh(\Pi a/2h_i)}{\sinh(\Pi b/2h_i)}\sqrt{\frac{\sinh^2(\Pi c/2h_i) - \sinh^2(\Pi b/2h_i)}{\sinh^2(\Pi c/2h_i) - \sinh^2(\Pi a/2h_i)}}, \tag{4.16}$$

$i = 2$, 3, 4, and 5.

Substituting Eqs. (4.4), (4.5), (4.8), and (4.11) to (4.14) into Eq. (4.1) results in

$$\begin{aligned}
\varepsilon_{\text{eff}}^{\text{CPW}} &= \frac{C}{C_0} \\
&= 1 + \frac{1}{2}(\varepsilon_{r1} - 1)\frac{K(k)}{K(k')}\frac{K(k'_1)}{K(k_1)} \\
&\quad + \frac{1}{2}(\varepsilon_{r2} - \varepsilon_{r1})\frac{K(k)}{K(k')}\frac{K(k'_2)}{K(k_2)} \\
&\quad + \frac{1}{2}(\varepsilon_{r3} - 1)\frac{K(k)}{K(k')}\frac{K(k'_3)}{K(k_3)} \\
&\quad + \frac{1}{2}(\varepsilon_{r4} - \varepsilon_{r3})\frac{K(k)}{K(k')}\frac{K(k'_4)}{K(k_4)} \\
&\quad + \frac{1}{2}(\varepsilon_{r5} - \varepsilon_{r4})\frac{K(k)}{K(k')}\frac{K(k'_5)}{K(k_5)}
\end{aligned} \tag{4.17}$$

From Eqs. (4.2) and (4.3) we obtain

$$v_{\rm ph}^{\rm CPW} = \frac{c'}{\sqrt{\varepsilon_{\rm eff}^{\rm CPW}}}, \tag{4.18}$$

$$Z_0^{\rm CPW} = \frac{30\pi}{\sqrt{\varepsilon_{\rm eff}^{\rm CPW}}} \cdot \frac{K(k)}{K(k')}\,\Omega, \tag{4.19}$$

where k and k' are given by Eqs. (4.6) and (4.7).

For a FW-CPW on a substrate with a single dielectric layer as shown in Figure 4.3(a), $\varepsilon_{r1} = \varepsilon_{r2} = \varepsilon_{r3} = \varepsilon_{r4} = 1$ and Eq. (4.17) reduces to

$$\varepsilon_{\rm eff}^{\rm CPW} = 1 + \frac{1}{2}(\varepsilon_{r5} - 1)\frac{K(k)}{K(k')}\frac{K(k'_5)}{K(k_5)}, \tag{4.20}$$

which is identical to the equation given in [3] and [4]. Figure 4.4(a) and (b) presents the computed $\varepsilon_{\rm eff}$ and Z_0, respectively.

For a FW-CPW sandwiched between two dielectric layers as shown in Figure 4.3(b), $\varepsilon_{r1} = \varepsilon_{r3} = \varepsilon_{r4} = 1$, and Eq. (4.17) reduces to

$$\varepsilon_{\rm eff}^{\rm CPW} = 1 + \frac{1}{2}(\varepsilon_{r2} - 1)\frac{K(k)}{K(k')}\frac{K(k'_2)}{K(k_2)} + \frac{1}{2}(\varepsilon_{r5} - 1)\frac{K(k)}{K(k')}\frac{K(k'_5)}{K(k_5)}. \tag{4.21}$$

Further, if the two dielectric layers have identical relative permittivity and thickness, then $\varepsilon_{r2} = \varepsilon_{r5} = \varepsilon_r$ and $h_2 = h_5 = h$. In that case Eq. (4.21) further simplifies to

$$\varepsilon_{\rm eff}^{\rm CPW} = 1 + (\varepsilon_r - 1)\frac{K(k)}{K(k')}\frac{K(k'_s)}{K(k_s)}, \tag{4.22}$$

where k and k' are given by Eqs. (4.6) and (4.7), respectively. The arguments k_s and k'_s are given by Eqs. (4.9) and (4.10), respectively, with h_1 replaced by h. Equation (4.22) is identical to that given in [5].

Finally, for a FW-CPW on an infinitely thick dielectric substrate, $h_5 \to \infty$ and $k_5 = k$. The effective dielectric constant from Eq. (4.20) is given by

$$\varepsilon_{\rm eff}^{\rm CPW} = \tfrac{1}{2}(\varepsilon_r + 1),$$

where ε_r is the relative permittivity of the substrate.

4.2.2 Dispersion and Characteristic Impedance from Full-Wave Analysis

The FW-CPW shown in Figure 4.3(a) has been analyzed using the spectral domain technique in [7]. The computed guide wavelength ratio, λ_0/λ_g, where

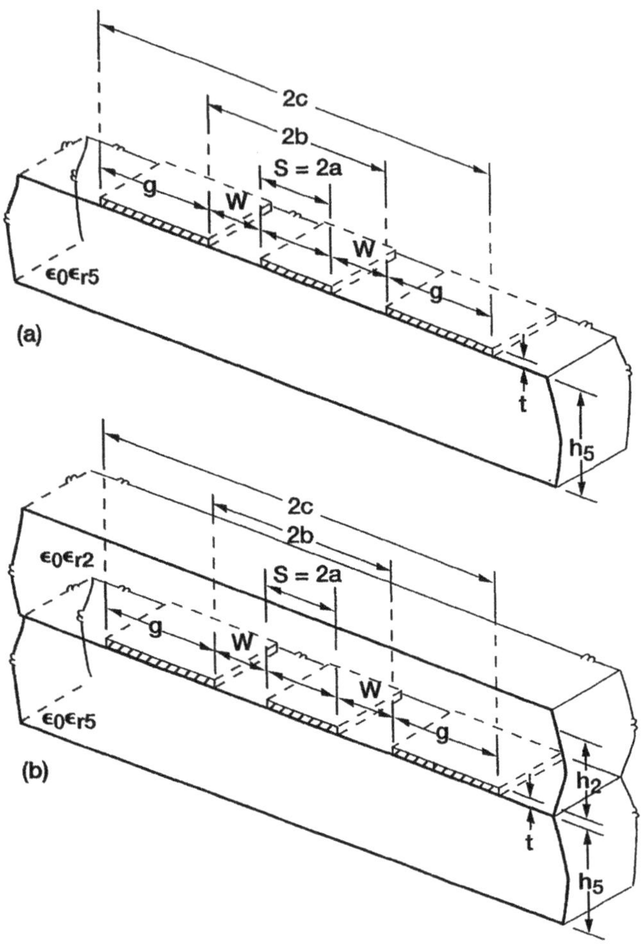

FIGURE 4.3 Schematic of (*a*) finite width coplanar waveguide (FW-CPW) and (*b*) sandwiched FW-CPW.

λ_g and λ_0 are the guide wavelength and free space wavelength, respectively, is shown in Figure 4.5(*a*). For small ground plane width, the energy is more closely bounded within the region of the substrate and consequently λ_0/λ_g is higher than the conventional CPW. As the ground plane width increases, the λ_0/λ_g converges to that of the conventional CPW. The characteristic impedance Z_0 defined as $Z_{0(\mathrm{air})}/\sqrt{\varepsilon_{\mathrm{eff}}}$, where $Z_{0(\mathrm{air})}$ is the characteristic impedance of the air filled structure, is shown in Figure 4.5(*b*). Since the FW-CPW has a higher $\varepsilon_{\mathrm{eff}}$, the Z_0 is lower than the conventional CPW. As the ground plane width increases, the Z_0 converges to that of the conventional CPW. The relative change in both λ_0/λ_g and Z_0 between extremes does not exceed a few percent.

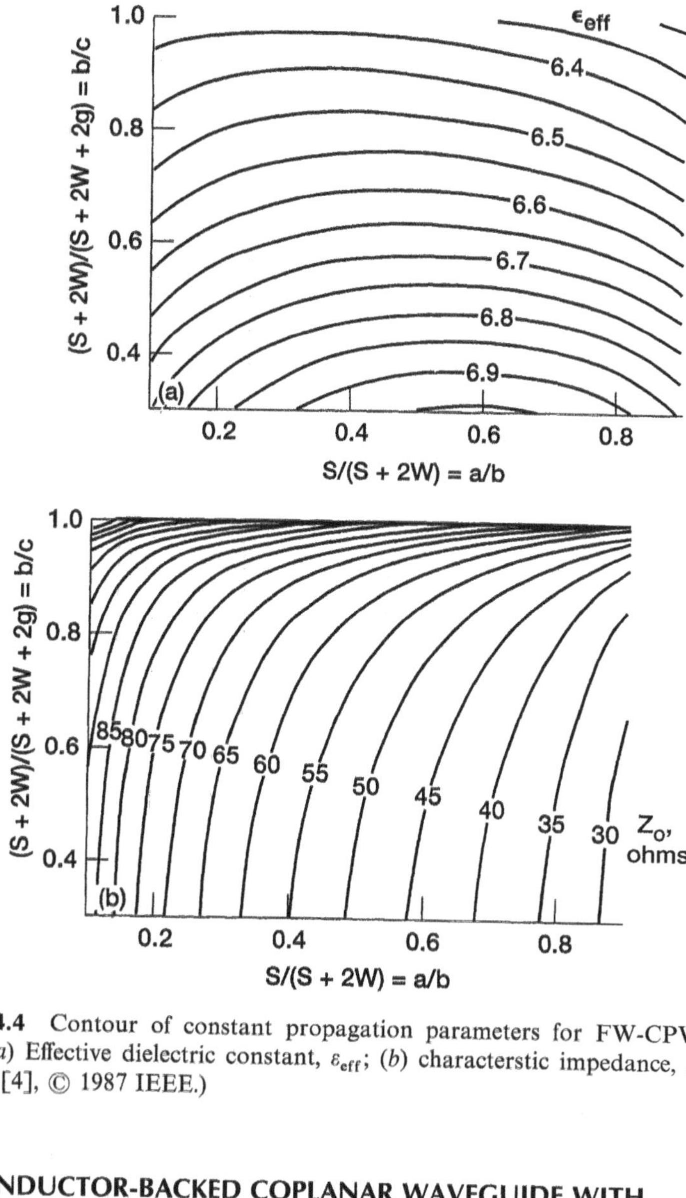

FIGURE 4.4 Contour of constant propagation parameters for FW-CPW, $\varepsilon_r = 13$, $h/b = 1$: (*a*) Effective dielectric constant, ε_{eff}; (*b*) characterstic impedance, Z_0. (From Reference [4], © 1987 IEEE.)

4.3 CONDUCTOR-BACKED COPLANAR WAVEGUIDE WITH FINITE-WIDTH GROUND PLANES ON A DIELECTRIC SUBSTRATE OF FINITE THICKNESS AND FINITE WIDTH

A conductor-backed coplanar waveguide with finite-width ground planes on a dielectric substrate of finite thickness and finite width (FW-CBCPW) is shown in Figure 4.6 [8]. The dominant CPW mode has its electric field components on this structure as sketched in Figure 4.6(*a*). In addition this structure can support a microstrip-like (MSL) propagating mode. This mode resembles a

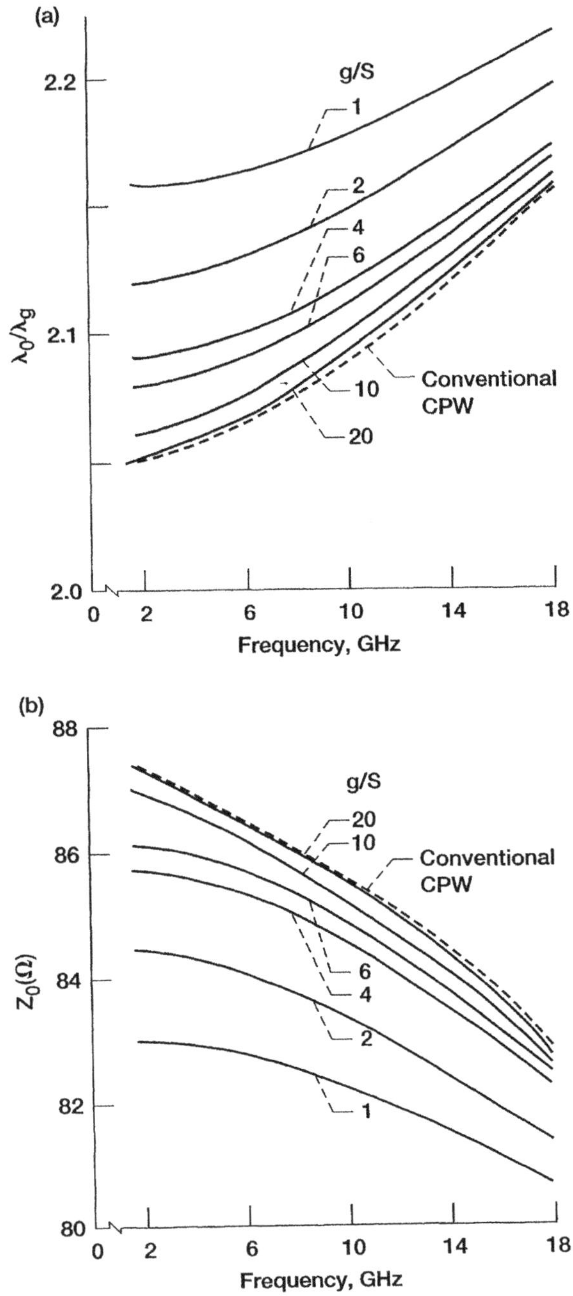

FIGURE 4.5 Computed characteristics of FWCPW with normalized ground plane width as parameter, $\varepsilon_r = 9.6$, $h = 0.07$ cm, $S = 0.05$ cm, $W = 0.1$ cm: (*a*) Guide wavelength ratio, λ_0/λ_g; (*b*) characteristic impedance, Z_0. (From Reference [7], with permission from AEÜ.)

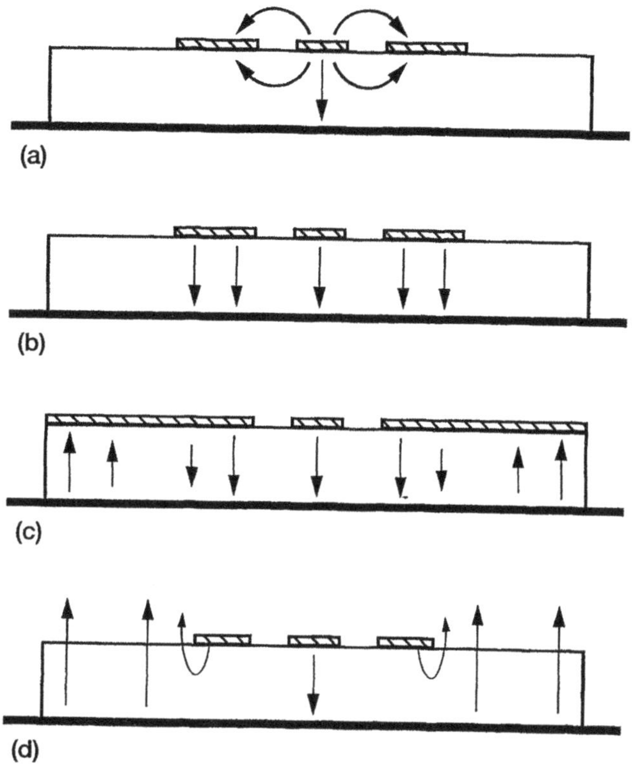

FIGURE 4.6 Schematic illustration of the modes on a FW-CBCPW: (*a*) CPW mode; (b) MSL mode; (c) first higher-order MSL mode; (*d*) image–guide-like mode.

parallel-plate transmission line mode and is shown in Figure 4.6(*b*). If the top ground planes are wide, then the next higher-order microstrip-like mode may exist as shown in Figure 4.6(*c*). If the dielectric substrate is even wider then the structure can support an image–guide-like propagating mode. However, for the initial study the structural parameters and the material constants are so chosen that the structure supports only the CPW mode and the MSL mode below 26.5 GHz [8].

For the purpose of analysis, modeling and computer simulation an equivalent CPW structure shown in Figure 4.7 is considered. The two perfect electric conductors (PEC) are located sufficiently far away from the guiding structure so as not to perturb the field distributions. This model is capable of providing the dispersion characteristics for a coplanar waveguide having top and bottom ground planes of finite width. Furthermore in this model, by setting both t_2 and h_2 equal to zero, the dispersion characteristics can be obtained for a coplanar waveguide with finite-width top ground planes and an infinitely wide lower ground plane.

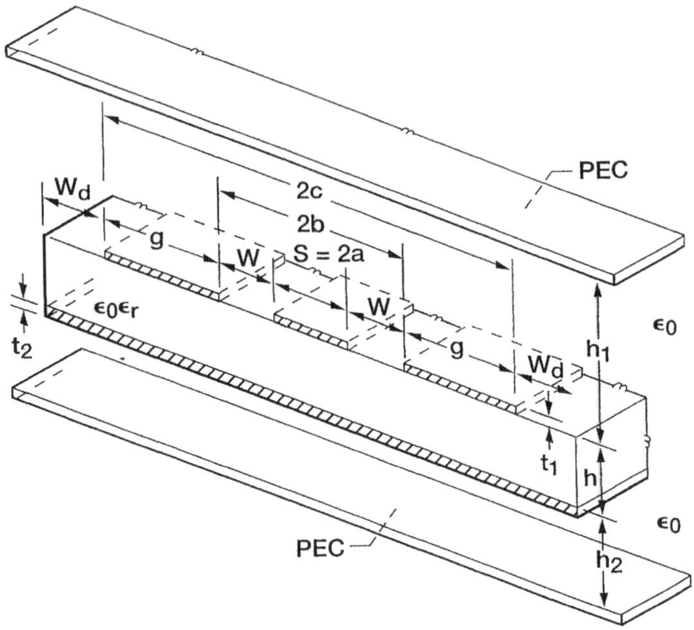

FIGURE 4.7 Schematic of the finite width conductor-backed coplanar waveguide (FW-CBCPW) with a top cover.

The structure in Figure 4.7 is analyzed using the full-wave mode-matching technique [8] and [9]. The computed normalized propagation constant β_g/β_0, where β_g and β_0 are the propagation constant in the guide and free space, respectively, for the FW-CPW with finite and infinitely wide lower ground plane [9] is presented in Figure 4.8. In the same figure the computed dispersion characteristics for the microstrip-like mode for the two cases of finite and infinitely wide lower ground plane, respectively, are also presented. The two sets of characteristics are computed over a very wide frequency range extending from about 1 GHz up to about 26 GHz. It is observed that at low frequencies, typically below 10 GHz, the phase constant of the MSL mode on a line with finite lower ground plane is significantly higher than with infinite lower ground plane. A possible reason for this to happen is the slower phase velocity on a line with finite lower ground plane because of smaller amount of fringing fields. On the other hand, the CPW mode has most of the energy concentrated around the slot regions. Hence the phase constants with finite and with infinite lower ground plane, respectively, differ by a constant amount over the above frequency range.

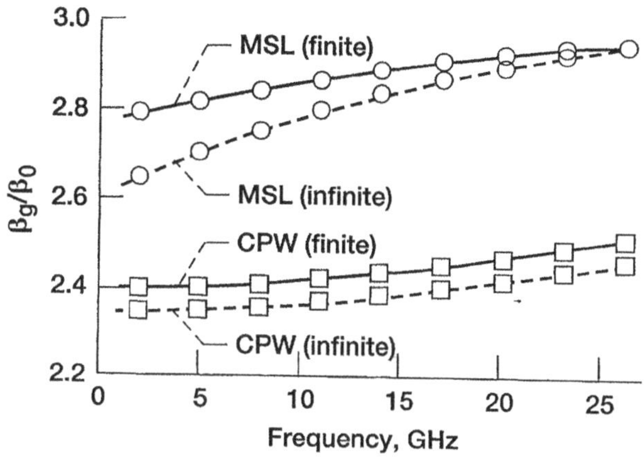

FIGURE 4.8 Computed normalized propagation constant β_g/β_0 as a function of frequency for MSL and CPW modes on FW-CBCPW with finite and infinite lower ground planes: $S = W = 0.508$ mm, $g = 1.0$ mm, $W_d = 0.0$ mm, $h = 0.2$ $h_1 = 0.635$ mm, $t_1 = 20\,\mu$m, $\sigma = 5.8 \times 10^7$ S/m (copper), and $\varepsilon_r = 10.2$. For infinite lower ground plane $h_2 = 0$, $t_2 = 0$, and for finite lower ground $h_2 = h_1$, $t_2 = t_1$. (From Reference [9], © 1993 IEEE.)

4.4 SIMPLE MODELS TO ESTIMATE FINITE GROUND PLANE RESONANCE IN CONDUCTOR-BACKED COPLANAR WAVEGUIDE

Consider the FW-CBCPW shown in Figure 4.9 and for the moment ignore the two notches cut in the top ground planes. The top ground planes are then viewed as two-dimensional resonators [10] and modeled as follows: first, as a patch antenna, second as an overmoded microstrip-like (MSL) transmission line, and finally using three-dimensional full-wave space domain integral equation (SDIE) technique.

When considering the ground planes as patch antennas, the effect of fringing fields are neglected and the resonant frequencies are computed as explained in [11] and [12]. When considering the ground planes as overmoded MSL, the dominant as well as the first higher-order modes are taken into consideration. Resonance occurs when the following condition is satisfied:

$$\beta_{\mathrm{MSL}} l = n\pi, \tag{4.23}$$

where β_{MSL} is the propagation constant of the dominant or higher-order MSL mode and l is the length of the line. In the case of SDIE techniques, the two-dimensional current distributions on the center strip conductor as well as the side planes are obtained. Knowing there current distributions, the resonance frequencies are determined.

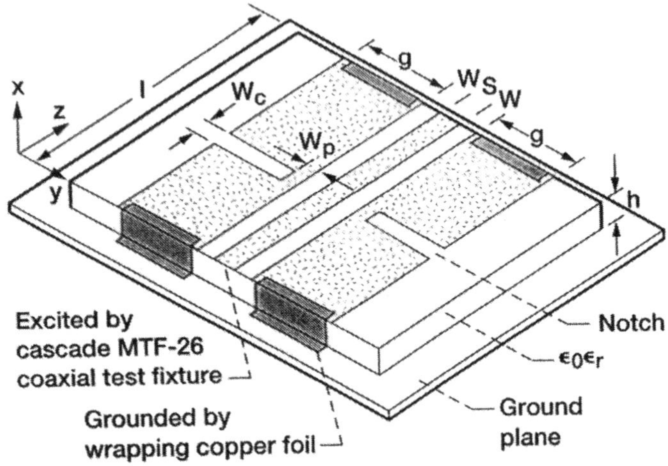

FIGURE 4.9 Schematic of the 50 Ω FW-CBCPW through line with two notches cut in the top ground plane: $S = W = 0.508$ mm, $g = 6.0$ mm, $l = 18.0$ mm, $W_p = 1.0$ mm, $W_c = 1.016$ mm, $h = 0.635$ mm, $\varepsilon_r = 10.2$

4.4.1 Experimental Validation

To validate the computed resonant frequencies, the top ground planes of the 50 Ω FW-CBCPW through line without notches are grounded by wrapping copper foils as shown in Figure 4.9. The FW-CBCPW is excited through a small opening between the foils by two Cascade MTF-26 coaxial test fixtures. The dimensions of the through line are presented in Figure 4.9. The reflection coefficient S_{11} and the transmission coefficient S_{21} are measured [10] on a calibrated network analyzer and the resonant frequencies are noted. Table 4.1 presents the measured as well as the computed resonant frequencies using three different models. The indexes m_y and n_z appearing in the first column of the table are the order of resonance. When $m_y = 0$ and $n_z = 1, 2, \ldots$, the resonance appears in the longitudinal z-direction with negligible field variation in the transverse y-direction. When $m_y = 1$ and $n_z = 1$, the resonance has one half-wave variation in both y- and z-directions. In comparing the resonant frequencies predicted by the patch antenna model and the MSL model with the measured resonant frequencies, the MSL model is observed to have much better agreement.

In a second experiment, the resonant frequencies are measured [10] for a FW-CBCPW having notches cut in the ground plane as shown in Figure 4.9. The dimensions W_p and W_c are also given in Figure 4.9. The measurements show that the slit causes a shift in the resonant frequencies. This experiment also proves that by providing notches in the ground plane, the MSL modes cannot be suppressed or eliminated.

TABLE 4.1 Measured and Modeled Resonant Frequencies for the FW-CBCPW Shown in Figure 4.9

Resonance Number and Indexes, $m_y n_z$	Measurement, GHz	Patch Antenna Model, GHz	Microstrip-like Model, $m_y = 0, 1$ $\beta_{MSL} l = n_z \pi$, GHz	Three-Dimensional Space Domain Integral Equation Model, GHz
1 (0.1)	2.71	2.61	2.74	2.76
2 (0.2)	5.38	5.22	5.40	5.44
3 (1.1)	7.65	8.23	7.71	7.75
4 (0.3)	8.01	7.83	8.04	8.08
5 (1.2)	9.07	9.39	9.21	9.22
6 (0.4)	10.61	10.44	10.67	10.71
7 (1.3)	10.94	11.06	11.27	11.08

Source: From Reference [10], © IEEE.

REFERENCES

[1] Y. Noguchi and N. Okamoto, "Analysis of Characteristics of the Coplanar Waveguide with Ground Planes of Finite Extent," *Trans. IECE., Japan*, Vol. 58-B, No. 12, pp. 679–680. Dec. 1975.

[2] R. E. DeBrecht, "Coplanar Balun Circuits for GaAs FET High-Power Push-Pull Amplifiers," in *1973 IEEE G-MTT Int. Microwave Symp. Dig.*, Boulder, CO, pp. 309–311, June 4–6, 1973.

[3] C. Veyres and V.F. Hanna, "Extension of the Application of Conformal Mapping Techniques to Coplanar Lines with Finite Dimensions," *Int. J. Electron.*, Vol. 48, No. 1, pp. 47–56, Jan. 1980.

[4] G. Ghione and C. U. Naldi, "Coplanar Waveguides for MMIC Applications: Effect of Upper Shielding, Conductor Backing, Finite-Extent Ground Planes, and Line-to-Line Coupling," *IEEE Trans. Microwave Theory Tech.*, Vol. 35, No. 3, pp. 260–267, March 1987.

[5] M. Cai, P. S. Kooi, M. S. Leong, and T. S. Yeo, "Symmetrical Coplanar Waveguide with Finite Ground Plane," *Microwave Optical Tech. Lett.*, Vol. 6, No. 3, pp. 218–220, March 1993.

[6] E. Chen and S. Y. Chou, "Characteristics of Coplanar Transmission Lines on Multilayer Substrates: Modeling and Experiments," *IEEE Trans. Microwave Theory Tech.*, Vol. 45, No. 6, pp. 939–945, June 1997.

[7] B. J. Janiczak, "Analysis of Coplanar Waveguide with Finite Ground Planes," AEÜ, Vol. 38, No. 5, pp. 341–342, May 1984.

[8] C.-C. Tien, C.-K. C. Tzuang, S. T. Peng, and C.-C. Chang, "Transmission Characteristics of Finite-Width Conductor-Backed Coplanar Waveguide," *IEEE Trans. Microwave Theory Tech.*, Vol. 41, No. 9, pp. 1616–1623, Sept. 1993.

[9] C.-C. Tien, C.-K.C. Tzuang and S.T. Peng, "Effect of Finite-Width Backside Plane on Overmoded Conductor-Backed Coplanar Waveguide," *IEEE Mircowave Guided Wave Lett.*, Vol. 3, No. 8, pp. 259–261, Aug. 1993.

[10] W.-T. Lo, C.-K. C. Tzuang, S. T. Peng, C.-C. Tien, C.-C. Chang, and J.-W. Huang, "Resonant Phenomena in Conductor-Backed Coplanar Waveguides (CBCPW's)," *IEEE Trans. Microwave Theory Tech.*, Vol. 41, No. 12, pp. 2099–2107, Dec. 1993.

[11] K. C. Gupta, R. Garg, and R. Chadha, *Computer-Aided Design of Microwave Circuits,* Dedham, MA: Artech House, 1981, Chap. 8.

[12] Y. T. Lo, D. Soloman, and W. F. Richards, "Theory and Experiment on Microstrip Antennas," *IEEE Trans. Antennas Propag.*, Vol. 27, No. 2, pp. 137–145, March 1979.

CHAPTER 5

Coplanar Waveguide Suspended inside a Conducting Enclosure

5.1 INTRODUCTION

In this chapter the characteristics of a coplanar waveguide (CPW) suspended inside a conducting metal enclosure are presented. By taking the enclosure into consideration, the inevitable package that is present in practice and its influence is studied and modeled.

Section 5.2 commences with a discussion on quasi-static TEM iterative techniques to analyze suspended CPW and presents computed as well as experimental results on effective dielectric constant ε_{eff} and characteristic impedance Z_0. These characteristics are compared with those obtained by full-wave analysis in Section 5.3. In addition to these the effects are presented of shielding and conductor thickness on the guide wavelength ratio λ_g/λ_0, where λ_g and λ_0 are the guide wavelength and the free space wavelength, respectively, and on Z_0. The section concludes with an estimate for the dispersion free bandwidth and the minimum width of a picosecond pulse that can propagate on the line without distortion. Experimental results are also presented to validate the analysis.

Section 5.4 considers a CPW suspended inside a nonsymmetrical split block type of housing. This type of construction has the advantage of simplifying assembly by soldering or epoxying the circuit to one-half of the housing. Critical mechanical tolerances and contacts at mounting grooves can be avoided by this method of assembly. The influences of a nonsymmetrical shielding enclosure on the dominant and higher-order modes, on the cutoff frequency of the first higher-order mode, and on Z_0 are presented. The substrate mounting recess dimensions are very critical at higher millimeter wave frequencies. Hence the impact of the groove dimensions on the cutoff frequency of the first higher order mode is presented.

Last, in Section 5.5, the λ_g/λ_0 and Z_0 of a CPW on a double-layer substrate and also of a CPW sandwiched between two dielectric substrates are presented. These additional layers are necessary for providing protection from mechanical or chemical damage to circuits exposed to the environment and also as an additional means to adjust the propagation characteristics.

The CPW in this chapter is assumed to be symmetric and on an isotropic homogeneous dielectric substrate of arbitrary thickness and relative permittivity. Further the substrate metalization and the metal enclosure are assumed to have perfect conductivity

5.2 QUASI-STATIC TEM ITERATIVE TECHNIQUE TO DETERMINE ε_{eff} AND Z_0 OF SUSPENDED CPW

A CPW suspended inside a conducting enclosure is shown in Figure 5.1(*a*). This structure has been analyzed by the use of relaxation method in [1] to [3]. In the relaxation method, the characteristics of the structure is determined by numerically solving the two-dimensional Laplace equation. In [1] and [2], the cross section of the structure is assumed to be symmetric about the *y*-axis. This simplifies the computational task, since it is sufficient to solve either the left half or the right half of the structure. Further, simplification is possible if the thickness t of the substrate metallization is assumed to be far less than the substrate thickness h, in which case it is set equal to zero.

5.2.1 Computed Quasi-static Characteristics and Experimental Validation

The effect of the metal side walls and the top and bottom metal covers on the propagation characteristics is first considered. Figure 5.2 shows the variation of the computed reciprocal of the effective dielectric constant ε_{eff} and the characteristic impendance Z_0 as a function of the normalized side wall separation [2]. While carrying out these computations, the heights of the metal top and bottom covers, the substrate thickness and dielectric constant, the strip width and the slot width, are all held fixed. As seen from this figure the characteristics saturates when the ratio $g/(W + S/2)$ is greater than 1.0.

Figure 5.3 shows the variation of the computed reciprocal of the ε_{eff} and the Z_0 as a function of the normalized height of the top and bottom metal covers [2]. While carrying out these computations, the side wall separation, the substrate thickness and dielectric constant, the strip width, and the slot width, are all held fixed. As seen from this figure the effect of H is small on the characteristics for h/H less than 0.142.

In Figure 5.4 the computed reciprocal ε_{eff} and the Z_0 are presented as function of the normalized slot width with the normalized center conductor strip width as a parameter [2]. The dimensions of the shielding enclosure, the

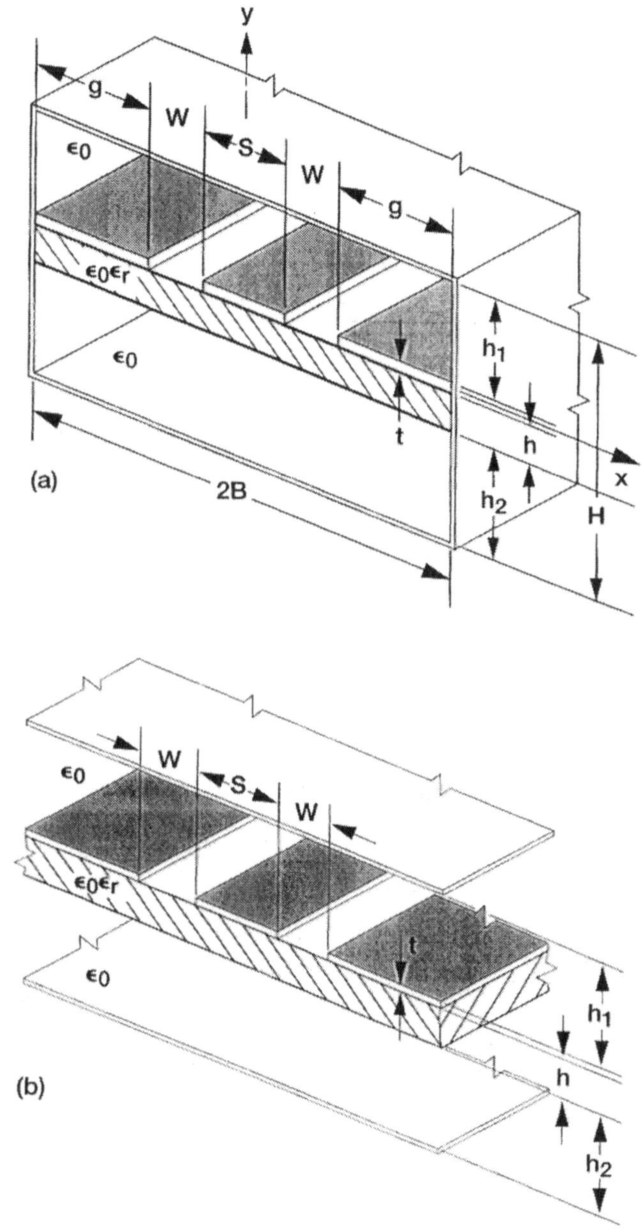

FIGURE 5.1 Shielded coplanar waveguide configurations: (*a*) Suspended; (*b*) covered; (*c*) grounded; (*d*) multilayer.

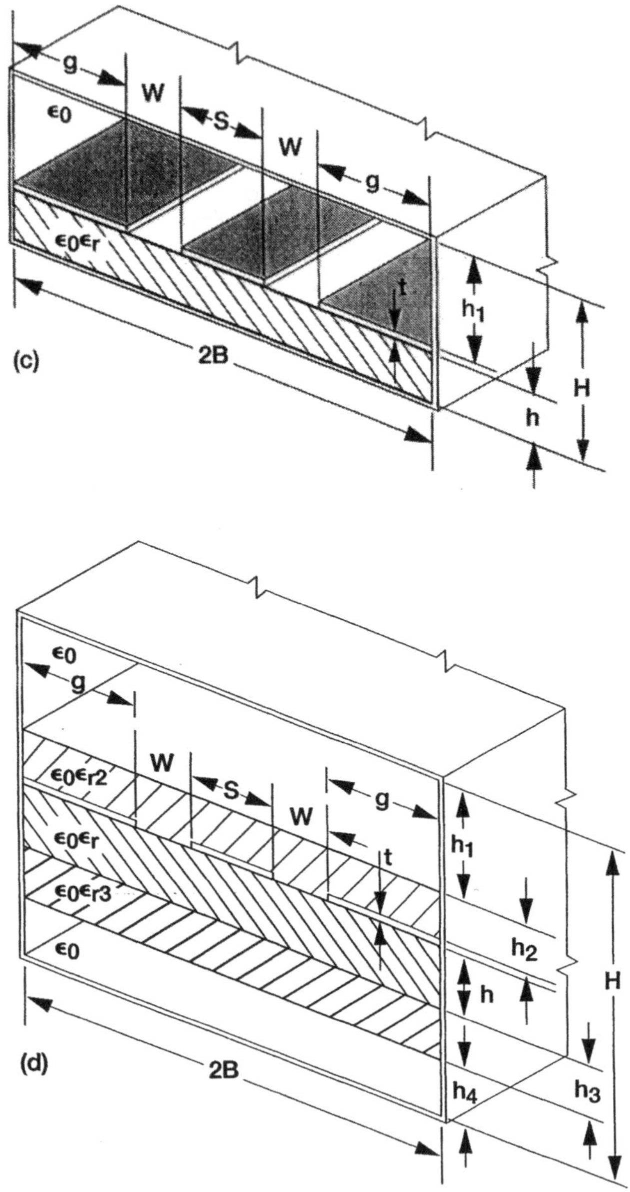

FIGURE 5.1 Shielded coplanar waveguide configurations: (*a*) Suspended; (*b*) covered; (*c*) grounded; (*d*) multilayer.

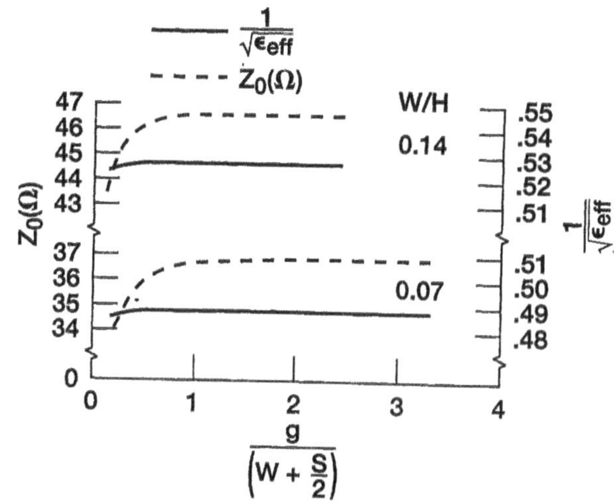

FIGURE 5.2 Computed reciprocal of the effective dielectric constant and characteristic impedance as a function of the side wall separation with the normalized slot width as a parameter: $H = 4.3\,\text{mm}$, $h = 0.61\,\text{mm}$, $\varepsilon_r = 9.4$, $h_1 = h_2$, $t = 0$, $S/H = 0.43$. (From Reference [2], © 1975 IEEE.)

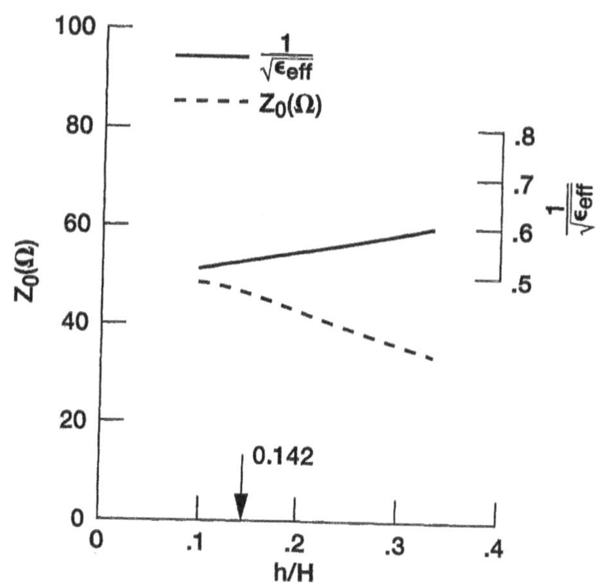

FIGURE 5.3 Computed reciprocal of the effective dielectric constant and characteristic impedance as a function of the top and bottom metal cover separation: $2B = 10.7\,\text{mm}$, $h = 0.61\,\text{mm}$, $\varepsilon_r = 9.4$, $h_1 = h_2$, $t = 0$, $S/2B = 0.18$, $W/2B = 0.06$. (From Reference [2], © 1975 IEEE.)

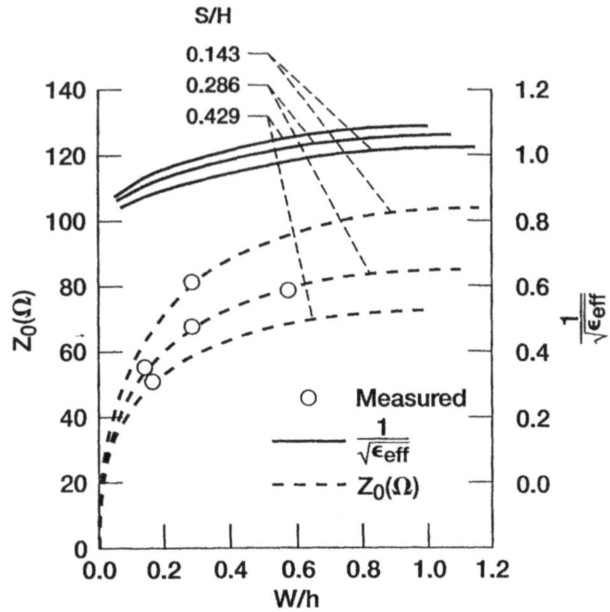

FIGURE 5.4 Computed reciprocal of the effective dielectric constant and characteristic impedance as a function of the normalized slot width with the normalized strip width as a parameter: Circular dots are experimental data; $2B = 10.7\,\text{mm}$, $h = 0.61\,\text{mm}$, $\varepsilon_r = 9.4$, $H = 4.3\,\text{mm}$, $h_1 = h_2$, $t = 0$. (From Reference [2], © 1975 IEEE.)

substrate thickness and dielectric constant are all held fixed while carrying out the computations. In Figure 5.5 the computed reciprocal ε_{eff} and Z_0 are presented as a function of the normalized center conductor strip width with the normalized slot width as a parameter [2].

To experimentally validate the computed results, a CPW is fabricated on an alumina substrate($\varepsilon_r = 9.4$) with gold-plated Cr-Au thin film. The characteristic impedance of the line is measured by a time domain reflectometer (TDR). The measured data are indicated by circular dots in Figures 5.4 and 5.5. The maximum difference between the measured and the calculated values is about 3 percent [2]. This is due to the uncertainty in the value of the substrate relative dielectric constant and also to experimental and measurement errors.

5.3 FREQUENCY-DEPENDENT NUMERICAL TECHNIQUES FOR DISPERSION AND CHARACTERISTIC IMPEDANCE OF SUSPENDED CPW

At frequencies below about 3 GHz, the quasi-static analysis provides sufficiently accurate results, however, as the frequency increases the deviation from quasi-static behavior is significant. Hence in this section the frequency

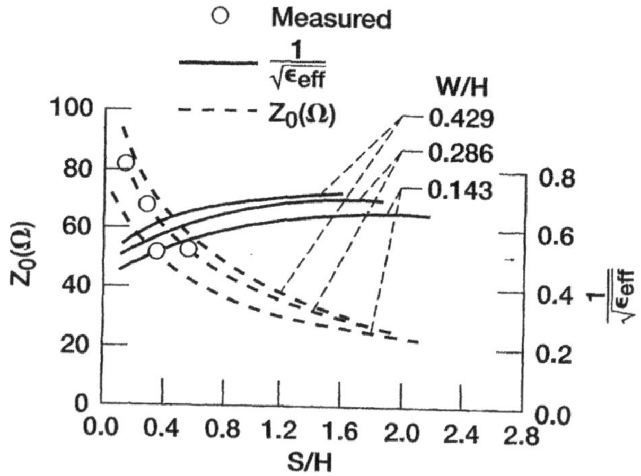

FIGURE 5.5 Computed reciprocal of the effective dielectric constant and characteristic impedance as a function of the normalized strip width with the normalized slot width as a parameter: Circular dots are experimental data; $2B = 10.7\,\text{mm}$, $h = 0.61\,\text{mm}$, $\varepsilon_r = 9.4$, $H = 4.3\,\text{mm}$, $h_1 = h_2$, $t = 0$. (From Reference [2], © 1975 IEEE.)

dependent characteristics will be discussed. The frequency dependent analysis of the structure shown in Figure 5.1(*a*) can be simplified by considering the even mode and the odd mode of propagation as explained in Chapter 2, Section 2.4.

In the literature several researchers have analyzed this structure using the hybrid-mode analysis [4], [5], the method of lines [6] to [8], integral equation technique [9], [10], transverse resonance method [11], [12], spectral domain method [13] to [18], modal analysis [19], and finite difference time domain method [20]. As an example, Figure 5.6 compares, the frequency dependent ε_{eff} defined as $(\lambda_0/\lambda_g)^2$ and Z_0 with the quasi-static ε_{eff} and Z_0 [4]. The deviation from quasi-static behavior is significant and clearly noticeable. The computations were carried out using the hybrid-mode technique.

5.3.1 Effect of Shielding on the Dispersion and Characteristic Impedance

Figure 5.7 shows the computed guide wavelength ratio λ_g/λ_0 as a function of the reciprocal of the free space wavelength, with the height of the shield as a parameter [13]. The computations were carried out using the spectral domain technique. It is observed that as the height of the enclosure increase, the dispersion characteristics approach that of an unshielded CPW [21]. Further, as the frequency increases, the dispersion curves converge upon each other, which is consistent with the decreasing effect of the shielding enclosure.

An useful variant of the above structure is the covered CPW. The covered CPW has the lateral wall separation $2B$ far greater than the height H of the

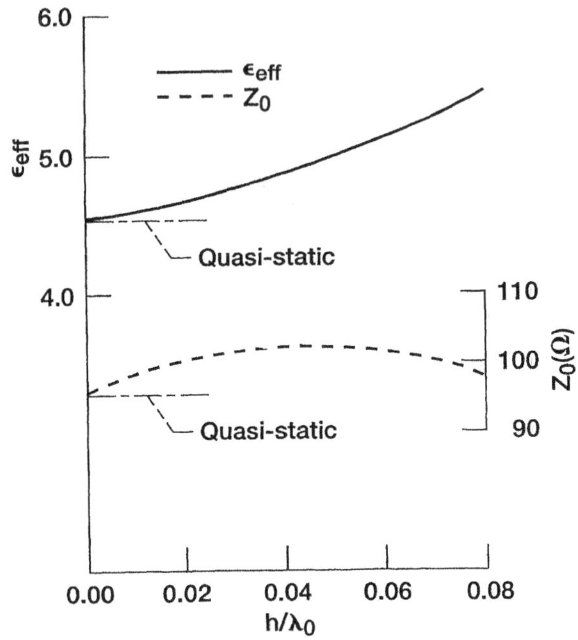

FIGURE 5.6 Comparison between the computed frequency dependent ε_{eff} and Z_0 with the quasi-static ε_{eff} and Z_0: $S/h = 0.4$, $(S + 2W)/h = 2.4$, $(S + 2W + 2g)/h = 15$, $(h_1 + t)/h = 14.5$, $h_2/h = 14.5$, $t = 0$, $\varepsilon_r = 9.6$. (From Reference [4], © IEE.)

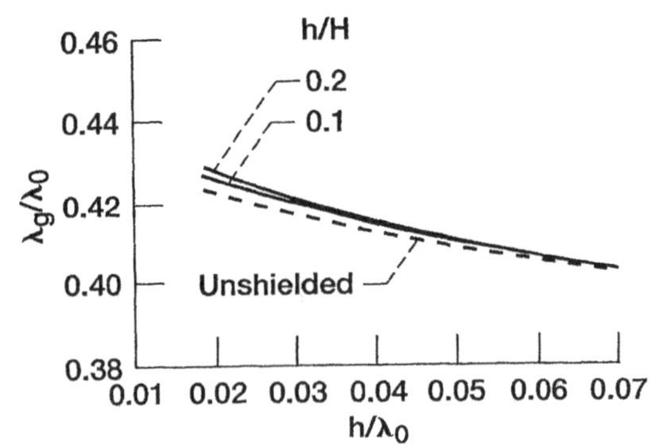

FIGURE 5.7 Computed guide wavelength ratio as a function of the reciprocal of the free space wavelength with the normalized height of the shield: $2B/H = 2$, $S/h = 1$, $W/h = 0.25$, $\varepsilon_r = 11$, $h_1 = h_2$, $t = 0$. (From Reference [13], © 1977 IEEE.)

shield. The covered CPW for all practical purpose is considered as a laterally open structure and is shown in Figure 5.1(*b*). The computed λ_g/λ_0 and Z_0 as a function of the reciprocal of the free space wavelength with the normalized slot width as a parameter [14] are shown in Figure 5.8(*a*) and (*b*). These characteristics are computed using the spectral domain technique.

5.3.2 Experimental Validation of Dispersion

In order to validate the computed λ_g/λ_0 for the covered CPW, the guide wavelength is measured for a CPW fabricated [14] on a alumina substrate of thickness 0.65 mm and ε_r equal to 9.7. The method relies on loosely coupled short-circuited resonating slots [14]. This method of measuring is not well suited for very narrow or very wide slots because, in the former case, the coupling probes have to be positioned extremely close to the slots, whereas in the later case the *Q*-factors of the resonances are rather low. The measured results [14] are shown by circular dots in Figure 5.8(*a*).

To validate the computed λ_g/λ_0 for the suspended CPW, the *S*-parameters are measured [7] for two shielded CPWs having identical strip and slot widths but different line lengths. The measurements are carried out by attaching two coaxial launchers with very similar characteristics to the lines and measuring the *S*-parameters on a personal computer controlled automatic network analyzer. After suitable algebraic manipulations, the *S*-parameters of the CPW with length equal to the difference between the two line lengths are obtained [24]. This procedure is conveniently handled by a software that resides on the personal computer. From the *S*-parameters the propagation constant γ of the CPW line is obtained. Measurements made on two CPWs with the following parameters: $\varepsilon_r = 10$, $\tan \delta_e = 0.002$, $h = 0.635$ mm, $h_1 = h_2 = 20$ mm, $2B = 12$ mm, $W = 0.5$ mm, and $S = 2$ mm and 4 mm, over the frequency range of 1 to 15 GHz are observed to be in good agreement with the computed values [7] using the method of lines.

5.3.3 Effect of Conductor Thickness on the Dispersion and Characteristic Impedance

The computed frequency dependent ε_{eff} and Z_0 as a function of the reciprocal of the free space wavelength with the conductor thickness as a parameter [4] for the shielded suspended CPW is shown in Figure 5.9. The computations are carried out using the hybrid-mode technique. The Z_0 presented here is based on the power-voltage definition. Both ε_{eff} and Z_0 decrease as the thickness of the conductor increases. The decrease in ε_{eff} is because of field lines concentrating in the air-filled slot region formed by the thick metal coating. The decrease in Z_0 is because of the corresponding fractional increase in the line capacitance. Also in Figure 5.9 is the quasi-static value to which the frequency dependent characteristic converges in the low-frequency limit.

The effect of the metal coating thickness on ε_{eff} and Z_0 of a shielded suspended CPW with different substrate dielectric constant ε_r [4] is shown in

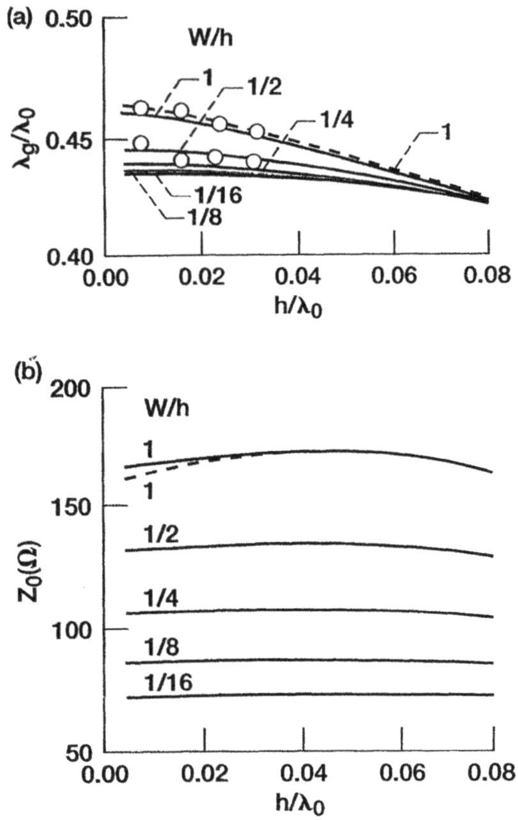

FIGURE 5.8 Computed guide wavelength ratio and characteristics impedance of covered CPW as a function of the reciprocal of the free space wavelength with the normalized slot width as a parameter: $h_1 = 20h$, $S/h = 0.5$, $h_2 = 20h$, $\varepsilon_r = 9.7$, $t = 0$. Circular dots are experimental data. Dash line is for $h_2 = h$. (*a*) Guide wavelength ratio; (*b*) characteristic impedance. (From Reference [14], copyright © IEE.)

Figure 5.10. At any fixed metal coating thickness the ε_{eff} increases as ε_r of the substrate increases; however, the Z_0 decreases as ε_r of the substrate increases. This is because the fractional line capacitance does not change essentially with the dielectric constant of the substrate.

5.3.4 Modal Bandwidth of a Suspended CPW

To understand modal bandwidth [16], first of all consider the shielded grounded CPW shown in Figure 5.1(*c*). The general behavior of the propagation constant β with frequency along this transmission line typically shows an increasing trend as illustrated by curve *A* in Figure 5.11 [16]. The free space propagation constant is denoted as β_0. The grounded CPW structure can also

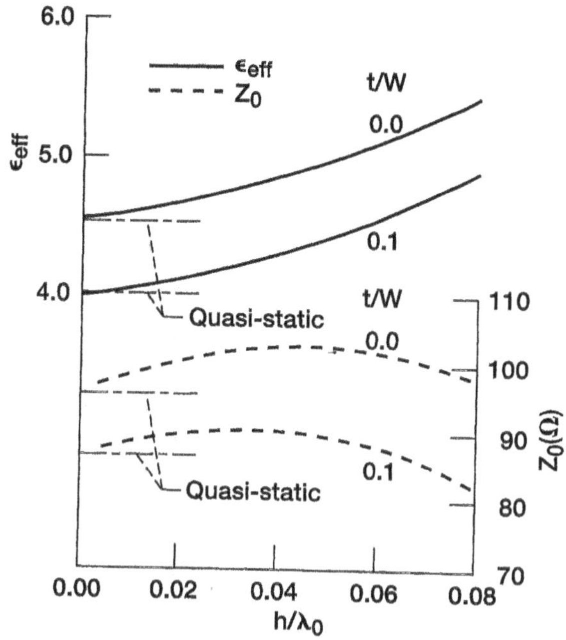

FIGURE 5.9 Computed ε_{eff} and Z_0 as a function of the reciprocal of the free space wavelength with the conductor thickness as a parameter: $\varepsilon_r = 9.6$, $S/h = 0.4$, $(S + 2W)/h = 2.4$, $(S + 2W + 2g)/h = 15$, $(h_1 + t)/h = 14.5$, $h_2/h = 14.5$. (From Reference [4], copyright © IEE.)

be considered as a shielded grounded slotline with an added center strip conductor. On this slotline the propagation constant for the dominant mode increases with frequency as shown by curve B in Figure 5.11. Further this dominant mode has a cutoff frequency of f_c. The added center strip conductor taken together with the two side strip conductors forms a two-conductor strip transmission line, such as a grounded CPW, and gives rise to a quasi-TEM propagation mode as the dominant mode. At low frequencies when quasi-TEM approximation is assumed, the normalized propagation constant is represented by $\beta_{TEM}/\beta_0 = \sqrt{\varepsilon_{eff}}$ as shown by curve C in Figure 5.11. At extremely high frequency, the electromagnetic fields are concentrated in the dielectric substrate and the normalized propagation constant eventually converges to $\beta/\beta_0 = \sqrt{\varepsilon_r}$. Between the low- and high-frequency extremes, the curve A appears as an asymptote to both the curves B and C.

In Figure 5.12 curve A_1 is the computed normalized propagation constant of a shielded grounded CPW [16]. Curves B_1 and C_1 are the computed normalized propagation constant of the shielded grounded slotline and the quasi-TEM propagation constant of the shielded grounded CPW, respectively [16]. The dimensions of shielded grounded CPW are also presented in Figure

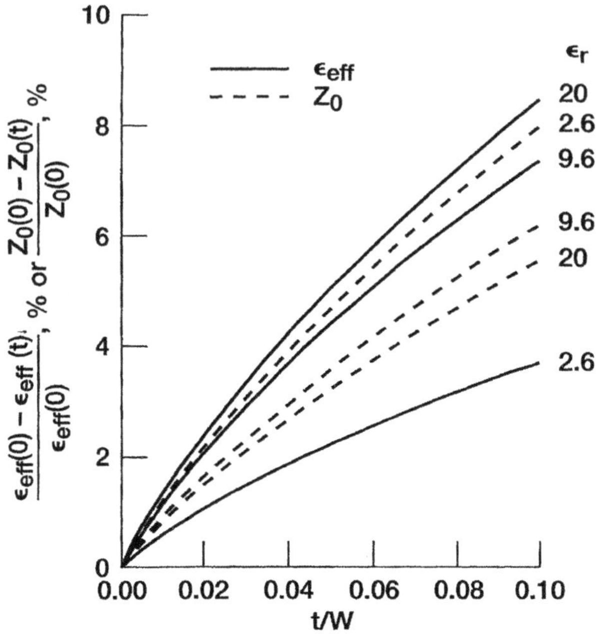

FIGURE 5.10 Computed percentage change in ε_{eff} and Z_0 of shielded suspended CPW as a function of the normalized conductor thickness with the substrate dielectric constant as a parameter: $S/h = 1.0$, $(S + 2W)/h = 3.0$, $(S + 2W + 2g)/h = 15.0$, $(h_1 + t)/h = 3.0$, $h_2/h = 3.0$, frequency = 0 GHz. (From Reference [4], copyright © IEE.)

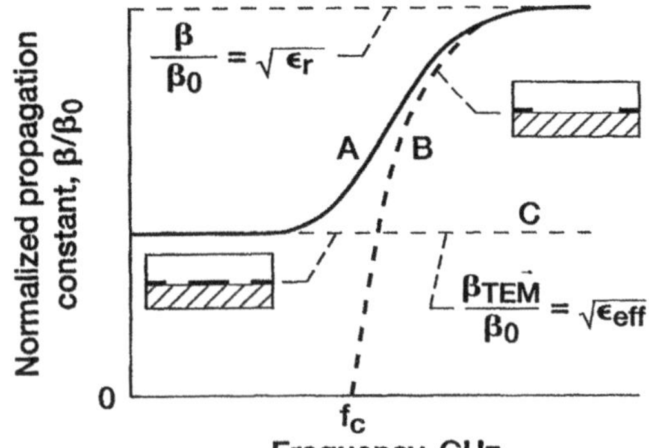

FIGURE 5.11 Typical dispersion characteristics of the grounded CPW dominant mode, which is considered as a superposition of the grounded slot line mode and the quasi-TEM mode of the grounded CPW. (From Reference [16], © 1992 IEEE.)

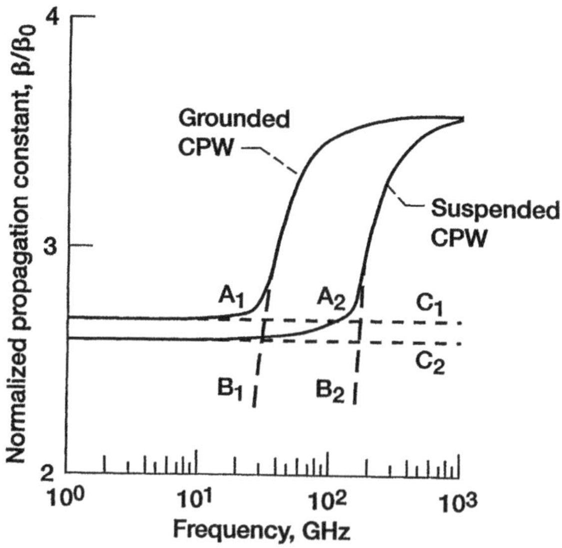

FIGURE 5.12 Computed dispersion characteristics of a shielded grounded CPW (A_1) and a shielded suspended CPW (A_2): $S = 0.14\,\text{mm}$, $2B = 2\,\text{mm}$, $h = 0.2\,\text{mm}$, $W = 10\,\text{mm}$, $\varepsilon_r = 12.9$, $t = 0$. For shielded grounded CPW, $h_1 = 2\,\text{mm}$, and for shielded suspended CPW, $h_1 = h_2 = 1\,\text{mm}$. (From Reference [16], © 1992 IEEE.)

5.12. This transmission line can be used up to about 30 GHz, that is, when β is 3 percent larger than β_{TEM} [16].

In Figure 5.12 curves A_2 and B_2 are the computed normalized propagation constants for the shielded suspended CPW and slotline, respectively [16]. Curve C_2 is the computed normalized quasi-TEM propagation constant for the shielded suspended CPW [16]. The dielectric constant, the substrate thickness, the center strip conductor width, and the slot width are identical to those of the shielded grounded CPW. The only difference is the 1 mm thick air layer below the dielectric substrate. In the case of the shielded suspended CPW, the useful frequency range approaches 100 GHz which is several times greater than the range of the shielded grounded CPW [16]. Thus, by providing a very thin air layer below the substrate, the dispersion free bandwidth can be significantly enhanced.

In Chapter 2, Section 2.4, it has been pointed out that the CPW structure can support both the odd mode and the even mode of propagation. In [17] it has been shown that the cutoff frequency of the first higher-order mode, that is, the even mode, is increased by suspending the dielectric substrate inside a conducting enclosure. Thus providing a thin air layer below the substrate also improves the modal bandwidth by increasing the frequency separation between the dominant and the first higher-order mode.

5.3.5 Pulse Propagation on a Suspended CPW

Consider the following Gaussian pulse as the input signal to a CPW [16]:

$$V(0, t) = V_0 \exp\left[-4 \ln 2 \left(\frac{t}{\tau}\right)^2\right], \tag{5.1}$$

where τ represents the full width at half maximum (FWHM) of the pulse. The pulse waveform at a propagation distance L is obtained by taking the inverse Fourier transform of the product of the Fourier transform of the input signal and the propagation factor, $\exp[-\gamma(f)L]$. Mathematically this can be expressed as [16]

$$V(L, t) = \mathscr{F}^{-1}\{\mathscr{F}[V(0, t)] \cdot \exp[-\gamma(f)L]\}, \tag{5.2}$$

where $\mathscr{F}$ denotes the Fourier transform and $\gamma(f)$ is the complex frequency-dependent propagation constant whose real and imaginary parts are the attenuation constant $\alpha(f)$ and the phase constant $\beta(f)$, respectively. For simplicity the attenuation constant is neglected, and only the phase constant is considered here. The phase constant $\beta(\mathrm{f})$ is given by the following closed form equation [16]:

$$\frac{\beta}{\beta_0} = \frac{\sqrt{\varepsilon_r} - \beta_{\mathrm{TEM}}/\beta_0}{1 + aF^{-b}} + \frac{\beta_{\mathrm{TEM}}}{\beta_0}, \tag{5.3}$$

where $F = f/f_{\mathrm{TE}}$, $f_{\mathrm{TE}} = c/4h\sqrt{\varepsilon_r - 1}$ is the cutoff frequency for the lowest order TE mode, c is the velocity of light in free space, β_{TEM} is the propagation constant assuming the quasi-TEM approximation, a and b are constants that depend on the type and dimensions of the transmission line. The values of a and b are obtained by curve fitting the dispersion data from numerical computations.

Figure 5.13 shows the computer simulation results of a Gaussian pulse propagation along a shielded suspended CPW [16]. The Gaussian pulse at the input has a FWHM of 20 ps with a 10 to 90 percent rise time of 15 ps. The frequency bandwidth of this pulse is approximately 40 GHz, at which point the spectral amplitude drops to 10 percent of the peak value. The shielded suspended CPW has a dispersion free bandwidth of about 100 GHz, as discussed in the previous section, and therefore the input pulse remains unaltered with the rise time still holding at 15 ps even after traveling a distance of 30 mm.

To obtain an expression for the minimum pulse width supported on such a transmission line, the Fourier transform of Eq. (5.1) is taken [16]:

$$V(0, \omega) = \sqrt{\frac{\pi}{4 \ln 2}}\; V_0 \tau \exp\left[-\frac{\omega^2 \tau^2}{16 \ln 2}\right]. \tag{5.4}$$

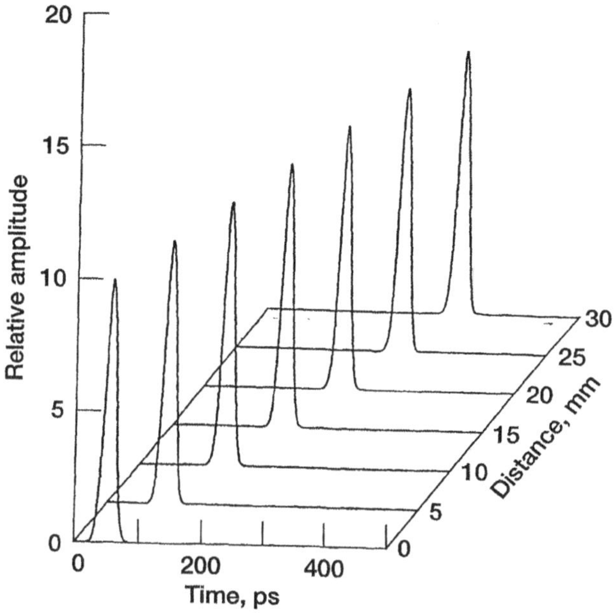

FIGURE 5.13 Computer simulation of a Gaussian pulse along a shielded suspended CPW. The FWHM of the pulse is 20 ps and the 10 to 90 percent rise time is 15 ps: $S = 0.14\,\text{mm}$, $2B = 2\,\text{mm}$, $h = 0.2\,\text{mm}$, $W = 10\,\text{mm}$, $\varepsilon_r = 12.9$, $t = 0$, $h_1 = h_2 = 1\,\text{mm}$. (From Reference [16], © 1992 IEEE.)

Let $f_{0.1}$ be the frequency at which the spectral amplitude is 10 percent of the peak value, then Eq. (5.4) gives [16]

$$f_{0.1} = \frac{2\sqrt{\ln 2 \cdot \ln 10}}{\pi \tau}. \tag{5.5}$$

Further let f_{TEM} be the frequency at which β in Eq. (5.3) is 3 percent larger than β_{TEM}. It is given by [16]

$$f_{\text{TEM}} = \left(\frac{\sqrt{\varepsilon_r} - 1.03\beta_{\text{TEM}}/\beta_0}{0.03a\beta_{\text{TEM}}/\beta_0} \right)^{-1/b} f_{\text{TE}}. \tag{5.6}$$

This frequency is considered as the upper limit of the usable frequency band. It thus defines the bandwidth of the CPW. Over this frequency range the dispersion is negligible. Since 99 percent of a pulse energy is contained in frequencies less than $f_{0.1}$, we may regard that the pulse will propagate without dispersion if $f_{0.1}$ does not exceed f_{TEM}. We thus obtain the minimum FWHM of the pulse as follows [16]:

$$\tau_{\text{min}} = \frac{2\sqrt{\ln 2 \cdot \ln 10}}{\pi f_{\text{TEM}}}. \tag{5.7}$$

The f_{TEM} for the shielded suspended CPW of Figure 5.12 is 76.2 GHz and τ_{min} is 10.6 ps.

5.3.6 Pulse Distortion—Experimental Validation

The picosecond pulse required for this experiment is generated optoelectronically. In the experiment a picosecond laser pulse is used to excite a photoconductive switch. The filtered output waveform of the switch is a Gaussian-like pulse and is coupled to the input of the CPW. The output pulse at the end of the CPW is coupled to the sampling head of a digital oscilloscope. The time resolution of the sampling head is about 8.8 ps. The Gaussian-like input pulse in the experiment has a rise time of 58 ps and a FWHM of 52 ps [16]. The shielded suspended CPW is fabricated on a DI-CLAD 810 dielectric substrate manufactured by Arlon and is 1.27 mm thick and has a dielectric constant of 10.5. The dimensions of the CPW are $S = 1.2$ mm, $W = 0.5$ mm, $2B = 20$ mm, $h_1 = h_2 = 2.7$ mm [16]. To support the dielectric substrate, 1 mm deep grooves are milled in the inner walls of the shielding enclosure. The measured FWHM of the pulse after traveling 60 mm is 65 ps [16]. The rise time of the leading edge of the pulse changed only slightly after propagating 60 mm [16]. This experiment demonstrates that a shielded suspended CPW is able to effectively minimize pulse distortion.

5.4 DISPERSION AND HIGHER-ORDER MODES OF A SHIELDED GROUNDED CPW

A shielded grounded CPW is shown in Figure 5.1(c). In the literature this structure is analyzed using the method of lines [8], the transverse resonance method [12], and the spectral domain method [15] and [16]. In [8] the dispersion characteristics are computed for the shielded grounded coplanar structure, assuming the same dimensions as those in [16] and a magnetic wall along the plane of symmetry. Figure 5.14 shows the computed characteristics.

The propagating mode on this structure is considered as a superposition of a TEM mode and slotline mode similar to a suspended CPW discussed earlier in Section 5.3.4. In Figure 5.14, the quasi-TEM dominant mode and the first higher-order mode are designated as curve A and curve C, respectively. The dominant mode of the resulting slotline structure when the center strip conductor of the CPW is removed is designated as curve B. The numerical results represented by curve A and curve B above are in good agreement [8] with the spectral domain computations [16].

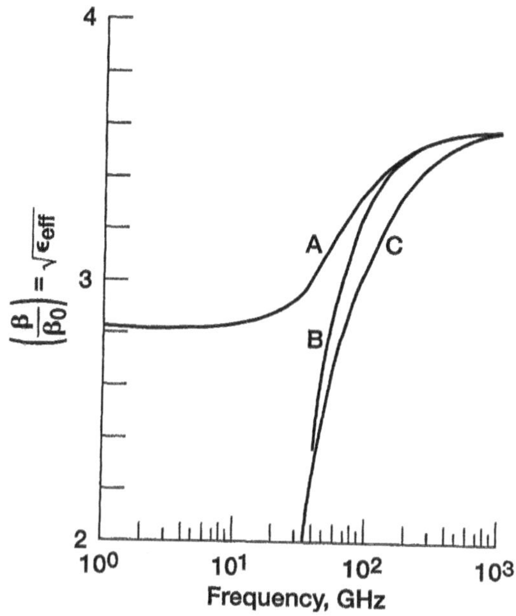

FIGURE 5.14 The quasi-TEM dominant mode (A) and the first higher-order mode (C) dispersion characteristics of a shielded grounded CPW: $S = 0.14$ mm, $2B = 2$ mm, $h = 0.2$ mm, $W = 0.28$ mm, $h_1 = 2$ mm, $\varepsilon_r = 12.9$, $t = 0$. (From Reference [8], copyright © IEE.)

5.5 DISPERSION, CHARACTERISTIC IMPEDANCE, AND HIGHER-ORDER MODES OF A CPW SUSPENDED INSIDE A NONSYMMETRICAL SHIELDING ENCLOSURE

A CPW suspended inside a nonsymmetrical shielding enclosure is shown in Figure 5.15. This structure is analyzed using the transverse resonance technique in [11], the spectral domain technique in [18] and the modal analysis method in [19]. The computed dispersion characteristics [11], for the dominant mode as well as the first two higher order modes are shown in Figure 5.16. The theoretical analysis takes into consideration the finite conductor thickness as well as the groove dimensions. However, their effect is very small on the dispersion characteristics in the frequency range considered. The characteristic is almost a straight line and therefore the dispersion is almost negligible over the frequency range considered. Figure 5.17(a) and (b) shows the computed Z_0 as a function of the normalized center conductor strip width with the normalized height of the shielding and normalized separation between the CPW ground planes as parameters, respectively [22]. From this figure it is observed that a wide range of characteristic impedances from about 25 Ω to over 200 Ω can be realized by appropriately choosing the geometry.

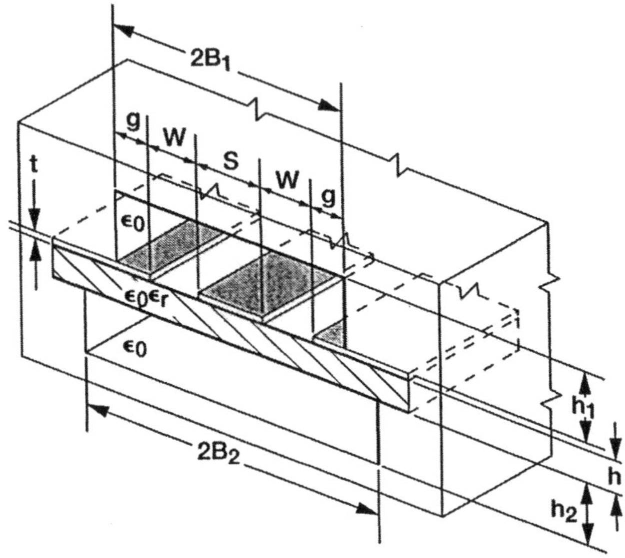

FIGURE 5.15 Schematic of a nonsymmetrically shielded coplanar waveguide.

The cutoff frequency for the first higher-order mode discussed earlier is computed as a function of the normalized center conductor strip width with the normalized height of the housing as a parameter. It is observed that the cutoff frequency is insensitive to either of these parameters [22] as shown in Figure 5.18. In fact it is approximately equal to the cutoff frequency of the fundamental mode of the rectangular waveguide of width $2B_2$.

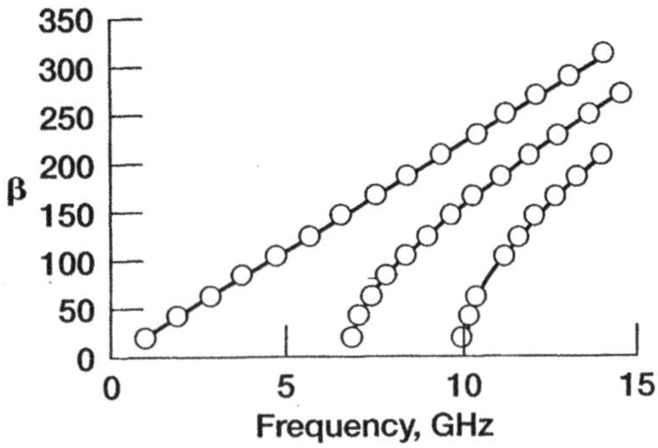

FIGURE 5.16 Computed and experimentally measured propagation constant as a function of the frequency for the dominant and the first two higher-order modes. Circular dots are experimental data; $2B_1 = 15.8$ mm, $2B_2 = 22.86$ mm, $h_1 = h_2 = B_2$, $h = 0.254$ mm, $t = 0.035$ mm, $S = 4.76$ mm, $W = 1.57$ mm, $\varepsilon_r = 2.2$. (From Reference [11], © 1989 IEEE.)

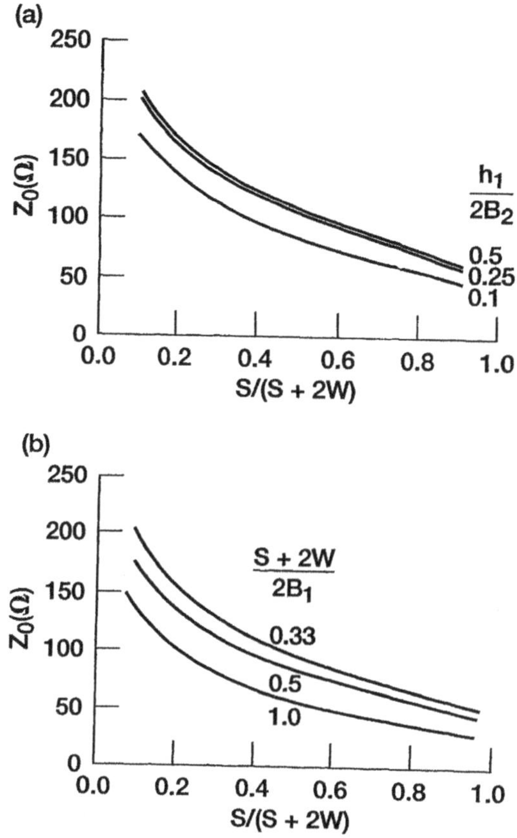

FIGURE 5.17 Computed characteristic impedance as a function of the center conductor strip width: $2B_1 = 15.8\,\text{mm}$, $2B_2 = 22.86\,\text{mm}$, $h_2 = 11.43\,\text{mm}$, $h = 0.254\,\text{mm}$, $t = 0.035\,\text{mm}$, $\varepsilon_r = 2.2$, $f = 4\,\text{GHz}$. (*a*) With h_1 as a parameter for $(S + 2W)/2B_1 = 0.5$; (*b*) with $S + 2W$ as a parameter for $h_1/2B_2 = 0.1$. (From Reference [22], © 1989 IEEE.)

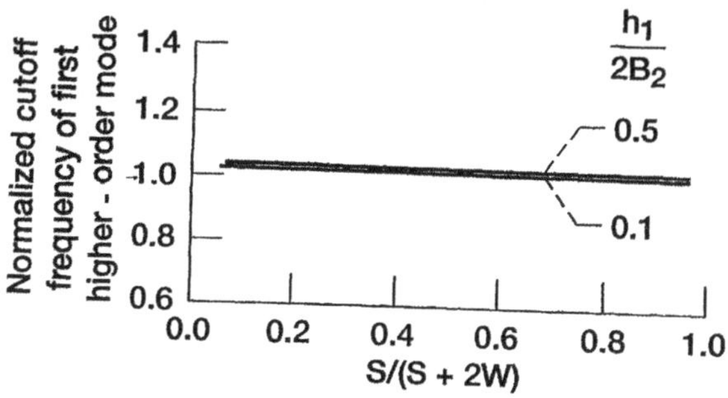

FIGURE 5.18 Computed normalized cutoff frequency of the first higher-order mode as a function of the center conductor strip width with the normalized height of the shield on a parameter: $2B_1 = 15.8\,\text{mm}$, $2B_2 = 22.86\,\text{mm}$, $h_2 = 11.43\,\text{mm}$, $h = 0.254\,\text{mm}$, $t = 0.035$, $\varepsilon_r = 2.2$, $(S + 2W)/2B_1 = 0.5$. (From Reference [22], © 1989 IEEE.)

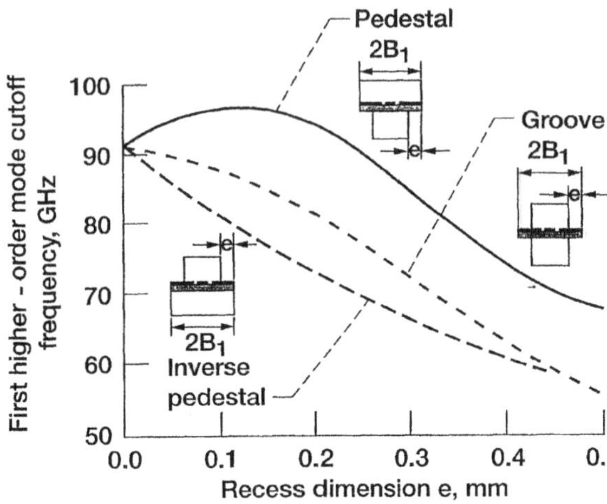

FIGURE 5.19 Computed cutoff frequency of the first higher order mode as a function of the recess dimension for different mounting structures: $2B_1 = 1.55\,\text{mm}$, $h_1 = 1.5\,\text{mm}$, $h_2 = 1.33\,\text{mm}$, $S = W = 0.2\,\text{mm}$, $h = 0.22\,\text{mm}$, $t = 50\,\text{mm}$, $\varepsilon_r = 3.75$. (From Reference [18], © 1995 IEEE.)

The effect of the mounting recess dimensions are significant at higher millimeter wave frequencies. As an example, in Figure 5.19 the computed cutoff frequency of the first higher-order mode is shown as a function of the recess depth [18]. It is also interesting to observe that the cutoff frequency depends on the type of mounting structure. For example, in the case of the groove and the inverse pedestal, the cutoff frequency decreases as the recess dimension increases. However, in the case of the pedestal, the cutoff frequency initially increases and beyond a certain recess dimension it decreases. Therefore there is an optimum choice of the pedestal dimensions for wide bandwidth.

5.5.1 Experimental Validation of the Dispersion Characteristics

To validate the computed results, a measurement technique utilizing a resonant cavity is used [11]. The cavity is formed by two rectangular waveguides that share a common broad wall which is formed by the suspended CPW substrate. The length L of the cavity is about five wavelengths at the center frequency, which is about 15 cms. Further the cavity is terminated by a short circuit at either ends. A two-port resonator is realized from this cavity by introducing two coaxial launchers. The coaxial launchers are oriented perpendicular to the suspended CPW substrate and enters the cavity from the two opposite broad walls as shown in Figure 5.20. The center conductor of the coaxial launchers are shorted to the cavity end walls and a part of the outer conductor is exposed to provided a weak magnetic coupling. The magnetic-field line trajectories at resonance are such that they loop around the center conductors of the launchers and are also normal to the plane of the suspended CPW substrate.

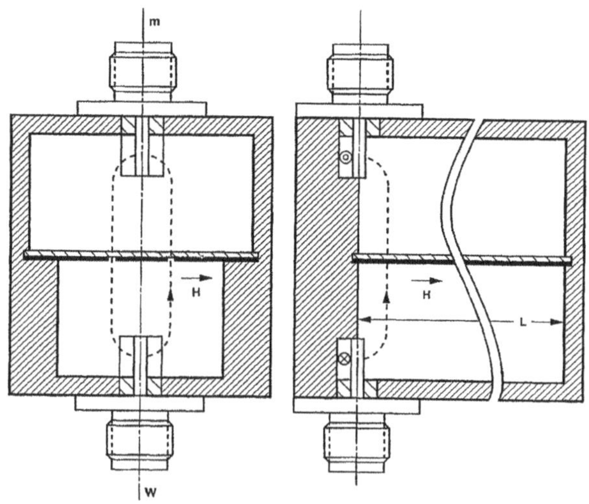

FIGURE 5.20 Cross-sectional view of the cavity showing magnetic-field lines for the dominant mode resonance. The magnetic wall symmetry is denoted by the plane *mw*.

Thus transverse magnetic fields are excited on the CPW circuit. This orientation of the magnetic fields allows a magnetic wall to be placed along the plane of symmetry analogous to a conventional CPW discussed in Chapter 2. The resonant frequencies corresponding to different mode number i and longitudinal order n are determined from the broadband swept frequency transmission measurements. The propagation constant β_i of mode i at frequency f_{ni} is determined from [11]

$$\frac{\beta_i}{k_o} = \frac{nc}{2Lf_{ni}}. \tag{5.8}$$

where $k_o = 2\pi/\lambda_0$ and λ_0 is the free space wavelength. The velocity of light in free space is denoted as c. The experimentally determined propagation constant is superimposed on the computed results in Figure 5.16 and is observed to be in good agreement [11].

5.6 DISPERSION AND CHARCTERISTIC IMPEDANCE OF SUSPENDED CPW ON MULTILAYER DIELECTRIC SUBSTRATE

A suspended CPW on a multilayer dielectric substrate is shown in Figure 5.1(*d*). The preceding generalized structure can be reduced to two useful practical structures, namely suspended CPW on a double-layer substrate and on a sandwiched substrate. These two structures are analyzed in [23], and their computed dispersion and characteristic impedance are presented here. Figures 5.21(*a*) and 5.22(*a*) show the λ_g/λ_0 and Z_0 as a function of the normalized

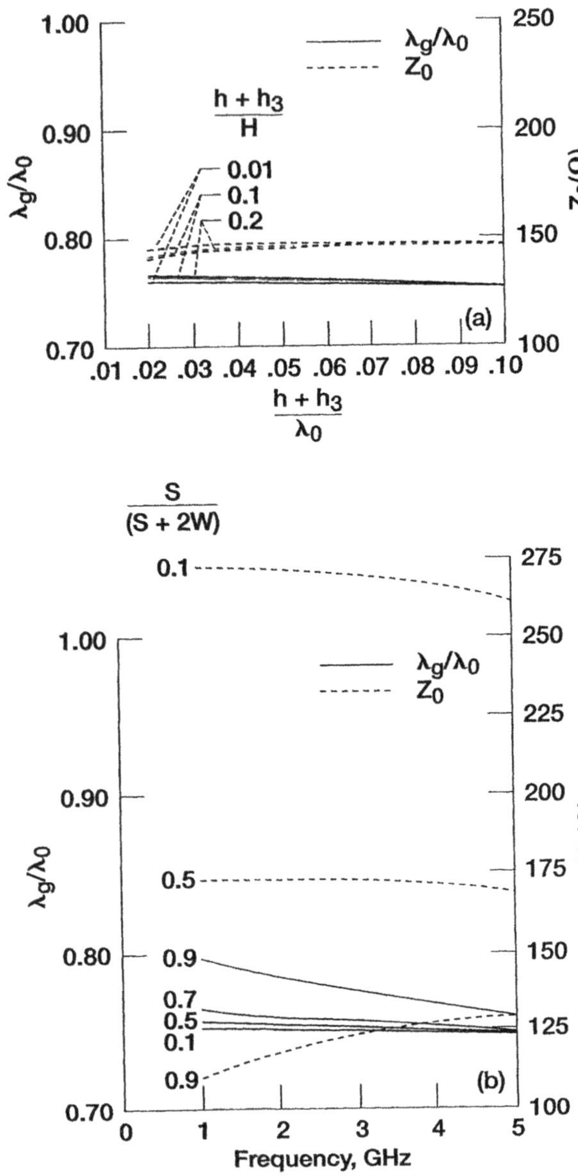

FIGURE 5.21 Computed guide wavelength ratio and characteristics impedance for CPW on suspended double-layer dielectric substrate: $W/(h + h_3) = 0.25$, $h = 3.2\,\text{mm}$, $h_3 = 5.0\,\text{mm}$, $B = H$, $h_1 = h_4$, $\varepsilon_r = 2.55$, $\varepsilon_{r3} = 2.62$, $t = 0$, $h_2 = 0$. (*a*) Function of the normalized substrate thickness with the normalized height of the shielding enclosure as a parameter, $S/(h + h_3) = 1.0$; (*b*) function of the frequency with the normalized slot width as a parameter, $(h + h_3)/H = 0.1$.

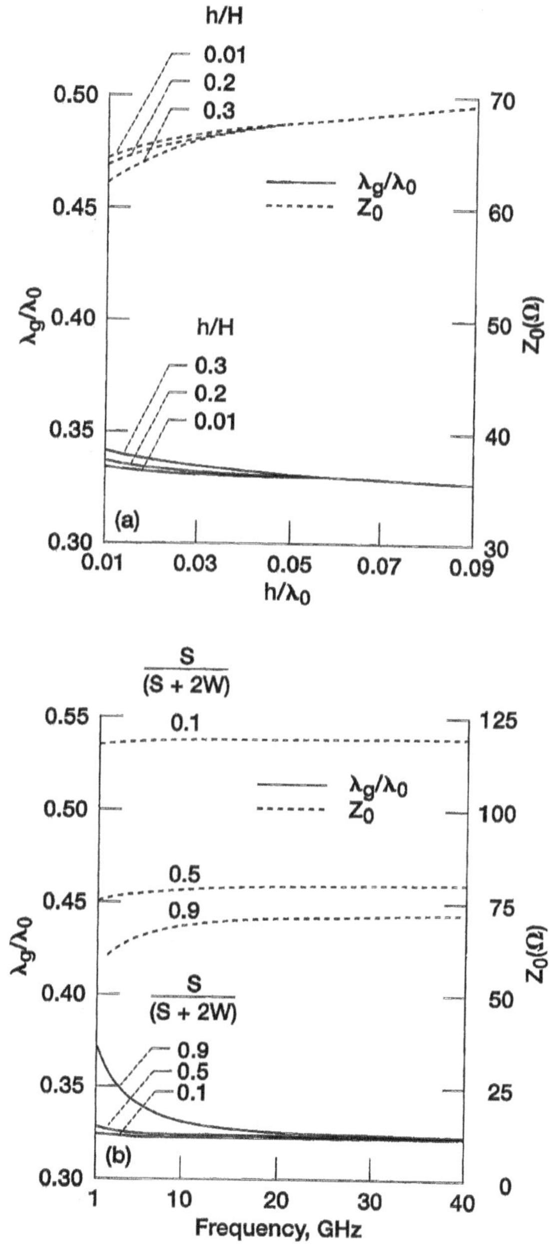

FIGURE 5.22 Computed guide wavelength ratio and characteristic impedance for CPW on suspended sandwiched dielectric substrate: $h_2 = h$, $W/h = 0.25$, $B = H$, $h_1 = h_4$, $h_3 = 0$, $t = 0$, $\varepsilon_r = \varepsilon_{r2} = 9.6$. (*a*) Function of the normalized substrate thickness with the normalized height of the shielding enclosure as a parameter, $S/h = 1.0$; (*b*) function of the frequency with the normalized slot width as a parameter, $h/H = 0.1$.

substrate thickness with the normalized height of the shielding enclosure as a parameter for the two structures, respectively. It is observed that the characteristics are almost insensitive to the height of the enclosure. In Figures 5.21(*b*) and 5.22(*b*), λ_g/λ_0 and Z_0 for the two structures are plotted as a function of the frequency with the normalized slot width as a parameter.

It is observed that as the slot width increases, the ratio $S/(S+2W)$ decreases; consequently the characteristic impedance increases and the guide wavelength ratio decreases. For large slot width, the λ_g/λ_0 is almost constant with frequency, indicating that the structure is dispersion free.

REFERENCES

[1] T. Hatsuda, "Computation of the Characteristics of Coplanar-Type Strip Lines by the Relaxation Method," *IEEE Trans. Microwave Theory Tech.*, Vol. 20, No. 6, pp. 413–416, June 1972.

[2] T. Hatsuda, "Computation of Coplanar-Type Strip-Line Characteristics by Relaxation Method and Its Application to Microwave Circuits," *IEEE Trans. Microwave Theory Tech.*, Vol. 23, No. 10, pp. 795–802, Oct. 1975.

[3] K. Koshiji, E. Shu, and S. Miki, "An Analysis of Coplanar Waveguides with Finite Conductor Thickness-Computation and Measurement of Characteristic Impedance," *Electron. Comm. in Japan*, Vol. 64-B, No. 8, pp. 69–78, 1981.

[4] T. Kitazawa and Y. Hayashi, "Quasi-static and Hybrid-Mode Analysis of Shielded Coplanar Waveguide with Thick Metal Coating," *IEE Proc.*, Vol. 134, Pt. H., No. 3, pp. 321–323 June 1987.

[5] T. N. Chang, "Hybrid-Mode Analysis of Shielded and Coupled Slotlines on a Suspended Substrate," *IEE Proc.*, Vol. 134, Pt. H, No. 3, pp. 327–329, June 1987.

[6] U. Schulz and R. Pregla, "A New Technique for the Analysis of the Dispersion Characteristics of Planar Waveguides Demonstrated for the Coplanar Line," *10th European Microwave Conf. Proc.*, Warsaw, Poland, pp. 331–335, Sept. 8–11, 1980.

[7] A. Stoeva, L. Urshev, and I. Angelov, "Theoretical and Experimental Determination of the Coplanar Line Dispersion Characteristics," *Microwave Optical Tech. Lett.*, Vol. 2, No. 4, pp. 132–135, April 1989.

[8] R. R. Kumar, S. Aditya, and D. Chadha, "Modes of a Shielded Conductor-Backed Coplanar Waveguide," *Electron. Lett.*, Vol. 30, No. 2 , pp. 146–148, Jan. 1994.

[9] Y. Fujiki, M. Suzuki, T. Kitazawa, and Y. Hayashi, "Higher-Order Modes in Coplanar-Type Transmission Lines," *Electron. Comm. in Japan*, Vol. 58-B, No. 2, pp. 74–81, 1975.

[10] P. K. Saha, "Dispersion in Shielded Planar Transmission Lines on Two-Layer Composite Structure," *IEEE Trans. Microwave Theory Tech.*, Vol. 25, No. 11, pp. 907–911, Nov. 1977.

[11] F. Alessandri, U. Goebel, F. Melai, and R. Sorrentino, "Theoretical and Experimental Characterization of Nonsymmetrically Shielded Coplanar Waveguides for Millimeter-Wave Circuits," *IEEE Trans. Microwave Theory Tech.*, Vol. 37, No. 12, pp. 2020–2026, Dec. 1989.

[12] F. Bouzidi, D. Bajon, H. Baudrand, and V. F. Hanna, "Influence of Grounding Lateral and Back Ground Planes in Transitions between Conductor Backed Coplanar Waveguide and Microstrip Lines," *23rd European Microwave Conf. Proc.*, Madrid, Spain, pp. 627–630, 1993.

[13] J. B. Davies and D. M.-Syahkal, "Spectral Domain Solution of Arbitrary Coplanar Transmission Line with Multilayer Substrate," *IEEE Trans. Microwave Theory Tech.*, Vol. 25, No. 2, pp. 143–146, Feb. 1977.

[14] R. H. Jansen, "Unified User-Oriented Computation of Shielded, Covered, and Open Planar Microwave and Millimeter—Wave Transmission-Line Characteristics," *IEE Proc. Microwaves, Optics Acoustics*, Vol. 3, No. 1, pp. 14–22, Jan. 1979.

[15] A. M. Pavio, "Hybrid Mode Technique Yields Waveguide Dispersion Analysis," *Microwave Sys. News*, Vol. 13, No. 4, pp. 106–111, April 1983.

[16] Y. Qian, E. Yamashita, and K. Atsuki, "Modal Dispersion Control and Distortion Suppression of Picosecond Pulses in Suspended Coplanar Waveguides," *IEEE Trans. Microwave Theory Tech.*, Vol. 40, No. 10, pp. 1903–1909, Oct. 1992.

[17] M. R. Lyons, J. P. K. Gilb, and C. A. Balanis, "Enhanced Dominant Mode Operation of Shielded Multilayer Coplanar Waveguide," *1993 IEEE MTT-S Int. Microwave Symp. Dig.*, Vol. 2, Atlanta, GA, pp. 943–946, June 14–18, 1993.

[18] T. Wang and K. Wu, "Effects of Various Suspended Mounting Schemes on Mode Characteristics of Coupled Slot lines Considering Conductor Thickness for Wideband MIC Applications," *IEEE Trans. Microwave Theory Tech.*, Vol. 43, No. 5, pp. 1106–1114, May 1995.

[19] S. Luo, R. Hinz, F. Williams, and V. K. Tripathi, "Effect of Rectangular Airbridge Tunnel on Microwave CPW Probe Design," *23rd European Microwave Conf. Proc.*, Madrid, Spain, pp. 621–623, Sept. 6–9, 1993.

[20] S. Xiao and R. Vahldieck, "An Improved 2D-FDTD Algorithm for Hybrid Mode Analysis of Quasi-Planar Transmission Lines," *1993 IEEE MTT-S Int. Microwave Symp. Dig.*, Vol. 1, Atlanta, GA, pp. 421–424, June 14–18, 1993.

[21] J. B. Knorr and K. D. Kuchler, "Analysis of Coupled Slots and Coplanar Strips on Dielectric Substrate," *IEEE Trans. Microwave Theory Tech.*, Vol. 23, pp. 541–548, July 1975.

[22] F. Alessandri, U. Goebel, F. Melai, and R. Sorrentino, "Theoretical and Experimental Characterization of Nonsymmetrically Shielded Coplanar Waveguides for Millimeter Wave Circuits," *1989 IEEE MTT-S Int. Microwave Symp. Dig.*, Vol. 3, Long Beach, CA, pp. 1219–1222, June 13–15, 1989.

[23] R. N. Simons, "Suspended Coupled Slotline Using Double Layer Dielectric," *IEEE Trans. Microwave Theory Tech.*, Vol. 29, No. 2, pp. 162–165, Feb. 1981.

[24] B. Bianco, M. Parodi, S. Ridella, and F. Selvaggi, "Launcher and Microstrip Characterization," *IEEE Trans. Instrument. Meas.*, Vol. 25, No. 4, pp. 320–323, Dec. 1976.

CHAPTER 6

Coplanar Striplines

6.1 INTRODUCTION

The coplanar stripline (CPS) consists of a dielectric substrate with two parallel strip conductors separated by a narrow gap. This basic structure is referred to as the conventional CPS. The advantages of CPS are as follows: (1) Both series as well as shunt mounting of devices is possible. (2) The CPS is a balanced transmission line. Hence its applications include balanced mixers and feed network work for printed dipole antennas. The main disadvantage of the CPS is that because it lacks a ground plane, the line can support besides the fundamental CPS mode two other parasitic modes, namely the TE_0 and TM_0 dielectric slab waveguide modes. These parasitic modes do not have a cutoff frequency. The TE_0 and TM_0 modes have their electric fields predominantly parallel and perpendicular to the dielectric–air interface, respectively. The electric field of the fundamental CPS mode is predominantly parallel to the dielectric-air interface and hence strongly couples to the TE_0 parasitic mode at discontinuities.

In Section 6.2 analytical expressions for effective dielectric constant and characteristics impedance for CPS and CPS variants are presented. In deriving the analytical expressions using conformal mapping technique, the CPS is assumed to be on an isotropic homogeneous dielectric of arbitrary thickness and relative permittivity. Further the metallization is assumed to have perfect conductivity and zero thickness. The variants considered are the asymmetric CPS (ACPS), the CPS with a infinitely wide ground plane (CPSWG), and the CPS with isolating ground planes (CPSIG). The advantage of the asymmetric CPS over conventional CPS is the flexibility to adjust the propagation parameters by changing the width of one of the strips while keeping the widths of the other strip and the gap fixed. The CPS with isolating ground planes has several advantages. First, line-to-line coupling is reduced because of the isolation provided by the ground planes. Second, the parasitic TE_0 dielectric slab waveguide mode discussed earlier is eliminated. Third, flexibility in obtaining lower characteristic impedance is added by the possibility of adjusting the outer slot width.

In Section 6.3 synthesis formulas to determine the slot width and the strip conductor width of conventional CPS are presented. Finally in Section 6.4 a few novel variants of the CPS, namely micro-coplanar stripline and CPS with a groove, are discussed.

6.2 ANALYTICAL EXPRESSIONS BASED ON QUASI-STATIC TEM CONFORMAL MAPPING TECHNIQUES TO DETERMINE EFFECTIVE DIELECTRIC CONSTANT AND CHARACTERISTIC IMPEDANCE

6.2.1 Coplanar Stripline on a Multilayer Dielectric Substrate

A coplanar stripline (CPS) sandwiched between multilayer dielectric substrates is shown in Figure 6.1. The strip and slot widths are indicates as W and S, respectively. On this structure a quasi-static TEM mode is assumed to propagate. For the quasi-static TEM mode, the effective dielectric constant $\varepsilon_{\mathrm{eff}}$, the phase velocity v_{ph}, and the characteristic impedance Z_0 are given by Eqs. (4.1), (4.2), and (4.3), respectively. Since these quantities are related to the

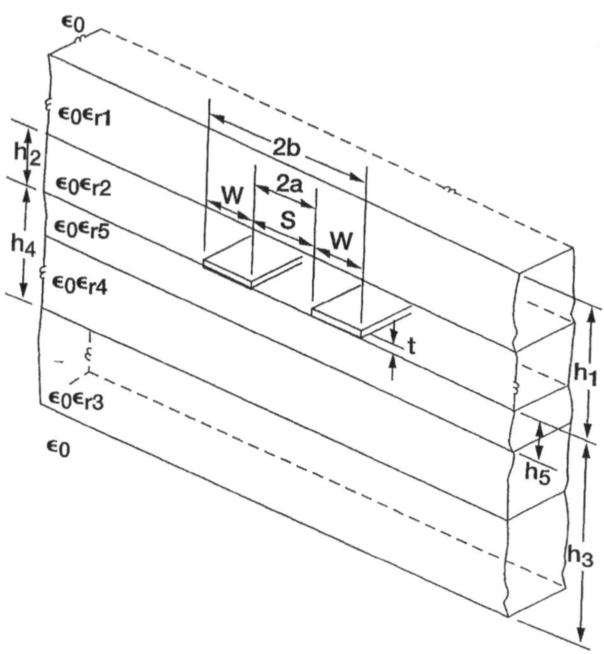

FIGURE 6.1 Schematic of a coplanar stripline (CPS) on a multi-layer dielectric substrate.

capacitance of structure, the first step is to determine the capacitance per unit length. This is accomplished by the method of superposition of partial capacitances as described in Section 4.2.1. Hence the capacitance per unit length of the CPS shown in Figure 6.1 is expressed as [1]

$$C_{\text{CPS}} = C_0 + C_1 + C_2 + C_3 + C_4 + C_5. \tag{6.1}$$

The configuration of these capacitances are similar to that shown in Figures 4.2(*a*) to (*f*). The capacitance C_0 per unit length of the CPS in the absence of all dielectric substrates is expressed as [1]

$$C_0 = \varepsilon_0 \frac{K(k)}{K(k')}, \tag{6.2}$$

where K is the complete elliptical integral of the first kind. The arguments k and k' are dependent on the geometry of the CPS and are given by [1]

$$k = \sqrt{1 - \left(\frac{a}{b}\right)^2} \tag{6.3}$$

and

$$k' = \sqrt{1 - k^2} = \frac{a}{b} = \frac{S}{S + 2W}. \tag{6.4}$$

The next step is to determine the capacitances C_1 to C_6. This can be accomplished by the conformal mapping technique as described Section 4.2.1 and through the application of Eq. (4.1), which yields the ε_{eff}. A more direct and simple procedure to obtain the ε_{eff} is through the application of the duality principle. The duality principle as described in [2] and [3] states that the ε_{eff} of complementary structures are identical. The CPS and the conventional CPW with infinitely wide top ground planes are considered as dual structures. Hence their effective dielectric constants are identical. Strictly this is true when the respective substrates are infinitely thick. However, the computed results in [4] indicate that the ε_{eff} of the two transmission lines are approximately the same for substrate thickness equal to or greater than five times the b dimension. In most applications involving microwave integrated circuits this requirement is met. Hence the expression for CPS effective dielectric constant $\varepsilon_{\text{eff}}^{\text{CPS}}$ is derived using the duality principle. Accordingly the dimension c in Eq. (4.17) is set equal to infinity, which results in the following expression [1]:

$$\begin{aligned}
\varepsilon_{\text{eff}}^{\text{CPS}} &= \frac{C_{\text{CPS}}}{C_0} \\
&= 1 + \frac{1}{2}(\varepsilon_{r1} - 1)\frac{K(k)K(k_1')}{K(k')K(k_1)}
\end{aligned}$$

$$+\frac{1}{2}(\varepsilon_{r2}-\varepsilon_{r1})\frac{K(k)K(k_2')}{K(k')K(k_2)}$$
$$+\frac{1}{2}(\varepsilon_{r3}-1)\frac{K(k)K(k_3')}{K(k')K(k_3)}$$
$$+\frac{1}{2}(\varepsilon_{r4}-\varepsilon_{r3})\frac{K(k)K(k_4')}{K(k')K(k_4)}$$
$$+\frac{1}{2}(\varepsilon_{r5}-\varepsilon_{r4})\frac{K(k)K(k_5')}{K(k')K(k_5)}. \tag{6.5}$$

Where ε_{ri} $(i = 1, 2, 3, 4, 5)$ is the relative permittivity of the dielectric substrate and K is the complete elliptical integral of the first kind. The arguments k and k' are given by Eqs. (6.3) and (6.4), respectively. The arguments k_i and k_i' are defined as follows:

$$k_i = \sqrt{1 - \frac{\sinh^2(\pi a/2h_i)}{\sinh^2(\pi b/2h_i)}}, \qquad i = 1, 2, 3, 4, 5. \tag{6.6}$$

and

$$k_i' = \sqrt{1 - k_i^2}, \tag{6.7}$$

where h_i is the dielectric substrate thickness as indicated in Figure 6.1.

The phase velocity and characteristic impedance from Eqs. (4.2), (4.3), and (6.5) are

$$v_{ph}^{CPS} = \frac{c'}{\sqrt{\varepsilon_{eff}^{CPS}}}, \tag{6.8}$$

$$Z_0^{CPS} = \frac{120\pi}{\sqrt{\varepsilon_{eff}^{CPS}}} \cdot \frac{K(k')}{K(k)}, \tag{6.9}$$

where c' is the velocity of light in free space.

6.2.2 Coplanar Stripline on a Dielectric Substrate of Finite Thickness

A coplanar stripline on a dielectric substrate of finite thickness is shown in Figure 6.2. The ε_{eff}^{CPS} for this transmission line is easily obtained by setting $\varepsilon_{r1} = \varepsilon_{r2} = \varepsilon_{r3} = \varepsilon_{r4} = 1$ in Eq. (6.5). With this substitution and simplification the new expression is as follows:

$$\varepsilon_{eff}^{CPS} = 1 + \frac{1}{2}(\varepsilon_{r5} - 1)\frac{K(k)}{K(k')}\frac{K(k_5')}{K(k_5)} \tag{6.10}$$

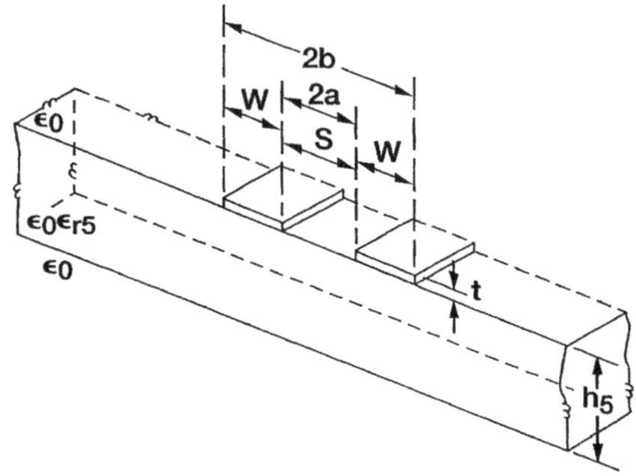

FIGURE 6.2 Schematic of a CPS on a dielectric substrate of finite thickness.

where K is the complete elliptical integral of the first kind. The arguments k, k', k_5, and k'_5 are given by Eqs. (6.3), (6.4), (6.6), and (6.7), respectively.

From duality principle discussed in the previous section, Eq. (6.10) is identical to that given in [5] for CPW. Therefore the computed $\varepsilon_{\text{eff}}^{\text{CPS}}$ is the same as $\varepsilon_{\text{eff}}^{\text{CPW}}$ which is shown in Figure 2.6(a) in Section 2.2.3.

The characteristic impedance Z_0^{CPS} computed from Eqs. (6.9) and (6.10) is shown in Figure 6.3. Since CPS and CPW are complementary structures, the

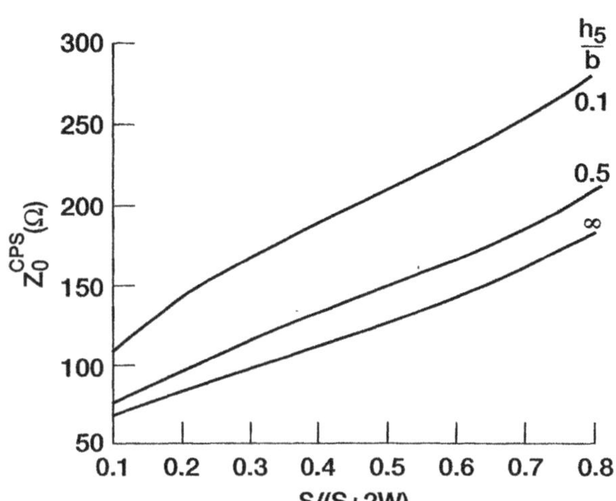

FIGURE 6.3 Computed characteristic impedance of CPS as a function of $S/(S + 2W)$, with the normalized substrate h_5/b as a parameter. $\varepsilon_{r5} = 10.0$, $t = 0$. (From Reference [5], copyright © IEE.)

product of Z_0^{CPS} and Z_0^{CPW} is a constant and is expressed as [2]

$$Z_0^{\text{CPS}} Z_0^{\text{CPW}} = \frac{\eta^2}{4\varepsilon_{\text{eff}}^{\text{CPS}}} = \frac{\eta^2}{4\varepsilon_{\text{eff}}^{\text{CPW}}}, \tag{6.11}$$

where η is the intrinsic impedance of the medium and is equal to $376.73\,\Omega \approx 120\pi$.

6.2.3 Asymmetric Coplanar Stripline on a Dielectric Substrate of Finite Thickness

In an asymmetric coplanar stripline (ACPS) one strip conductor is wider than the other, as shown in Figure 6.4. The strip widths W_1 and W_2 and the separation S between them are indicated in the same figure. The capacitance per unit length of the line in the absence of the dielectric substrate is given by the expression [6]

$$C_0^{\text{ACPS}} = 2\varepsilon_0 \frac{K(k')}{K(k)}, \tag{6.12}$$

where ε_0 is the permittivity of free space and K is the complete elliptic integral of the first kind. The arguments k and k' are [6]

$$k = \sqrt{\frac{(2a)(b_1 + b_2)}{(a + b_1)(a + b_2)}}, \tag{6.13}$$

$$k' = \sqrt{1 - k^2} = \sqrt{\frac{(b_1 - a)(b_2 - a)}{(a + b_1)(a + b_2)}}, \tag{6.14}$$

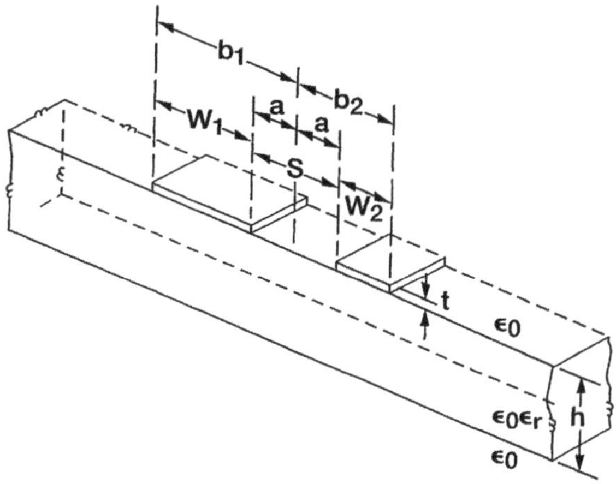

FIGURE 6.4 Schematic of asymmetric coplanar stripline (ACPS) on a dielectric substrate of finite thickness.

where

$$a = S/2$$
$$b_1 = a + W_1$$
$$b_2 = a + W_2.$$

The characteristic impedance of the line in the absence of the dielectric substrate is defined as [6]

$$Z_{0(\text{air})}^{\text{ACPS}} = \left(\frac{L}{C_0^{\text{ACPS}}}\right)^{1/2}, \tag{6.15}$$

where L is the inductance per unit length of the line and is given by

$$L = \frac{1}{(c')^2 C_0^{\text{ACPS}}} \tag{6.16}$$

where c' is the velocity of light in free space. Substituting Eqs. (6.12) and (6.16) into Eq. (6.15) and simplifying gives [6]

$$Z_{0(\text{air})}^{\text{ACPS}} = 60\pi \frac{K(k)}{K(k')}. \tag{6.17a}$$

In the presence of the dielectric substrate, the expression above has to be modified to take into account the ACPS effective dielectric constant $\varepsilon_{\text{eff}}^{\text{ACPS}}$. Thus the characteristic impedance in the presence of a dielectric substrate is

$$Z_0^{\text{ACPS}} = \frac{60\pi}{\sqrt{\varepsilon_{\text{eff}}^{\text{ACPS}}}} \frac{K(k)}{K(k')}. \tag{6.17b}$$

The $\varepsilon_{\text{eff}}^{\text{ACPS}}$ is given below for two cases, namely infinitely thick substrate and finite thickness substrate. The $\varepsilon_{\text{eff}}^{\text{ACPS}}$ in the case of an infinitely thick substrate is [7]

$$\varepsilon_{\text{eff}}^{\text{ACPS}} = \frac{(\varepsilon_r + 1)}{2}, \tag{6.18}$$

where ε_r is the relative permittivity of the dielectric substrate.

The $\varepsilon_{\text{eff}}^{\text{ACPS}}$ in the case of a finite thickness substrate is [6]

$$\varepsilon_{\text{eff}}^{\text{ACPS}} = 1 + \frac{\varepsilon_r - 1}{2} \frac{K(k)}{K(k')} \frac{K(k_2')}{K(k_2)} \tag{6.19}$$

where ε_r is the relative permittivity of the dielectric substrate, K is the complete elliptic integral of the first kind and the arguments k and k' are given by Eqs. (6.13) and (6.14), respectively. The arguments k_2 and k_2' are [6]

$$k_2 = \sqrt{\frac{\{\exp[2\pi(b_1+a)/h]-\exp[2\pi(b_1-a)/h]\}\{\exp[2\pi(b_1+b_2)/h]-1\}}{\{\exp[2\pi(b_1+b_2)/h]-\exp[2\pi(b_1-a)/h]\}\{\exp[2\pi(b_1+a)/h]-1\}}}, \quad (6.20)$$

$$k_2' = \sqrt{1-k_2^2}, \quad (6.21)$$

where h is the substrate thickness.

Equation 6.17(b) reduces to Eq. (6.9) when the two strip conductors are symmetric, that is, when $b_1 = b_2 = b$. To demonstrate that this happens, the following transformation rules for elliptic integrals are considered:

$$K(k) = (1 + k_s)K(k_s), \quad (6.22)$$

$$K(k') = \frac{1 + k_s}{2} K(k_s'), \quad (6.23)$$

where the arguments k_s and k_s' correspond to the symmetric case and

$$k_s' = \sqrt{1 - k_s^2}, \quad (6.24)$$

$$k = \frac{2\sqrt{k_s}}{1 + k_s}, \quad (6.25)$$

$$k' = \frac{1 - k_s}{1 + k_s}. \quad (6.26)$$

On substituting $b_1 = b_2 = b$ into Eqs. (6.13) and (6.15), we obtain

$$k = \frac{2\sqrt{a/b}}{1 + (a/b)}, \quad (6.27)$$

$$k' = \frac{1 - (a/b)}{1 + (a/b)}. \quad (6.28)$$

By equating Eqs. (6.27) and (6.28) to Eqs. (6.25) and (6.26) respectively, we obtain

$$k_s = \frac{a}{b}, \quad (6.29)$$

$$k_s' = \sqrt{1 - \left(\frac{a}{b}\right)^2}. \quad (6.30)$$

The next step is to replace k_s and k_s' in Eqs. (6.22) and (6.23), respectively, by the quantities above and then to substitute Eqs. (6.22) and (6.23) into Eq. (6.17b). Equation (6.17b), when simplified, reduces to Eq. (6.9)

6.2.4 Coplanar Stripline with Infinitely Wide Ground Plane on a Dielectric Substrate of Finite Thickness

A coplanar stripline with infinitely wide ground plane (CPSWG) is shown in Figure 6.5. The strip width W and the separation S between the strip and the ground are also indicated in the same figure. The capacitance per unit length of the line in the absence of the dielectric substrate is [6]

$$C_0^{\mathrm{CPSWG}} = 2\varepsilon_0 \frac{K(k_3')}{K(k_3)}, \tag{6.31}$$

where K is the complete elliptic integral of the first kind and the arguments k_3 and k_3' are obtained by $b_1 \to \infty$ and $b_2 = b$ in Eqs. (6.13) and (6.14), respectively. Thus

$$k_3 = \sqrt{\frac{2a}{a+b}} \tag{6.32}$$

$$k_3' = \sqrt{1 - k_3^2} = \sqrt{\frac{(b-a)}{(b+a)}}, \tag{6.33}$$

where the strip and slot widths are $2a$ and $(b - a)$, respectively.

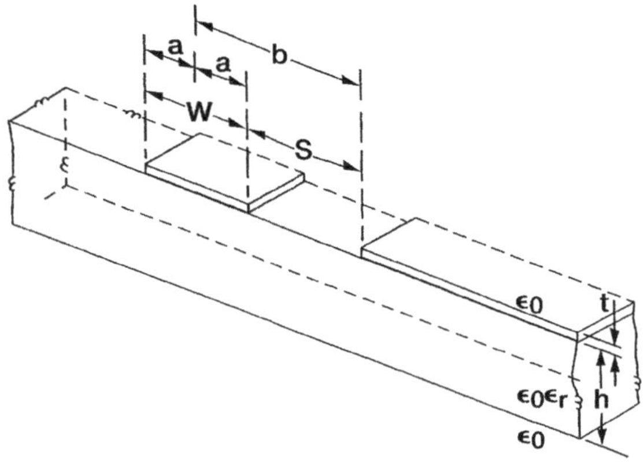

FIGURE 6.5 Schematic of coplanar stripline with infinitely wide ground plane (CPSWG) on a dielectric substrate of finite thickness.

The characteristic impedance in the absence of the dielectric substrate is derived as in Section 6.2.3, and it is written as

$$Z_{0(\mathrm{air})}^{\mathrm{CPSWG}} = 60\pi \frac{K(k_3')}{K(k_3)}. \tag{6.34}$$

In the presence of a dielectric substrate, the expression above is modified by the effective dielectric constant $\varepsilon_{\mathrm{eff}}^{\mathrm{CPSWG}}$. The $\varepsilon_{\mathrm{eff}}^{\mathrm{CPSWG}}$ determined by the method of superposition of partial capacitances is [6]

$$\varepsilon_{\mathrm{eff}}^{\mathrm{CPSWG}} = 1 + \frac{\varepsilon_r - 1}{2} \frac{K(k_3)}{K(k_3')} \frac{K(k_4')}{K(k_4)}, \tag{6.35}$$

where ε_r is the relative permittivity of the dielectric substrate and K is the complete elliptic integral of the first kind. The arguments k_3 and k_3' are given by Eqs. (6.32) and (6.33), respectively. The arguments k_4 and k_4' are given by [6]

$$k_4 = \sqrt{\frac{\exp(2\pi a/h) - 1}{\exp[\pi(b + a)/h] - 1}}, \tag{6.36}$$

$$k_4' = \sqrt{1 - k_4^2} = \sqrt{\frac{\exp[\pi(b + a)/h) - \exp(2\pi a/h)}{\exp[\pi(b + a)/h] - 1}}, \tag{6.37}$$

where h is the dielectric substrate thickness. The corresponding characteristic impedance is given by

$$Z_0^{\mathrm{CPSWG}} = \frac{60\pi}{\sqrt{\varepsilon_{\mathrm{eff}}^{\mathrm{CPSWG}}}} \frac{K(k_3')}{K(k_3)}. \tag{6.38}$$

6.2.5 Coplanar Stripline with Isolating Ground Planes on a Dielectric Substrate of Finite Thickness

This configuration of the coplanar stripline has additional ground planes symmetrically located on either side of the balanced pair of signal lines as shown in Fig. 6.6. This variant of the conventional CPS is known as the coplanar stripline with isolating ground planes (CPSIG) [8]. The effective dielectric constant $\varepsilon_{\mathrm{eff}}^{\mathrm{CPSIG}}$ is given by the expression [8]

$$\varepsilon_{\mathrm{eff}}^{\mathrm{CPSIG}} = 1 + \left(\frac{\varepsilon_r - 1}{2}\right) \frac{K(k_1)}{K(k_1')} \frac{K(k')}{K(k)}, \tag{6.39}$$

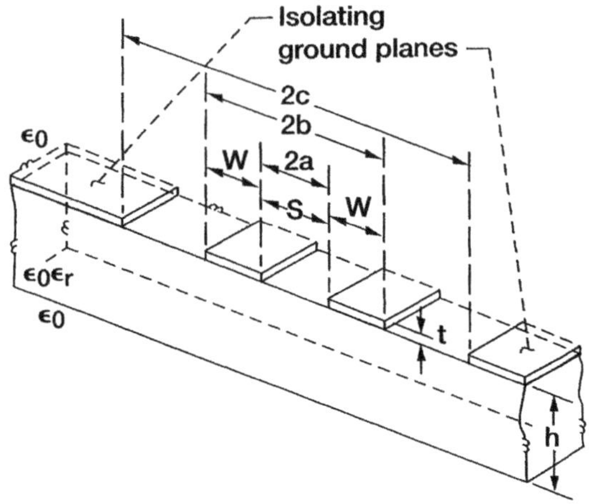

FIGURE 6.6 Schematic of coplanar stripline with isolating ground planes on a dielectric substrate of finite thickness.

where ε_r is the relative permittivity of the dielectric substrate and K is the complete elliptic integral of the first kind. The arguments k, k', k_1, and k_1' are given by [8]

$$k = \frac{a}{b}\sqrt{\frac{1-(b/c)^2}{1-(a/c)^2}}, \tag{6.40}$$

$$k' = \sqrt{1-k^2}, \tag{6.41}$$

$$k_1 = \frac{\sinh(\pi a/2h)}{\sinh(\pi b/2h)}\sqrt{\frac{1-\dfrac{\sinh^2(\pi b/2h)}{\sinh^2(\pi c/2h)}}{1-\dfrac{\sinh^2(\pi a/2h)}{\sinh^2(\pi c/2h)}}}, \tag{6.42}$$

$$k_1' = \sqrt{1-k_1^2}, \tag{6.43}$$

where h is the dielectric substrate thickness. The strip and slot widths are b-a and $2a$, respectively. The separation between the ground planes is $2c$. The characteristic impedance is given by [8]

$$Z_0^{\text{CPSIG}} = \frac{120\pi}{\sqrt{\varepsilon_{\text{eff}}^{\text{CPSIG}}}}\frac{K(k)}{K(k')}. \tag{6.44}$$

6.3 COPLANAR STRIPLINE SYNTHESIS FORMULAS TO DETERMINE THE SLOT WIDTH AND THE STRIP CONDUCTOR WIDTH

A CPS with zero thickness conductors on a dielectric of relative permittivity ε_r is shown in Figure 6.7. In this figure, the strip and slot widths are denoted as W and S respectively. For this CPS accurate closed form synthesis formulas are available in the literature [9]. These formulas are based on function approximation and curve fitting quasi-static analysis results. The application range of these formulas encompass most CPS circuit design, including monolithic microwave integrated circuits. The accuracy of these formulas is better than 1 percent when compared with the spectral domain [10] results. The expression for the slot width S will be first presented.

Synthesis Formula for Slot Width S

When

$$\frac{W}{h} \leqslant \frac{10}{1 + \ln \varepsilon_r} \quad \text{and} \quad \frac{S}{h} \leqslant \frac{10}{3(1 + \ln \varepsilon_r)}, \tag{6.45}$$

the separation or the slot width S can be expressed as

$$S = W \times G(\varepsilon_r, h, Z_0, W) \tag{6.46}$$

with

$$G = \left[\frac{1}{8}\exp\left(\frac{60\pi^2}{Z_0\varepsilon_{\text{eff}}^{0.5}}\right) - \frac{1}{2}\right]^{-1} \quad \text{for } Z_0 \leqslant \frac{60\sqrt{2}\pi}{(\varepsilon_r + 1)^{0.5}}, \tag{6.47a}$$

$$G = \frac{1}{4}\exp\left(\frac{Z_0\varepsilon_{\text{eff}}^{0.5}}{120}\right) + \exp\left(-\frac{Z_0\varepsilon_{\text{eff}}^{0.5}}{120}\right) - 1 \quad \text{for } Z_0 > \frac{60\sqrt{2}\pi}{(\varepsilon_r + 1)^{0.5}}, \tag{6.47b}$$

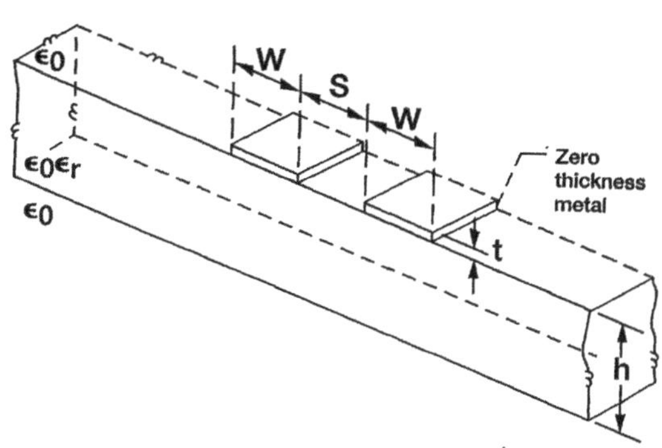

FIGURE 6.7 Coplanar stripline configuration.

where the effective dielectric constant ε_{eff} is given by

$$\varepsilon_{eff} = \varepsilon_{eff}(\varepsilon_r, h, Z_0, W) = T_1 \times \left(1 + \frac{\varepsilon_r - 1}{Z_0(\varepsilon_r + 1)^{0.5}} T_2\right) \tag{6.48}$$

with

$$T_1 = 1 + \tanh\left\{\frac{Z_0^2 \varepsilon_r^7}{720\pi^3(\varepsilon_r + 1)^8} \times \exp\left[\left(1 + 0.0004\varepsilon_r Z_0 \frac{W}{h}\right) \ln\left(\frac{W}{h}\right)\right]\right\} \tag{6.49}$$

and

$$T_2 = 84.85 \ln\left(2 \frac{1+g}{1-g}\right) \qquad \text{for } 0.841 \leqslant g \leqslant 1, \tag{6.50a}$$

$$T_2 = \frac{837.5}{\ln\left(2 \dfrac{1 + (1 - g^4)^{0.25}}{1 - (1 - g^4)^{0.25}}\right)} \qquad \text{for } 0 < g \leqslant 0.841, \tag{6.50b}$$

$$g = \left\{\frac{\exp\left(\dfrac{\pi(1+Q)W}{2h}\right) - \exp\left(\dfrac{\pi W}{2h}\right)}{\exp\left(\dfrac{\pi(2+Q)W}{2h}\right) - 1}\right\}^{0.5}, \tag{6.51}$$

$$Q = G|_{\varepsilon_{eff} = \frac{\varepsilon_r + 1}{2}}. \tag{6.52}$$

Synthesis Formula for Strip Conductor Width W

When

$$\frac{S}{h} \leqslant \frac{10}{3(1 + \ln \varepsilon_r)} \quad \text{and} \quad \frac{W}{h} \leqslant \frac{10}{1 + \ln \varepsilon_r} \tag{6.53}$$

the strip conductor width W can be expressed as

$$W = \frac{S}{G(\varepsilon_r, h, Z_0, W)} \tag{6.54}$$

with

$$G = \left[\frac{1}{8} \exp\left(\frac{60\pi^2}{Z_0 \varepsilon_{eff}^{0.5}}\right) - \frac{1}{2}\right]^{-1} \qquad \text{for } Z_0 \leqslant \frac{60\sqrt{2}\pi}{(\varepsilon_r + 1)^{0.5}}, \tag{6.55a}$$

$$G = \frac{1}{4}\exp\left(\frac{Z_0\varepsilon_{\mathrm{eff}}^{0.5}}{120}\right) + \exp\left(-\frac{Z_0\varepsilon_{\mathrm{eff}}^{0.5}}{120}\right) - 1 \qquad \text{for } Z_0 > \frac{60\sqrt{2}\pi}{(\varepsilon_r + 1)^{0.5}}. \tag{6.55b}$$

In the expressions above, $\varepsilon_{\mathrm{eff}}$ is given by

$$\varepsilon_{\mathrm{eff}} = \varepsilon_{\mathrm{eff}}(\varepsilon_r, h, Z_0, S) = T_1 \times \left(1 + \frac{\varepsilon_r - 1}{Z_0(\varepsilon_r + 1)^{0.5}}\, T_2\right) \tag{6.56}$$

with

$$T_1 = \operatorname{sech}\left\{\frac{\pi\varepsilon_r^5}{(\varepsilon_r + 1)^6}\left(\frac{60}{Z_0}\right)^2 \times \exp\left[\left(1 + \varepsilon_r\left(\frac{Z_0}{121}\right)^4 \frac{S}{Qh}\right)\ln\left(\frac{S}{Qh}\right)\right]\right\} \tag{6.57}$$

and

$$T_2 = 84.85\ln\left(2\,\frac{1+g}{1-g}\right) \qquad \text{for } 0.841 \leqslant g \leqslant 1, \tag{6.58a}$$

$$T_2 = \frac{837.5}{\ln\left(2\,\dfrac{1 + (1 - g^4)^{0.25}}{1 - (1 - g^4)^{0.25}}\right)} \qquad \text{for } 0 < g \leqslant 0.841, \tag{6.58b}$$

$$g = \left\{\frac{\exp\left(\dfrac{\pi(1 + 1/Q)S}{2h}\right) - \exp\left(\dfrac{\pi S}{2Qh}\right)}{\exp\left(\dfrac{\pi(1 + 2/Q)S}{2h}\right) - 1}\right\}^{0.5}, \tag{6.59}$$

$$Q = G|_{\varepsilon_{\mathrm{eff}} = \frac{\varepsilon_r + 1}{2}}. \tag{6.60}$$

6.4 NOVEL VARIANTS OF THE COPLANAR STRIPLINE

6.4.1 Micro-coplanar Stripline

The geometry of the micro-coplanar stripline (MCPS) [11] is shown in Figure 6.8. In the MCPS a very thin dielectric spacer layer separates the two strip conductors. The spacer layer and the strip conductors are supported from below by a dielectric substrate.

The MCPS has several advantages, some of which are discussed here: (1) The dimensions of the strip conductors as well as the spacer layer are on the order of few microns for a 50 Ω transmission line. Thus the transmission line is very compact, resulting in small amount of parasitics when combined with active devices. (2) The strip conductors are on two levels, thus making vertical as well as horizontal integration possible. This feature is useful in applications,

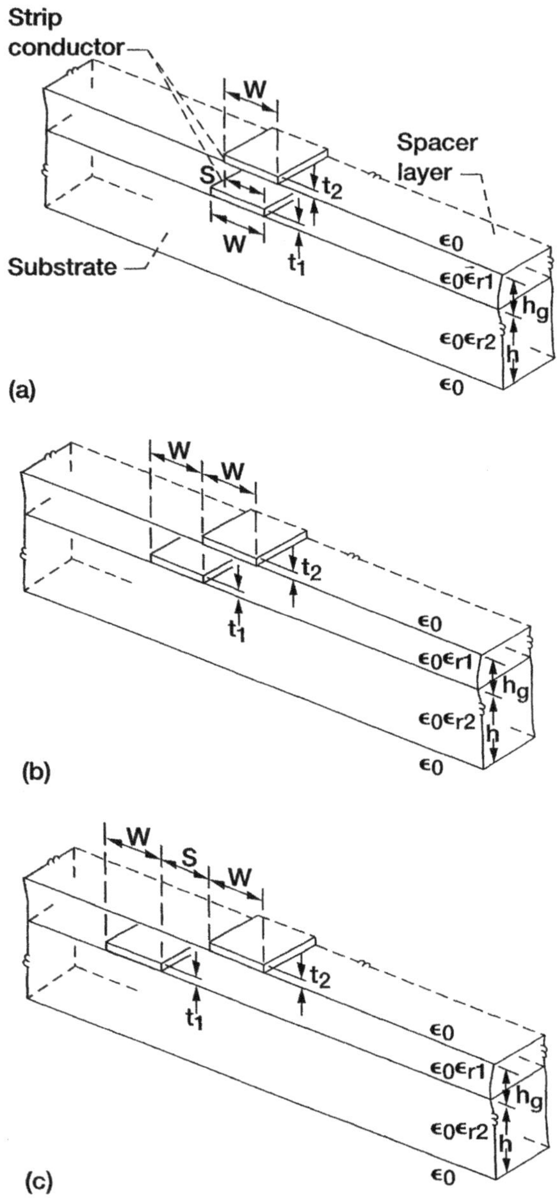

FIGURE 6.8 Micro-coplanar stripline (MCPS) configuration. (*a*) Strips overlap; (*b*) strip edges are even; (*c*) strips are separated.

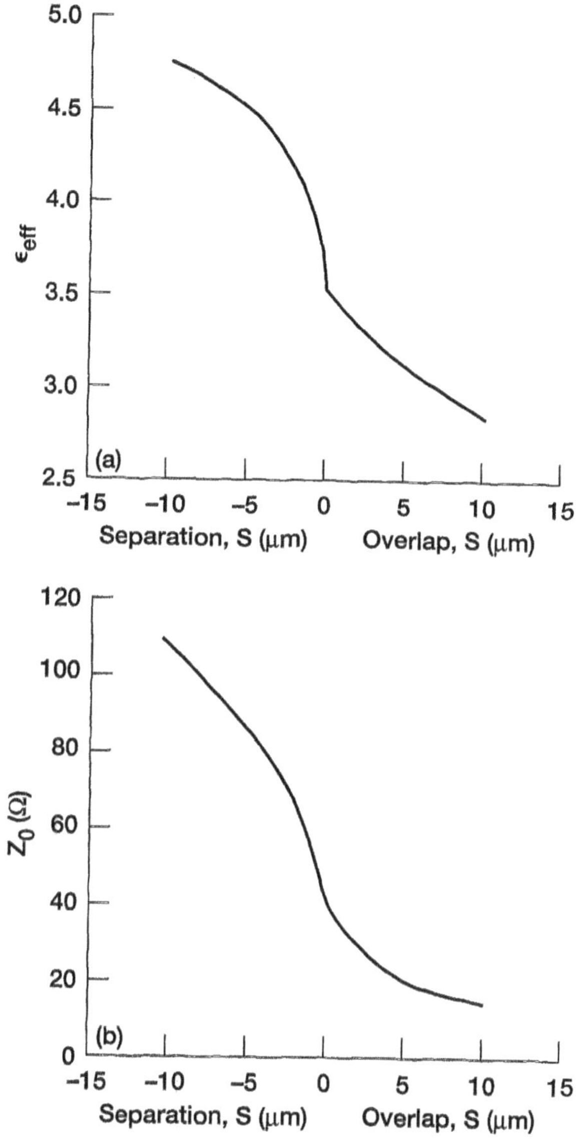

FIGURE 6.9 Micro-coplanar stripline: (*a*) Effective dielectric constant and (*b*) characteristic impedance. $h = 200\,\mu m$, $h_g = 1.35\,\mu m$, $\varepsilon_{r1} = 3.1$, $\varepsilon_{r2} = 11.7$, $W = 10\,\mu m$, $t_1 = t_2 = 0.45\,\mu m$.

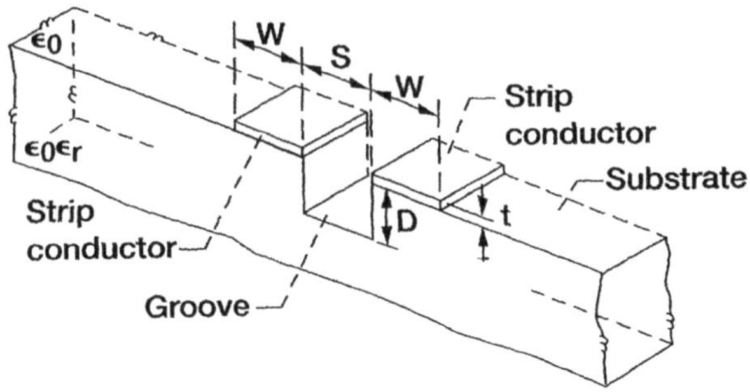

FIGURE 6.10 Coplanar stripline with a Groove.

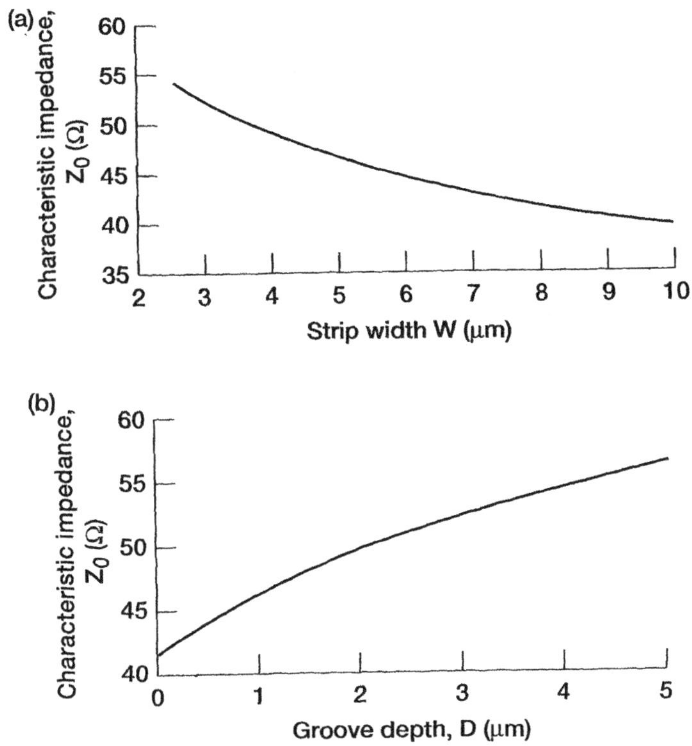

FIGURE 6.11 Computed characteristic impedance: (*a*) Conventional CPS, $\varepsilon_r = 35.2$, $S = 5\,\mu m$, and $t = 3\,\mu m$; (*b*) CPS with a groove, $\varepsilon_r = 35.2$, $S = 5\,\mu m$, $W = 8\,\mu m$, $t = 3\,\mu m$. (From Reference [17] © 1980 IEEE.)

such as mixers using broadside couplers with tight coupling and wide bandwidth. (3) The spacer layer is very thin, hence making it possible to realize large capacitances in a small area. This feature is particularly desirable in the design of compact low-pass and band-pass filters. (4) The two strip conductors form a balanced line, making the MCPS an ideal line for feeding integrated antennas, such as patches, bow ties, and dipoles.

The MCPS is modeled using the well-known FDTD technique [12], [13], [14]. The model assumes a high resistivity silicon wafer ($\varepsilon_{r2} = 11.7$) as the substrate and spin-on-glass (SOG) [15] ($\varepsilon_{r1} = 3.1$) as the spacer layer. In addition the model takes into consideration the finite thickness of the substrate, spacer layer, and the strip conductors. The computed ε_{eff} and Z_0 are presented in Figure 6.9. The figure shows that ε_{eff} and Z_0 are small when the two strip conductors fully overlap each other. As the overlap decreases and the strip conductors separate, the ε_{eff} and Z_0 increase. Thus a very wide range of impedances can be realized.

6.4.2 Coplanar Stripline with a Groove

A CPS with a groove of depth D in the dielectric substrate between the strip conductors is shown in Figure 6.10. This variant of the CPS was first investigated in connection with travelling wave electrooptic modulators on $LiNbO_3$ ($\varepsilon_r \approx 35$) substrate [16]. For a desired characteristic impedance Z_0, separation S between the two strips and strip conductor thickness t, this variant of the CPS has a much wider strip width W than a conventional CPS. This characteristic is evident when the computed Z_0 in Figure 6.11(*a*) for conventional CPS is compared with Figure 6.11(*b*) for CPS with a groove [17]. As an example, for a Z_0, S, and t equal to $50\,\Omega$, $5\,\mu$m, and $34\,\mu$m, respectively, for conventional CPS W is less than $3.75\,\mu$m while for the variant with a groove depth D of about $2\,\mu$m, W is $8\,\mu$m. Hence the CPS with a groove has potentially lower attenuation due to conductor losses. A second advantage of the groove is the lower ε_{eff}. Hence in traveling wave electrooptic modulators the velocity mismatch between the microwave signal and the optical signal is reduced, resulting in a wider bandwidth.

REFERENCES

[1] E. Chen and S. Y. Chou, "Characteristics of Coplanar Transmission Lines on Multilayer Substrates: Modeling and Experiments," *IEEE Trans. Microwave Theory Tech.*, Vol. 45, No. 6, pp. 939–945, June 1997.

[2] W. J. Getsinger, "Circuit Duals on Planar Transmission Media," *1983 IEEE MTT-S Int. Microwave Symp. Dig.*, pp. 154–156, Boston, MA, May 31–June 3, 1983.

[3] B. D. Popovic and A. Nesic, "Some Extensions of the Concept of Complementary Electromagnetic Structures," *IEE Proc.*, Vol. 132, Pt.H, No. 2, pp. 131–137, April 1985.

[4] N. H. Zhu, Z. Q, Wang, and W. Lin, "On the Accuracy of Analytical Expressions for Calculating the Parameters of Coplanar Strips on a Finitely Thick Substrate," *Microwave Optical Tech. Lett.*, Vol. 8, No. 3, pp. 160–164, Feb. 1995.

[5] G. Ghione and C. Naldi, "Analytical Formulas for Coplanar Lines in Hybrid and Monolithic MICs," *Electron. Lett.*, Vol. 20, No. 4, pp. 179–181, Feb. 1984.

[6] G. Ghione, "A CAD-Oriented Analytical Model for the Losses of General Asymmetric Coplanar Lines in Hybrid and Monolithic MICs," *IEEE Trans. Microwave Theory Tech.*, Vol. 41, No. 9, pp. 1499–1510, Sept. 1993.

[7] S. S. Bedair, "Characteristics of Some Asymmetrical Coupled Transmissions Lines," *IEEE Trans. Microwave Theory Tech.*, Vol. 32, No. 1, pp. 108–110, Jan. 1984.

[8] J. S. McLean and T. Itoh, "Analysis of a New Configuration of Coplanar Stripline," *IEEE Trans. Microwave Theory Tech.*, Vol. 40, No. 4, pp. 772–774, April 1992.

[9] T. Q. Deng, M. S. Leong, P. S. Kooi, and T. S. Yeo, "Synthesis Formulas for Coplanar Lines in Hybrid and Monolithic MICs," *Electron. Lett.*, Vol. 32, No. 24, pp. 2253–2254, Nov. 1996.

[10] J. B. Knorr and K.-D. Kuchler, "Analysis of Coupled Slots and Coplanar Strips on Dielectric Substrate," *IEEE Trans. Microwave Theory Tech.*, Vol. 23, No. 7, pp. 541–548, July 1975.

[11] K. Goverdhanam, R. N. Simons, and L. P. B. Katehi, "Micro-Coplanar Striplines — New Transmission Media for Microwave Applications," *1998 IEEE MTT-S Int. Microwave Symp. Dig.*, Vol. 2, pp. 1035–1038, Baltimore, MD, June 7–12, 1998.

[12] K. Goverdhanam, R. N. Simons, and L. P. B. Katehi, "Coplanar Stripline Propagation Characteristics and Bandpass Filter," *IEEE Microwave Guided Wave Lett.*, Vol. 7, No. 8, pp. 214–216, Aug. 1997.

[13] K. Goverdhanam, R. N. Simons, and L. P. B. Katehi, "Coplanar Stripline Components for High-Frequency Applications," *IEEE Trans. Microwave Theory Tech.*, Vol. 45, No. 10, pp. 1725–1729, Oct. 1997.

[14] R. N. Simons, N. I. Dib, and L. P. B. Katehi, "Modeling of Coplanar Stripline Discontinuities," *IEEE Trans. Microwave Theory Tech.*, Vol. 44, No. 5, pp. 711–716, May 1996.

[15] Accuglass 512 Spin-on-Glass (SOG), Product Bulletin, Allied-Signal Inc., Planarization and Diffusion Products, Milpitas, CA 95035.

[16] H. Haga, M. Izutsu, and T. Sueta, "$LiNbO_3$ Travelling-Wave Light Modulator/Switch with an Etched Groove," *IEEE J. Quantum Electron.*, Vol. 22, No. 6, pp. 902–906, June 1986.

[17] J. C. Yi, S. H. Kim, and S. S. Choi, "Finite-Element Method for the Impedance Analysis of Travelling-Wave Modulators," *IEEE J. Lightwave Tech.*, Vol. 8, No. 6, pp. 817–822, June 1980.

CHAPTER 7

Microshield Lines and Coupled Coplanar Waveguide

7.1 INTRODUCTION

This chapter is broadly divided into two types of transmission lines, microshield lines and coupled coplanar waveguides. Section 7.2 presents analytical equations to compute the characteristic impedance of microshield lines of various shapes. Section 7.3 presents analytical equations to compute the even- and odd-mode characteristics of edge coupled coplanar waveguides without a lower ground plane. Section 7.4 is devoted to conductor-backed edge coupled coplanar waveguides. Last, Section 7.5 presents the broadside coupled coplanar waveguides.

7.2 MICROSHIELD LINES

A microshield line [1] consists of a center strip conductor with two adjacent coplanar ground planes supported on a dielectric membrane and backed by a metalized shielding cavity. Figure 7.1(*a*) gives an example of a rectangular microshield line. In some applications an additional shielding metal cover may be present on the top side. Figure 7.1(*b*) shows a technique for fabricating a microshield line by attaching together two silicon wafers. The first wafer supports a membrane with a coplanar waveguide on the top surface. This wafer is anisotropically etched from below to form a cavity and the walls of the cavity are metalized. Since the membrane is very thin, the overlap capacitances between the top ground planes and cavity side walls are very large. Hence at microwave frequencies the overlap region behaves as a virtual short circuit to the propagating mode. The second wafer is metalized on the top surface and form the bottom wall of the cavity. This structure is therefore considered as an evolution of the microstrip line and the channelized coplanar waveguide discussed in Section 3.5, Chapter 3.

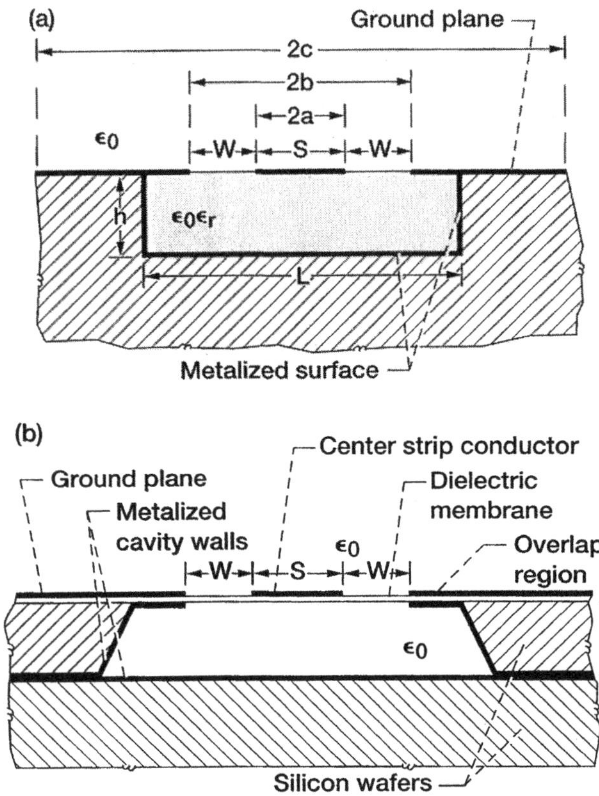

FIGURE 7.1 Rectangular microshield line: (*a*) Analytical model; (*b*) practical realization.

In the microshield line the effect of the membrane on the electromagnetic wave propagation is negligible, and the wave number k is approximately the same in the regions above and below the center strip. The propagation constant γ of the microshield line therefore satisfies the condition

$$\gamma^2 + k^2 = 0 \tag{7.1}$$

for a true TEM-wave [2]. This suggest that the wave propagation on the microshield line can be modeled as a pure TEM-mode. The dimensions of the rectangular cavity in the example above are such that no higher-order modes are excited at the desired operating frequency.

In the sections that follow an expression for the total capacitance of microshield lines with lower cavity of different shapes are presented. The total capacitance per unit length of the microshield line is the sum of the capacitance of the upper air region and the lower shielding cavity region. These capacitances are determined by a sequence of conformal mappings. In the conformal mapping technique all the metallic conductors of the structure are assumed to have negligible thickness and perfect conductivity. The width of the ground planes on either sides of the center strip conductor are considered to be infinite. The air-dielectric interface in the slot regions on either sides of the center strip conductor is assumed to behave as a perfect magnetic wall. Knowing the capacitance of the transmission line, the propagation parameters, such as effective dielectric constant ε_{eff} and characteristic impedance Z_0, are determined. Microshield lines with a rectangular shape, a V-shape, an elliptic shape, and a circular shape shielding cavities are considered in the subsequent sections.

7.2.1 Rectangular Shaped Microshield Line

A microshield line with a center strip conductor of width $2a$ symmetrically located between two finite ground planes separated by a distance $2b$ and backed by a rectangular shaped metalized cavity is shown in Figure 7.1(a). The width and depth of the cavity at any transverse section is L and h, respectively. The cavity is assumed to be filled with a dielectric of relative permittivity ε_r. To determine the ε_{eff} and Z_0 of the line, the capacitance of the structure is obtained using the Schwartz-Christoffel transformation. The total capacitance C_t per unit length is the sum of the capacitance C_a of the upper half-plane consisting of the air region and the capacitance C_{CR} of the lower half-plane consisting of the rectangular cavity. That is,

$$C_t = C_a + C_{\text{CR}}. \tag{7.2}$$

The capacitance C_a is determined in a manner similar to that in Section 4.2.1 and is one-half of that given by (4.5). Hence

$$C_a = 2\varepsilon_0 \frac{K(k_1')}{K(k_1)}, \tag{7.3}$$

where $K(k_1)$ and $K(k_1')$ are the complete elliptic integrals of the first kind and the arguments k and k' are dependent on the geometry of the line and are given by

$$k_1 = \frac{c}{b}\sqrt{\frac{b^2 - a^2}{c^2 - a^2}}, \tag{7.4}$$

$$k_1' = \sqrt{1 - k^2} = \frac{a}{b}\sqrt{\frac{c^2 - b^2}{c^2 - a^2}}, \tag{7.5}$$

where the widths a, b, and c are as indicated in Figure 7.1(a). If the ground planes extend to infinity on either sides and in the presence of a top metal cover at a height h_1, the capacitance C_a is determined in a manner similar to that in Section 2.2.1 and is given by

$$C_a = 2\varepsilon_0 \frac{K(k_1)}{K(k_1')}, \tag{7.6}$$

where

$$k_1 = \frac{\tanh(\pi S/4h_1)}{\tanh[\pi(S+2W)/4h_1]}, \tag{7.7}$$

$$k_1' = \sqrt{1-k_1^2}. \tag{7.8}$$

The capacitance C_{CR} is given by [3], [4] as

$$C_{CR} = 2\varepsilon_0\varepsilon_r \frac{K(k_2)}{K'(k_2)} = 2\varepsilon_0\varepsilon_r \frac{K(t_a/t_b)}{K'(t_a/t_b)} \tag{7.9}$$

where t_a, t_b, and t_c are related to the geometry of the structure through the following set of equations:

$$\frac{F(\arcsin(t_a/t_c), t_c)}{K(t_c)} = \frac{2a}{L} \tag{7.10}$$

$$\frac{F(\arcsin(t_b/t_c), t_c)}{K(t_c)} = \frac{2b}{L}, \tag{7.11}$$

$$t_c = \sqrt{1 - \left(\frac{\{\exp(2h\pi/L)\} - 2}{\{\exp(2h\pi/L)\} + 2}\right)^4} \quad \text{for } 0 < \frac{L}{2h} < 1, \tag{7.12}$$

$$t_c = \left(\frac{\{\exp(L\pi/2h)\} - 2}{\{\exp(L\pi/2h)\} + 2}\right)^2 \quad \text{for } 1 < \frac{L}{2h} < \infty. \tag{7.13}$$

In the equations above $K(t_c)$ is the complete elliptic integral of the first kind, and $F(\arcsin(t_a/t_c), t_c)$ and $F(\arcsin(t_b/t_c), t_c)$ are the incomplete elliptic integrals of the first kind written in the Jacobi's notation. The functions $F(\arcsin(t_a/t_c), t_c)$ and $F(\arcsin(t_b/t_c), t_c)$ are evaluated for the values of t_a/t_c and t_b/t_c using the analytical formulas [4] presented below. While evaluating t_a/t_c, the subscript x as well as the factor x in the equations below are set equal to a. In a similar manner t_b/t_c is evaluated with $x = b$.

For $0 \leqslant 2x/L \leqslant \delta_1 = [1/S(t_c)] \arcsin(t_c\zeta)$,

$$\frac{t_x}{t_c} = \sin\left\{\frac{1}{t_c}\sin\left[\frac{2x}{L}S(t_c)\right]\right\}. \tag{7.14}$$

For $\delta_1 \leqslant 2x/L \leqslant \delta_2 = 1 - [1/S(t_c)] \operatorname{arc\,sinh}(t_c\zeta/\sqrt{1-t_c^2})$,

$$\frac{t_x}{t_c} = \sin\left\{\frac{\pi}{4} + \frac{1}{2\zeta}\left(\frac{1-\psi^2}{t_c^2} - \frac{1}{2}\right)\right\}, \tag{7.15}$$

$$\psi = \sqrt{1 - \frac{t_c^2}{2}(1-\zeta\xi) - t_c\zeta S(t_c)\left\{\frac{2x}{L} - \delta_1\right\}}. \tag{7.16}$$

For $\delta_2 \leqslant 2x/L \leqslant 1$,

$$\frac{t_x}{t_c} = \cos\left\{\frac{\sqrt{1-t_c^2}}{t_c}\sinh\left[\left(1 - \frac{2x}{L}\right)S(t_c)\right]\right\}, \tag{7.17}$$

where

$$\zeta = \frac{\pi}{4} - \frac{\xi}{4}, \tag{7.18}$$

$$\xi = \sqrt{\pi^2 - 8}, \tag{7.19}$$

$$S(t_c) = \arcsin(t_c\zeta) + \operatorname{arc\,sinh}(t_c\zeta/\sqrt{1-t_c^2}) + \frac{1}{\zeta t_c}\left\{\sqrt{1 - \frac{t_c^2}{2}(1-\zeta\xi)} - \sqrt{1 - \frac{t_c^2}{2}(1+\zeta\xi)}\right\}. \tag{7.20}$$

The effective dielectric constant ε_{eff} and the characteristic impedance Z_0 of the line, as defined in Section 2.2.1, are given by

$$\varepsilon_{\text{eff}} = \frac{C_t}{C_a + C_{CR}\ (\text{with } \varepsilon_r = 1)}, \tag{7.21}$$

$$Z_0 = \frac{1}{c'[C_a + C_{CR}\ (\text{with } \varepsilon_r = 1)]\sqrt{\varepsilon_{\text{eff}}}}, \tag{7.22}$$

where c' is the speed of light in free space. For a membrane supported line, the $\varepsilon_{\text{eff}} \approx 1.0$, and hence

$$Z_0 = \frac{1}{c'[C_a + C_{CR}\ (\text{with } \varepsilon_r = 1)]}. \tag{7.23}$$

Figure 7.2(*a*) shows the computed Z_0 as a function of the slot width W with the center strip width S as a parameter [3]. Figure 7.2(*b*) shows the computed Z_0 as a function of the normalized ground plane width c/b with the strip width S as a parameter [3]. From this figure it is inferred that the finite extent ground planes have negligible effect on Z_0 as long as $c/b > 2$.

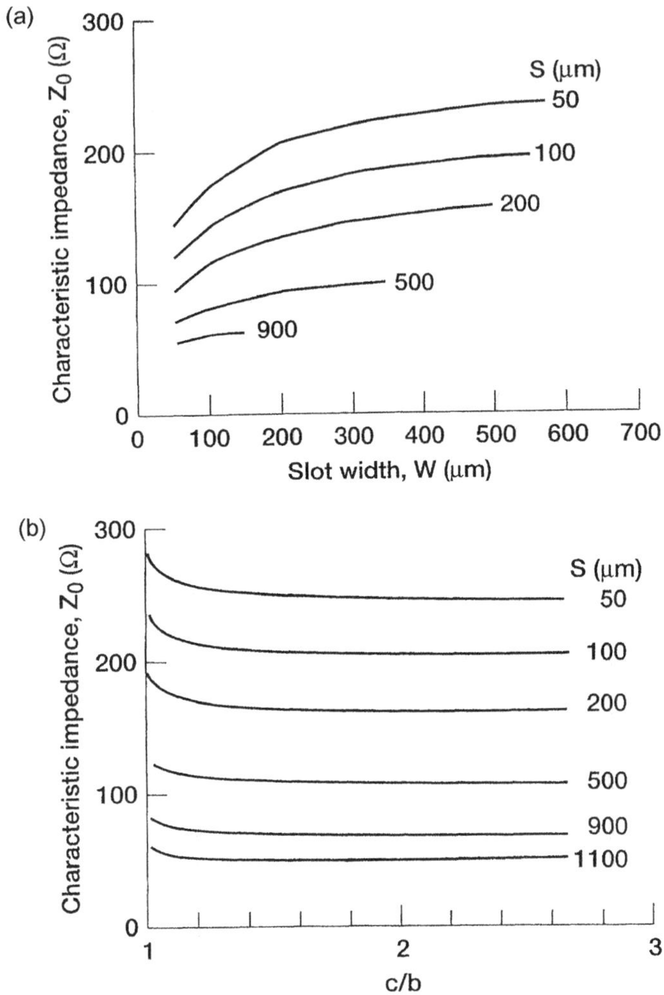

FIGURE 7.2 Computed characteristic impedance of a microshield line: (*a*) As a function of the slot width, $L = 1200\ \mu\text{m}$, $h = 400\ \mu\text{m}$, $c = 1800\ \mu\text{m}$; (*b*) as a function of the normalized ground plane width, $L = 1200\ \mu\text{m}$, $h = 400\ \mu\text{m}$, $b = 600\ \mu\text{m}$. (From Reference [3], © IEEE 1992.)

7.2.2 V-Shaped Microshield Line

An asymmetric microshield line with a center strip conductor of width S separated from two adjacent ground planes by a distance W_1 and W_2, respectively, and backed by a V-shaped metalized cavity is shown in Figure 7.3. At any transverse section along the line, the strip conductor, the ground planes and the V-shaped cavity walls form an inverted isosceles triangle with a base of width L and an apex angle of 2θ. The cavity is assumed to be filled with a

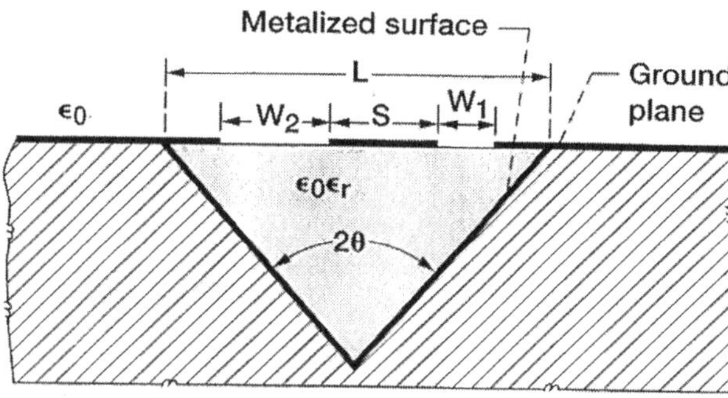

FIGURE 7.3 V-shaped microshield line.

dielectric of relative permittivity ε_r. To determine the ε_{eff} and Z_0 of the line, the capacitance of the structure is first obtained using the Schwartz-Christoffel transformation. The total capacitance C_t per unit length is the sum of the capacitance C_a of the upper half-plane consisting of the air region and the capacitance C_{CV} of the lower half-plane consisting of the V-shaped cavity. That is,

$$C_t = C_a + C_{CV}. \tag{7.24}$$

The capacitance C_a is given by [5]

$$C_a = \varepsilon_0 \frac{K(k_1)}{K'(k_1)}, \tag{7.25}$$

where $K(k)$ is the complete elliptic integral of the first kind and the arguments k and k' are dependent on the geometry of the line. They are given by [5]

$$k_1 = \sqrt{\frac{S(S + W_1 + W_2)}{(S + W_1)(S + W_2)}}, \tag{7.26}$$

$$k_1' = \sqrt{1 - k_1^2}. \tag{7.27}$$

The capacitance C_{CV} is given by [5]

$$C_{CV} = \varepsilon_0 \varepsilon_r \frac{K(k_2)}{K'(k_2)}, \tag{7.28}$$

where

$$k_2 = \sqrt{\frac{2t_E(t_D - t_C)}{(t_E + t_D)(t_E - t_C)}}. \tag{7.29}$$

The function t_x (x equals to C, D, or E) depends on the value of Z_x/Z_B (x equals to C, D, or E) and is evaluated using the equations given below [5]:

$$\frac{Z_E}{Z_B} = \frac{W_2 + W_1 + S}{L}, \tag{7.30}$$

$$\frac{Z_C}{Z_B} = \frac{W_2 - W_1 - S}{L}, \tag{7.31}$$

$$\frac{Z_D}{Z_B} = \frac{W_2 - W_1 + S}{L}, \tag{7.32}$$

$$\psi = 1 + \frac{\pi}{4} - \sqrt{2} + \frac{\sqrt{2}}{1 - (2\theta/\pi)}\left[1 - \left(1 - \frac{1}{\sqrt{2}}\right)\left(\frac{1 - (\pi/4)}{\sqrt{2} - 1}\right)^{1-(2\theta/\pi)}\right]. \tag{7.33}$$

For $0 \leqslant Z_x/Z_B \leqslant [1 + (\pi/4) - \sqrt{2}]/\psi$,

$$t_x = \sin\left(\psi \frac{Z_x}{Z_B}\right). \tag{7.34}$$

For

$$\frac{1 + (\pi/4) - \sqrt{2}}{\psi} \leqslant \frac{Z_x}{Z_B} \leqslant 1 - \frac{1}{\psi[1 - (2\theta/\pi)]}\left[\frac{1 - (\pi/4)}{\sqrt{2} - 1}\right]^{1-(2\theta/\pi)}$$

$$t_x = \sin\left\{1 + \frac{\pi}{4} - \sqrt{2}\left[1 - \frac{1 - (2\theta/\pi)}{\sqrt{2}}\left(\psi\frac{Z_x}{Z_B} - 1 - \frac{\pi}{4} + \sqrt{2}\right)\right]^{1/[1-(2\theta/\pi)]}\right\} \tag{7.35}$$

For $1 - \{1/\psi[1 - (2\theta/\pi)]\}\{[1 - (\pi/4)]/(\sqrt{2} - 1)\}^{1-(2\theta/\pi)} \leqslant Z_x/Z_B \leqslant 1$,

$$t_x = \cos\left\{\left[\psi\left(1 - \left(\frac{2\theta}{\pi}\right)\right)\left(1 - \frac{Z_x}{Z_B}\right)\right]^{1/[1-(2\theta/\pi)]}\right\}. \tag{7.36}$$

The ε_{eff} and Z_0 of the line are given by [5]

$$\varepsilon_{\text{eff}} = \frac{C_t}{C_a + C_{CV}\ (\text{with } \varepsilon_r = 1)} \tag{7.37}$$

$$= 1 + q(\varepsilon_r - 1), \tag{7.38}$$

where the filling factor q is [5]

$$q = \frac{\dfrac{K(k_2)}{K'(k_2)}}{\dfrac{K(k_1)}{K'(k_1)} + \dfrac{K(k_2)}{K'(k_2)}} \tag{7.39}$$

and

$$Z_0 = \frac{1}{c'[C_a + C_{CV}\,(\text{with } \varepsilon_r = 1)]\sqrt{\varepsilon_{eff}}} \tag{7.40}$$

and where c' is the velocity of light in free space. After substituting for C_a and C_{CV} and simplifying [5],

$$Z_0 = \frac{120\pi}{\sqrt{\varepsilon_{eff}}} \frac{1}{\dfrac{K(k_1)}{K'(k_1)} + \dfrac{K(k_2)}{K'(k_2)}}. \tag{7.41}$$

The computed ε_{eff} and Z_0 [5] are presented in Figure 7.4. The analysis and computed characteristics of coupled coplanar waveguides with V-shaped microshields can be found in [6] and [7].

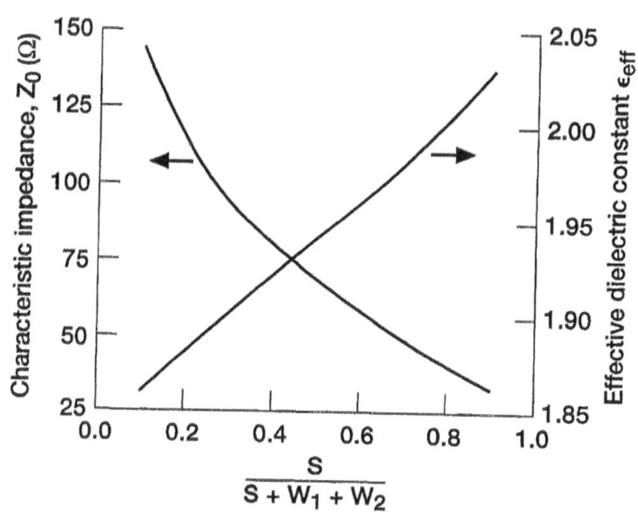

FIGURE 7.4 Computed characteristic impedance and effective dielectric constant of a V-shaped microshield line, $W_1 = W_2$, $\theta = 30°$, $\varepsilon_r = 2.55$, $L = S + W_1 + W_2$. (From Reference [5], © IEEE 1995.)

7.2.3 Elliptic Shaped Microshield Line

An elliptic shaped microshield line is shown in Figure 7.5. The capacitance C_a of the upper half-plane for a structure with finite-size ground planes and in the absence of a top metal cover is given by Eq. (7.3). When a top metal cover is present and the ground planes are infinite, C_a is given by Eq. (7.6). The capacitance of the lower half-plane consisting of the elliptic shaped cavity filled with a dielectric of relative permittivity ε_r is [8]

$$C_{CE} = \pi\varepsilon_0\varepsilon_r \left[\ln\frac{b + h_2}{a} + \ln(1 - B') \right]^{-1}, \tag{7.42}$$

where

$$B' = \frac{(\sqrt{b^2 - h_2^2} - a)^2(b - h_2)}{2h_2 a}. \tag{7.43}$$

The total capacitance C_t per unit length is

$$C_t = C_a + C_{CE}. \tag{7.44}$$

The ε_{eff} and Z_0 of the line are given by Eqs. (7.21) and (7.22) with C_{CR} replaced by C_{CE}. An example of the computed Z_0 [8] as a function of the slot width for a particular geometry is shown in Figure 7.6.

7.2.4 Circular Shaped Microshield Line

A microshield line with a circular shaped cavity is shown in Figure 7.7. In a manner analogous to the elliptic shaped microshield line the capacitance C_a is determined from Eq. (7.3) or Eq. (7.6). By letting $b = h_2$ in Eq. (7.42), the

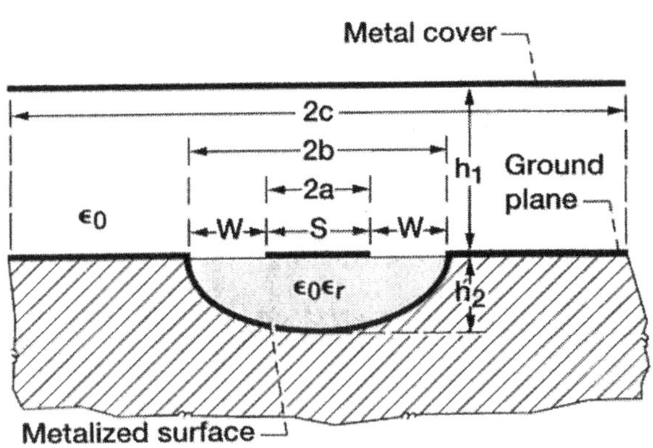

FIGURE 7.5 Elliptic-shaped microshield line.

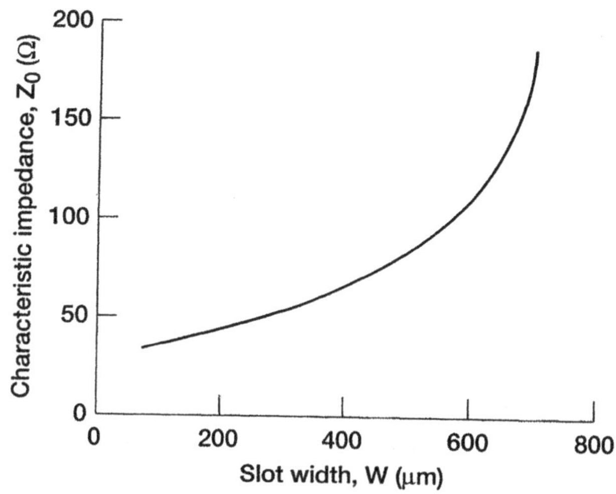

FIGURE 7.6 Computed Z_0 as a function of the slot width, $\varepsilon_r = 2.55$, $h_1 = 3.0\,\text{mm}$, $h_2 = 400\,\mu\text{m}$, $b = 692.82\,\mu\text{m}$. (From Reference [8], © IEEE 1994.)

capacitance C_{CC} of the lower half-plane is determined and is given by [8]

$$C_{CC} = \frac{\pi\varepsilon_0\varepsilon_r}{\ln(2b/a)}. \tag{7.45}$$

The total capacitance C_t per unit length is

$$C_t = C_a + C_{CC}. \tag{7.46}$$

The ε_{eff} and Z_0 of the line are given by Eqs. (7.21) and (7.22) with C_{CR} replaced

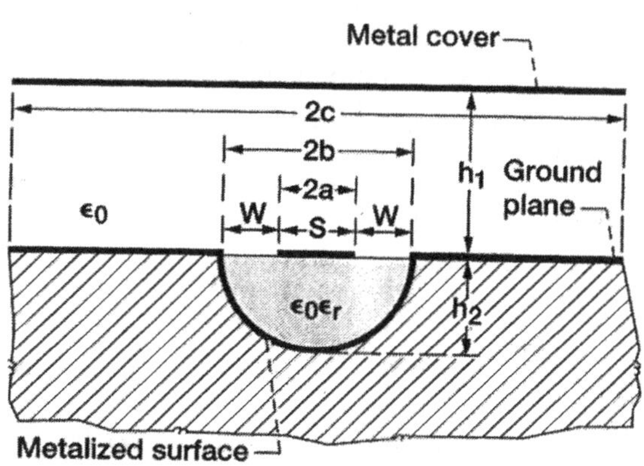

FIGURE 7.7 Circular shaped microshield line.

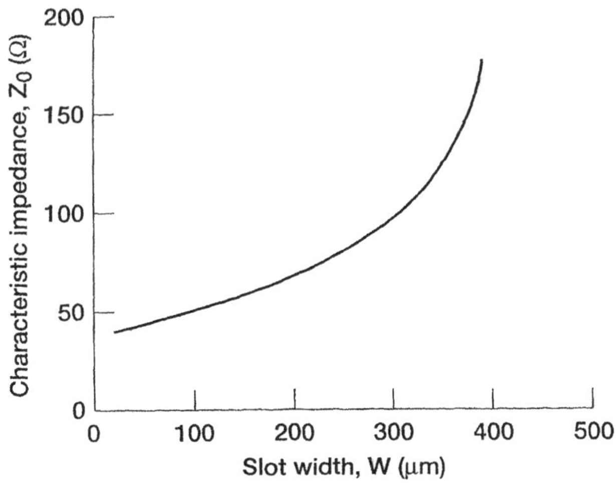

FIGURE 7.8 Computed Z_0 as a function of the slot width, $\varepsilon_r = 2.55$, $h_1 = 3.0\,\text{mm}$, $h_2 = 400\,\mu\text{m}$, $b = 400\,\mu\text{m}$. (From Reference [8], © IEEE 1994.)

by C_{CC}. An example of the computed Z_0 [8] as a function of the slot width for a particular geometry is shown in Figure 7.8.

7.3 EDGE COUPLED COPLANAR WAVEGUIDE WITHOUT A LOWER GROUND PLANE

When two transmission lines are placed in close proximity, there is a strong interaction between their fields and power is coupled from one line to the other. In general, the amount of coupling is dependent on the distance of separation between the lines and the interaction length. An edge coupled coplanar waveguide with two parallel coupled strip conductors symmetrically located between two ground planes is shown in Figure 7.9. This structure can support two modes of propagation, the even mode and the odd mode. These modes are illustrated in Figure 7.10(*a*) and (*b*). In general, any arbitrary excitation of the coupled lines in Figure 7.9 can be considered as a superposition of two appropriate amplitudes of even mode and odd mode. In the succeeding sections expressions for the effective dielectric constant as well as the characteristic impedance for the even mode and the odd mode obtained using the quasi-static conformal mapping technique will be presented.

7.3.1 Even Mode

For the even excitation a magnetic wall is placed along the plane of symmetry in Figure 7.10(*a*). It then suffices to restrict the analysis to the right half of the

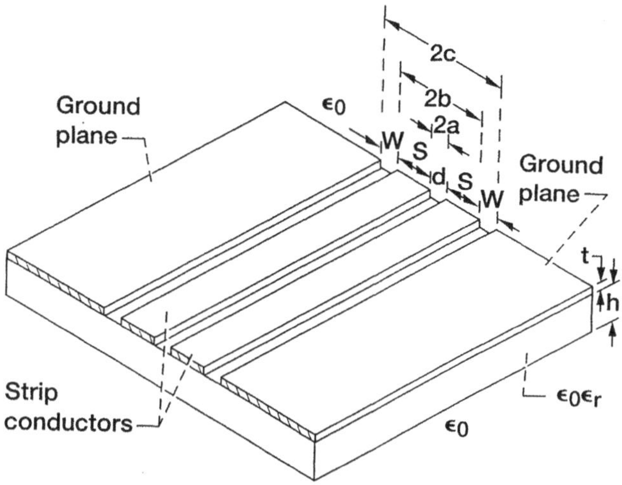

FIGURE 7.9 Edge coupled coplanar waveguides.

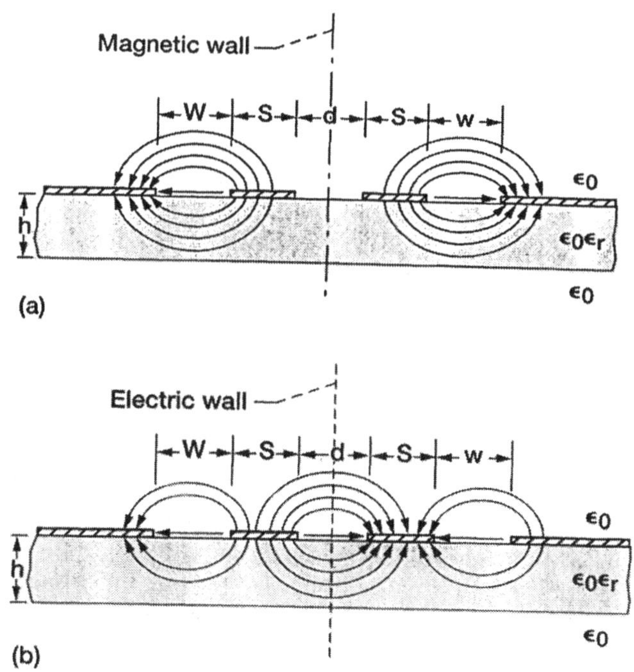

FIGURE 7.10 Excitation of coupled coplanar waveguides: (*a*) Even mode; (*b*) odd mode.

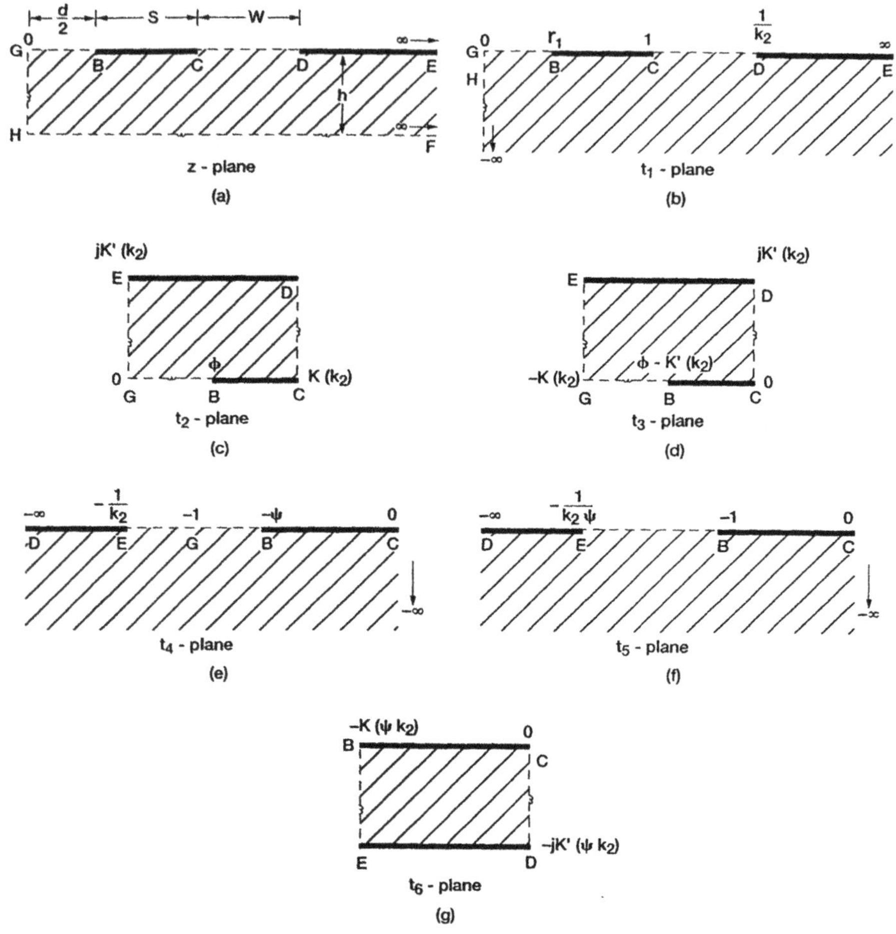

FIGURE 7.11 Conformal mapping transformation for the calculation of capacitance for even mode. The mapping functions are (*a*) *z*-plane; (*b*) t_1-plane, $t_1 = \sinh(\pi z/2h)/\sinh\{\pi[(d/2) + S]/2h\}$; (c) t_2-plane, $t_1 = \mathrm{sn}(t_2/k_2)$; (*d*) t_3-plane, $t_3 = t_2 - K(k_2)$; (*e*) t_4-plane, $t_4 = \mathrm{sn}(t_3/k_2)$; (*f*) t_5-plane, $t_5 = t_4/\psi$; (*g*) t_6-plane, $t_5 = \mathrm{sn}(t_6/k_2\psi)$. (From Reference [9], © Ann. Telecommun. 1984.)

structure. Through a sequence of conformal mapping steps [9], the right half of the structure is mapped into an ideal parallel plane structure for which the capacitance per unit length is easily obtained. The mapping functions and the transformed configurations are presented in Figure 7.11(*a*) to (*g*). In this figure, the mapping function

$$t = \mathrm{sn}(\zeta, k) \tag{7.47}$$

is an elliptic sine function of a complex variable ζ to modulus k. The following elliptic sine function identities are also helpful in the transformation process:

$$\mathrm{sn}(0, k) = 0 \tag{7.48}$$

$$\mathrm{sn}(K(k), k) = 1, \tag{7.49}$$

$$\mathrm{sn}(K(k) + jK'(k), k) = \frac{1}{k} \tag{7.50a}$$

$$\mathrm{sn}(-K(k) + jK'(k), k) = -\frac{1}{k} \tag{7.50b}$$

$$\mathrm{sn}(jK'(k), k) = \pm\infty, \tag{7.51}$$

where $K(k)$ and $K'(k)$ are the complete elliptic integrals of the first kind and k' is the complementary modulus. These integrals and modulus are related as follows:

$$K'(k) = K(k'), \tag{7.52}$$

$$k' = \sqrt{1 - k^2}. \tag{7.53}$$

The total even-mode capacitance $C_{t,e}$ per unit length is the sum of the partial capacitances $C_{a,e}$ and $C_{d,e}$, as explained in Section 2.2.1. That is,

$$C_{t,e} = C_{a,e} + C_{d,e}, \tag{7.54}$$

where $C_{a,e}$ is the partial capacitance in the absence of the dielectric layer and $C_{d,e}$ is the partial capacitance assuming that the electric field exists only in the dielectric layer of thickness h and relative dielectric constant $\varepsilon_r - 1$. The partial capacitance $C_{a,e}$ is given by [9]

$$C_{a,e} = 2\varepsilon_0 \frac{K(\delta k_1)}{K'(\delta k_1)}, \tag{7.55}$$

where

$$\delta = \left\{ \frac{(1 - r^2)}{(1 - k_1^2 r^2)} \right\}^{1/2}, \tag{7.56}$$

$$r = \frac{d}{d + 2S}, \tag{7.57}$$

$$k_1 = \frac{d + 2S}{d + 2S + 2W}, \tag{7.58}$$

and $K(\delta k_1)$ and $K'(\delta k_1)$ are the complete elliptic integrals of the first kind with modulus δk_1. The partial capacitance $C_{\mathrm{d,e}}$ is given by [9]

$$C_{\mathrm{d,e}} = \varepsilon_0(\varepsilon_r - 1)\frac{K(\psi k_2)}{K'(\psi k_2)}, \tag{7.59}$$

where

$$\psi = \left\{\frac{(1 - r_1^2)}{(1 - k_2^2 r_1^2)}\right\}^{1/2}, \tag{7.60}$$

$$r_1 = \frac{\sinh(\pi d/4h)}{\sinh\{(\pi/2h)[(d/2) + S]\}}, \tag{7.61}$$

$$k_2 = \frac{\sinh\{(\pi/2h)[(d/2) + S]\}}{\sinh\{(\pi/2h)[(d/2) + S + W]\}}, \tag{7.62}$$

and $K(\psi k_2)$ and $K'(\psi k_2)$ are the complete elliptic integrals of the first kind with modulus ψk_2. The even-mode effective dielectric constant $\varepsilon_{\mathrm{eff,e}}$ based on the definition in Section 2.2.1 is

$$\varepsilon_{\mathrm{eff,e}} = \frac{C_{\mathrm{t,e}}}{C_{\mathrm{a,e}}}, \tag{7.63}$$

$$= 1 + \frac{1}{2}(\varepsilon_r - 1)\frac{K(\psi k_2)}{K'(\psi k_2)}\frac{K'(\delta k_1)}{K(\delta k_1)}. \tag{7.64}$$

The even-mode phase velocity $v_{\mathrm{ph,e}}$ and characteristic impedance $Z_{0,\mathrm{e}}$ based on the definition in Section 2.2.1 are

$$v_{\mathrm{ph,e}} = c'(\varepsilon_{\mathrm{eff,e}})^{-1/2} \tag{7.65}$$

$$Z_{0,\mathrm{e}} = \frac{1}{c' C_{\mathrm{a,e}}\sqrt{\varepsilon_{\mathrm{eff,e}}}} \tag{7.66}$$

$$= \frac{60\pi}{\sqrt{\varepsilon_{\mathrm{eff,e}}}}\frac{K'(\delta k_1)}{K(\delta k_1)}, \tag{7.67}$$

where c' is the velocity of light in free space.

7.3.2 Odd Mode

For the odd excitement an electric wall is placed along the plane of symmetry in Figure 7.10(*b*). It then suffices to restrict the analysis to the right half of the structure. Through a sequence of conformal mapping steps [9], the right half

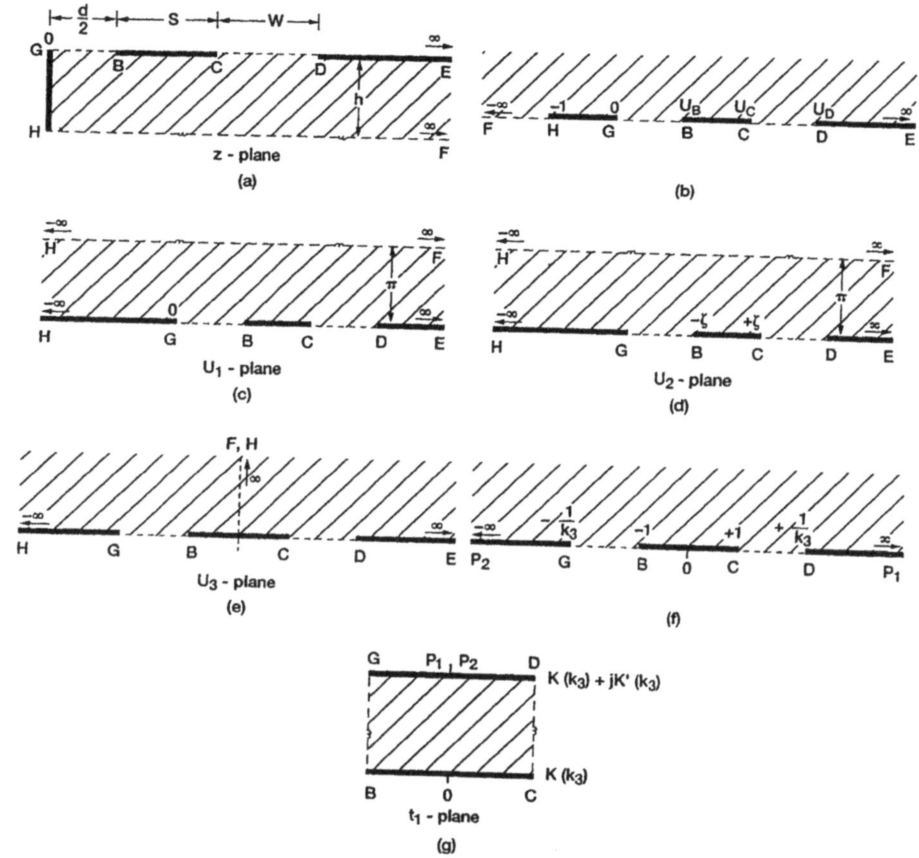

FIGURE 7.12 Conformal mapping transformation for the calculation of capacitance for odd mode. The mapping functions are (*a*) *z*-plane; (*b*) $U = \sinh^2(\pi z/2h)$; (*c*) U_1-plane, $U_1 = \log(1 + U)$; (*d*) U_2-plane, $U_2 = U_1 - (U_{1C} + U_{1B})/2$ and $\zeta = (U_{1C} - U_{1B})/2$; (*e*) U_3-plane, $U_3 = \sinh(U_2/2)$; (*f*) $t = f(U_3)$; (*g*) t_1-plane, $t = \mathrm{sn}(t_1/k_3)$. (From Reference [9], © Ann. Telecommun. 1984.)

of the structure is mapped into an ideal parallel-plate structure for which the capacitance per unit length is easily obtained as in the even-mode case. The mapping functions and the transformed configurations are presented in Figure 7.12(*a*) to (*g*).

The total odd-mode capacitance $C_{t,o}$ per unit length is the sum of the partial capacitances $C_{a,o}$ and $C_{d,o}$. That is,

$$C_{t,o} = C_{a,o} + C_{d,o}, \tag{7.68}$$

where $C_{a,o}$ is the partial capacitance in the absence of the dielectric layer and $C_{d,o}$ is the partial capacitance, assuming that the electric field exists only in the dielectric layer of thickness h and relative dielectric constant $\varepsilon_r - 1$. The partial capacitance $C_{a,o}$ is given by [9]

$$C_{a,o} = 2\varepsilon_0 \frac{K(\delta)}{K'(\delta)}, \tag{7.69}$$

where δ is given by Eq. (7.56) and $K(\delta)$ and $K'(\delta)$ are the complete elliptic integrals of the first kind with modulus δ. The capacitance $C_{d,o}$ is given by [9]

$$C_{d,o} = 2\varepsilon_0(\varepsilon_r - 1)\frac{K(k_3)}{K'(k_3)}, \tag{7.70}$$

where

$$k_3 = \frac{C_{11}(1 + \kappa C_{12})}{(C_{12} + \kappa C_{11}^2)}, \tag{7.71}$$

$$C_{11} = \frac{1}{2}\left[\left\{\frac{1 + C_{13}}{1 + C_{14}}\right\}^{1/4} - \left\{\frac{1 + C_{14}}{1 + C_{13}}\right\}^{1/4}\right], \tag{7.72}$$

$$C_{13} = \sinh^2\left[\frac{\pi}{2h}\left(\frac{d}{2} + S\right)\right], \tag{7.73}$$

$$C_{14} = \sinh^2\left(\frac{\pi d}{4h}\right), \tag{7.74}$$

$$C_{12} = \frac{1}{2}\left[\frac{(1 + C_{15})^{1/2}}{[(1 + C_{13})(1 + C_{14})]^{1/4}} - \frac{[(1 + C_{13})(1 + C_{14})]^{1/4}}{(1 + C_{15})^{1/2}}\right], \tag{7.75}$$

$$C_{15} = \sinh^2\left[\frac{\pi}{2h}\left(\frac{d}{2} + S + W\right)\right], \tag{7.76}$$

$$\kappa = \frac{1}{(C_{12} - \chi)}\left[-1 - \frac{C_{12}\chi}{C_{11}^2} - \left\{\left(\frac{C_{12}^2}{C_{11}^2} - 1\right)\left(\frac{\chi^2}{C_{11}^2} - 1\right)\right\}^{1/2}\right], \tag{7.77}$$

$$\chi = -\frac{1}{2}\left[\{(1 + C_{13})(1 + C_{14})\}^{1/4} - \{(1 + C_{13})(1 + C_{14})\}^{-1/4}\right]. \tag{7.78}$$

The odd-mode effective dielectric constant $\varepsilon_{\text{eff,o}}$ is

$$\varepsilon_{\text{eff,o}} = \frac{C_{t,o}}{C_{a,o}} \tag{7.79}$$

$$= 1 + (\varepsilon_r - 1)\frac{K(k_3)}{K'(k_3)}\frac{K'(\delta)}{K(\delta)}. \tag{7.80}$$

The odd-mode phase velocity $v_{ph,o}$ and characteristic impedance $Z_{0,o}$ are

$$v_{ph,o} = c'(\varepsilon_{eff,o})^{-1/2}, \tag{7.81}$$

$$Z_{0,o} = \frac{1}{c' C_{a,o} \sqrt{\varepsilon_{eff,o}}} \tag{7.82}$$

$$= \frac{60\pi}{\sqrt{\varepsilon_{eff,o}}} \frac{K'(\delta)}{K(\delta)}, \tag{7.83}$$

where c' is the velocity of light in free space.

7.3.3 Computed Even- and Odd-Mode Characteristic Impedance and Coupling Coefficient

The computed even- and odd-mode characteristic impedances for the edge-coupled CPWs [9] are presented in Figure 7.13. In this figure, for a given

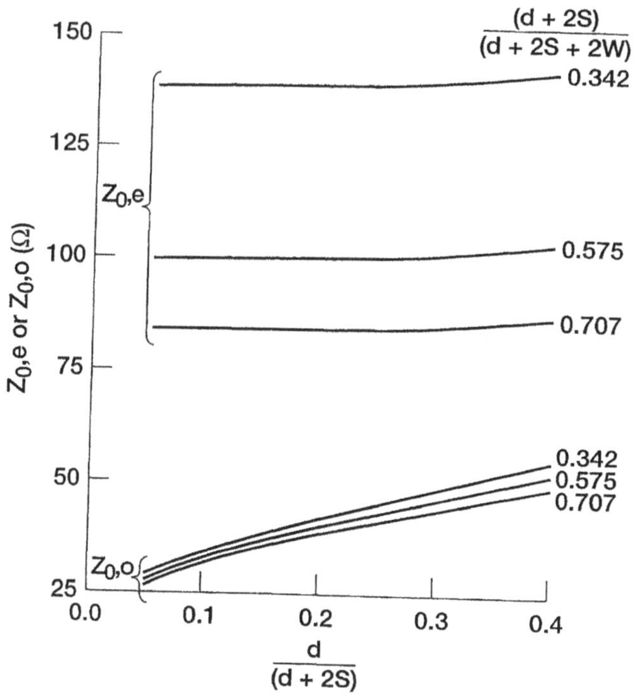

FIGURE 7.13 Computed even- and odd-mode characteristic impedance as a function of $d/(d + 2S)$ with $[(d/2) + S]/h = 0.5$ and $\varepsilon_r = 9.9$. (From Reference [9], © Ann. Telecommun. 1984.)

substrate and for a fixed W and $d + 2S$ as d increases, S proportionately decreases. The decrease in S reduces both the even- and odd-mode capacitances and effective dielectric constants of the structure. The reduction in these parameters is small for the even mode, since the two strip conductors are isolated, as shown in Figure 7.10(*a*). Therefore $Z_{0,e}$ marginally increases as $d/(d + 2S)$ increases in Figure 7.13. In the case of the odd mode, the reduction in these parameters is large because the two strip conductors are tightly coupled as shown in Figure 7.10(*b*). Therefore $Z_{0,o}$ increases as $d/(d + 2S)$ increases in Figure 7.13.

Further, for a fixed d and $d + 2S$, as W increases, the ratio

$$(d + 2S)/(d + 2S + 2W)$$

decreases. The increase in W reduces both the even- and odd-mode capacitances and effective dielectric constants of the structure. The reduction in these parameters is large for the even mode, since the fields are concentrated across the slots of width W, as shown in Figure 7.10(*a*). Therefore $Z_{0,e}$ increases significantly as $(d + 2S)/(d + 2S + 2W)$ decreases. In the case of a odd mode, the reduction in the above parameters is small because the fields are concentrated mainly across the separation d as shown in Figure 7.10(*b*). Therefore $Z_{0,o}$ increases by a very small amount as $(d + 2S)/(d + 2S + 2W)$ decreases.

The voltage coupling coefficient K is defined as [9]

$$K = \frac{Z_{0,e} - Z_{0,o}}{Z_{0,e} + Z_{0,o}}. \tag{7.84}$$

The computed K [9] is shown in Figure 7.14. In this figure $[(d/2) + S]/h = 0$ corresponds to the case of a substrate with infinite thickness. Further, for small $[(d/2) + S]/h$ ratios such as 0.01 and 0.1, the curves are indistinguishable. In general, for a given substrate and for a fixed W and $d + 2S$, the coupling decreases as d increases.

7.4 CONDUCTOR-BACKED EDGE COUPLED COPLANAR WAVEGUIDE

In this section expressions for computing the even- and odd-mode effective dielectric constant and characteristic impedance of a conductance-backed edge coupled coplanar waveguide obtained using the quasi-static conformal mapping technique [10] will be presented. The structure is shown in Figure 7.15. As in Section 7.3, a magnetic wall and an electric wall is placed along the plane of symmetry for the even and odd mode of excitation, respectively, and one-half of the structure is isolated. Through a sequence of conformal mapping steps, the total capacitance per unit length is obtained. From the capacitance the effective dielectric constant and characteristic impedance are obtained. Expressions for these quantities are given in the following sections.

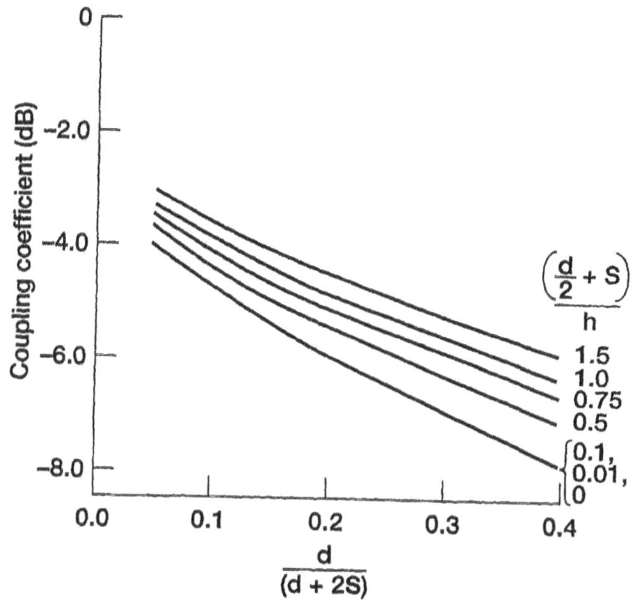

FIGURE 7.14 Computed coupling coefficient as a function of $d/(d + 2S)$ with $(d + 2S)/(d + 2S + 2W) = 0.342$ and $\varepsilon_r = 9.9$. (From Reference [9], © Ann. Telecommun. 1984.)

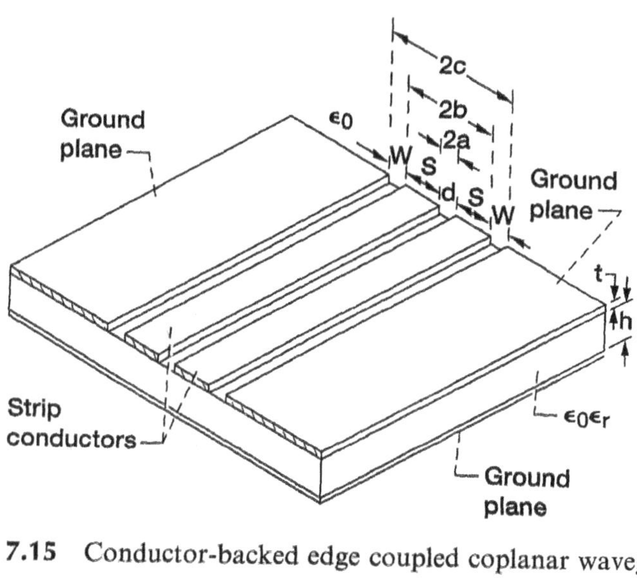

FIGURE 7.15 Conductor-backed edge coupled coplanar waveguides.

7.4.1 Even Mode

The even mode effective dielectric constant $\varepsilon_{\mathrm{eff,e}}$ is given by [10]

$$\varepsilon_{\mathrm{eff,e}} = \frac{\left[2\varepsilon_r \dfrac{K(k_e)}{K'(k_e)} + \dfrac{K(\delta k_1)}{K'(\delta k_1)}\right]}{\left[2\dfrac{K(k_e)}{K'(k_e)} + \dfrac{K(\delta k_1)}{K'(\delta k_1)}\right]}, \tag{7.85}$$

and the characteristic impedance $Z_{0,\mathrm{e}}$ is given by [10]

$$Z_{0,\mathrm{e}} = \frac{120\pi}{\sqrt{\varepsilon_{\mathrm{eff,e}}}\left[2\dfrac{K(k_e)}{K'(k_e)} + \dfrac{K(\delta k_1)}{K'(\delta k_1)}\right]}, \tag{7.86}$$

where δ and k_1 are given by Eqs. (7.56) and (7.58), respectively. Further $K(k_e)$ and $K'(k_e)$ are the complete elliptic integrals of the first kind with modulus k_e. The modulus k_e is given by

$$k_e = \phi_1 \frac{-(\phi_1^2 - \phi_2^2)^{1/2} + (\phi_1^2 - \phi_3^2)^{1/2}}{\phi_3(\phi_1^2 - \phi_2^2)^{1/2} + \phi_2(\phi_1^2 - \phi_3^2)^{1/2}}, \tag{7.87}$$

where

$$\phi_1 = \frac{1}{2}\cosh^2\left[\frac{\pi}{2h}\left(\frac{d}{2} + S + W\right)\right], \tag{7.88}$$

$$\phi_2 = \sinh^2\left[\frac{\pi}{2h}\left(\frac{d}{2} + S\right)\right] - \phi_1 + 1, \tag{7.89}$$

$$\phi_3 = \sinh^2\left(\frac{\pi d}{4h}\right) - \phi_1 + 1. \tag{7.90}$$

7.4.2 Odd Mode

The odd mode effective dielectric constant $\varepsilon_{\mathrm{eff,o}}$ is given by [10]

$$\varepsilon_{\mathrm{eff,o}} = \frac{\left[2\varepsilon_r \dfrac{K(k_o)}{K'(k_o)} + \dfrac{K(\delta)}{K'(\delta)}\right]}{\left[2\dfrac{K(k_o)}{K'(k_o)} + \dfrac{K(\delta)}{K'(\delta)}\right]}, \tag{7.91}$$

and the characteristic impedance $Z_{0,o}$ is given by [10]

$$Z_{0,o} = \frac{120\pi}{\sqrt{\varepsilon_{\mathrm{eff},o}}\left[2\dfrac{K(k_o)}{K'(k_o)} + \dfrac{K(\delta)}{K'(\delta)}\right]}, \tag{7.92}$$

where δ is given by Eq. (7.56). Further $K(k_o)$ and $K'(k_o)$ are the complete elliptic integrals of the first kind with modulus k_o. The modulus k_o is given by

$$k_0 = \phi_4 \frac{-(\phi_4^2 - \phi_5^2)^{1/2} + (\phi_4^2 - \phi_6^2)^{1/2}}{\phi_6(\phi_4^2 - \phi_5^2)^{1/2} + \phi_5(\phi_4^2 - \phi_6^2)^{1/2}}, \tag{7.93}$$

where

$$\phi_4 = \frac{1}{2}\sinh^2\left[\frac{\pi}{2h}\left(\frac{d}{2} + S + W\right)\right], \tag{7.94}$$

$$\phi_5 = \sinh^2\left[\frac{\pi}{2h}\left(\frac{d}{2} + S\right)\right] - \phi_4, \tag{7.95}$$

$$\phi_6 = \sinh^2\left(\frac{\pi d}{4h}\right) - \phi_4. \tag{7.96}$$

The voltage coupling coefficient K is defined as in Eq. (7.84), and the computed values can be found in [10]. The compound even- and odd-mode $\varepsilon_{\mathrm{eff}}$ and Z_0 for a conductor-backed edge coupled coplanar waveguide placed inside a metal enclosure are given in [11]. These characteristics are obtained by solving the propagation parameters using the method of lines.

7.4.3 Even- and Odd-Mode Characteristics with Elevated Strip Conductors

A coplanar waveguide with an elevated center strip conductor is shown in Figure 2.20. In [12] the characteristics of a conductor-backed coplanar waveguide with two elevated and coupled strip conductors are presented. The two strip conductors are elevated by inserting a thin dielectric layer between the conducting strips and the substrate. For the purpose of analysis, the structure is placed inside a shielding metal enclosure and the propagation characteristics are obtained by numerically solving Laplace's equations along a cross sectional plane.

7.5 BROADSIDE COUPLED COPLANAR WAVEGUIDE

This section presents the coupling characteristics between two coplanar waveguides that are stacked one on top of another. The two coplanar waveguides

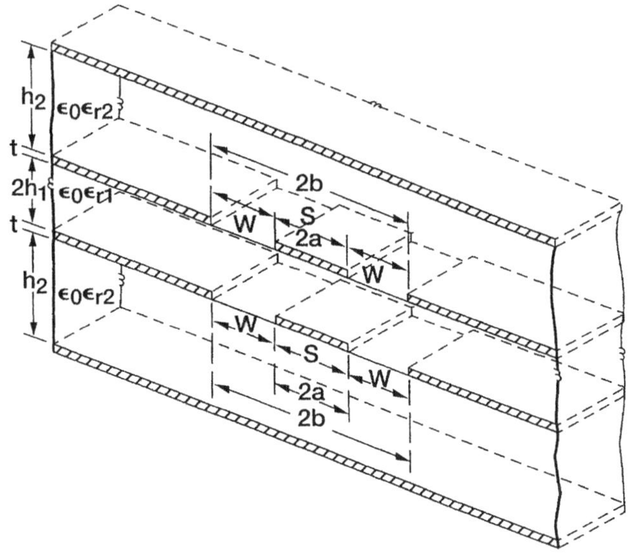

FIGURE 7.16 Schematic of broadside coupled coplanar waveguides.

are separated from each other by a thin dielectric sheet. A pair of ground planes that acts as a shield is placed symmetrically above and below the coupled coplanar waveguides, respectively, as shown in Figure 7.16. This structure is analyzed in a manner similar to the edge coupled coplanar waveguide by considering the even and odd modes of excitation. For each of these modes analytical expressions are presented to determine the effective dielectric constant and characteristic impedance based on quasi-static conformal mapping techniques. The assumption made to simplify the analysis is similar to those described in Section 2.2.1.

7.5.1 Even Mode

For the even excitation a magnetic wall is placed along the plane of symmetry in Figure 7.17(a) It then suffices to restrict the analysis of the upper half of the structure. The total even-mode capacitance per unit length C_e is derived as explained in Section 2.2.1. This total capacitance is expressed as the sum of two capacitances C_{e1} and C_{e2}, where C_{e1} and C_{e2} are the capacitances per unit length contributed by half the substrate region of height h_1 and the superstrate region of height h_2, respectively. That is, [13]

$$C_e = C_{e1} + C_{e2}, \tag{7.97}$$

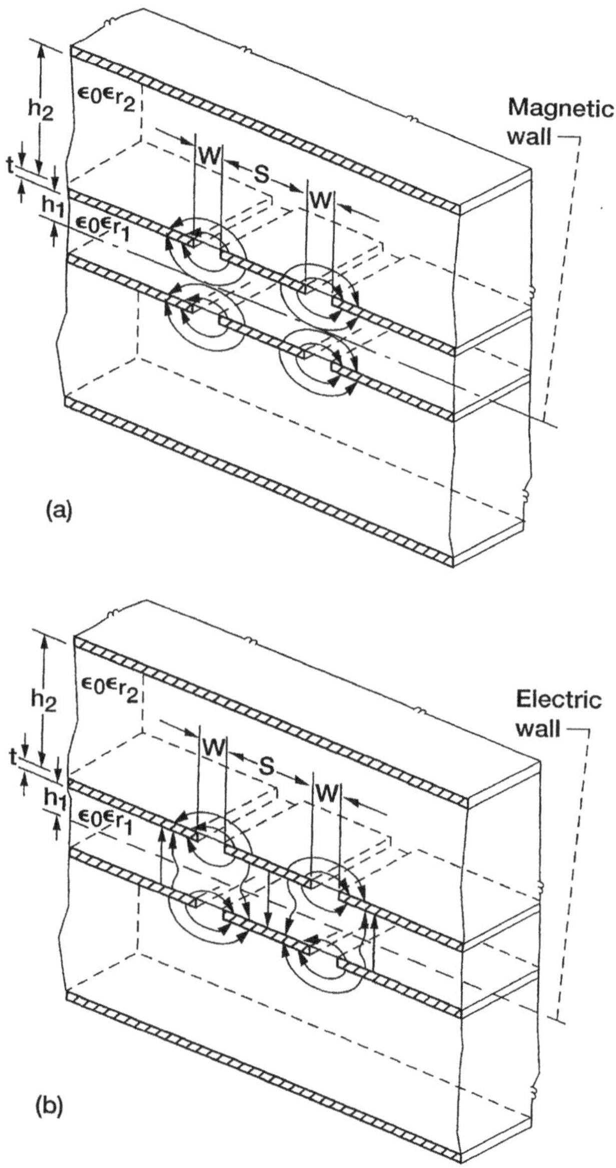

FIGURE 7.17 Excitation of broadside coupled coplanar waveguides: (*a*) Even mode; (*b*) odd mode.

where

$$C_{e1} = 2\varepsilon_0\varepsilon_{r1}\frac{K(k_{e1})}{K(k'_{e1})}, \tag{7.98}$$

$$k_{e1} = \frac{\sinh(\pi S/4h_1)}{\sinh(\pi(S+2W)/4h_1)}, \tag{7.99}$$

$$k'_{e1} = \sqrt{1-k_{e1}^2}, \tag{7.100}$$

$$C_{e2} = 2\varepsilon_0\varepsilon_{r2}\frac{K(k_{e2})}{K(k'_{e2})}, \tag{7.101}$$

$$k_{e2} = \frac{\tanh(\pi S/4h_2)}{\tanh(\pi(S+2W)/4h_2)}, \tag{7.102}$$

$$k'_{e2} = \sqrt{1-k_{e2}^2}, \tag{7.103}$$

and $K(k_{ei})$ and $K(k'_{ei})$, $(i = 1, 2)$ are the complete elliptic integral of the first kind and its complement. Substituting Eqs. (7.98) and (7.101) into Eq. (7.97), the total even-mode capacitance per unit length is

$$C_e = 2\varepsilon_0\left\{\varepsilon_{r1}\frac{K(k_{e1})}{K(k'_{e1})} + \varepsilon_{r2}\frac{K(k_{e2})}{K(k'_{e2})}\right\}. \tag{7.104}$$

The even-mode effective dielectric constant $\varepsilon_{eff,e}$ is defined as [13]

$$\varepsilon_{eff,e} = \frac{C_e}{C_{e(air)}}, \tag{7.105}$$

where $C_{e(air)}$ is the even-mode capacitance per unit length with the dielectric replaced by air, that is,

$$C_{e(air)} = 2\varepsilon_0\left\{\frac{K(k_{e1})}{K(k'_{e1})} + \frac{K(k_{e2})}{K(k'_{e2})}\right\}. \tag{7.106}$$

Substituting Eqs. (7.104) and (7.106) into Eq. (7.105), we have

$$\varepsilon_{eff,e} = \varepsilon_{r2} + q_e(\varepsilon_{r1} + \varepsilon_{r2}), \tag{7.107}$$

where

$$q_e = \frac{\dfrac{K(k_{e1})}{K(k'_{e1})}}{\dfrac{K(k_{e1})}{K(k'_{e1})} + \dfrac{K(k_{e2})}{K(k'_{e2})}}. \tag{7.108}$$

The even-mode characteristic impedance $Z_{0,e}$ is defined as [13]

$$Z_{0,e} = \frac{1}{c'\sqrt{C_e C_{e(air)}}}, \tag{7.109}$$

where c' is the velocity of light in free space. Substituting Eqs. (7.104) and (7.106) into Eq. (7.109), we have

$$Z_{0,e} = \frac{60\pi}{\sqrt{\varepsilon_{eff,e}}} \cdot \frac{1}{\left[\frac{K(k_{e1})}{K(k'_{e1})} + \frac{K(k_{e2})}{K(k'_{e2})}\right]}. \tag{7.110}$$

7.5.2 Odd Mode

For the odd excitation an electric wall is placed along the plane of symmetry in Figure 7.17(*b*). The analysis is then restricted to the upper half of the structure. The total odd-mode capacitance per unit length C_o is derived as explained in Section 3.2.3. The total capacitance is expressed as the sum of two capacitances C_{01} and C_{02}, where C_{01} and C_{02} are the capacitances per unit length contributed by half the substrate region of height h_1 and the superstrate region of height h_2, respectively. That is [13],

$$C_o = C_{o1} + C_{o2}, \tag{7.111}$$

where

$$C_{oi} = 2\varepsilon_0 \varepsilon_{ri} \frac{K(k_{oi})}{K(k'_{oi})} \qquad (i = 1, 2), \tag{7.112}$$

and

$$k_{oi} = \frac{\tanh(\pi S/4h_i)}{\tanh(\pi(S + 2W)/4h_i)} \qquad (i = 1, 2), \tag{7.113}$$

and $K(k_{oi})$ and $K(k'_{oi})$, $i = 1, 2$, are the complete elliptic integral of the first kind and its complement. Substituting Eq. (7.112) into Eq. (7.111) yields the total odd-mode capacitance power unit length

$$C_o = 2\varepsilon_0 \left[\varepsilon_{r1} \frac{K(k_{01})}{K(k'_{01})} + \varepsilon_{r2} \frac{K(k_{02})}{K(k'_{02})}\right]. \tag{7.114}$$

The odd-mode effective dielectric constant $\varepsilon_{eff,0}$ is defined as [13]

$$\varepsilon_{eff,0} = \frac{C_o}{C_{o(air)}}, \tag{7.115}$$

where $C_{\mathrm{o(air)}}$ is the odd-mode capacitance per unit length with the dielectric replaced by air, that is,

$$C_{\mathrm{o(air)}} = 2\varepsilon_0 \left\{ \frac{K(k_{\mathrm{o1}})}{K(k'_{\mathrm{o1}})} + \frac{K(k_{\mathrm{o2}})}{K(k'_{\mathrm{o2}})} \right\}. \tag{7.116}$$

Substituting Eqs. (7.114) and (7.116) into Eq. (7.115) obtains

$$\varepsilon_{\mathrm{eff,o}} = \varepsilon_{r2} + q_{\mathrm{o}}(\varepsilon_{r1} - \varepsilon_{r2}), \tag{7.117}$$

where

$$q_{\mathrm{o}} = \frac{\dfrac{K(k_{\mathrm{o1}})}{K(k'_{\mathrm{o1}})}}{\dfrac{K(k_{\mathrm{o1}})}{K(k'_{\mathrm{o1}})} + \dfrac{K(k_{\mathrm{o2}})}{K(k'_{\mathrm{o2}})}}. \tag{7.118}$$

The odd-mode characteristic impedance $Z_{0,\mathrm{o}}$ is defined as [13]

$$Z_{0,\mathrm{o}} = \frac{1}{c'\sqrt{C_{\mathrm{o}} C_{\mathrm{o(air)}}}}, \tag{7.119}$$

where c' is the velocity of light in free space. Substituting Eqs. (7.114) and (7.116) into Eq. (7.119) obtains

$$Z_{0,\mathrm{o}} = \frac{60\pi}{\sqrt{\varepsilon_{\mathrm{eff,o}}}} \cdot \frac{1}{\left[\dfrac{K(k_{\mathrm{o1}})}{K(k'_{\mathrm{o1}})} + \dfrac{K(k_{\mathrm{o2}})}{K(k'_{\mathrm{o2}})} \right]}. \tag{7.120}$$

7.5.3 Computed Even- and Odd-Mode Effective Dielectric Constant, Characteristic Impedance, Coupling Coefficient, and Mode Velocity Ratio

The computed even- and odd-mode effective dielectric constant [13] for the broadside coupled coplanar waveguide as a function of the strip conductor width is shown in Figure 7.18(*a*). For a fixed substrate thickness and slot width, the $\varepsilon_{\mathrm{eff,o}}$ increases while $\varepsilon_{\mathrm{eff,e}}$ decreases as the strip conductor width S increases. The computed even- and odd-mode characteristic impedance [13] as a function of the strip conductor width S is shown in Figure 7.18(*b*). For a fixed substrate thickness and slot width, the $Z_{0,\mathrm{e}}$ and $Z_{0,\mathrm{o}}$ decreases as S increases. The broadside coupled coplanar waveguides have also been analyzed using the spectral domain method [14]. The numerical results show that the effective dielectric constants are almost insensitive to change in frequency, thus confirming that the quasi-static results given above are adequate for circuit design [14].

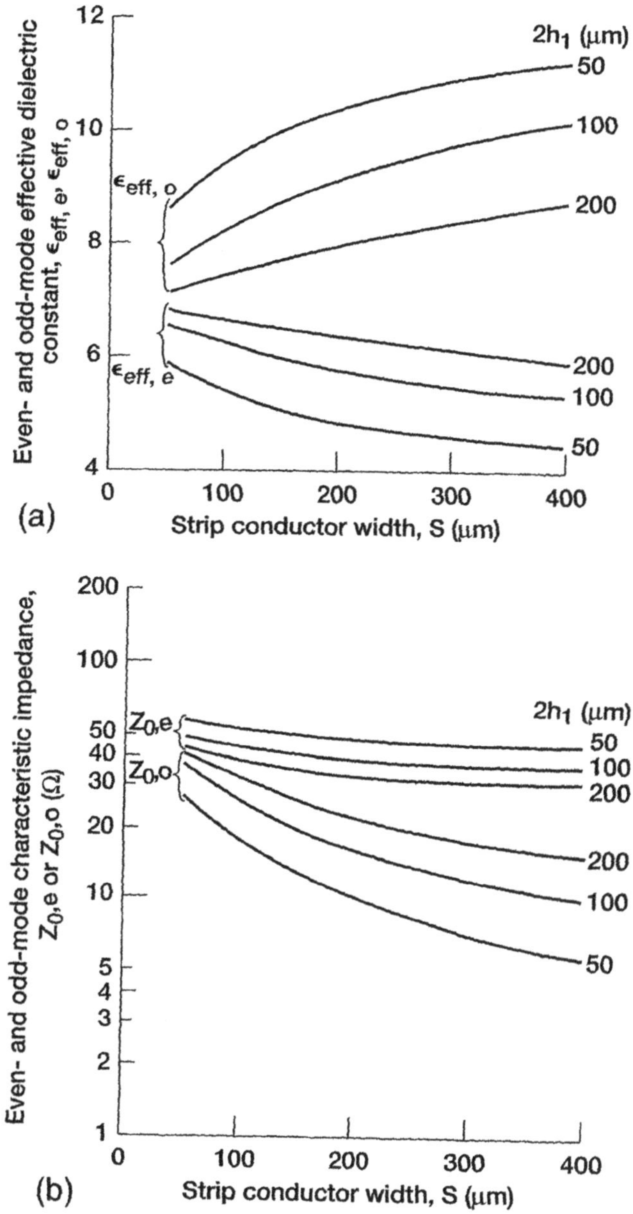

FIGURE 7.18 Computed even- and odd-mode (*a*) effective dielectric constant and (*b*) characteristic impedance, $\varepsilon_{r1} = 12.9$, $\varepsilon_{r2} = 1.0$, $W = 20\ \mu m$, $h_2 = 1$ mm. (From Reference [13], © IEEE 1989.)

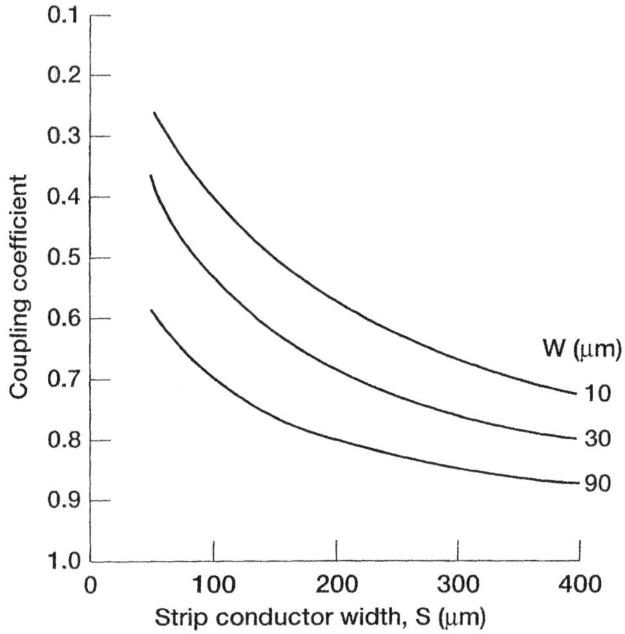

FIGURE 7.19 Computed voltage coupling coefficient as a function of the strip conductor width, $\varepsilon_{r1} = 12.9$, $\varepsilon_{r2} = 1.0$, $2h_1 = 50\,\mu\text{m}$, $h_2 = 5.0\,\text{mm}$. (From Reference [13], © IEEE 1989.)

The voltage coupling coefficient K is defined as [13]

$$K = \frac{Z_{0,e} - Z_{0,o}}{Z_{0,e} + Z_{0,o}}. \tag{7.121}$$

The computed voltage coupling coefficients [13] as a function of the strip conductor width S is shown in Figure 7.19. For a fixed substrate thickness and slot width, the coupling increases as S increases.

The even- and odd-mode velocity ratio is defined as [13]

$$\frac{v_e}{v_o} = \sqrt{\frac{\varepsilon_{eff,o}}{\varepsilon_{eff,e}}} \tag{7.122}$$

The compared velocity ratio [13] as a function of the strip conductor width is shown in Figure 7.20. For a fixed substrate thickness and slot width, the velocity ratio increases as S increases. From Figures 7.19 and 7.20 it is observed that as S increases both K as well as v_e/v_o increases. The computed characteristics for open broadside coupled coplanar waveguides with a dielectric overlay can be found in [15].

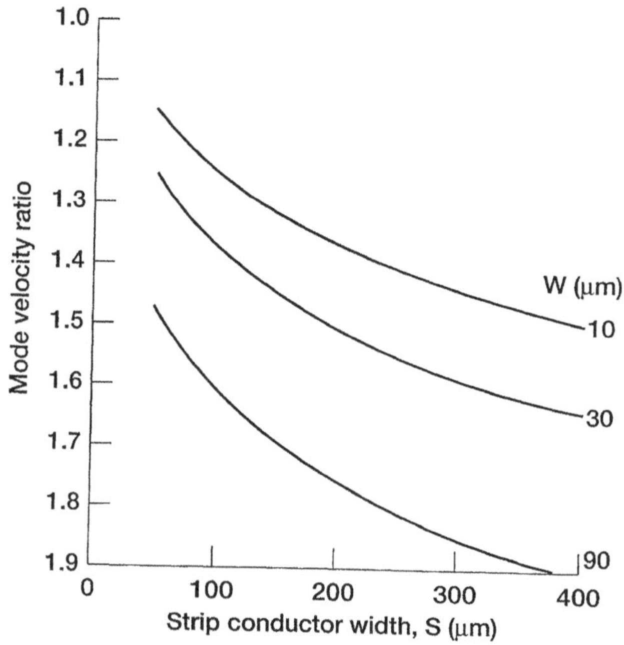

FIGURE 7.20 Computed even- and odd-mode velocity ratio, $\varepsilon_{r1} = 12.9$, $\varepsilon_{r2} = 1.0$, $2h_1 = 50\,\mu m$, $h_2 = 5.0\,mm$. (From Reference [13], © IEEE 1989.)

REFERENCES

[1] N. I. Dib, W. P. Harokopus, L. P. B. Katehi, C. C. Ling, and G. M. Rebeiz, "Study of a novel planar transmission line," 1991 IEEE MTT-S *Int. Microwave Symp. Dig.*, Vol. 2, Boston, MA, pp. 623–626, 1991.

[2] S. Ramo, J. R. Whinnery, and T. V. Duzer, *Fields and Waves in Communication Electronics*, 3rd ed., New York: Wiley, 1994, p. 411.

[3] N. I. Dib and L. P. B. Katehi, "Impedance Calculation for the Microshield Line," *IEEE Microwave Guided Wave Lett.*, Vol. 2, No. 10, pp. 406–408, Oct. 1992.

[4] K.-K. M. Cheng and I. D. Robertson, "Quasi-TEM Study of Microshield Lines with Practical Cavity Sidewall Profiles," *IEEE Trans. Microwave Theory Tech.*, Vol. 43, No. 12, pp. 2689–2694, Dec. 1995.

[5] K.-K. M. Cheng and I. D. Robertson, "Simple and Explicit Formulas for the Design and Analysis of Asymmetrical V-Shaped Microshield Line," *IEEE Trans. Microwave Theory Tech.*, Vol. 43, No. 10, pp. 2501–2504, Oct. 1995.

[6] N. Yuan, C. Ruan, W. Lin, J. He, and C. He, "Coplanar Coupled Lines: The Effects of the Presence of the Lateral Ground Planes, Upper and Lower Ground Planes, and the V-Shaped Microshield Ground Walls," *IEE Proc. Microwave Antennas Propag.*, Vol. 142, No. 1, pp. 63–66, Feb. 1995.

[7] E. Costamagna and A. Fanni, "Inhomogeneous Dielectric Transmission Line Geometries Analysed by Combining Conformal Mapping-Finite Difference Procedures," *8th Mediterranean Electrotechnical Conf. Proc.* (Melecon '96), Vol. III, Bari, Italy, pp. 1393–1396, May 13–16, 1996.

[8] N. Yuan, C. Ruan, and W. Lin, "Analytical Analyses of V, Elliptic, and Circular-Shaped Microshield Transmission Lines," *IEEE Trans. Microwave Theory Tech.*, Vol. 42, No. 5, pp. 855–858, May 1994.

[9] V. F. Hanna and D. Thebault, "Analyse des coupleurs directifs coplanaires," *Annales des Telecomm.*, Vol. 39, No. 7–8, pp. 299–306, July–Aug. 1984.

[10] V. F. Hanna, "Parameters of Coplanar Directional Couplers with Lower Ground Plane," *15th European Microwave Conf. Proc.*, Paris, France, pp. 820–825, Sept. 9–13, 1985.

[11] R. R. Kumar, S. Aditya, and D. Chadha, "Characteristics of a Shielded Dielectric-Loaded Edge-Coplanar Waveguide," *Microwave Optical Tech. Lett.*, Vol. 11, No. 11, pp. 279–281, April 1996.

[12] S. Raju, B. N. Nityanandan, and V. Abhaikumar, "Analysis of a Shielded Elevated Coupled Coplanar Waveguide Using a Modified Finite Element Method," *Int. J. Microwave Millimeter-Wave Computer-Aided Engineering*, Vol. 6, No. 5, pp. 305–318, 1996.

[13] S. S. Bedair and I. Wolff, "Fast and Accurate Analytic Formulas for Calculating the Parameters of a General Broadside-Coupled Coplanar Waveguide for (M)MIC Applications," *IEEE Trans. Microwave Theory Tech.*, Vol. 37, No. 5, pp. 843–850, May 1989.

[14] C. Nguyen, "Dispersion Characteristics of the Broadside-Coupled Coplanar Waveguide," *IEEE Trans. Microwave Theory Tech.*, Vol. 41, No. 9, pp. 1630–1633, Sept. 1993.

[15] F. Tefiku, E. Yamashita, and J. Funada, "Novel Directional Couplers Using Broadside-Coupled Coplanar Waveguides for Double-Sided Printed Antennas," *IEEE Trans. Microwave Theory Tech.*, Vol. 44, No. 2, pp. 275–282, Feb. 1996.

CHAPTER 8

Attenuation Characteristics of Conventional, Micromachined, and Superconducting Coplanar Waveguides

8.1 INTRODUCTION

In this chapter the attenuation characteristics of conventional, micromachined, and superconducting coplanar waveguides are presented. Section 8.2 presents closed form equations based on the conformal mapping technique, the mode-matching method, and the modified matched asymptotic expansion technique in order to compute the attenuation constant of a coplanar waveguide. In Section 8.3 the influence of the coplanar waveguide geometry on the attenuation characteristics is discussed. Recently silicon (Si) and silicon-germanium (SiGe) based integrated circuits have emerged for RF/microwave applications. Hence Section 8.4 discusses the attenuation characteristics of coplanar waveguides on CMOS grade low-resistivity as well as high-resistivity silicon wafers. Further, microelectromechanical systems (MEMS) based switches and actuators are being explored for potential phased-array antenna applications. Therefore the attenuation characteristics of coplanar waveguides on micromachined silicon wafers are presented in Section 8.5. Filter circuits based on superconducting coplanar waveguides have demonstrated potential for application in wireless and satellite communications systems. Hence the attenuation characteristics of high-temperature superconducting coplanar waveguides are discussed in Section 8.6

8.2 CLOSED FORM EQUATIONS FOR CONVENTIONAL CPW ATTENUATION CONSTANT

Coplanar waveguide (CPW) circuits fabricated using standard photolithography process on copper-clad soft plastic-like sheets have a conductor thickness t several times the skin depth δ. Typically this thickness t is on the order of 17 μm, which is the result of rolling a half ounce of copper over an area of one square feet [1]. Hence, for deriving an expression for the CPW attenuation constant, the conductors are considered to be electrically very thick and with conductivity the same as that of bulk copper. For the case of thick conductors a simple closed form expression for CPW attenuation constant is derived in [2] using conformal mapping technique.

In monolithic microwave integrated circuits the conductor patterns are formed by evaporating or by sputtering precious metal such as gold on the surface of a dielectric substrate. These deposition processes yield conducting films with thickness on the order of a micron. The thickness is then built up to about 2.5 μm by an electroplating process. The conductor thickness t in this case is said to be thin and in some cases may be less than or equal to the skin depth δ at the operating frequency. When $t \leqslant \delta$, the fields penetrate into the conductors and the conductivity inside the conductors can be accounted for by introducing a complex dielectric constant. The CPW attenuation constant for this case is determined using accurate full wave analysis, such as the mode-matching method [3]. The disadvantage of the mode-matching method is that it requires significant amount of computer resources to carry out the computations. However, closed form expressions that approximate the results of the mode-matching method are also available [4].

Besides thickness the cross section of the CPW conductors in MMICs may deviate from the ideal rectangular shape. The shape is strongly dependent on the fabrication process involved. In most cases the shape would be a trapezoid with 60° or 70° angles. The CPW attenuation constant for this case is derived by the modified matched asymptotic expansion technique [5], [6]. In this technique an approximate current density in first assumed on the CPW conductors and an equation for the CPW attenuation constant is derived. This equation involves a contour integral. Instead of evaluating this integral out to the edge of the conductors where the fields are singular, the limits of the integral are taken at some distance just before the edge. This short length before the edge is defined as the stopping distance Δ [6]. Once Δ is determined for a given edge shape as a function of t/δ, a closed form expression is derived for the CPW conductor loss in [6]. In the subsections that follow closed form equations based on the foregoing numerical methods as well as measurement-based design equations are presented, and the corresponding computed attenuation constant is compared with measured data.

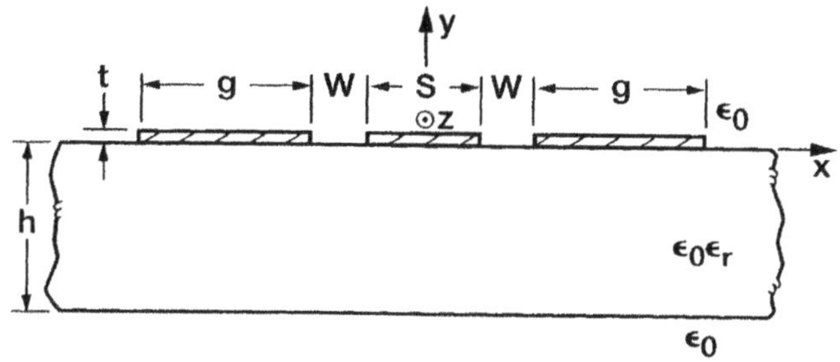

FIGURE 8.1 Coplanar waveguide with finite thickness conductors.

8.2.1 Conformal Mapping Method

Consider the CPW structure shown in Figure 8.1 with g tending to infinity. In the conformal mapping method [2] the total CPW attenuation is expressed as the sum of the attenuation due to dielectric losses in the substrate and the attenuation due to the conductor losses in the strip and the ground planes. The attenuation constant due to the dielectric loss is expressed as

$$\alpha_d = \frac{\pi}{\lambda_0} \frac{\varepsilon_r}{\sqrt{\varepsilon_{eff}}} q \tan \delta_e \text{ Nepers/meter,} \tag{8.1}$$

where λ_0 is the free space wavelength in meters, ε_r is the relative permittivity of the substrate, $\tan \delta_e$ is the dielectric loss tangent, ε_{eff} is the effective dielectric constant defined in Eq. (2.37) and is expressed as

$$\varepsilon_{eff} = 1 + q(\varepsilon_r - 1), \tag{8.2}$$

and q is the filling factor and is given by

$$q = \frac{1}{2} \frac{K(k_1)}{K(k'_1)} \frac{K(k'_0)}{K(k_0)}. \tag{8.3}$$

The terms $K(k_1)$ and $K(k_0)$ are the complete elliptic integrals of the first kind with moduli k_1 and k_0 which are given by Eqs. (2.3) and (2.35), respectively,

$$k_1 = \frac{\sinh(\pi S/4h_1)}{\sinh[\pi(S + 2W)/4h_1]}, \tag{8.4}$$

$$k_0 = \frac{S}{S + 2W}. \tag{8.5}$$

The terms k_1' and k_0' are the complementary moduli given by

$$k_1' = \sqrt{1 - k_1^2}, \tag{8.6}$$

$$k_0' = \sqrt{1 - k_0^2}. \tag{8.7}$$

An expression for the attenuation constant α_c due to conductor loss in the center strip conductor and the ground planes of the CPW is given below. The assumption made in deriving this expression is that the thickness t of the CPW conductors is far greater than the skin depth δ in the metal. Typically t would be greater than 5δ. The attenuation constant α_c is given by [2]

$$\alpha_c = \frac{R_c + R_g}{2Z_0} \quad \text{Nepers/meter} \tag{8.8}$$

where R_c is the series resistance in ohms per unit length of the center strip conductor and is given by

$$R_c = \frac{R_s}{4S(1 - k_0^2)K^2(k_0)}\left[\pi + \ln\left(\frac{4\pi S}{t}\right) - k_0 \ln\left(\frac{1 + k_0}{1 - k_0}\right)\right], \tag{8.9}$$

R_g is the distributed series resistance in ohms per unit length of the ground planes and is given by

$$R_g = \frac{k_0 R_s}{4S(1 - k_0^2)K^2(k_0)}\left[\pi + \ln\left(\frac{4\pi(S + 2W)}{t}\right) - \frac{1}{k_0} \ln\left(\frac{1 + k_0}{1 - k_0}\right)\right], \tag{8.10}$$

and Z_0 is the coplanar waveguide characteristic impedance given by Eq. (2.38). The term R_s is the skin effect surface resistance given by

$$R_s = \frac{1}{\delta\sigma} \quad \text{ohms}, \tag{8.11}$$

where σ is the conductivity of the conductor in Siemens/meter, δ is the skin depth given by

$$\delta = \sqrt{\frac{2}{\omega\mu\sigma}} \quad \text{meters}, \tag{8.12}$$

with

$\omega = 2\pi f$ as the angular frequency in radians,
$\mu = \mu_0\mu_r$, as the permeability of the conductor,
$\mu_0 = 4\pi \times 10^{-7}$ Henry/meter.

The total attenuation α is

$$\alpha = (\alpha_c + \alpha_d) \quad \text{Nepers/meters.} \tag{8.13}$$

8.2.2 Mode-Matching Method and Quasi-TEM Model

Consider the CPW structure shown in Figure 8.1 with g tending to infinity. In the mode-matching method [3] the fields in the substrate and in the air regions on top and bottom are initially represented as a sum of all the longitudinal-section modes LSE_{xm} and LSH_{xm} with unknown amplitudes. Next, the propagation constants k_x and k_z are obtained by a two-step procedure. In the first step, k_{xm} in the plane of the metallization is determined for each LSE_{xm} and LSH_{xm} mode by satisfying the field continuity at the vertical boundaries and by solving the transverse resonance condition. In the second step, the field continuity at the horizontal planes is used and leads to a system of homogeneous equations. The eigenvalue of this system of equations is the complex propagation constant k_z of the CPW. From the propagation constant the phase constant β and the attenuation constant α can be obtained. Although the mode-matching method is very accurate, it is not suitable for computer-aided design (CAD) software because it requires considerable amount of computer resources. Hence closed form equations that approximate the mode-matching results are obtained in [4].

In deriving the closed form equations, the CPW is represented by a distributed equivalent circuit model consisting of a resistance R and an inductance L in series with a shunt capacitance C and a conductance G. Such a model assumes a quasi-TEM mode of propagation on the CPW. This is a valid assumption and holds good even at millimeter-wave frequencies as discussed in Section 2.2.6 and shown in Table 2.3. The advantage of the model above is that it fits well into common network analysis software and takes into consideration both the finite thickness as well as the non-ideal conductivity of the CPW conductors. Since the closed form expressions are very lengthy, they are not reproduced here, but interested readers can refer to [4].

8.2.3 Matched Asymptotic Technique and Closed Form Expressions

This method takes into consideration both the finite thickness as well as the non-ideal conductivity of the CPW conductors. In addition the shape of the conductor cross section needs not be rectangular, but instead can be trapezoidal. Thus the sides of the conductor which are normally assumed to be vertical can now be considered to be inclined at an angle θ as shown in Figure 8.2. In this figure the average center strip width and the ground plane separation are indicated as $2a$ and $2b$, respectively.

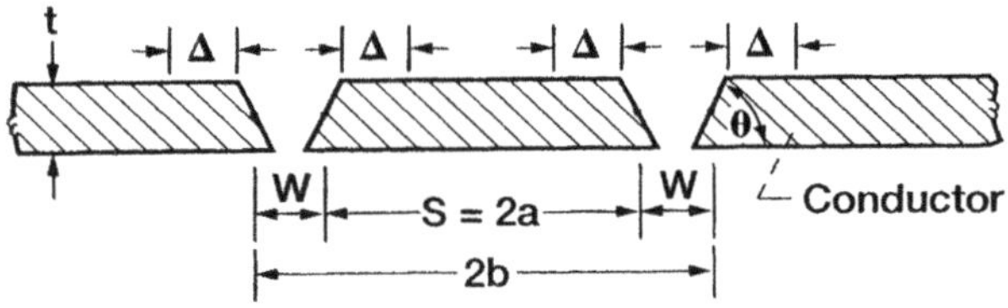

FIGURE 8.2 Coplanar waveguide conductor geometry showing the stopping distance Δ and edge profile θ.

The attenuation constant α_c due to conductor loss is given by [6]

$$\alpha_c = \frac{R_{sm} b^2}{16 Z_0 [K^2(k)](b^2 - a^2)} \left\{ \frac{1}{a} \ln\left(\frac{2a}{\Delta} \frac{(b-a)}{(b+a)} \right) + \frac{1}{b} \ln\left(\frac{2b}{\Delta} \frac{(b-a)}{(b+a)} \right) \right\} \quad \text{Nepers/meter,} \tag{8.14}$$

where

$$Z_0 = \frac{30\pi}{\sqrt{\varepsilon_{eff}}} \frac{K(k')}{K(k)} \text{ ohms,} \tag{8.15}$$

$$k = \frac{a}{b}, \tag{8.16}$$

$$k' = \sqrt{1 - k^2}, \tag{8.17}$$

$$K(k) = \text{complete elliptic integral of the first kind,} \tag{8.18}$$

$$\varepsilon_{eff} = \frac{\varepsilon_r + 1}{2}, \tag{8.19}$$

$$\varepsilon_r = \text{relative permittivity of the substrate,} \tag{8.20}$$

$$R_{sm} = \omega \mu_c t \text{ Imaginary} \left(\frac{\cot(k_c t) + \csc(k_c t)}{k_c t} \right), \tag{8.21}$$

$$\omega = 2\pi f, \tag{8.22}$$

$$\mu_c = \mu_0 \mu_{rc} \tag{8.23a}$$

$$= \mu_0 (\mu_{rc} \approx 1 \text{ for copper or gold}), \tag{8.23b}$$

$$\varepsilon_c = \varepsilon_c' - j \frac{\sigma_c}{\omega} \tag{8.24a}$$

$$= \varepsilon_0 \varepsilon_{rc} - j\,\frac{\sigma_c}{\omega} \tag{8.24b}$$

$$= \varepsilon_0 - j\,\frac{\sigma_c}{\omega}\ (\varepsilon_{rc} \approx 1 \text{ for copper or gold}), \tag{8.24c}$$

$$\sigma_c = \text{conductivity of the metal in Siemens/meter}, \tag{8.25}$$

$$k_c = \omega\sqrt{\mu_c \varepsilon_c} \tag{8.26a}$$

$$= \omega\sqrt{\mu_0\left(\varepsilon_0 - j\,\frac{\sigma_c}{\omega}\right)}$$

$$= \omega\sqrt{\mu_0\varepsilon_0}\sqrt{1 - j\,\frac{\sigma_c}{\omega\varepsilon_0}} \tag{8.26b}$$

$$= \tilde{k}_0\sqrt{-j\,\frac{\sigma_c}{\omega}}, \quad \text{since } \frac{\sigma_c}{\omega\varepsilon_0} \gg 1, \tag{8.26c}$$

$$\tilde{k}_0 = \omega\sqrt{\mu_0\varepsilon_0} \tag{8.27a}$$

$$= \frac{2\pi}{\lambda_0}, \tag{8.27b}$$

$$\lambda_0 = \text{free space wavelength}, \tag{8.28}$$

$$\Delta = \text{stopping distance}. \tag{8.29}$$

The stopping distance Δ for a given conductor thickness t, skin depth δ and edge profile θ, is given in Table 8.1.

In Figure 8.3, the computed attenuation constant α_c due to conductor loss [6] using Eq. (8.14) is compared with the experimentally measured valves of Williams and Marks [7] over the frequency range of 0.2 to 20.0 GHz. The measured and the modeled results are in very good agreement at frequencies above 0.5 GHz. The reason for the discrepancy below 0.5 GHz is because Eq. (8.15) for the characteristics impedance Z_0 of a CPW is not valid. In deriving Eq. (8.15), the assumption made is that the conductors are perfectly conducting and infinitely thin. As a consequence the field penetration inside the conductors is negligibly small. This assumption is not valid when $t \leqslant \delta$ and the fields penetrate inside the conductors. The characteristic impedance Z_c taking field penetration into account is given by [8]

$$Z_c = Z_0\left(1 + \frac{Q}{j\tilde{k}_0\sqrt{\varepsilon_{eff}}Z_0}\,\frac{2}{1 + \sqrt{1 + \dfrac{4Q}{j\tilde{k}_0\sqrt{\varepsilon_{eff}}Z_0}}}\right) \text{ ohms}, \tag{8.30}$$

TABLE 8.1 Computed t/Δ as a Function of $t/2\delta$ for edge Profile $\theta = 45°$ and $90°$

	t/Δ			t/Δ	
$t/2\delta$	90°	45°	$t/2\delta$	90°	45°
0.03	9.18	6.57	1.87	266.73	306.15
0.04	9.18	6.57	2.00	244.95	288.35
0.05	9.19	6.57	2.18	221.57	271.18
0.06	9.19	6.58	2.29	210.25	265.33
0.10	9.25	6.62	2.51	200.43	264.15
0.14	9.45	6.77	2.76	189.28	274.11
0.25	11.76	8.43	3.00	178.57	288.74
0.50	33.97	26.72	3.55	170.73	317.89
0.64	61.90	49.41	4.00	168.50	327.08
0.71	81.32	65.51	4.53	171.59	327.42
0.79	108.83	89.23	4.74	172.81	326.99
0.87	138.42	115.75	5.0	174.33	327.01
0.94	169.39	144.36	6.0	185.89	336.93
1.0	200.50	173.95	7.0	193.43	357.86
1.07	235.98	209.06	8.0	195.96	383.06
1.12	258.21	232.17	9.0	196.58	410.96
1.15	270.22	245.19	9.49	196.94	426.04
1.22	299.73	279.65	10.49	198.36	461.40
1.32	324.12	314.30	12.25	203.04	543.07
1.50	329.88	342.89	13.0	205.58	587.90
1.63	312.74	339.00	14.0	209.24	659.63
1.73	293.30	326.77	16.0	217.25	862.14

Source: From Reference [6], © IEEE 1995.

where

$$Q = \frac{(Z_s + Z_m)b^2}{16K^2(k)(b^2 - a^2)} \left\{ \frac{1}{a} \ln\left(\frac{2a}{\Delta} \frac{(b-a)}{(b+a)} \right) + \frac{1}{b} \ln\left(\frac{2b}{\Delta} \frac{(b-a)}{(b+a)} \right) \right\}, \tag{8.31}$$

$$Z_s = -j \sqrt{\frac{\mu_0}{\varepsilon_c' - j\dfrac{\sigma_c}{\omega}}} \cos(k_c t) \tag{8.32a}$$

$$= -j \sqrt{\frac{\mu_0}{\varepsilon_0 - j\dfrac{\sigma_c}{\omega}}} \cot(k_c t) \tag{8.32b}$$

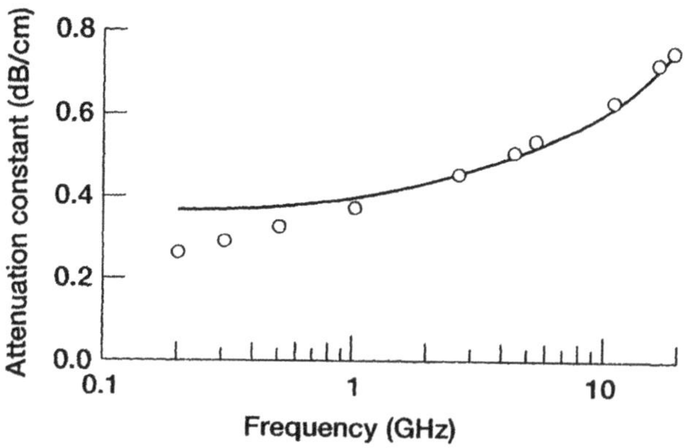

FIGURE 8.3 Comparison of computed and measured attenuation constant due to conductor loss as a function of the frequency for a CPW with $t = 1.61\,\mu$m, $a = 35.6\,\mu$m, $b = 84.6\,\mu$m, $\varepsilon_r = 12.9$, $g \to \infty$, and $\sigma_c = 3.602 \times 10^7$ S/meter (the circles are measured attenuation). (Reference [6], © IEEE 1995.)

$$= -j\eta \frac{1}{\sqrt{1 - j\dfrac{\sigma_c}{\omega\varepsilon_0}}} \cot(k_c t) \tag{8.32c}$$

$$= \frac{-j\eta}{\sqrt{-j\dfrac{\sigma_c}{\omega\varepsilon_0}}} \cot(k_c t), \quad \text{since } \frac{\sigma_c}{\omega\varepsilon_0} \gg 1, \tag{8.32d}$$

$$\eta = \sqrt{\frac{\mu_0}{\varepsilon_0}}, \tag{8.33}$$

$$Z_m = -j\sqrt{\frac{\mu_0}{\varepsilon_c' - j\dfrac{\sigma_c}{\omega}}} \csc(k_c t) \tag{8.34a}$$

$$= \frac{-j\eta}{\sqrt{-j\dfrac{\sigma_c}{\omega\varepsilon_0}}} \csc(k_c t). \tag{8.34b}$$

As a caution, it should be mentioned that analogous to Eq. (8.15) for Z_0, Eq. (8.30) for Z_c also has a low-frequency limit below which the equation is not valid. This frequency limit is typically on the order of few MHz [8].

The effect of conductor thickness t on the computed attenuation constant α_c for a CPW [6] at a fixed frequency of 20 GHz is shown in Figure 8.4. As expected, the attenuation increases very rapidly once the conductor thickness t becomes less than about two to three times the skin depth δ.

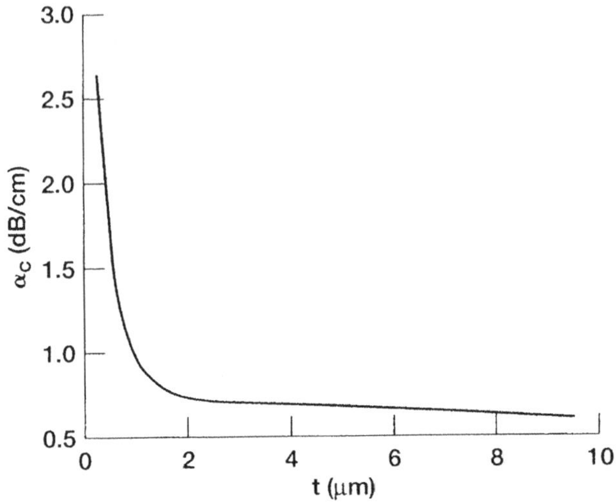

FIGURE 8.4 Computed attentuation constant due to conductor loss as a function of the conductor thickness t for a CPW with $a = 35.6\,\mu\text{m}$, $b = 84.6\,\mu\text{m}$, $\varepsilon_r = 12.9$, $\sigma_c = 3.602 \times 10^7$ S/meter, $g \to \infty$, frequency = 20.0 GHz, and $\delta = 0.593\,\mu\text{m}$. (Reference [6], © IEEE 1995.)

Figures 8.5 and 8.6 present the computed attenuation constant α_c for two CPWs [6] with t equal to 3 and 1.61 μm, respectively. As a worst-case scenario the two CPWs are assumed to have an edge profile θ equal to 45°. These figures show that the attenuation constant α_c for θ equal to 45° is not significantly greater than the attenuation constant α_c for θ equal to 90° when t is equal to 1.61 μm. However, when t is equal to 3 μm, there is about 7 percent increase in the attenuation at 40 GHz. Therefore for CPWs with thick conductors the additional attenuation due to edge profile is significant.

The stopping distance Δ given in Table 8.1 is derived for an isolated strip edge. Hence, if the CPW center strip conductor becomes very narrow or if the slot width becomes very small, the stopping distance Δ may be in error. To discover the range of validity, the results of this method (α_c) are compared in [6] with the data (α) obtained using the rigorous spectral-domain/perturbation method [9]. This comparison is graphically illustrated in Figure 8.7. It is observed that when t/a and $t/(b - a)$ approach 0.3 and 0.1, respectively, the error is in the range of 5 to 10 percent [6] between the two sets of results.

8.2.4 Measurement-Based Design Equations

In [10] the measured attenuation characteristics of CPWs fabricated on GaAs ($\varepsilon_r = 12.85$), InP($\varepsilon_r = 12.4$) and high-resistivity silicon ($\rho \geqslant 2500\,\Omega$-cm and $\varepsilon_r = 11.9$) wafers are used to derive a closed form equation to predict the

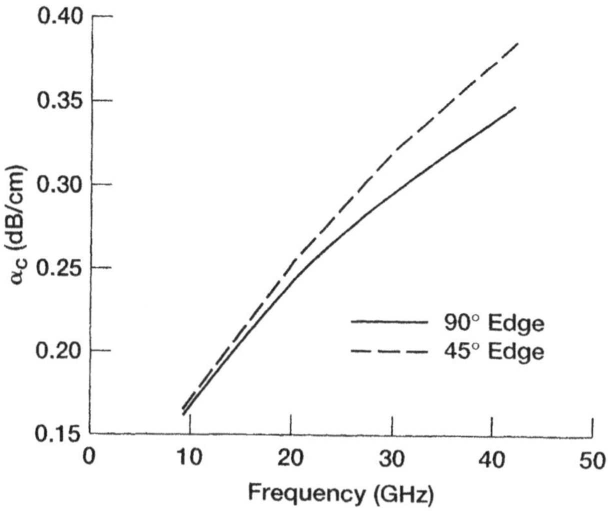

FIGURE 8.5 Computed attentuation constant due to conductor loss as a function of frequency for a CPW with 90° and 45° edges and $t = 3\,\mu m$, $a = 5\,\mu m$, $b = 25\,\mu m$, $\varepsilon_r = 12.9$, $g \to \infty$, and $\sigma_c = 3.0 \times 10^7$ S/meter. (Reference [6], © IEEE 1995.)

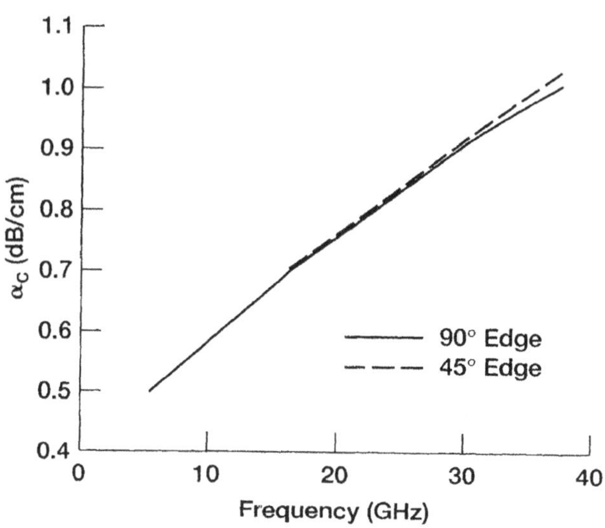

FIGURE 8.6 Computed attentuation constant due to conductor loss as a function of frequency for a CPW with 90° and 45° edges and $t = 1.61\,\mu m$, $a = 35.6\,\mu m$, $b = 84.6\,\mu m$, $\varepsilon_r = 12.9$, $g \to \infty$, and $\sigma_c = 3.602 \times 10^7$ S/meter. (Reference [6], © IEEE 1995.)

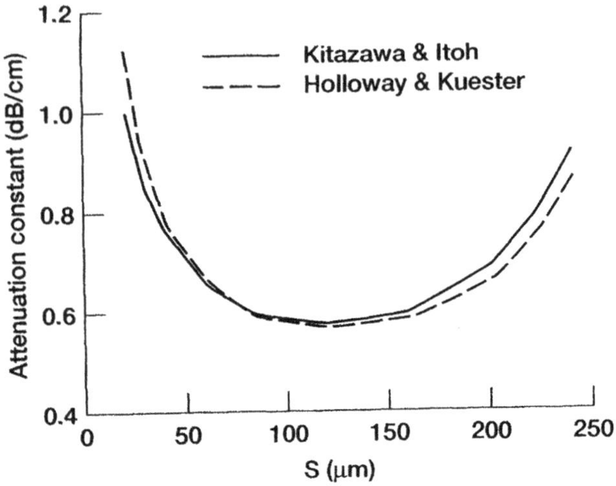

FIGURE 8.7 Computed attentuation constant due to conductor loss as a function of center strip conductor width S for a CPW with $t = 3.0\,\mu$m, $b = 150.0\,\mu$m, $\varepsilon_r = 12.8$, $g \to \infty$, and $\sigma_c = 5.882 \times 10^7$ S/meter. (Reference [6], © IEEE 1995.)

attenuation constant α. The equation for attenuation constant is as follows:

$$\alpha = pf^q \quad \text{dB/cm}, \tag{8.35}$$

where

$$p = \sqrt{\frac{\varepsilon_r + 1}{2}} \left[\frac{45.152}{(SW)^{0.41} e^{2.127\sqrt{t}}} \right], \tag{8.36}$$

$$q = 0.183(t + 0.464) - 0.095k_t^{2.484}(t - 2.595), \tag{8.37}$$

$$S = \text{strip width in } \mu\text{m}, \tag{8.38a}$$

$$W = \text{slot width in } \mu\text{m}, \tag{8.38b}$$

$$t = \text{conductor thickness in } \mu\text{m}, \tag{8.38c}$$

$$f = \text{frequency in GHz}, \tag{8.38d}$$

$$k_t = \frac{S + \Delta t}{S + 2W - \Delta t}, \tag{8.39}$$

$$\Delta t = \frac{1.25t}{\pi} \left[1 + \ln \frac{4\pi S}{t} \right]. \tag{8.40}$$

The ranges of validity of the equations above are

$$0.5 < t < 3\mu\text{m}, \tag{8.41a}$$

$$0.2 < \frac{S}{S + 2W} < 0.7, \tag{8.41b}$$

$$10 < S < 80\,\mu\text{m}, \tag{8.41c}$$

$$1 < f < 40\,\text{GHz}. \tag{8.41d}$$

In the experimental CPWs the propagation parameters are made independent of the ground plane width g and the substrate height h by choosing g to be $4S$ and h greater than $3(S+2W)$. In [10] it is shown that the α predicted by Eq. (8.35) is in good agreement with the measured α in [11] for CPW lines on a GaAs wafer and on a InP wafer ($t = 0.5\mu$m).

8.2.5 Accuracy of Closed Form Equations

In [10] an estimate of the accuracy of each of the four closed form equations described in Sections 8.2.1 through 8.2.4 is demonstrated by comparing them with a single set of measured data. The measurements in [10] are performed using a HP8510C automatic network analyzer and a microwave probe station. The HP8510C is calibrated using on-wafer coplanar waveguide (CPW) through-reflect-line (TRL) standards and four delay lines to cover the frequency range of 1 to 40 GHz. During calibration the HP8510C is computer controlled through the National Institute of Standards and Technology (NIST) software [12], and it acquires the scattering parameters (S-parameters) pertaining to each of the CPW standards and delay lines. The software de-embeds the CPW attenuation constant (α_{meas}) from the measured S-parameters as explained in [13]. This experiment is repeated three times, and the average $\bar{\alpha}_{\text{meas}}$ is calculated. This process is then extended to all 17 combinations of S and W included in the study. These combinations are representative of those encountered in typical MMIC design.

An estimate of the accuracy is obtained by calculating the normalized difference ($\Delta\alpha$) between the average measured attenuation constant ($\bar{\alpha}_{\text{meas}}$) and the calculated attenuation constant (α_{calc}), that is,

$$\Delta\alpha = \frac{\bar{\alpha}_{\text{meas}} - \alpha_{\text{calc}}}{\bar{\alpha}_{\text{meas}}}. \tag{8.42}$$

TABLE 8.2 Average Root-Mean-Square Error between Measured and Calculated Attenuation across the Frequency Band of 1–40 GHz for CPW Lines Fabricated on GaAs with Metal Thickness of 1.58 μm ($\sigma = 4.1 \times 10^7$ S/m Used to Calculate Attenuation)

$S(\mu m)$	$W(\mu m)$	Collin [2]	Holloway-Kuester [6]	Heinrich [4]	Ponchak et al. [10]
10	4.5	0.177	0.248	0.154	0.018
20	6	0.124	0.177	0.088	0.018
30	8	0.111	0.150	0.076	0.034
40	10	0.080	0.111	0.046	0.028
50	11	0.074	0.093	0.042	0.031
10	9	0.085	0.143	0.087	0.022
20	16	0.083	0.062	0.040	0.037
30	22	0.111	0.035	0.070	0.055
40	28	0.141	0.057	0.106	0.069
50	35	0.119	0.050	0.082	0.031
60	40	0.184	0.105	0.165	0.070
70	46	0.215	0.143	0.209	0.082
80	51	0.224	0.153	0.222	0.069
10	20	0.088	0.076	0.054	0.043
20	35	0.123	0.044	0.068	0.045
30	50	0.163	0.081	0.123	0.058
40	65	0.216	0.142	0.200	0.082
average	error	0.136	0.110	0.108	0.047

Source: From Reference [10], © IEEE 1999.

The $\Delta\alpha$ is determined at 201 points across the frequency interval of 1 to 40 GHz and the average-root-mean-square error ($\Delta\alpha_{rms}$) is calculated, that is,

$$\Delta\alpha_{rms} = \sqrt{\frac{(\Delta\alpha_1)^2 + (\Delta\alpha_2)^2 + \cdots + (\Delta\alpha_{201})^2}{201}}. \tag{8.43}$$

The $\Delta\alpha_{rms}$ for all 17 combinations and the average error for each of the four closed form equations [10] are presented in Table 8.2

According to [10], the attenuation constant predicted by Collin's method [2] underestimates the valves at low frequencies. This is attributed to the thick metal assumption, that is, $t \gg d$. The average error for this method is on the order of 13.6 percent. On the other hand, Heinrich's method [4] and Holloway's and Kuester's technique [5] do not assume the metal to be thick. Hence the difference between the measured and calculated attenuation constant is nearly independent of frequency. The average error for these two methods are on the order of 10.8 percent and 11.0 percent, respectively. The Ponchak et al.

equations [10] result in an average error on the order of 4.7 percent. Thus among the above four closed-form equations, the method described in [10] has the least average error.

8.3 INFLUENCE OF GEOMETRY ON COPLANAR WAVEGUIDE ATTENUATION

8.3.1 Attenuation Constant Independent of the Substrate Thickness and Dielectric Constant

In [14] the attenuation constant α of two coplanar waveguides fabricated on a thin quartz substrate (165 μm) and on a thick semi-insulating GaAs wafer (525 μm) is measured. These coplanar waveguides have identical center strip conductor, slot, and ground plane width. The measurements show that the attenuation in dB/wavelength is about the same for both coplanar waveguides, even though the substrate thickness and dielectric constant are different. This suggests that α is independent of substrate thickness and dielectric constant.

8.3.2 Attenuation Constant Dependent on the Aspect Ratio

In [9] the attenuation constant α is computed as a function of the characteristic impedance Z_0 with the ground plane separation $(S+2W)$ as a parameter. The results from [9] are shown in Figure 8.8. The figure shows that for a fixed

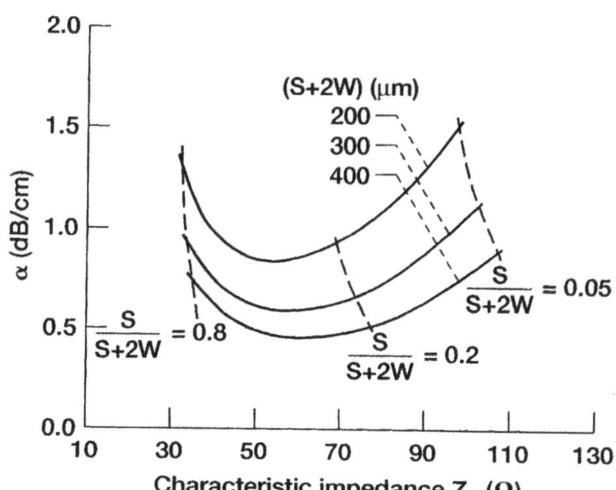

FIGURE 8.8 Computed coplanar waveguide attentuation constant as a function of the characteristic impedance with the aspect ratio and ground plane separation as parameters, $t = 3.0\,\mu$m, $\varepsilon_r = 12.8$, $\tan\delta_e = 0.0006$, $g \rightarrow \infty$, and $\rho = 1.7 \times 10^6\,\Omega$-cm. (Reference [9], © IEEE 1991.)

ground plane separation, the attenuation constant initially decreases and later increases as the characteristic impedance increases or the aspect ratio $S/(S+2W)$ decreases. It is interesting to note that minimum attenuation occurs for characteristic impedence close to 50 Ω.

8.3.3 Attenuation Constant Varying with the Elevation of the Center Strip Conductor

The attenuation characteristics when the center strip conductor is elevated above the plane of the substrate are discussed in Section 2.7.

8.4 ATTENUATION CHARACTERISTICS OF COPLANAR WAVEGUIDE ON SILICON WAFER

8.4.1 High-Resistivity Silicon Wafer

Attenuation Constant as a Function of Wafer Resistivity The measured attenuation constant α of coplanar waveguide on silicon wafer as a function of frequency with the silicon wafer resistivity as a parameter is reported in [15]. The resistivity of the wafers are in the range of 400 to 30,000 Ω-cm and the conductors of the coplanar waveguide are separated from the silicon wafer by a thin layer of silicon dioxide of thickness t_1, for reasons explained later. In the slot regions the silicon dioxide is etched away as shown in Figure 8.9. The measured attenuation constant is summarized in Table 8.3, where it is observed that there is no significant improvement in the attenuation constant beyond 2500 Ω-cm wafer resistivity. In [16] it is shown that the measured attenuation constant of a coplanar waveguide on a silicon wafer of resistivity 2500 Ω-cm is about the same as that of a coplanar waveguide on a semi-insulating GaAs wafer. Since it has been proved that semi-insulating GaAs wafers are excellent materials for monolithic microwave integrated circuits (MMICs), these data

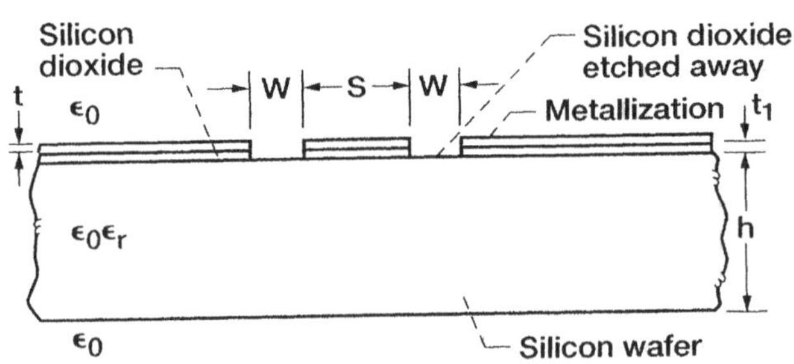

FIGURE 8.9 Coplanar waveguide on a silicon wafer with a thin interfacial layer of silicon dioxide.

TABLE 8.3 Measured Attenuation of CPW on High-Resistivity Silicon Wafers

Resistivity Ω-cm	Attenuation (dB/cm)		
	10 GHz	20 GHz	30 GHz
400–720	1.08	1.23	1.39
2500–3300	0.62	0.77	0.91
5000	0.58	0.72	0.87
5000–10,000	0.50	0.70	0.86
>30,000	0.45	0.64	0.85

Note: $S = 100\,\mu m$, $W = 50\,\mu m$, $t = 2.5\,\mu m$ (gold), $t_1 = 1000$ Å (SiO_2), $Z_0 \approx 50\,\Omega$.

indicate that silicon wafers with resistivity greater than 2500 Ω-cm are also viable substrates for MMIC applications.

Conductors in direct Contact with the Silicon Wafer When the coplanar waveguide conductors are in direct contact with the high-resistivity silicon (HR-Si) wafer, the signal suffers the least attenuation [17]. However, a dc leakage current can flow between the CPW center strip conductor and the ground planes, since the HR-Si is not a perfect insulator [17]. The leakage current is not a concern in the case of passive devices such as couplers and filters. But leakage can be a serious problem when active devices are integrated. The measured attenuation constant α [16] for typical coplanar waveguide geometries used in MMIC design when the conductors are in direct contact with the silicon wafer are summarized in Table 8.4.

TABLE 8.4 Measured Attenuation of CPW on 2500 Ω-cm Silicon Wafer

Frequency (GHz)	Attenuation (dB/cm)	
	$S = 10\ \mu m$, $W = 9\ \mu m$	$S = 80\ \mu m$, $W = 51\ \mu m$
5	2.59	0.6
10	3.18	0.74
20	3.88	0.88
30	4.59	1.06
40	5.29	1.20

Note: $h = 410\,\mu m$, $t = 0.02/1.45\,\mu m$ (Ti/Au), $Z_0 = 50\,\Omega$.

TABLE 8.5 Measured Attenuation of CPW on HR-Si Wafer with a Thick Interface Oxide Layer

Frequency (GHz)	Attenuation (dB/cm)
2.5	1.6
5.0	2.0
10.0	2.7
15.0	3.6
20.0	4.3

Note: $S = 8\,\mu m$, $W = 5\,\mu m$, $g = 40\,\mu m$, $t = 2.4\,\mu m$ (Cu), $t_1 = 6.6\,\mu m$ (SiO_2), $\rho > 1000\,\Omega$-cm, $Z_0 = 60\,\Omega$.

Conductors Separated from the Silicon Wafer by a Thin Insulating Layer A simple solution to block the leakage current is to deposit a thin insulator, such as silicon dioxide (SiO_2, $\varepsilon_r = 3.9$) or silicon nitride (Si_3N_4, $\varepsilon_r = 7.5$), between the silicon wafer and the metal conductors. In such a structure because of defects in the oxide layer, fixed positive charges are present at a distance of few tens of angstroms from the semiconductor-insulator interface [18]. These fixed positive charges set up an electron populated inversion layer in a p-type silicon or an accumulation layer in a n-type silicon at the interface [17], and [19]. This charge layer provides a low-resistance path for the RF fields in the coplanar waveguide slots. Consequently the coplanar waveguide attenuation increases substantially [17], [19] to [21] when an oxide or a nitride layer is present. In addition the interface can be charged or discharged depending upon the surface potential, hence the attenuation varies as a function of the dc bias voltage [19]. The charge layer can be interrupted by etching away the SiO_2 or Si_3N_4 layer from the slot regions as shown in Figure 8.9 [17] and [19]. After this etch step, the coplanar waveguide attenuation is almost independent of the dc bias and is about the same as when the metal is in direct contact with the high-resistivity silicon [17].

Conductors Separated from the Silicon Wafer by a Very Thick Insulating Layer In the coplanar waveguide structures considered in the previous sections the thickness of the oxide or nitride layer is few thousand angstroms while the slot width is several microns wide. This proximity causes significant interaction to take place between the RF fields in the slot and the interface charge sheet. However, when the oxide or nitride layer is very thick on the order of several microns and the slot width is less than the oxide thickness ($W < t_1$), the interaction of the RF fields with the interface charge sheet is small due to the large distance of separation. This is evident from the attenuation measurements reported in [22]. The results from [22] are summarized in Table 8.5. It is interesting to note that the attenuation constant given in Table 8.5 for a coplanar waveguide with very thick oxide layer is about the same as when the coplanar waveguide conductors are in direct contact with the high-resistivity silicon wafer (see the second column of Table 8.4).

8.4.2 Low-Resistivity Silicon Wafer

In the previous sections it was shown that for low attenuation the resistivity of the silicon wafer should be greater than 2500 Ω-cm. However, these high-resistivity silicon wafers are more expensive than the standard CMOS grade silicon wafers. Further there is also the possibility of them not being fully compatible with the standard silicon IC process. The standard process uses CMOS grade wafers with resistivity typically on the order of 0.5 to 20.0 Ω-cm and aluminium for interconnects. Hence in this section a technique to lower the attenuation of coplanar waveguide on standard CMOS grade silicon wafers is discussed. A simple solution to isolate the RF fields in the coplanar waveguide slots from the low-resistivity wafer is to use a thick insulating layer of SiO_2 or polyimide. To lower further the surface resistance of the coplanar waveguide conductors, a metal with conductivity higher than that of aluminium is chosen. The measured attenuation constants for coplanar waveguide with different insulating material thickness is summarized in Table 8.6. Two observations can be made from this table. First, as the polyimide thickness increases, the attenuation constant decreases [23]. Second, when the slot width is less than the SiO_2 layer thickness, the attenuation constant is insensitive to the substrate resistivity [22]. This is evident by comparing the attenuation constant of a coplanar waveguide on a low-resistivity silicon wafer (the last row of Table 8.6) with the attenuation constant of an identical coplanar waveguide on a high-resistivity silicon wafer (see Table 8.5).

8.5 ATTENUATION CHARACTERISTICS OF COPLANAR WAVEGUIDE ON MICROMACHINED SILICON WAFER

8.5.1 Microshield Line

A microshield line [26] to [28] is shown in Figure 7.1(*b*). In the microshield line the center strip conductor and the upper ground planes are supported by a dielectric membrane. The dielectric membrane is formed by depositing a tri-layer of $SiO_2/Si_3N_4/SiO_2$ on a silicon wafer. The shielding cavity below the membrane is formed by micromachining the silicon wafer and coating the surface with a metal, such as gold. The bottom wall of the cavity is formed by the metallized surface of a second silicon wafer that is placed below. The thickness of the strip conductor, the ground planes and the metal coating is about two to three times the skin depth at the operating frequency. A transition between a microshield line and a grounded coplanar waveguide is essential for on-wafer characterization of microshield lines. A typical transition is discussed

TABLE 8.6 Attenuation of CPW on CMOS Grade Silicon Wafer with an Interfacial Dielectric Layer

Substrate Resistivity, Ω-cm	Insulating Material Thickness, μm	Metal for CPW Conductor and thickness, μm	CPW Strip Width, μm	CPW Slot Width, μm	Z_0, Ω	Attenuation (dB/mm) 4.0 GHz	10.0 GHz	20.0 GHz	40.0 GHz	Reference
n-type Si 0.5–2.5	Polyimide 8.83	Au 1.5	10	4.5	50	—	0.3	—	0.75	[23]
n-type Si 0.5–2.5	Polyimide 14.59	Au 1.5	20	6	50	—	0.22	—	0.62	[23]
n-type Si 0.5–2.5	Polyimide 20.15	Au 1.5	30	8	50	0.11	0.17	0.285	0.55	[23]
p-type Si 20	Polyimide 10	Al 4.0	12	18	—	0.2	—	—	—	[24]
Si 5–20	Polyimide 10	Ti/Au 0.05/0.3	34	8	50	0.4	—	—	1.2	[19]
p-type Si 8	Oxidized porous silicon 20	Au 2.5	100	30	50	0.1	0.3	—	—	[25]
Si 10	SiO_2 6.3	Cu 2.7	8	5	53	0.22	0.325	0.5	—	[22]

later inSection 10.4.1. Using two such transitions, a set of RF probes and the NIST software [12] the effective dielectric constant and attenuation constant can be determined. In Table 8.7 the geometry and the measured attenuation constant [28] of typical microshield lines are summarized. The table shows that the attenuation constant is as low as 0.03 dB/mm at 40.0 GHz.

8.5.2 Coplanar Waveguide with V-Shaped Grooves

A coplanar waveguide with V-shaped micromachined grooves and with finite width ground planes [29], [30] is shown in Figure 8.10. In this line the silicon material below the two slots is removed by a wet chemical etchant such as ethylenediamene pyrocatechol (EDP). As a result a V-shaped groove is formed below the slots with side walls dictated by the 54.74° angle of the ⟨111⟩ silicon crystal planes [31]. The groove size G is defined as the slot width W plus the lateral undercut U on either side. The depth of the center of the V-shaped groove depends on the dimension G and the angle of 54.74°. A $SiO_2/Si_3N_4/SiO_2$ protective layer of thickness t_1 is present between the conductors and the silicon wafer. In Table 8.8 the geometry and the measured attenuation constant [29] of typical coplanar waveguides with V-shaped grooves are summarized. In addition for the purpose of comparison, the measured attenuation constant of conventional coplanar waveguides with identical strip and slot width is presented. This table shows that the effective dielectric constant and the attenuation constant decreases as material below the slots is etched away. Furthermore a lower effective dielectric constant results in a lower line capacitance that increases the characteristic impedance Z_0.

8.5.3 Coplanar Waveguide Suspended by a Silicon Dioxide Membrane over a Micromachined Wafer

In this section, first, a technique to lower the attenuation of a coplanar waveguide on a standard CMOS grade silicon wafer through the use of micromachining techniques is explained. Second, the measured attenuation constant of the coplanar waveguide before and after micromachining the wafer is presented.

A coplanar waveguide suspended by a silicon dioxide membrane over a micromachined silicon wafer [32], [33] is shown in Figures 8.11(a) and (b). The line is fabricated in a commercial 2 μm CMOS n-well process on a 460 μm thick ⟨100⟩ silicon wafer. The three steps involved in the fabrication process are briefly explained below. First, the silicon dioxide layer for the membrane and the aluminum conductor strips for the coplanar waveguide are formed over the silicon wafer. Second, through the openings in silicon dioxide layer small cavities are created in the silicon wafer adjacent to the conductors by reactive ion etching process. This is an isotropic etch, and hence the grooves propagate outward radially beneath the coplanar waveguide conductors and connect the cavities on either sides of the coplanar waveguide. This is followed by the third

TABLE 8.7 Measured Effective Dielectric Constant and Attenuation of Microshield Line

Strip Width S, μm	Slot Width W, μm	Height H, μm	Metallization Thickness t, μm	Characteristic Impedance Z_0, Ω	Effective Dielectric Constant, ε_{eff}	Attenuation in dB/mm		
						10.0 GHz	20.0 GHz	40.0 GHz
250	25	355	1.2	75	1.15	0.045	0.055	0.06
190	55	355	1.2	100	1.08	0.02	0.025	0.03

Note: Width of lower shielding cavity is 1800 μm. Membrane is SiO_2 (7000 Å)/Si_3N_4 (3000 Å)/SiO_2 (4000 Å).

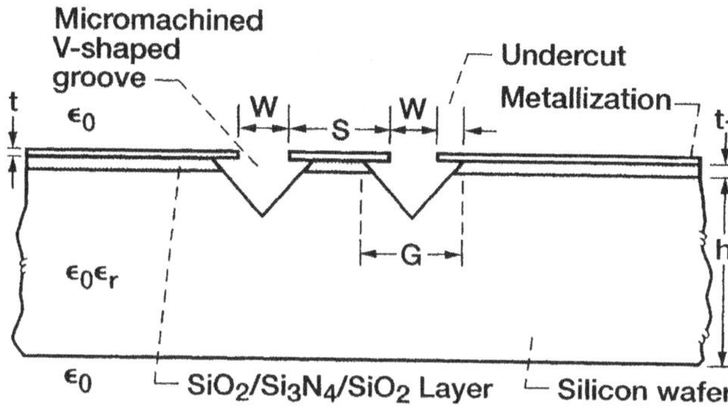

FIGURE 8.10 Coplanar waveguide with V-shaped micromachined grooves.

and last step in which the wafer is immersed in a chemical etchant. This is an anisotropic etch and follows the crystalline structure of the ⟨100⟩ wafer forming a V-shaped groove. The walls of the groove slope at an angle of 54.74° from the surface plane and are aligned to the ⟨111⟩ crystallographic plane of the silicon wafer.

The measured attenuation constant before and after micromachining the wafer [32] is presented in Table 8.9. Clearly, there is a very significant improvement in the attenuation constant after the micromachining process. However, the improvement is limited by the relatively high series resistance of the strip conductors because of the very small thickness of the metal layers. In a commercial CMOS process, the thickness of the metal layer is fixed and is not a design parameter. Hence further improvement in attenuation constant can be obtained by increasing the strip conductor widths.

8.6 ATTENUATION CONSTANT FOR SUPERCONDUCTING COPLANAR WAVEGUIDES

In this section the closed form expressions presented for computing the attenuation constant α_c of superconducting coplanar waveguide [34] include conductor losses. These expressions are valid for arbitrary conductor thickness and are derived in a manner analogous to that for normal metal as in Section 8.2.3.

8.6.1 Stopping Distance

The first step is to determine the stopping distance Δ. The value of the stopping distance is determined numerically and is a function only of the edge profile θ, the strip thickness t, and the material properties of the conductors. The edge profile θ is assumed to be 90°; however, θ can take any arbitrary value between

TABLE 8.8 Measured Attenuation of Si-Micromachined Coplanar Waveguide at *W*-Band (94 GHz)

Strip Width, μm	Slot Width, μm	Parameter	Conventional Coplanar Waveguide	Micromachined Coplanar Waveguide Undercut				
				2 μm	4 μm	6 μm	8 μm	10 μm
20	45	ε_{eff}	4.938	4.165	3.560	2.520	—	—
		α(dB/cm)	3.65	3.2	2.7	2.3	—	—
		Z_0, (Ω)	71.	77.31	83.62	99.39	—	—
40	80	ε_{eff}	5.57	4.625	4.0	3.494	3.189	2.973
		α(dB/cm)	1.95	1.8	1.75	1.65	1.6	1.5
		Z_0 (Ω)	64	70.2	75.5	80.8	84.58	87.59
50	45	ε_{eff}	5.408	4.83	4.34	4.081	3.845	3.698
		α(dB/cm)	2.2	1.6	1.2	0.8	—	—
		Z_0, (Ω)	57	60.31	63.53	65.62	67.6	68.93

Note: Silicon substrate thickness and resistivity are 500 μm and $>2000\,\Omega$-cm. $SiO_2/Si_3N_4/SiO_2$ protective layer is 1.5 μm thick. Metallization is 1.7 to 2.0 μm thick.

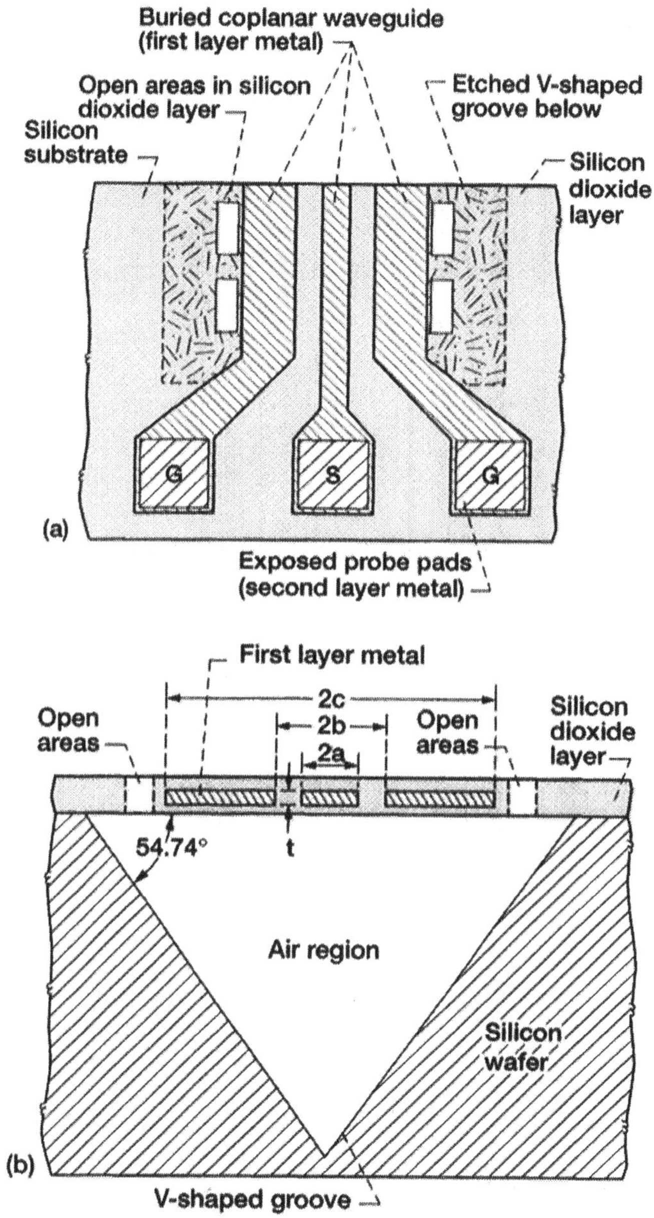

FIGURE 8.11 Coplanar waveguide suspended by a silicon dioxide membrane over a micromachined substrate: (*a*) Top view; (*b*) cross section.

TABLE 8.9 Measured Attenuation of Coplanar Waveguide Suspended by a Silicon Dioxide Membrane over a Micromachined Wafer

Strip Width, $2a$ (μm)	Slot Width, $b-a$ (μm)	Ground plane Width, $c-b$ (μm)	Conductor Thickness t (μm)	ε_{eff}	Z_0 (Ω)	Attenuation (dB/cm)							
						Before Micromachining				After Micromachining			
						10 GHz	20 GHz	30 GHz	40 GHz	10 GHz	20 GHz	30 GHz	40 GHz
24	3	50	0.6	2.18	50	17.0	25.5	31.75	38.0	2.5	3.0	3.5	4.0

45° and 90°, as described in Section 8.2.3 for normal metals. The thickness t is also arbitrary and typically can range from 0.05 to 0.5 μm. The material properties of the conductors are taken into account through the complex conductivity $\bar{\sigma}$ which is defined as [34]

$$\bar{\sigma} = \sigma_1 - j\sigma_2 \quad \text{S/meter}, \tag{8.44}$$

where σ_1 and σ_2 are the real and imaginary part respectively of the conductivity. The real and imaginary part represent the normal and super fluid, respectively, in a two-fluid model for a superconductor. The real part of the conductivity can be included in the expression for the normal fluid skin depth δ_{nf} defined as [34]

$$\delta_{\text{nf}} = \sqrt{\frac{2}{\mu_0 \omega \sigma_1}} \quad \text{meters}, \tag{8.45}$$

where

$\mu_0 = 4\pi \times 10^{-7}$ Henry/meter
$\omega = 2\pi f$
f = frequency of operation in c/s.

Likewise the imaginary part of the conductivity can be included in the expression for the superconducting magnetic penetration depth λ_p which is defined as [34]

$$\lambda_p = \sqrt{\frac{1}{\mu_0 \omega \sigma_2}} \quad \text{meters}. \tag{8.46}$$

The λ_p is also referred to as the London penetration depth. Typically for a good superconductor $\lambda_p \ll \delta_{\text{nf}}$ and the ratio $t/2\delta_{\text{nf}}$ is $\leqslant 0.1$. This is evident from Table 8.10, which presents the mesured material characteristics of superconducting $YBa_2Cu_3O_{7-\delta}$ thin films [34]. This material represents a general class of high-T_c superconductors.

In Figure 8.12 universal curves [34] for the numerically determined stopping distance Δ are presented as a function of the normalized thickness $t/2\lambda_p$ and the ratio $t/2\delta_{\text{nf}}$ as a parameter for a superconducting strip with an edge profile $\theta = 90$. Notice that the stopping distance Δ is sensitive to the ratio $t/2\lambda_p$ but insensitive to small values of the parameter $t/2\delta_{\text{nf}}$, typically in the range of $0 < t/2\delta_{\text{nf}} \leqslant 0.2$. The parameter $t/2\delta_{\text{nf}}$ for a good superconductor is $\leqslant 0.1$, from Table 8.10. Hence we can set the ratio $t/2\delta_{\text{nf}} \approx 0$ and use just one of the curves shown in Figure 8.12 for the stopping distance Δ of a strip fabricated from a good superconductor. The data in Figure 8.12 are tabulated in Table 8.11 [37] to facilitate computation of the attenuation constant as explained in the next section.

TABLE 8.10 Material Characteristics of $YBa_2Cu_3O_{7-\delta}$ Thin Film Grown by Pulsed-Laser Deposition on $LaAlO_3$ ($T_c \approx 91$ K)

t (μm)	λ_p (μm)	$t/2\lambda_p$	σ_1 (S/m)	Frequency (GHz)	δ_{nf} (μm)	$t/2\delta_{nf}$
			Temperature = 76 K			
0.05	0.316	0.0791	1.72×10^6	3	7.00	0.0036
				10	3.84	0.0065
				20	2.71	0.0092
				30	2.22	0.0113
				40	1.92	0.0130
			Temperature = 30 K			
0.05	0.205	0.122	2.13×10^6	3	6.30	0.0040
				10	3.45	0.0072
				25	2.18	0.0115
0.32		0.7805	2.63×10^6	3	5.66	0.0282
				10	3.10	0.0516
				25	1.96	0.0815
0.5		1.2195	3.13×10^6	3	5.19	0.0481
				10	2.84	0.0879
				25	1.80	0.1390

8.6.2 Closed Form Equations

After the stopping distance Δ has been determined, the attenuation constant of a superconducting coplanar waveguide can be computed using [34]

$$\alpha_c \approx \frac{R_{sm} b^2}{16 Z_0 [K^2(a/b)](b^2 - a^2)} \left\{ \frac{1}{a} \ln\left(\frac{2a(b-a)}{\Delta(b+a)}\right) + \frac{1}{b} \ln\left(\frac{2b(b-a)}{\Delta(b+a)}\right) \right\}, \tag{8.47}$$

where

$2a$ = coplanar waveguide center strip conductor width S
$2b$ = coplanar waveguide ground plane separation equal to $(S + 2W)$
$K(a/b)$ = complete elliptic integral of the first kind
Z_0 = coplanar waveguide characteristic impedance given by Eq. (8.15):

$$R_{sm} = \mu_0 \omega t \text{ Imaginary} \left(\frac{\cot(k_c t) + \csc(k_c t)}{k_c t} \right) \tag{8.48}$$

which is a modified surface impedance for a superconductor strip of thickness t.

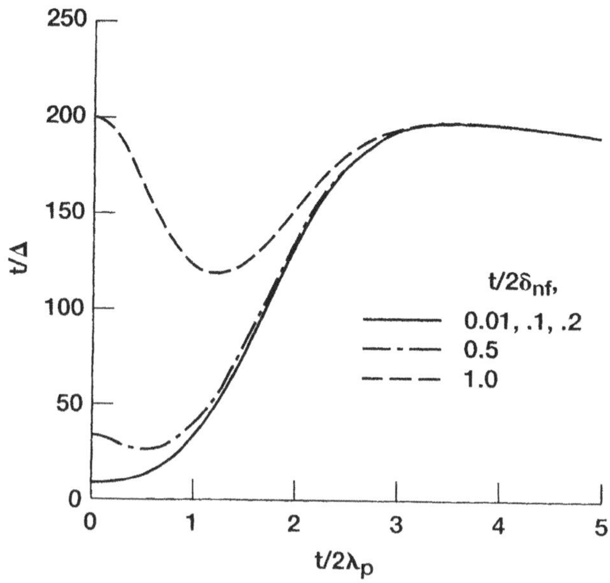

FIGURE 8.12 Computed stopping distance as a function of $t/2\lambda_p$ for various values of $t/2\delta_{nf}$. (Reference [34], © IEEE 1999.)

The complex wavenumber k_c in the superconductor at a frequency ω is given by [34]

$$k_c^2 = \left(\frac{1}{\lambda_p}\right)^2 + 2j\left(\frac{1}{\delta_{nf}}\right)^2. \tag{8.49}$$

The losses occurring both in the center strip conductor as well as the two ground planes of the coplanar waveguide are included in the expression for the attenuation constant given in Eq. (8.47).

The quality factor (Q) of a superconducting coplanar waveguide resonator is given by [34]

$$Q = \frac{\beta}{2\alpha}, \tag{8.50}$$

where β is the phase constant and is expressed as

$$\beta \approx \left(\frac{\omega}{c}\right)\sqrt{\varepsilon_{eff}} \text{ rad/m}, \tag{8.51}$$

where c is the velocity of light in vacuum and ε_{eff} is the effective dielectric

TABLE 8.11 Stopping Distance t/Δ as a Function of $t/2\lambda_p$ for Various Values of $t/2\delta_{nf}$

	t/Δ				
$t/2\lambda_p$	$t/2\delta_{nf}=1.0$	$t/2\delta_{nf}=0.5$	$t/2\delta_{nf}=0.2$	$t/2\delta_{nf}=0.1$	$t/2\delta_{nf}=0.01$
0.0000	200.50	33.968	10.336	9.2570	9.1794
0.10000	198.70	33.266	10.267	9.2685	9.1974
0.20000	193.49	31.465	10.275	9.4667	9.4088
0.30000	185.47	29.289	10.706	10.095	10.054
0.40000	175.50	27.493	11.742	11.285	11.254
0.50000	164.54	26.594	13.447	13.096	13.072
0.60000	153.54	26.840	15.865	15.585	15.565
0.70000	143.33	28.300	19.052	18.822	18.805
0.80000	134.49	30.965	23.072	22.877	22.864
0.90000	127.43	34.814	27.992	27.823	27.812
1.0000	122.33	39.823	33.865	33.719	33.708
1.1000	119.23	45.967	40.726	40.601	40.588
1.2000	118.06	53.202	48.576	48.454	48.454
1.3000	118.64	61.455	57.370	57.278	57.262
1.4000	120.79	70.608	67.013	66.925	66.918
1.5000	124.26	80.500	77.354	77.279	77.271
1.6000	128.80	90.928	88.198	88.131	88.125
1.7000	134.15	101.66	99.312	99.261	99.251
1.8000	140.06	112.45	110.45	110.39	110.40
1.9000	146.27	123.05	121.37	121.32	121.33
2.0000	152.57	133.24	131.84	131.82	131.80
2.1000	158.75	142.83	141.68	141.63	141.66
2.2000	164.66	151.68	150.75	150.70	150.73
2.3000	170.16	159.69	158.94	158.94	158.92
2.4000	175.16	166.80	166.21	166.23	166.19
2.5000	179.62	173.01	172.54	172.55	172.52
2.6000	183.50	178.32	177.96	177.99	177.95
2.7000	186.81	182.80	182.51	182.50	182.50
2.8000	189.57	186.49	186.27	186.27	186.26
2.9000	191.82	189.47	189.30	189.29	189.30
3.0000	193.58	191.82	191.70	191.69	191.70
3.1000	194.93	193.62	193.53	193.53	193.53
3.2000	195.90	194.94	194.88	194.87	194.88
3.3000	196.54	195.86	195.81	195.81	195.81
3.4000	196.91	196.44	196.40	196.39	196.40
3.5000	197.05	196.73	196.70	196.70	196.70
3.6000	196.99	196.79	196.77	196.77	196.77
3.7000	196.77	196.66	196.65	196.66	196.65
3.8000	196.43	196.38	196.37	196.36	196.37
3.9000	195.98	195.98	195.98	195.99	195.98
4.0000	195.46	195.49	195.49	195.50	195.49
4.1000	194.89	194.94	194.94	194.94	194.94
4.2000	194.28	194.34	194.35	194.38	194.35
4.3000	193.64	193.72	193.72	193.69	193.72
4.4000	192.99	193.07	193.08	193.05	193.08
4.5000	192.34	192.42	192.43	192.46	192.43
4.6000	191.70	191.78	191.78	191.75	191.78
4.7000	191.06	191.14	191.15	191.17	191.15
4.8000	190.45	190.52	190.53	190.56	190.53
4.9000	189.85	189.92	189.93	189.94	189.93
5.0000	189.28	189.35	189.35	189.40	189.35

Source: From Reference [37].

TABLE 8.12 Measured Attenuation Constant of $YBa_2Cu_3O_{7-\delta}$ Superconducting 50 Ω Coplanar Waveguides on $LaAlO_3$ Substrate ($T_c \approx 91$ K)

t (μm)	T (K)	Measured λ_p (μm)	Measured σ_1 (S/m)	Strip Width (μm)	Slot Width (μm)	Measured Attenuation (dB/cm) 10 GHz	20 GHz	30 GHz	40 GHz
0.05	76	0.316	1.72×10^6	21	40	0.1	0.5	1.15	2.0
				53	101	0.05	0.25	0.75	1.4
				105	200	—	0.2	0.35	0.65

constant approximated as

$$\varepsilon_{\text{eff}} \approx \frac{\varepsilon_r + 1}{2} \tag{8.52}$$

and ε_r is the dielectric constant of the substrate material.

8.6.3 Comparison with Numerical Calculations and Measured Results

The attenuation constant computed for a high-temperature superconducting coplanar waveguide using the closed form Eq. (8.47) and using the mode-matching method [3], [35] are shown to be in excellent agreement in [34]. In addition the measured attenuation constant and the computed attenuation constant using the closed form Eq. (8.47) for three high-temperature superconducting coplanar waveguide structures are shown to be in good agreement in [34]. These measured attenuation constants are representative of a general class of high-temperature superconducting materials, and they are summarized in Table 8.12. Additional charactersitics and applications of high-temperature superconducting coplanar waveguides can be found in Section 12.4 and [36], respectively.

REFERENCES

[1] Data Sheet RT/duroid 5880, Glass Microfiber Reinforced Polytetrafluoroethylene Composite, Rogers Corporation, Chandler, AZ 85226.

[2] R. E. Collin, *Foundations for Microwave Engineering*, 2nd ed., New York: McGraw-Hill, 1992, pp. 178–179.

[3] W. Heinrich, "Full-Wave Analysis of Conductor losses on MMIC Transmission Lines," *IEEE Trans. Microwave Theory Tech.*, Vol. 38, No. 10, pp. 1468–1472, Oct. 1990.

[4] W. Heinrich, "Quasi-TEM Description of MMIC Coplanar Lines Including Conductor-Loss Effects," *IEEE Trans. Microwave Theory Tech.*, Vol. 41, No. 1, pp. 45–52, Jan. 1993.

[5] C. L. Holloway and E. F. Kuester, "Edge Shape Effects and Quasiclosed Form Expressions for the Conductor Loss in Microstrip," *Radio Science*, Vol. 29, No. 3, pp. 539–559, May-June 1994.

[6] C. L. Holloway and E. F. Kuester, "A Quasi-Closed Form Expression for the Conductor Loss of CPW Lines, with an Investigation of Edge Shape Effects," *IEEE Trans. Microwave Theory Tech.*, Vol. 43, No. 12, pp. 2695–2701, Dec. 1995.

[7] D. Williams and R. Marks, NIST, Boulder, CO., personal communication with authors of [6].

[8] C. L. Holloway, R. R. DeLyser, and E.F. Kuester, "Field Penetration Corrections for the Characteristic Impedance of Planar Structures," 1997 *IEEE Antennas and Propagation Society Int. Symp. Dig.*, Vol. 1, pp. 186–189, Montreal, Canada, July 13–18, 1997.

[9] T. Kitazawa and T. Itoh, "Propagation Characteristics of Coplanar-Type Transmission Lines with Lossy Media," *IEEE Trans. Microwave Theory Tech.*, Vol. 39, No. 10, pp. 1694–1700, Oct. 1991.

[10] G. E. Ponchak, M. Matloubian, and L. P. B. Katehi, "A Measurement-Based Design Equation for the Attenuation of MMIC-Compatible Coplanar Waveguides," *IEEE Trans. Microwave Theory Tech.*, Vol. 47, No. 2, pp. 241–243, Feb. 1999.

[11] W. H. Haydl, J. Braunstein, T. Kitazawa, M. Schlechtweg, P. Tasker, and L. F. Eastman, "Attenuation of Millimeterwave Coplanar Lines on Gallium Arsenide and Indium Phosphide over the Range 1 to 60 GHz," *1992 IEEE Int. Microwave Symp. Dig.*, Vol. 1, pp. 349–352, Albuquerque, New Mexico, June 1–5, 1992.

[12] NIST De-embedding Software, Program DEEMBED, Revision 4.04, 1994.

[13] R. B. Marks, "A Multiline Method of Network Analyzer Calibration," *IEEE Trans. Microwave Theory Tech.*, Vol. 39, No. 7, pp. 1205–1215, July 1991.

[14] J. Papapolymerou, J. East, and L. P. B. Katehi, "GaAs versus Quartz FGC Lines for MMIC Applications," *IEEE Trans. Microwave Theory Tech.*, Vol. 46, No. 11, pp. 1790–1793, Nov. 1998.

[15] S. R. Taub and P. G. Young, "Attenuation and ε_{eff} of Coplanar Waveguide Transmission Lines on Silicon Substrates," *11th Annual Benjamin Franklin Symp. Dig.*, pp. 8–11, King of Prussia, PA, May 1, 1993.

[16] G. E. Ponchak, A. N. Downey, and L. P. B. Katehi, "High Frequency Interconnects on Silicon Substrates," *1997 IEEE Radio Frequency Integrated Circuits (RFIC) Symp. Dig.*, pp. 101–104, Denver, CO, June 8–11, 1997.

[17] Y. Wu, H. S. Gamble, B. M. Armstrong, V. F. Fusco, and J. A. C. Stewart, "SiO_2 Interface Layer Effects on Microwave Loss of High-Resistivity CPW Line," *IEEE Microwave Guided Wave Lett.*, Vol. 9, No. 1, pp. 10–12, Jan. 1999.

[18] D. K. Schroder, *Semiconductor Material and Device Characterization*, New York, Wiley, 1990, ch. 6, p. 244.

[19] C. Warns, W. Menzel, and H. Schumacher, "Transmission Lines and Passive Elements for Multilayer Coplanar Circuits on Silicon," *IEEE Trans. Microwave Theory Tech.*, Vol. 46, No. 5, pp. 616–622, May 1998.

[20] A. C. Reyes, S. M. El-Ghazaly, S. J. Dorn, M. Dydyk, D. K. Schroder, and H. Patterson, "Coplanar Waveguides and Microwave Inductors on Silicon Substra-

tes," *IEEE Trans. Microwave Theory Tech.*, Vol. 43, No. 9, pp. 2016–2022, Sept. 1995.

[21] A. L. Caviglia, R. C. Potter, and L. J. West, "Microwave Performance of SOI n-MOSFET's and Coplanar Waveguides," *IEEE Electron Devices Lett.*, Vol. 12, No. 1, pp. 26–27, Jan. 1991.

[22] J. N. Burghartz, D. C. Edelstein, K. A. Jenkins, and Y. H. Kwark,"Spiral Inductors and Transmission Lines on Silicon Technology Using Copper-Damascene Interconnects and Low-Loss Substrates," *IEEE Trans. Microwave Theory Tech.*, Vol. 45, No. 10, pp. 1961–1968, Oct. 1997.

[23] G. E. Ponchak and L. P. B. Katehi, "Measured Attenuation of Coplanar Waveguide on CMOS Grade Silicon Substrates with Polyimide Interface Layer," *Electron. Lett.*, Vol. 34, No. 13, pp. 1327–1329, June 1998.

[24] B.-K. Kim, B.-K. Ko, and K. Lee, "Monolithic Planar Inductor and Waveguide Structures on Silicon with Performance Comparable to Those in GaAs MMIC," *Int. Electron Devices Meet. Tech. Dig.*, pp. 717–720, Washington, DC, Dec. 10–13, 1995.

[25] C.-M. Nam and Y.-S. Kwon, "Coplanar Waveguide on Silicon Substrate with Thick Oxidized Porous Silicon (OPS) Layer," *IEEE Microwave Guided Wave Lett.*, Vol. 8, No. 11, pp. 369–371, Nov. 1998.

[26] L. P. B. Katehi, "Novel Transmission Lines for the Submillimeter-Wave Region," *Proc. IEEE.*, Vol. 80, pp. 1771–1787, Nov. 1992.

[27] L. P. B. Katehi, G. M. Rebeiz, T. M. Weller, R. F. Drayton, H.-J. Cheng, and J. F. Whitaker, "Micromachined Circuits for Millimeter- and Sub-millimeter-Wave Applications," *IEEE Antennas Propagation Mag.*, Vol. 35, No. 5, pp. 9–17, Oct. 1993.

[28] T. M. Weller, L. P. B. Katehi, and G. M. Rebeiz, "High Performance Microshield Line Components," *IEEE Trans. Microwave Theory Tech.*, Vol. 43, No. 3, pp. 534–543, March 1995.

[29] K. J. Herrick, T. A. Schwarz, and L. P. B. Katehi, "Si-Micromachined Coplanar Waveguides for Use in High-Frequency Circuits," *IEEE Trans. Microwave Theory Tech.*, Vol. 46, No. 6, pp. 762–768, June 1998.

[30] K. J. Herrick, J.-G. Yook, and L. P. B. Katehi, "Microtechnology in the Development of Three-Dimensional Circuits," *IEEE Trans. Microwave Theory Tech.*, Vol. 46, No. 11, pp. 1832–1844, Nov. 1998.

[31] M. Madou, *Fundamentals of Microfabrication*, Boca Raton, FL: CRC Press, 1997, pp. 174–175.

[32] V. Milanovic, M. Gaitan, E. D. Bowen, and M. E. Zaghloul, "Micromachined Microwave Transmission Lines in CMOS Technology," *IEEE Trans. Microwave Theory Tech.*, Vol. 45, No. 5, pp. 630–635, May 1997.

[33] V. Milanovic, M. Gaitan, E. D. Bowen, N. H. Tea, and M. E. Zaghloul, "Design and Fabrication of Micromachined Passive Microwave Filtering Elements in CMOS Technology," *1997 Int. Conf. on Solid-State Sensors and Actuators*, Vol. 2, pp. 1007–1010, Chicago, IL, June 16–19, 1997.

[34] J. C. Booth and C. L. Holloway, "Conductor Loss in Superconducting Planar Structures: Calculations and Measurements," *IEEE Trans. Microwave Theory Tech.*, Vol. 47, No. 6, pp. 769–774, June 1999.

[35] W. Heinrich, "Mode-Matching Approach for Superconducting Planar Transmission Lines Including Finite Conductor Thickness," *IEEE Microwave Guided Wave Lett.*, Vol. 1, No. 10, pp. 294–296, Oct 1991.

[36] R. N. Simons, "High-Temperature Superconducting Coplanar Waveguide Microwave Circuits and Antennas" in *Advances in High-T_c Superconductors, Material Science Form*, Vols. 137–139, J. J. Pouch, S. A. Alterovitz, R. R. Romanofsky, and A. F. Hepp, eds., Aedermannsdorf, Switzerland: Trans Tech Publications, 1993, pp. 759–782.

[37] C. L. Holloway, personal communication.

CHAPTER 9

Coplanar Waveguide Discontinuities and Circuit Elements

9.1 INTRODUCTION

This chapter presents the characteristics of coplanar waveguide (CPW) discontinuities and circuits. These are typical discontinuities and are an intergral part of practical CPW circuits. Therefore a good understanding of their circuit behavior is essential for design success. The CPW discontinuities in Sections 9.2 through 9.7 include open end, short end, series gap in the center strip conductor, step change in the width of the center strip conductor, and right angle bend.

Air-bridges are frequently employed in CPW circuits to suppress the propagation of the parasitic coupled slot line mode. Hence, different types of air-bridges and their characteristics are presented in Section 9.8. To realize CPW components such as stub filters, power dividers, and hybrid couplers, we require T-junctions. Hence Section 9.9 is devoted to two types of T-junctions and their characteristics. Sections 9.10 through 9.13 present the characteristics of CPW spiral inductors, interdigital and metal-insulator-metal (MIM) capacitors, series and shunt stubs, and shunt inductors, respectively.

9.2 COPLANAR WAVEGUIDE OPEN CIRCUIT

A coplanar waveguide (CPW) open circuit [1] is formed by ending the center strip conductor of width S a short distance before the slot ends, thereby creating a gap g_1 as shown in Figure 9.1. An electric field exists at the open circuit between the terminated center conductor and the surrounding ground

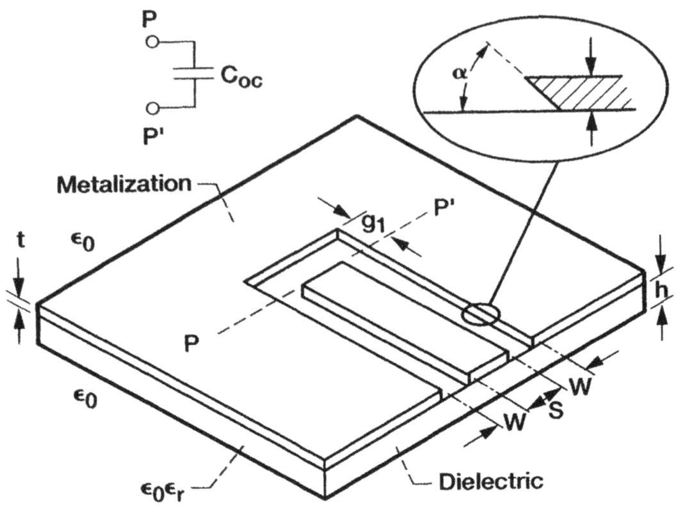

FIGURE 9.1 Coplanar waveguide open circuit.

conductor and hence gives rise to a capacitive reactance. This reactance is seen at the plane P–P' coincident with the open end of the center conductor. Thus the apparent position of the open circuit is beyond the physical end of the center conductor. This length extension is denoted as l_{oc}. The open circuit capacitance C_{oc} is a parallel combination of the capacitance due to the fringing fields across the gap g_1 and those across the slot W. The gap-dependent capacitance varies proportionately as $1/g_1$. The slot-dependent capacitance is constant. The capacitance C_{oc} is related to l_{oc} by

$$C_{oc} = \frac{\tan(\beta l_{oc})}{\omega Z_0}, \tag{9.1}$$

where β is the phase constant and is given by

$$\beta = \frac{2\pi f}{c}\sqrt{\varepsilon_{eff}}, \tag{9.2}$$

and where

ε_{eff} = CPW effective dielectric constant
f = frequency in Hz
c = velocity of light in free space in meters/second
Z_0 = CPW characteristic impedance and

$$\omega = 2\pi f. \tag{9.3}$$

When l_{oc} is small compared to a wavelength, the term $\tan(\beta l_{oc})$ is approximately equal to βl_{oc} and hence Eq. (9.1) can be rewritten as

$$C_{oc} = \left(\frac{\beta}{\omega Z_0}\right) l_{oc}. \tag{9.4}$$

The capacitance C per unit length in F/meter for an ideal lossless CPW is given by

$$C = \frac{\beta}{\omega Z_0}. \tag{9.5}$$

Hence, Eq. (9.4) can be written as

$$\frac{C_{oc}}{C} = l_{oc}. \tag{9.6}$$

9.2.1 Approximate Formula for Length Extension When the Gap Is Large

The l_{oc} has been determined as a function of the gap g_1 using finite difference method (FDM) in the frequency domain in [2], [3]. These studies show that when the dimensions of the CPW are very small compared to the guide wavelength, the dispersion is negligible and l_{oc} remains nearly constant and independent of the frequency. In general, l_{oc} is a function of the strip conductor width S, slot width W and the open end gap width g_1. However, for a fixed S and W, as g_1 increases, l_{oc} monotonically decreases and eventually saturates. This limit is reached when $g_1 \geqslant S + 2W$. Hence, for a minimum parasitic C_{oc} or l_{oc}, it is sufficient to choose $g_1 = (S + 2W)$. A simple design rule that provides this minimum l_{oc} when $g_1 \geqslant S + 2W$ and $0.2 \leqslant S/(S + 2W) \leqslant 0.8$ is [2]

$$l_{oc} = \frac{S + 2W}{4}. \tag{9.7}$$

This equation holds good for most semiconductor substrates including GaAs. In addition it has less than 1 percent error for a line with $Z_0 = 50\,\Omega$ and for all other line geometries a deviation up to 20 percent is observed [2].

9.2.2 Closed Form Equation for Open End Capacitance When the Gap Is Narrow

An analytical formula for C_{oc} under the narrow gap assumption, that is, when $g_1 \leqslant 0.1(S + 2W)$, has been derived in [4]. The derivation in [4] is based on variational technique and makes use of analytical Green's function and exact

TABLE 9.1 CPW Open Circuit Capacitance

h, μm	ε_r	Strip Width, μm	Slot Width, μm	ε_{eff}	Gap g_1, μm	C_{oc}(fF)[a]	C_{oc}(fF)[b]
200	12.9	20	20	4.8088	20	2.1	1.31
		160	20	5.3205	20	16	13.3
		20	160	6.0636	160	2.9	5.4

[a] From Bromme and Jansen [5].
[b] Using Eq. (9.9).

integration formulas to arrive at a closed form expression for capacitance. The computed results in [4] show that the ratio of the open end capacitance with and without a dielectric substrate ($C_{oc}/C_{oc(air)}$) is within 5 percent of the CPW ε_{eff} when $S/(S+2W) \geqslant 0.25$ and $h/(S+2W) \geqslant 1.0$. Hence in [4] C_{oc} is approximated as

$$C_{oc} = C_{oc(air)}\varepsilon_{eff}, \tag{9.8}$$

where

$$C_{oc(air)} = \frac{2\varepsilon_0}{\pi}\left\{(S+W)\left[\frac{\ln(\eta+\sqrt{1+\eta^2})}{\eta} + \ln\left(\frac{\sqrt{1+\eta^2}+1}{\eta}\right) - \frac{1}{3}\left(\frac{1}{1+\sqrt{1+\eta^2}} + \frac{1}{\eta+\sqrt{1+\eta^2}}\right)\right] - \left(S+\frac{2}{3}W\right)\right\}, \tag{9.9}$$

$$\eta = \frac{g_1}{S+W}. \tag{9.10}$$

The C_{oc} values determined using Eq. (9.9) are compared with the values from [5] in Table 9.1. It is observed that for $g_1 \leqslant 0.1(S+2W)$, the two sets of values are in fair agreement.

9.2.3 Radiation Loss

The radiation loss from a CPW open circuit has been numerically studied using the space domain integral equation method in [6], [7]. The computed results show that the radiation loss increases as the gap width g_1 and the center strip conductor width S increases. As an example, for a CPW with $S = 0.5$ mm, $W = 0.25$ mm, on a substrate with $h = 0.635$ mm and $\varepsilon_r = 2.1$, the radiation loss factor which is given by $10\log(1-|\Gamma|^2)$, where $|\Gamma|$ is the magnitude of the reflection coefficient, increases from 2.8 to 3.0 percent as $g_1/S+2W$ increases from 0.25 to 1.25 [6].

9.2.4 Effect of Conductor Thickness and Edge Profile Angle

The length extension l_{oc} has been determined as a function of the conductor thickness t with the edge profile angle α as a parameter using the hybrid finite-element method in [8]. The edge profile angle α is indicated in the inset in Figure 9.1. The computed l_{oc} is for a fixed $\alpha = 90°$, and when t increases from 0 to 5 μm, it is observed to increase by only 1 percent [2]. Therefore l_{oc} is considered to be independent of t. Furthermore, when $\alpha = 45°$ or 135°, the computed l_{oc} when t increases from 0 to 5 μm is observed to decrease or increase, respectively, by only a few percent. Hence l_{oc} is insensitive to both conductor thickness and edge profile.

9.3 COPLANAR WAVEGUIDE SHORT CIRCUIT

A coplanar waveguide (CPW) short circuit [1] is formed by abruptly ending the slots as shown in Figure 9.2. In which case a RF current flows around the end of the slots, and therefore magnetic energy is stored behind the termination. This magnetic energy gives rise to an inductive reactance that is modeled as L_{sc} and located at the plane P–P'. Thus the apparent position of the short circuit is beyond the physical end of the slots. This length extension is denoted as l_{sc}.

9.3.1 Approximate Formula for Length Extension

The l_{sc} has been determined as a function of the slot width W using finite difference method (FDM) in the frequency domain in [2], [3]. Analogous to the CPW open circuit termination, the l_{sc} is independent of the frequency when

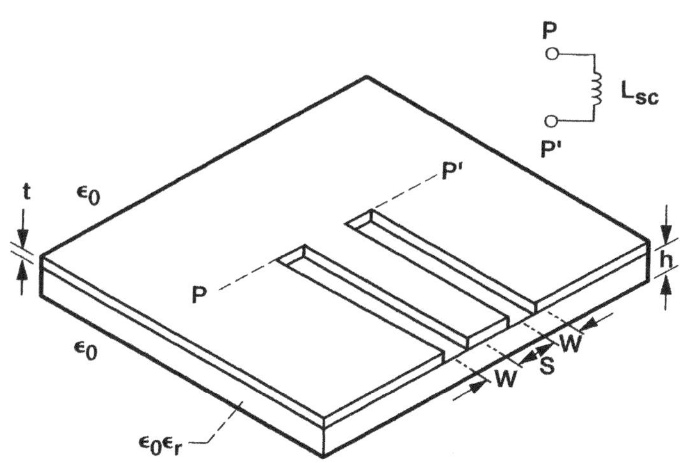

FIGURE 9.2 Coplanar waveguide short circuit.

TABLE 9.2 CPW Short-Circuit Length Extension and Inductance

h, μm	ε_r	Strip Width, μm	Slot Width, μm	ε_{eff}	Z_0, Ω	l_{sc}/h [a]	l_{sc}/h [b]	L(pH) [c]	L(pH) [d]
635	9.7	127	127	—	—	0.1	0.075	—	—
			254	—	—	0.15	0.125	—	—
			381	—	—	0.21	0.175	—	—
			508	—	—	0.25	0.225	—	—
			635	—	—	0.29	0.275	—	—
635	9.7	635	127	—	—	0.22	0.175	—	—
			254	—	—	0.31	0.225	—	—
			381	—	—	0.37	0.275	—	—
			508	—	—	0.44	0.325	—	—
			635	—	—	0.49	0.375	—	—
200	12.9	40	40	5.6	45.8	—	—	16	5.2
		40	160	6.1	75.6	—	—	50	36.2
		160	40	5.7	31.2	—	—	19	6.2

[a] From Jansen [9].
[b] Using Eq. (9.11).
[c] From Bromme and Jansen [5].
[d] Using Eq. (9.12).

the dispersion is small. A simple design rule that provides l_{sc} is given by [2]

$$l_{sc} \approx \frac{S + 2W}{8}. \tag{9.11}$$

Since l_{sc} is sensitive to the conductor thickness t as explained in the next section, Eq. (9.11) holds good only when $t < W/3$. In general, the values for l_{sc} are about one-half of those obtained for an identical CPW with an open circuit termination [2]. Hence the parasitic effects associated with a short circuit are smaller than those for an open circuit termination. The l_{sc} values determined using Eq. (9.11) are compared with the values from the graph in [9] for $h/\lambda_0 = 0.002$, where λ_0 is the free space wavelength, and presented in Table 9.2. It is observed that for small S typically $S/h \leqslant 0.2$, the two sets of values are fairly close.

9.3.2 Closed Form Equations for Short-Circuit Inductance

A closed form equation for L_{sc} based on magnetostatic analysis of a finite-length two-wire transmission line is derived in [10] and is given below:

$$L_{sc} = \left(\frac{2}{\pi}\right)\varepsilon_0\varepsilon_{eff}(a + b)Z_0^2\left\{1 - \frac{1}{[\cosh(60\pi^2/Z_0\sqrt{\varepsilon_{eff}})]}\right\}, \tag{9.12}$$

where

ε_{eff} = CPW effective dielectric constant
Z_0 = CPW characteristic impedance
$2a = S$
$2b = S + 2W$

In a manner analogous to an open circuit termination discussed in Section 9.2, the length extension is given by

$$l_{sc} = \frac{L_{sc}}{L}, \tag{9.13}$$

where L is the inductance per unit length in Henry/meter for an ideal loss less CPW and is given by

$$L = \frac{\beta Z_0}{\omega}, \tag{9.14}$$

where β and ω are defined in Eq. (9.2) and (9.3), respectively. The L_{sc} values determined using Eq. (9.12) are compared with the values from [5] in Table 9.2. It is observed that Eq. (9.12) greatly underestimates the inductance.

9.3.3 Effect of Conductor Thickness and Edge Profile Angle

The length extension l_{sc} has been determined as a function of the conductor thickness t with the edge profile angle α as the parameter using the hybrid finite-element method in [8]. The edge profile angle is indicated in the inset in Figure 9.1. For $\alpha = 90$, the computed l_{sc}, when t increases from 0 to 5 μm is observed to monotonically decrease by as much as 24 percent [8]. Furthermore, for α either equal to 45° or 135°, the computed l_{sc}, when t increases from 0 to 5 μm is also observed to decrease. However, the decrease is as much as 36 percent [8]. Hence l_{sc} is very sensitive to both the conductor thickness as well as edge profile angle.

9.4 COPLANAR WAVEGUIDE MIM SHORT CIRCUIT

In the CPW short circuit described in Section 9.3, the center strip conductor and the ground planes are in direct contact, and hence they are at the same potential. This poses a problem in designing circuits with active devices such as MESFETs that require different voltages on the gate and drain terminals. This difficulty can be overcome by including a metal-insulator-metal (MIM)

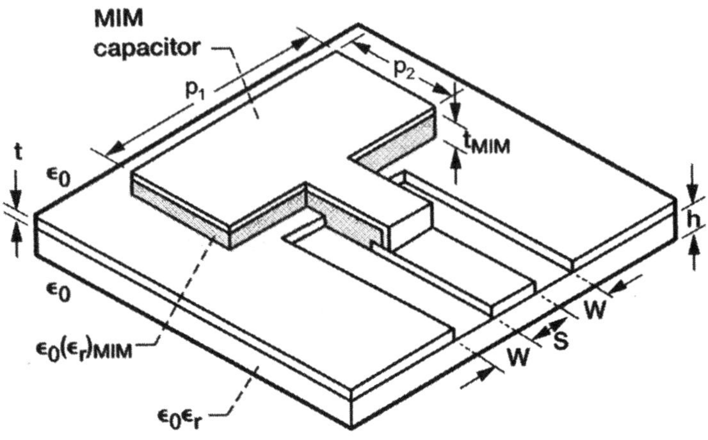

FIGURE 9.3 Coplanar waveguide MIM short circuit.

capacitor as an integral part of the short circuit as shown in Figure 9.3. The capacitor acts as a dc block for the bias voltage and as an ideal short circuit for signals at microwave frequencies. The length extension $l_{sc(MIM)}$ is given by [2]

$$l_{sc(MIM)} = \frac{L_{sc}}{L} - \frac{1}{\omega^2 L C_{MIM}}, \tag{9.15}$$

where L_{sc}/L is approximated by Eq. (9.11), L is given by Eq. (9.14), C_{MIM} can be determined from the well-known parallel-plate capacitance formula

$$C_{MIM} = (\varepsilon_r)_{MIM} \frac{p_1 p_2}{t_{MIM}}. \tag{9.16}$$

The quantities p_1, p_2, t_{MIM}, and $(\varepsilon_r)_{MIM}$ in Eq. (9.16) are indicated in Figure 9.3. Typically in a MIM capacitor, the dielectric material is silicon nitride for which $(\varepsilon_r)_{MIM}$ and t_{MIM} are about 6 and 0.2 μm, respectively. Circuit design considerations dictate that C_{MIM} should be as small as possible to keep the size small. However, for good short-circuit performance C_{MIM} should be sufficiently large. Hence C_{MIM} should satisfy the condition [2]

$$\frac{1}{\omega C_{MIM}} \leqslant \frac{1}{50} Z_0, \tag{9.17}$$

where Z_0 is the characteristic impedance of the CPW.

9.5 SERIES GAP IN THE CENTER STRIP CONDUCTOR OF A COPLANAR WAVEGUIDE

A series gap of length g in the center conductor of a CPW [1] is shown in Figure 9.4. In general, the widths S_1 and S_2 of the center strip conductor need not be equal. Hence the gap is modeled as a lumped π-network consisting of two unequal fringing capacitances C_1 and C_2 and a coupling capacitance C_3. This type of discontinuity has been theoretically investigated using, an unified strip/slot three dimensional full-wave electromagnetic simulator and three-dimensional finite difference method in [5] and [11], respectively. The computed values of C_1, C_2, and C_3 as a function of the gap g [11] are shown in Figures 9.4(*b*), (*c*), and (*d*). Furthermore closed form equations to calculate the fringing and coupling capacitances have been obtained in [12] by curve fitting the numerical results of [11]. The results above show that the coupling capacitance decreases proportionally as $1/g$. The fringing capacitances for $S_1 = S_2$ increase from the CPW line capacitance for $g = 0$, to the open circuit saturation capacitance for large g. The effect of metallization thickness t on the coupling capacitance C_3 is shown in Figure 9.4(*e*). The figure shows that the coupling capacitances, in the case of very small gap, is sensitive to the conductor thickness.

9.6 STEP CHANGE IN THE WIDTH OF CENTER STRIP CONDUCTOR OF A COPLANAR WAVEGUIDE

A step change in width of the center strip conductor of a CPW [1] is shown in Figure 9.5. The step discontinuity perturbs the normal CPW electric and magnetic fields and gives rise to additional reactances. These additional reactances are assumed to be lumped and located at the plane of the strip discontinuity.

The step discontinuity has been analyzed using finite difference time domain (FDTD) technique in [13] and modeled as a shunt capacitor C_s. The influence of this capacitance is to effectively lengthen the center strip conductor of the lower impedance CPW toward the higher-impedance CPW. Further, for two CPWs with identical aspect ratio $S/(S + 2W)$ on the same substrate, C_s is larger for the CPW with the wider center strip conductor.

The step discontinuity has also been analyzed using hybrid finite element method [14], [15] and modeled as a T-network consisting of two series inductances L_{s1} and L_{s2} and a shunt capacitance C_s. The computed series inductance and shunt capacitance as a function of the step change in the width of the center strip conductor, with the metallization thickness t as a parameter, are shown in Figure 9.6 (*a*) and (*b*), respectively. In Figure 9.6 (*a*) the series inductance L_s is the sum of L_{s1} and L_{s2}. It is observed that for a fixed metallization thickness as the step ratio changes both L_s and C_s increase. Besides, at a fixed step ratio as the metallization thickness increases, L_s

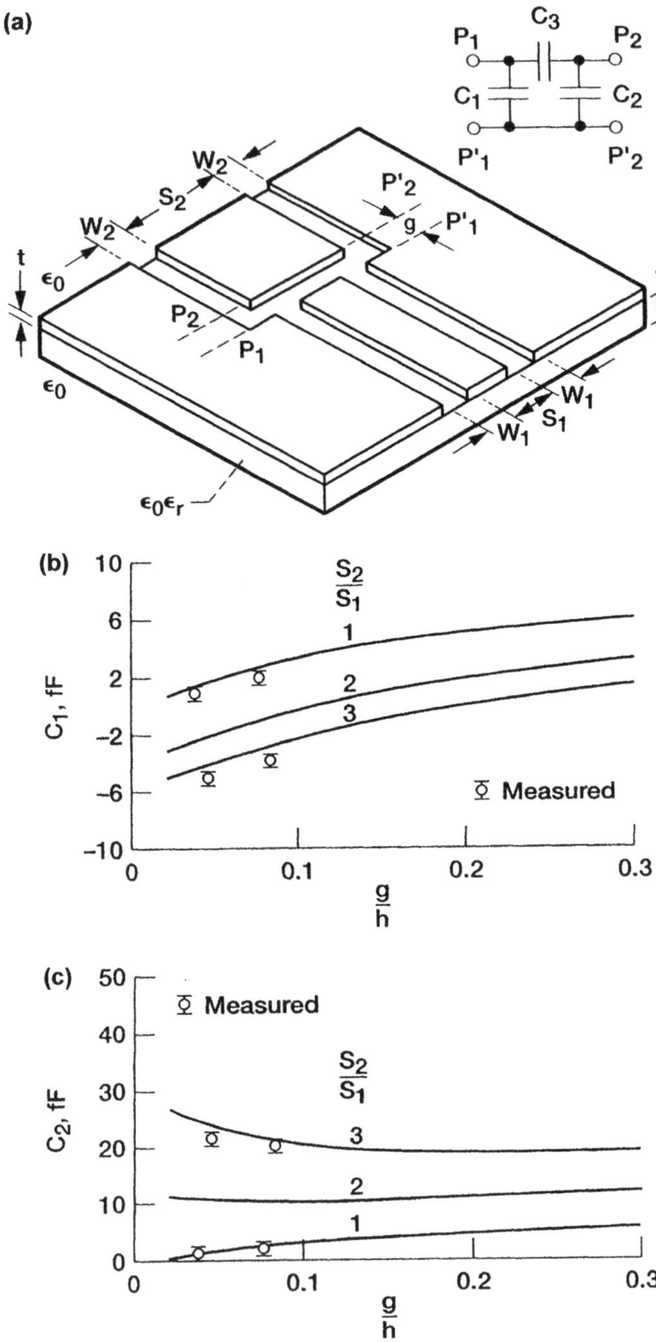

FIGURE 9.4 Series gap in the center strip conductor of a coplanar waveguide: (*a*) circuit layout; (*b*) computed fringing capacitance C_1; (*c*) computed fringing capacitance C_2; (*d*) computed coupling capacitance C_3; (*e*) effect of metallization thickness t on the coupling capacitance C_3. (Reference [11], ©IEEE 1990.)

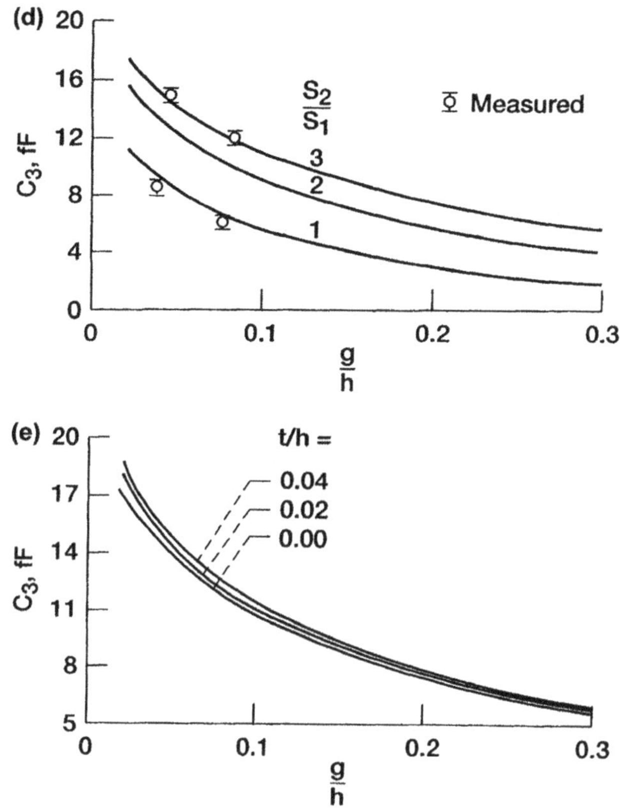

FIGURE 9.4 Continued.

decreases while C_s increases. However, the percentage decrease in L_s is greater than the percentage increase in C_s.

9.7 COPLANAR WAVEGUIDE RIGHT ANGLE BEND

A coplanar waveguide right angle bend with chamfered corners [16], [17] is shown in Figure 9.7. Air-bridges are placed near the CPW bends in order to suppress the parasitic slot line mode. However, the air-bridges add parasitic capacitance to the bend and degrade its performance. In addition the bend reactance is also responsible for performance degradation. Chamfering the corners provides a simple method to compensate for both parasitic capacitance as well as bend reactance. In Figure 9.7 the chamfer in the ground plane in defined by the ratio of the distances, m/m_{max}, and takes a value between 0 and 1. The strip corner is chamfered by a proportional amount given by $mS/(S + W)$. For a given S/W, the optimum chamfer $(m/m_{max})_{opt}$ for minimum

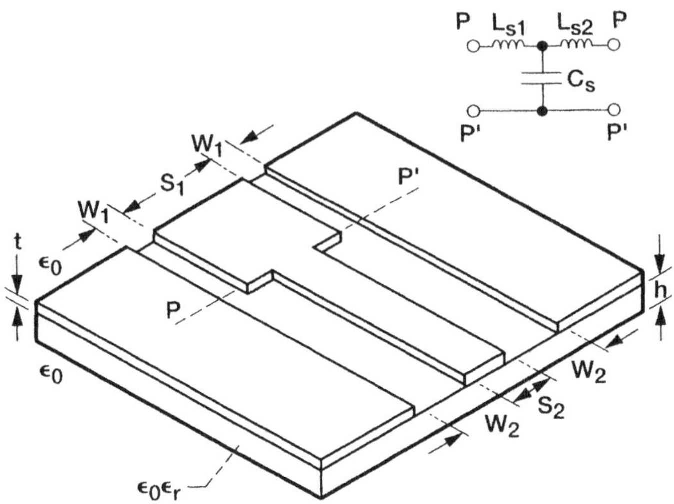

FIGURE 9.5 Step change in the width of the center strip conductor of a coplanar waveguide.

return loss is [17]

$$\left(\frac{m}{m_{\max}}\right)_{\text{opt}} = \begin{cases} 0.2102 \ln\left(\dfrac{S}{W}\right) + 0.7677 & \text{for } 0 \leqslant S/W \leqslant 2, \\ 1 & \text{for } S/W > 2.5. \end{cases} \tag{9.18}$$

In developing this equation, it is assumed that the substrate is GaAs ($\varepsilon_r = 12.9$) and that the thickness h is 635 μm. In addition the length L_a and height h_a above the substrate, of the air-bridges, are assumed to be 40 μm and 3 μm, respectively.

Besides chamfering, compensation for the parasitic capacitance and bend reactance can also be provided by a step change in the width of the center strip conductor [18] or by a dielectric overlay [1]. These techniques are shown in Figures 9.8 (*a*) and (*b*). Not much information is available in the literature on the bend shown in Figure 9.8 (*a*). On the other hand, the CPW bend shown in Figure 9.8(*b*) is fabricated on a 3.175 mm thick RT-5880 Duroid substrate. The CPW strip and slot widths are 4.5212 and 0.254 mm, respectively, resulting in a characteristic impedance of 50 Ω. The dielectric overlay on top of the inner slot is formed by an alumina piece measuring 12.5 × 12.5 × 0.635 mm. Experimental results demonstrate that both techniques provide compensation over a broadband.

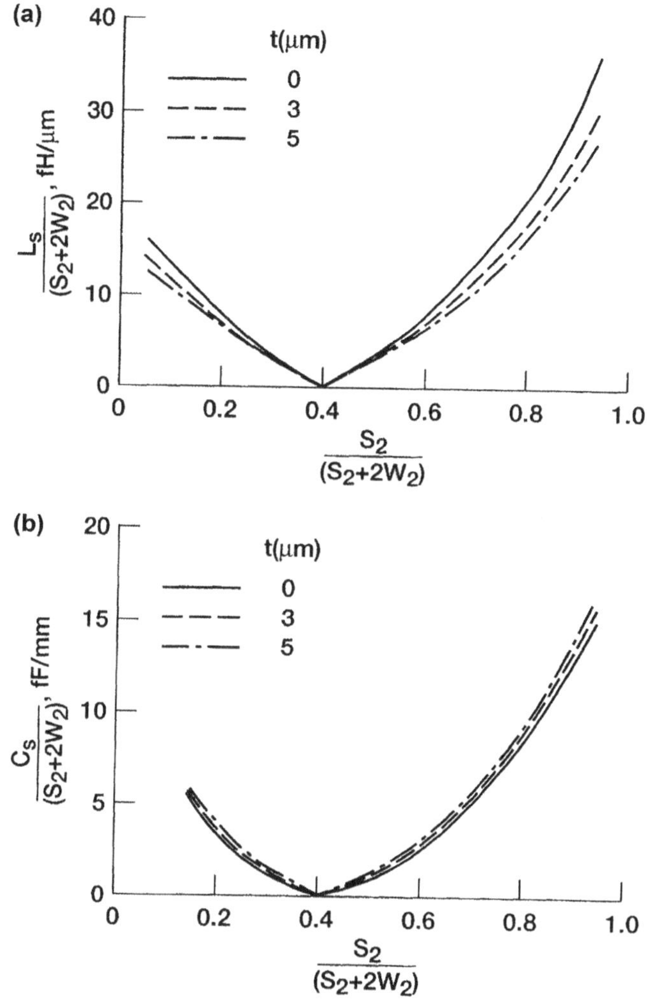

FIGURE 9.6 Computed equivalent circuit element values as a function of the step change in width of the center strip conductor with metallization thickness as a parameter, $\varepsilon_r = 12.9$, $h = 350\,\mu\text{m}$, $S_1 + 2W_1 = S_2 + 2W_2 = 150\,\mu\text{m}$, $S_1/(S_1 + 2W_1) = 0.4$): (*a*) Total series inductance ($L_s = L_{s1} + L_{s2}$); (*b*) shunt capacitance. (Reference [15], ©IEE 1998; Reference [14], ©IEEE 1997.)

9.8 AIR-BRIDGES IN COPLANAR WAVEGUIDE

Air-bridges are used in coplanar waveguide (CPW) based circuits to suppress the parasitic coupled slot line mode and also to provide a dc bias to active devices. There are two types of air-bridges [19], [20] that are commonly used in CPW circuits, and their construction and characteristics are described below.

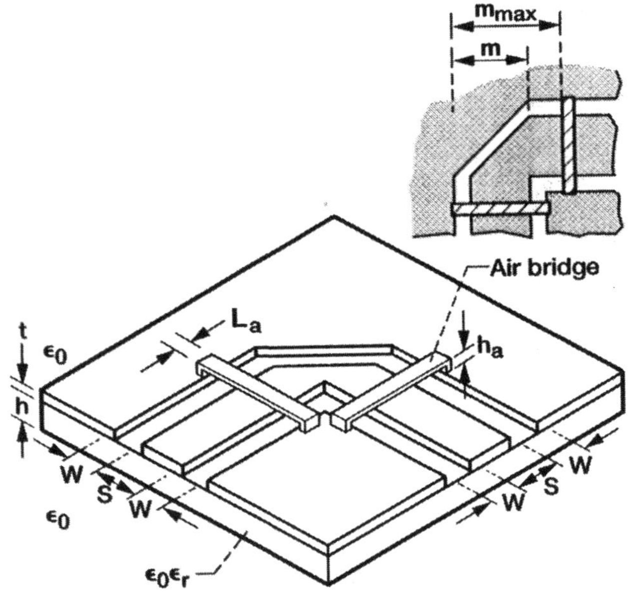

FIGURE 9.7 Coplanar waveguide right angle bend with chamfered corners and air-bridges.

9.8.1 Type A Air-Bridge

In the air-bridge of Type A, the bridge is formed by an elevated metal strip that connects the ground planes on either side of the center strip conductor as shown in Figure 9.9(*a*). In this type of air-bridge, the thin metal film for the CPW center strip conductor is fabricated using evaporation, while the bridge is fabricated by electroplating. Typically the thickness t_m of the metal film and t_a of the bridge are about 0.4 and 3.0 μm, respectively [19]. The cross section of the bridge is shown in the inset in Figure 9.9(*a*).

9.8.2 Type B Air-Bridge

The air-bridge of Type B consists of an elevated metal strip that bridges the center strip conductors of two back-to-back CPW open circuits as shown in Figure 9.9(*b*). The thin metal film for the ground plane below the bridge is fabricated using evaporation while the bridge is fabricated using electroplating. The thickness of the conductors are same as in Type A air-bridge [19]. The cross section of the bridge is shown in the inset in Figure 9.9(*b*).

9.8.3 Air-Bridge Characteristics

The Type A and Type B air-bridges have been studied using full wave frequency domain TLM method in [21]. The computed magnitude and phase

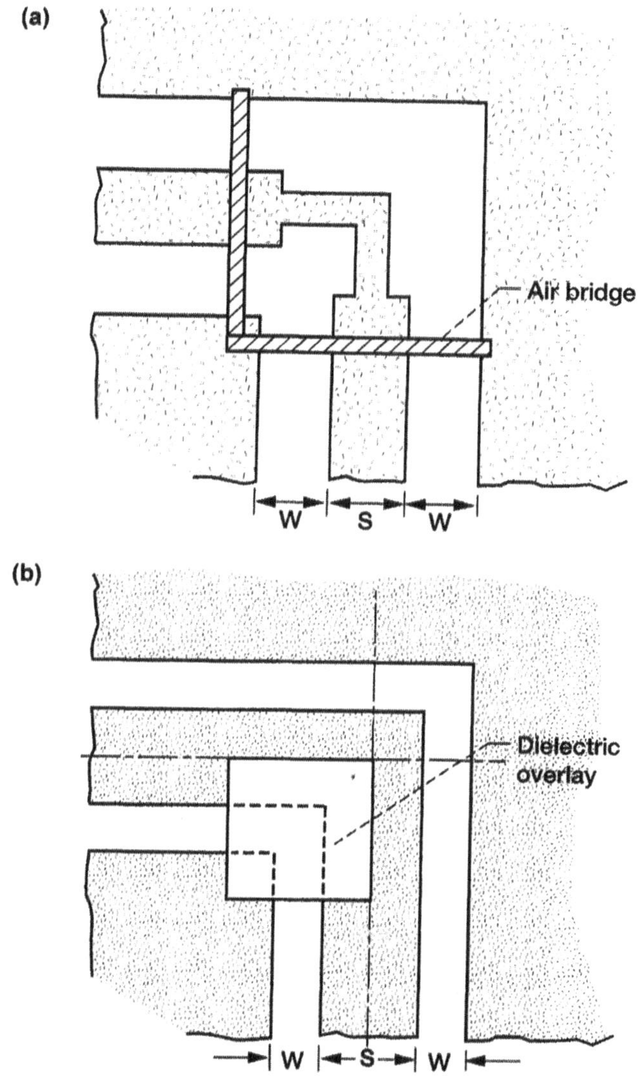

FIGURE 9.8 Coplanar waveguide right angle bend with (*a*) step compensation and air-bridges and (*b*) a dielectric overlay.

of the reflection coefficient S_{11} for Type A and Type B air-bridges are shown as a function of the frequency in Figure 9.10(*a*) and (*b*). The height h_a and length L_a of both types of air-bridges are held fixed at 3 and 30 μm, respectively, while carrying out the computations and are typical dimensions of a practical air-bridge. The computed results show that the magnitude of S_{11} is small and almost identical for both types of air-bridges. However, the phase of S_{11} for Type A and Type B air-bridges are opposite, and this indicates that

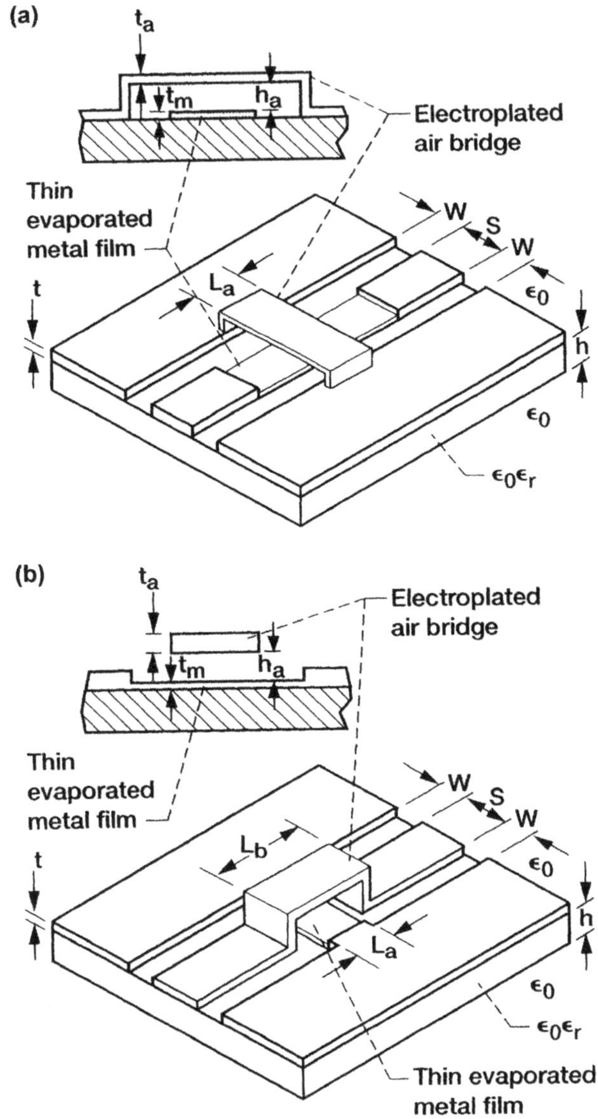

FIGURE 9.9 Air-bridges for coplanar waveguide: (*a*) Type A; (*b*) Type B.

they behave as a capacitor and as a inductor, respectively. The excess capacitance in the case of Type A air-bridge is because of the parasitic parallel-plate capacitor formed by the air bridge and the center strip conductor [20]. In the case of Type B air-bridge, the current distributions is different on the CPW and the section of the line formed by the air-bridge. This change in current distribution gives rise to an inductance [20]. Figure 9.10(*a*) and (*b*) also

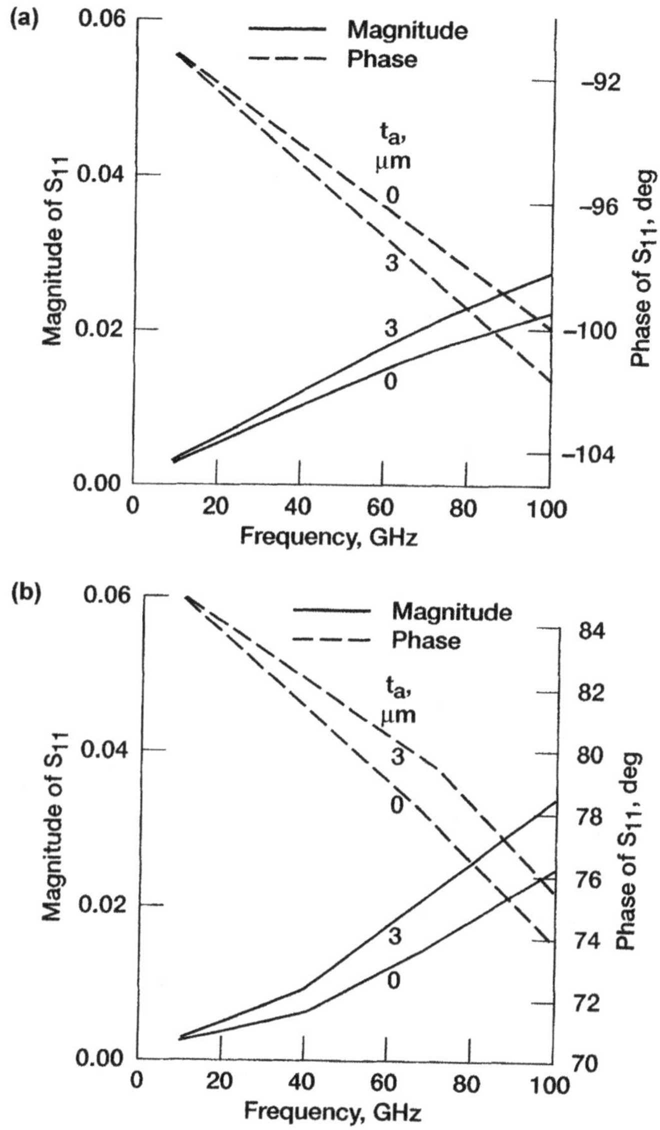

FIGURE 9.10 Computed reflection coefficient as a function of the frequency with the air-bridge thickness as a parameter, $W = 10\,\mu m$, $S = 15\,\mu m$, $h = 100\,\mu m$, $\varepsilon_r = 12.9$, $t = t_m = t_a$, $L_a = 30\,\mu m$, $h_a = 3\,\mu m$): (*a*) Type A; (*b*) Type B ($L_b = 45\,\mu m$). (Reference [21], ©IEEE 1992.)

shows that the magnitude of S_{11} for both type of air bridges are marginally higher, when the air bridge thickness t_a is increased from 0 to 3 μm.

The computed magnitude and phase of S_{11} as a function of the height h_a and length L_a of the air-bridges are shown in Figures 9.11 and 9.12, respectively. It is observed from Figure 9.11 that S_{11} changes marginally for both type of air-bridges when the height h_a is increased beyond 3 μm. Figure 9.12 shows that S_{11} changes linearly for both type of air-bridges when the length L_a increases beyond 30 μm.

9.8.4 Air-Bridge Discontinuity Characteristics

A CPW step discontinuity in close proximity of a Type A air bridge [22] is shown in Figure 9.13. The characteristic impedances Z_{01} and Z_{02} of the CPW on either side of the step are 50 and 45 Ω, respectively. The computed magnitude and phase of the reflection coefficient S_{11} as a function of the distance of separation d_1 between the step and the air-bridge are shown in Figure 9.14. It is interesting to observe that the actual location of the air bridge with respect to the step has virtually no effect on S_{11}. The change in phase is due to the change in the length of the CPW between the step and the air bridge [21].

A step discontinuity in the air bridge of Type B [22] is shown in Figure 9.15. The characteristic impedances Z_{01} and Z_{02} of the CPW on either side of the air-bridge are 70 and 50 Ω, respectively. The computed magnitude and phase of the reflection coefficient S_{11} and the transmission coefficient S_{12} as a function of the distance d_2 from the end of the air-bridge are shown in Figure 9.16. It is observed that the location of the step has little effect on S_{11} and S_{12}.

9.9. COPLANAR WAVEGUIDE T-JUNCTION

In the design of coplanar waveguide components such as, stub filters, power dividers and branch-line couplers, T-junctions are necessary. Since a T-junction includes several discontinuities, mode convertion takes place within the circuit. Thus power is converted from the desired CPW mode to the parasitic coupled slotline mode as will be shown later. A simple method to suppress the excitation of the coupled slotline mode is by maintaining electrical continuity between the ground planes of the circuit. Normally this is accomplished by bond wires or air-bridges. There are two types of CPW T-junctions that employ air-bridges [19], [23]. Their construction as well as their characteristics will be described in the following sections.

9.9.1 Conventional T-Junction

A CPW conventional T-junction is shown in Figure 9.17. In this structure, the three CPW center strip conductors meet to form the junction while the

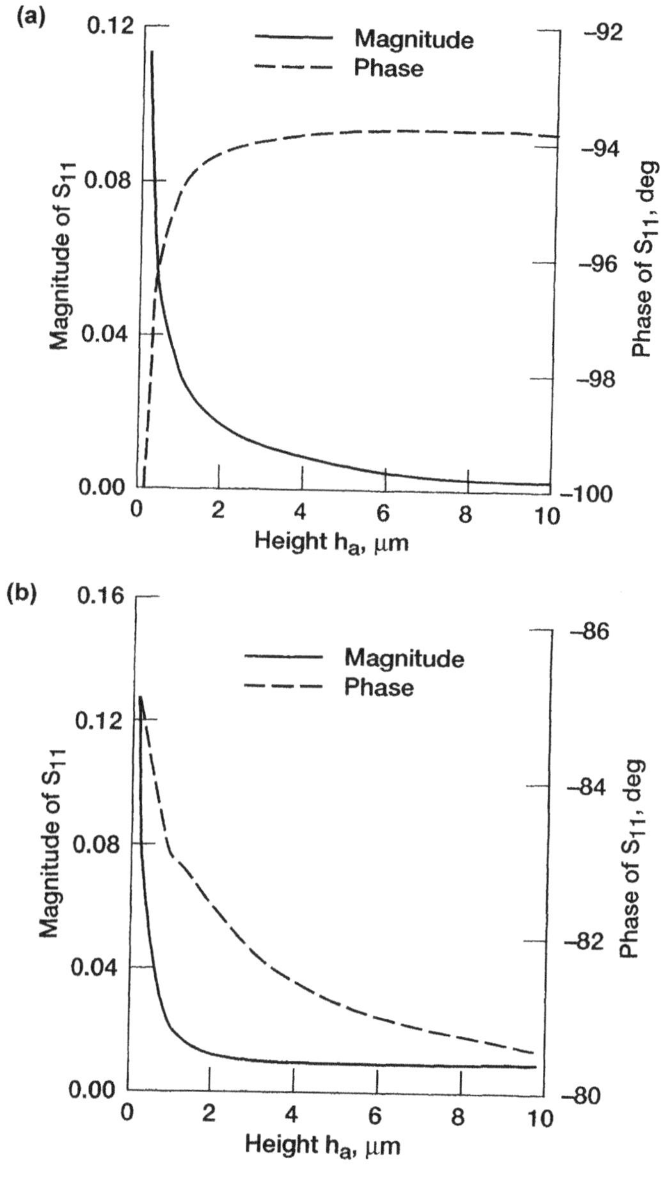

FIGURE 9.11 Computed reflection coefficient as a function of the air-bridge height, $W = 10\,\mu\text{m}$, $S = 15\,\mu\text{m}$, $h = 100\,\mu\text{m}$, $\varepsilon_r = 12.9$, $t = t_m = t_a = 3\,\mu\text{m}$, $L_a = 30\,\mu\text{m}$, frequency = 40.0 GHz: (*a*) Type A; (*b*) Type B ($L_b = 45\,\mu\text{m}$). (Reference [21], ©IEEE 1992.)

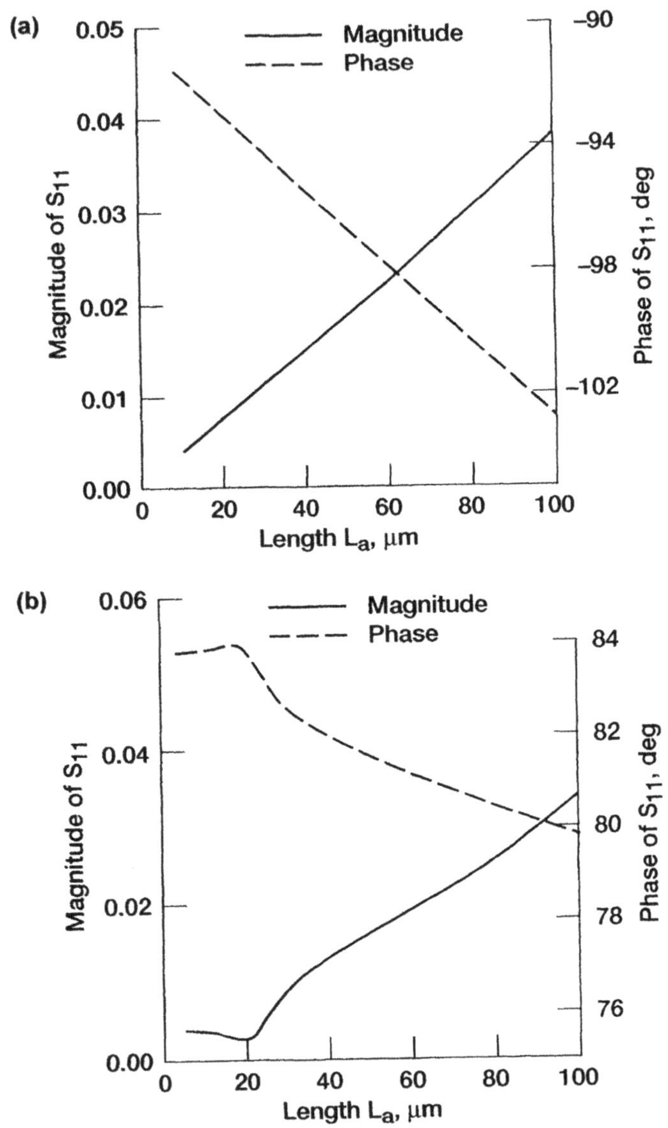

FIGURE 9.12 Computed reflection coefficient as a function of the air-bridge length, $W = 10\,\mu m$, $S = 15\,\mu m$, $h = 100\,\mu m$, $\varepsilon_r = 12.9$, $t = t_m = t_a = 3\,\mu m$, $h_a = 3\,\mu m$, frequency = 40.0 GHz): (*a*) Type A; (*b*) Type B ($L_b = L_a + 15.0\,\mu m$). (Reference [21], ©IEEE 1992.)

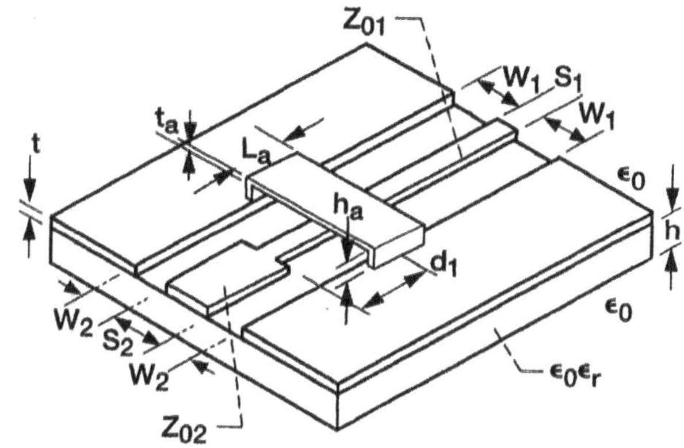

FIGURE 9.13 A CPW step discontinuity in close proximity of a Type A air-bridge.

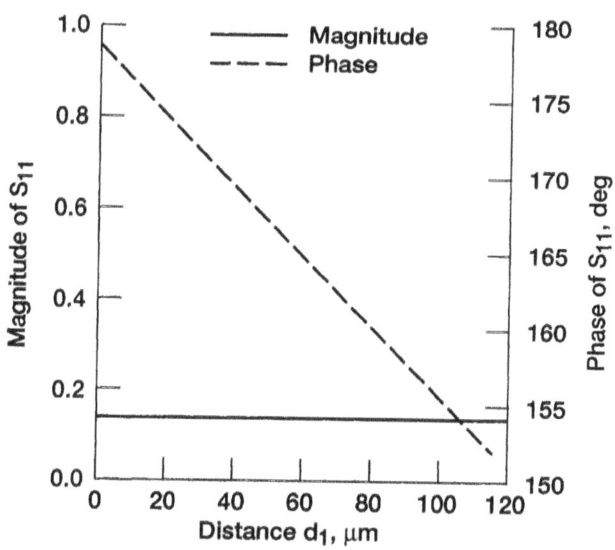

FIGURE 9.14 Computed reflection coefficient as a function of the distance of separation, $W_1 = 12\,\mu$m, $S_1 = 12\,\mu$m, $W_2 = 8\,\mu$m, $S_2 = 20\,\mu$m, $h = 100\,\mu$m, $\varepsilon_r = 12.9$, $h_a = 3\,\mu$m, $L_a = 30\,\mu$m , $t = t_a = 3\,\mu$m, frequency = 40 GHz, metallization $\sigma = 10{,}000$ S/mm). (Reference [22], ©IEEE 1993.)

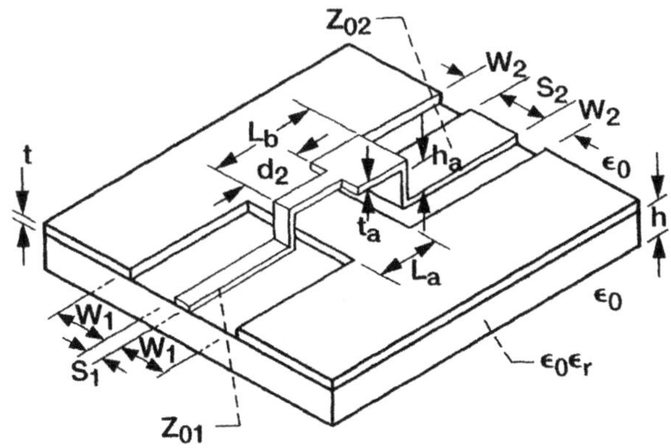

FIGURE 9.15 A step discontinuity in the air-bridge of Type B.

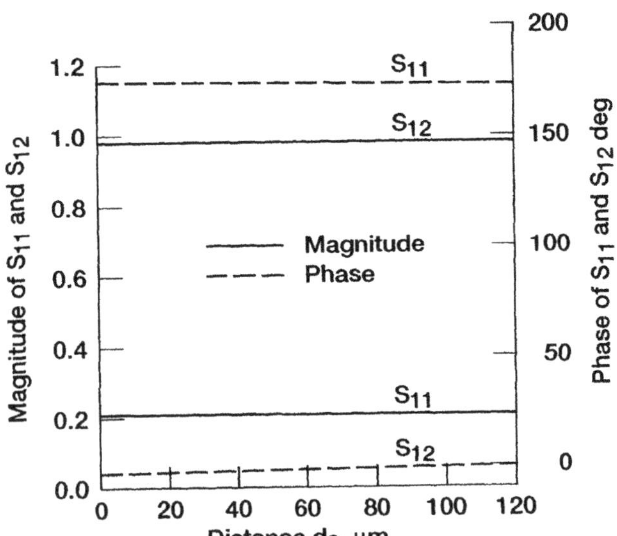

FIGURE 9.16 Computed reflection and transmission coefficient as a function of the distance, $W_1 = 12.5\ \mu m$, $S_1 = 10\ \mu m$, $W_2 = 7\ \mu m$, $S_2 = 21\ \mu m$, $h = 100\ \mu m$, $\varepsilon_r = 9.8$, $h_a = 3\ \mu m$, $L_b = 45\ \mu m$, $L_a = 30\ \mu m$, $t = t_a = 3\ \mu m$, frequency = 40 GHz, metallization $\sigma = 10{,}000$ S/mm). (Reference [22], ©IEEE 1993.)

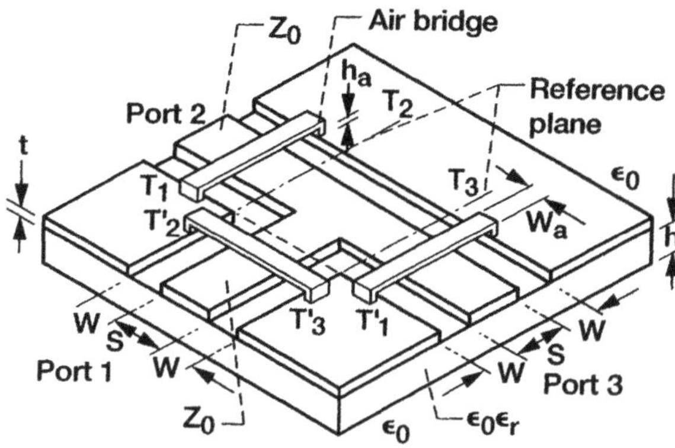

FIGURE 9.17 A conventional T-junction.

continuity between ground planes at the junction is maintained either by bond wires or by air-bridges. Typically, the bond wires are 0.7 mil in diameter and the dimensions of the air bridges are as described in Section 9.8.

9.9.2 Air-Bridge T-Junction

In this circuit the CPW center strip conductors are elevated and meet to form a T-junction that resembles an air-bridge. The respective ground planes are extended to intersect and form a rectangular metal patch below the elevated strip conductors. The air-bridge T-junction is shown in Figure 9.18.

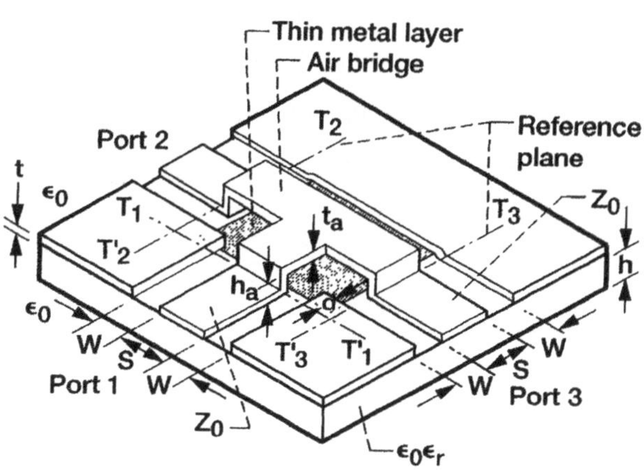

FIGURE 9.18 An air-bridge T-junction.

9.9.3 Mode Conversion in CPW T-Junctions

Mode conversion is a serious problem in CPW circuits with discontinuities and can result in excess insertion/return loss. A simple method to alleviate this problem is to insert air-bridges very close to the discontinuities as pointed out earlier. Here we will discuss the efficacy of this method when used in conjunction with CPW T-junctions shown in Figures 9.17 and 9.18. A good estimate of the effectiveness of this method is the amount of power converted from an incident coplanar waveguide mode at port 1 to a coupled slotline mode which emerges from port 2 eventually. This convertion can be expressed as a special transmission coefficient denoted as $S_{21(\mathrm{CPW} \rightarrow \mathrm{slotline})}$. Through the use of integral equation technique, the $S_{21(\mathrm{CPW} \rightarrow \mathrm{slotline})}$ has been computed for both types of CPW T-junctions in [24]. For the purpose of design two examples are considered for each type of T-junction in [24]. The two examples correspond to CPW with center strip conductor and slot width both of which are either wide or narrow, respectively. The computed $S_{21(\mathrm{CPW} \rightarrow \mathrm{slotline})}$ as a function of the frequency is shown in Figure 9.19. By comparing the $S_{21(\mathrm{CPW} \rightarrow \mathrm{slotline})}$ for both types of T-junction one clearly notices that the mode conversion is reduced to a greater extent in the case of an air-bridge T-junction. Further the reduction in mode conversion is up to a factor of five better for a junction formed with narrow strip and slot widths than with wide strip and slot widths.

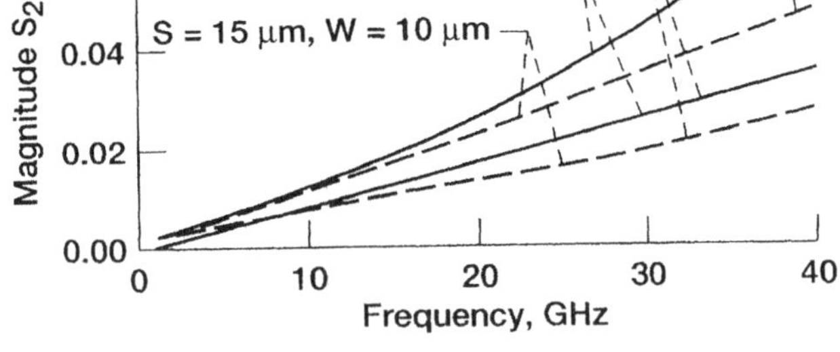

FIGURE 9.19 Computed magnitude of transmission coefficient $S_{21(\mathrm{CPW} \rightarrow \mathrm{slotline})}$ as a function of frequency for both type CPW T-junctions on a substrate of thickness $h = 410\,\mu$m and $\varepsilon_r = 12.9$, $t = 0\,\mu$m, $t_a = 0\,\mu$m; line geometry $S = 75\,\mu$m, $W = 50\,\mu$m, $Z_0 = 50\,\Omega$, $h_a = 3\,\mu$m (solid line), $S = 15\,\mu$m, $W = 10\,\mu$m, $Z_0 = 50\,\Omega$, $h_a = 3\,\mu$m ($5 \times S_{21(\mathrm{CPW} \rightarrow \mathrm{slotline})}$ dash line). (Reference [24], ©IEEE 1993.)

9.9.4 CPW T-Junction Characteristics

The CPW T-junctions shown in Figures 9.17 and 9.18 have been modeled using integral equation technique and complex image technique in [24] and [25]. In addition a grounded CPW T-junction has been modeled using quasi-static spectral domain method in [26]. In an ideal nonsymmetric CPW T-junction if the characteristic impedance of the coplanar waveguide at port 1 and ports 2 and 3, are Z_{in} and Z_{out}, respectively, then the magnitude of the reflection coefficient at port 1 ignoring junction effects is

$$|S_{11}| = \frac{|Z_{out} - 2Z_{in}|}{|Z_{out} + 2Z_{in}|} \tag{9.19}$$

and the magnitude of the transmission coefficient between ports 1 and 2 and between ports 1 and 3 is

$$|S_{21}| = |S_{31}| = \sqrt{\frac{1 - |S_{11}|^2}{2}}. \tag{9.20}$$

For a symmetric CPW T-junction, $Z_{in} = Z_{out}$, and hence

$$|S_{11}| = \tfrac{1}{3} \tag{9.21}$$

and

$$|S_{21}| = |S_{31}| = \tfrac{2}{3}. \tag{9.22}$$

The numerically modeled, as well as measured, magnitude of S_{11} and S_{21} as a function of the frequency for a symmetric conventional T-junction and for an air-bridge T-junction [24] is shown in Figure 9.20(*a*) and (*b*). This figure shows that when the frequency is small, the magnitude of S_{11} and S_{21} approach the values given by Eqs. (9.21) and (9.22). However, when the frequency increases, the magnitude of S_{11} for both junctions increases due to the parasitic capacitive reactance of the bridge structures.

A lumped element equivalent circuit model for a conventional T-junction and an air-bridge T-junction are shown in Figure 9.21(*a*) and (*b*). In these models, the inductances L_1, L_2, and L_3 arise because of current flow interruptions. This effect is significant since most of the current flows along the slot edges of the CPW as shown by three-dimensional finite difference method (3D-FDM) in [27]. The capacitance C arises through additional charge accumulation due to field concentration at the corners as shown by 3D-FDM in [11]. The capacitance C_a in the case of conventional T-junction is the parasitic capacitance associated with the air-bridge. As examples, the element values determined numerically [26], [27] for both types of symmetric T-junctions are listed in Table 9.3. In the case of conventional T-junction, the

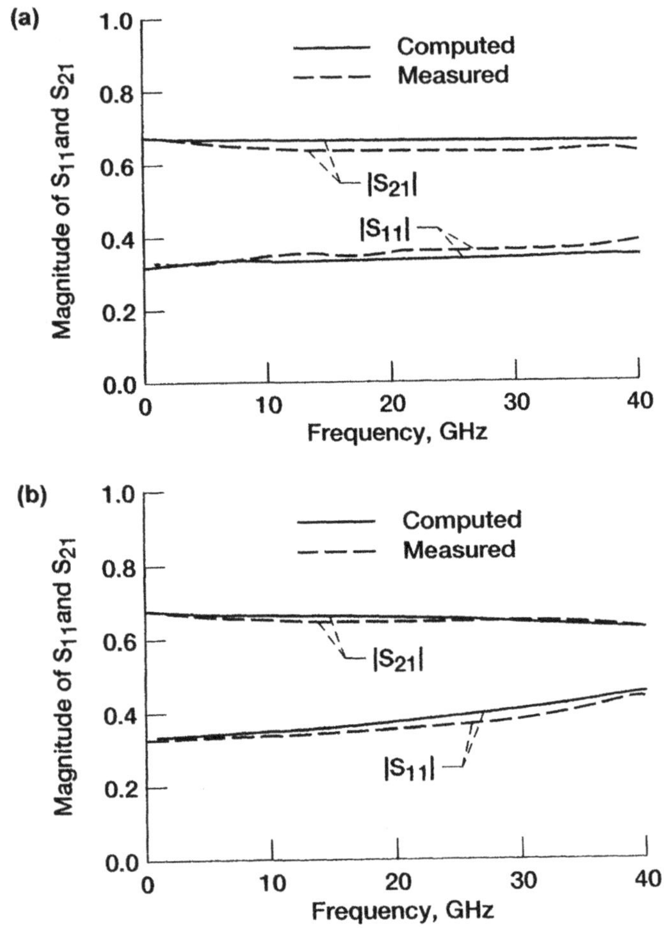

FIGURE 9.20 Computed magnitude of reflection and transmission coefficients for a symmetric structure as a function of the frequency, $S = 75\ \mu\text{m}$, $W = 50\ \mu\text{m}$, $Z_0 = 50\ \Omega$, $h_a = 3\ \mu\text{m}$, $h = 410\ \mu\text{m}$, $\varepsilon_r = 12.9$, $t = 0\ \mu\text{m}$, $t_a = 0\ \mu\text{m}$: (*a*) Conventional T-junction; (*b*) air-bridge T-junction. (Reference [24], ©IEEE 1993.)

capacitance C_a of the air-bridge is approximated by that of a parallel-plate capacitor.

9.10 COPLANAR WAVEGUIDE SPIRAL INDUCTOR

Lumped circuit elements such as spiral inductors have the potential to reduce the overall circuit size of MMIC balanced mixers [28] and frequency doublers

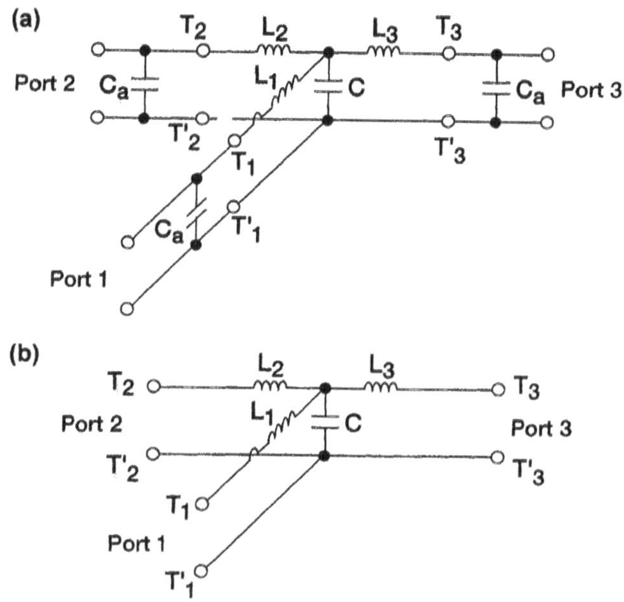

FIGURE 9.21 Lumped element equivalent circuit model: (*a*) Conventional T-junction; (*b*) air-bridge T-junction.

[29]. The size reduction in mixers and doublers is accomplished by replacing the distributed hybrid couplers and phase shifters by lumped element versions. As an example, a V-band (50.0 to 75.0 GHz) lumped 180° hybrid requires only 0.26 mm^2 of chip area, while a Lange coupler occupies about 3 mm^2 chip area that is larger by a factor of 10 [28].

A typical spiral inductor with two and a half turns in a CPW environment [29] is shown in Figure 9.22. This spiral inductor for the purpose of circuit simulation can be represented by a lumped equivalent circuit model [29] as shown in Figure 9.23. The inductance L and the parasitic capacitances C_{p1},

TABLE 9.3 Computed Equivalent Circuit Element Values for Symmetric T-Junctions

S, μm	W, μm	h, μm	ε_r	d, μm	h_a, μm	W_a, μm	t_a, μm	t, μm	C, fF	C_a, fF	L_1, pH	L_2, pH	L_3, pH
Air-bridge T-junction [27]													
75	50	400	12.9	50	2.5	—	0	0	141	—	13	74	74
Conventional T-junction in a conductor back CPW [26]													
40	20	200	12.5	—	2.0	20	—	—	5.83	3.5	3.57	19.59	19.59

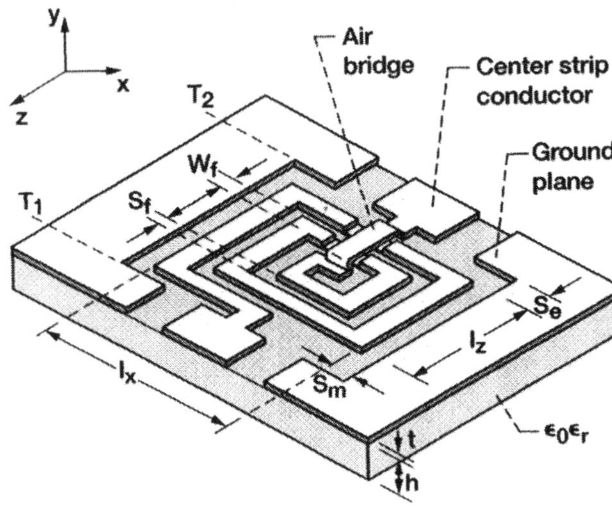

FIGURE 9.22 Coplanar waveguide spiral inductor.

C_{p2}, and C_g in the model can be obtained using quasi-static finite difference method of analysis as described in [29] and [11], respectively. The resistance $R_f(f)$ in the equivalent circuit is given by [29]

$$R_f(f) = \begin{cases} R_{dc} + \{[R_f(1\ \text{GHz})\cdot\sqrt{f_g}] - R_{dc}\}\cdot\left(\dfrac{f}{f_g}\right) & \text{for } f < f_g, \\ R_f(1\ \text{GHz})\cdot\sqrt{f} & \text{for } f > f_g, \end{cases} \tag{9.23}$$

where the resistance $R_f(f)$, R_{dc}, and $R_f(1\ \text{GHz})$ are in ohms, and the frequency f and f_g are in GHz. The frequency f_g is given by [29]

$$f_g = 39.131\frac{(\rho/\rho_{Cu})}{t}\ \text{GHz}, \tag{9.24}$$

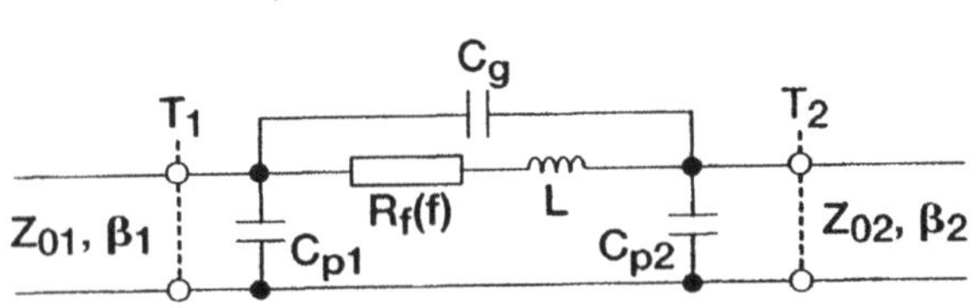

FIGURE 9.23 Lumped element equivalent circuit model for the spiral inductor.

TABLE 9.4 Geometrical Dimensions, Lumped Equivalent Circuit Model Element Values, and λ/4 Resonance Frequency of Three Rectangular Coplanar Spiral Inductors on a semi-insulating GaAs Substrate ($\varepsilon_r = 12.9$)

Parameters	Inductor Number		
	1	2	3
Number of turns, N	2.5	3.5	4.5
Line width, W_f (μm)	25	25	25
Gap width, S_f (μm)	5	5	5
Inductor size, l_x, l_z (μm)	185	240	300
Gap to ground, S_m (μm)	50	50	50
Gap to ground, S_e (μm)	50	50	50
Turn length l_{turn} (mm)	1.245	2.055	3.155
Model parameters			
L (nH)	0.7013	1.455	2.813
C_{p1} (fF)	41.97	60.72	81.34
C_{p2} (fF)	19.03	23.36	29.65
C_g (fF)	21.38	22.16	32.67
R_f (1 GHz) (Ω)	0.334	0.563	0.838
R_{dc} (Ω)	0.434	0.716	1.110
$f(\lambda/4)$ (GHz)	30	17	12

Source: From Reference [29], ©IEEE 1993.

where

ρ = resistivity per unit length

ρ_{Cu} = resistivity of copper

t = metallization thickness in μm.

In Table 9.4 the physical parameters as well as the lumped equivalent circuit model element values for spiral inductors with different number of turns [29] are presented as examples. One notices from this table that as the number of turns increases, the inductance L increases. In Figure 9.24 the measured and simulated transmission coefficient S_{21} for the spiral inductors [29] in the above table are presented. It is observed first that at a fixed frequency, as the number of turns increases, the S_{21} decreases. Second, there is good agreement between measured and simulated S_{21}. Measurements show that for an inductor on a GaAs substrate with $W_f = 3\,\mu$m, $S_f = 5\,\mu$m and $N = 7.5$, the L is about 5.16 nH. When N is increased to 14.5, the L increases to 28 nH [43].

9.11 COPLANAR WAVEGUIDE CAPACITORS

Coplanar waveguide filter circuits, lumped hybrid couplers, dc blocking circuits, tuning elements in impedance matching network, and RF bypass circuits

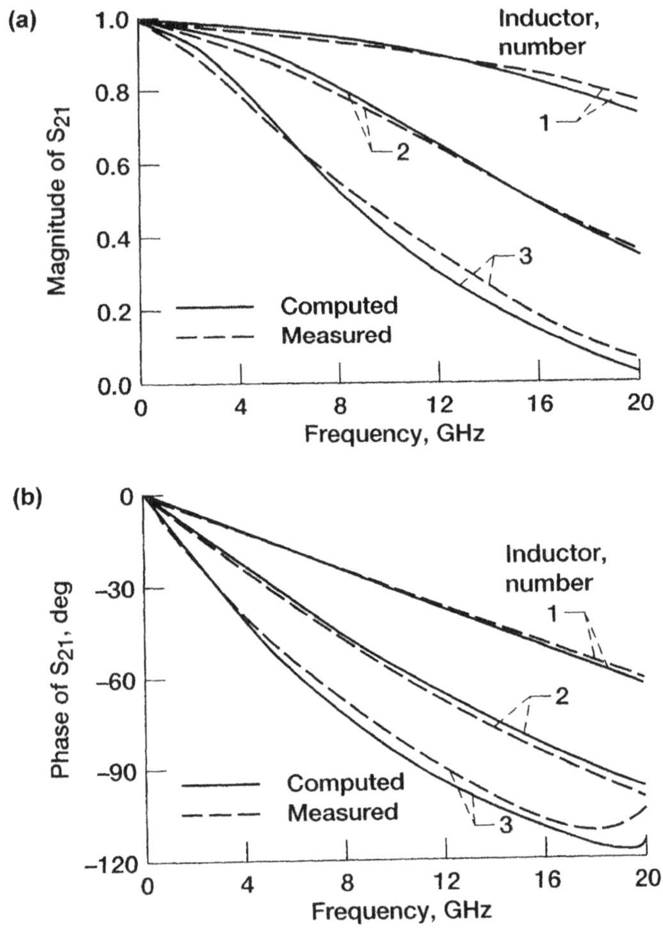

FIGURE 9.24 Calculated and measured transmission coefficients of coplanar spiral inductors: (*a*) Magnitude; (*b*) phase. (Reference [29], ©IEEE 1993.)

are a few examples that require capacitors. These capacitors are of two types, interdigital and metal-insulator-metal (MIM). In this section the construction, the equivalent circuit models, and the numerically simulated results for both types of capacitors are presented.

9.11.1 Interdigital Capacitor

In Figure 9.25 the circuit layout for a typical interdigital capacitor [30] having five fingers is shown. The air-bridges at the two ends of the circuit are of Type B as explained in Section 9.8.2, and they are required for the suppression of the parasitic coupled slotline mode. The interdigital capacitor for the purpose of

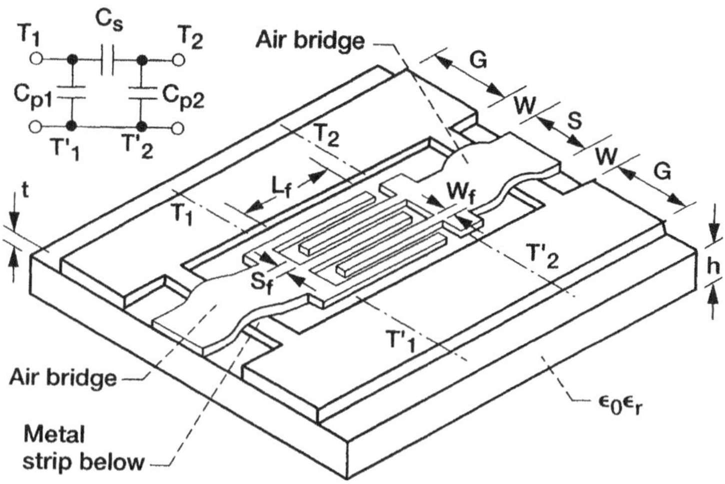

FIGURE 9.25 Coplanar waveguide interdigital capacitor and inset shows the lumped element equivalent circuit model.

circuit simulation can be represented by a π-equivalent circuit model shown in the inset in Figure 9.25. In this model, the series capacitance C_s is the desired capacitance, while the shunt capacitances C_{p1} and C_{p2} are the parasitic capacitances between the interdigital structure and the CPW ground planes. The C_s, C_{p1}, and C_{p2} can be calculated using three-dimensional finite difference method as explained in [11], [30]. The calculated as well as the measured C_s, C_{p1}, and C_{p2} for several interdigital capacitors as a function of the number of fingers n_f, finger width W_f, finger separation S_f, finger length L_f, and substrate parameters are presented in Table 9.5. The experimental values in the table are

TABLE 9.5 Measured and Modeled Coplanar Waveguide Interdigital Capacitor Values

							C_{p1} (fF)		C_{p2} (fF)		C_s (fF)	
ε_r	h, μm	t, μm	n_f	W_f, μm	S_f, μm	L_f, μm	Calculated	Measured	Calculated	Measured	Calculated	Measured
12.9	400	3	4	17	3	100	9.65	11	9.65	11	40.1	41
12.9	400	3	10	12	2	100	9.9	11	9.9	11	113.3	116
12.9	400	3	4	17	3	200	19.2	20	19.2	20	73	76
9.8	635	5	5	38	25	200	22.94	23	10.78	11	55	56
9.8	635	5	7	38	25	200	21.78	22	11.64	12	85	86

Source: From Reference [11], ©IEEE 1990.

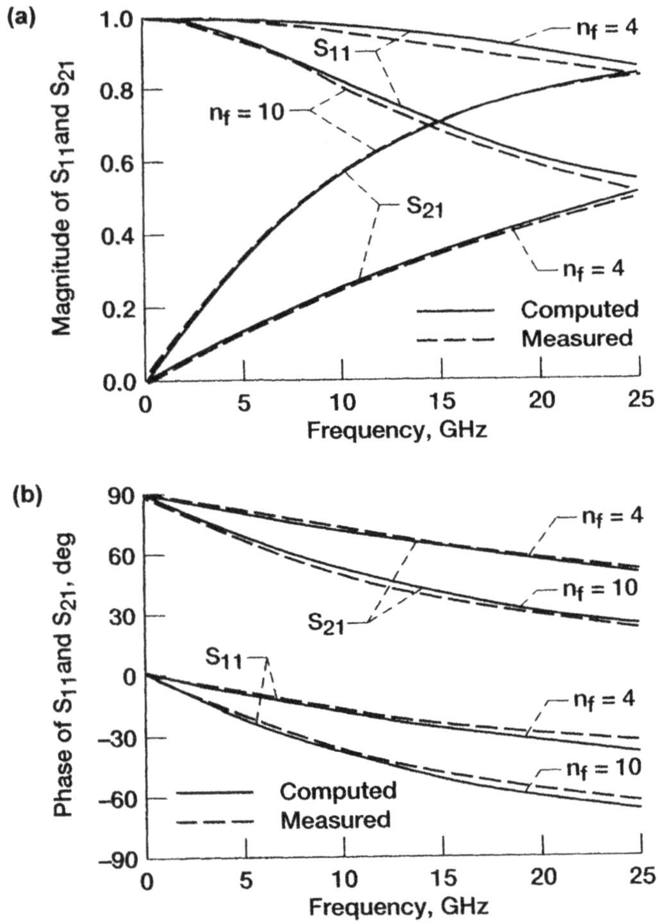

FIGURE 9.26 Calculated and measured reflection and transmission coefficient of coplanar waveguide interdigital capacitors, $\varepsilon_r = 12.9$, $h = 400\ \mu m$, $L_f = 100\ \mu m$. Case 1: $n_f = 4$, $W_f = 17\ \mu m$, $S_f = 3\ \mu m$); case 2: $n_f = 10$, $W_f = 12\ \mu m$, $S_f = 2\ \mu m$. (*a*) Magnitude; (*b*) phase. (Reference [11], ©IEEE 1990.)

obtained from the measured scattering parameters. This step is carried out by a software routine which optimizing the π-equivalent circuit model element values till a good fit is established between the calculated and measured scattering parameters. Figure 9.26(*a*) and (*b*), presents examples of the calculated and measured magnitude and phase of the reflection and transmission coefficient for capacitors in Table 9.5 with the smallest and largest C_s values [11]. It is important to note that the capacitances C_{p1} and C_{p2} do not depend on the size and the number of fingers. This is due to the fact that the location of the CPW ground planes can be varied to minimize their effect. As a result it is possible to realize large values of C_s without increases C_{p1} and C_{p2}. In practice, this is accomplished by increasing the number of fingers as well as the distance of separation between the intedigitated structure and the CPW ground planes.

9.11.2 Series Metal-Insulator-Metal Capacitor

The circuit layout of a series metal-insulator-metal (MIM) capacitor [30] is shown in Figure 9.27. The coplanar waveguide (CPW) center strip conductor and slot widths are indicated as S and W, respectively. The capacitor is considered to be lumped and appears as a series element in the center strip conductor of the CPW. The inset in Figure 9.27 shows the cross section of the capacitor. In the cross section the thickness of the top and bottom electrodes are indicated as t_{top} and t_{bot}, respectively. Further the electrodes are shown separated by a thin dielectric film of thickness t_c and dielectric constant ε_{rc}. In most cases, the width S_{top} and S_{bot} are identical and same as S. Further for tight coupling, the electrodes overlap each other, and the overlap length L_c can take a value anywhere from 0.5 S to 4.0 S [30].

The lumped element equivalent circuit model for the series MIM capacitor is shown in Figure 9.28. As a typical example consider a series MIM capacitor with the following parameters: $S = 100\ \mu\text{m}$, $W = 75\ \mu\text{m}$, $h = 450\ \mu\text{m}$, $\varepsilon_r = 12.9$, $Z_{0(CPW)} = 50\ \Omega$, $t_c = 0.2\ \mu\text{m}$, $t_{top} = 3.0\ \mu\text{m}$, $t_{bot} = 0.48\ \mu\text{m}$, $\varepsilon_{rc} = 7.45$, and $L_c = 300\ \mu\text{m}$. For this example the series capacitance C_s can be calculated using the well-known parallel-plate capacitor formula as follows:

$$C_s = \frac{\varepsilon_0 \varepsilon_{rc} S L_c}{t_c}. \tag{9.25}$$

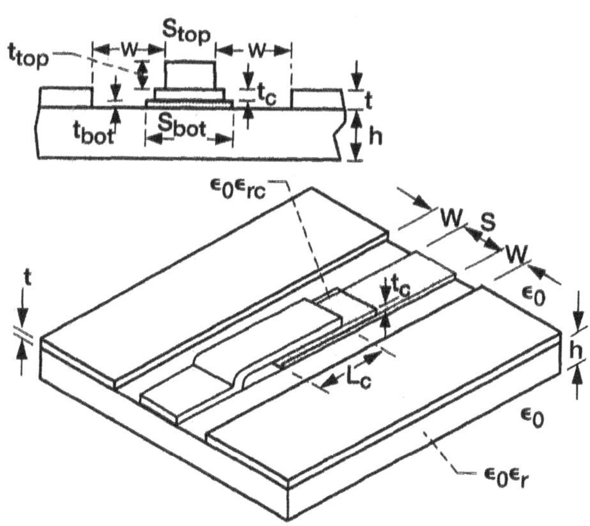

FIGURE 9.27 Coplanar waveguide series metal-insulator-metal capacitor. Inset gives the cross section.

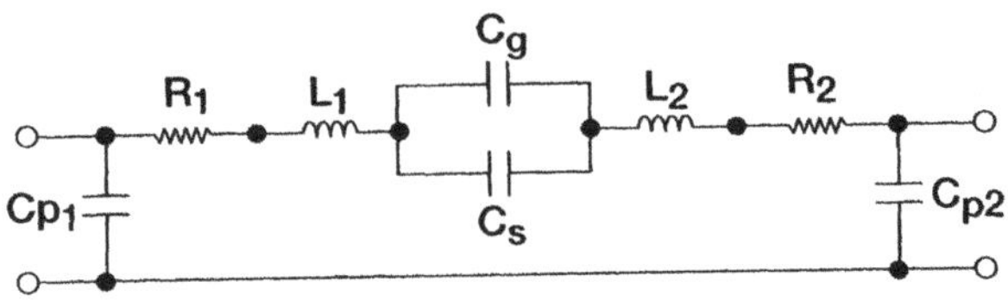

FIGURE 9.28 Lumped element equivalent circuit model for the series MIM capacitor.

By substituting the parameters, it is found that C_s is about 9.9 pF. The capacitance C_g is the parasitic capacitance between the open end of the top electrode and the center strip conductor below the dielectric film. This element is independent of the overlap length L_c and is dependent on the electrode width and thickness as explained in Section 9.5. The C_g computed for the example above has a value small compared to C_s and is about 0.18 fF [30]. The parasitic resistance and inductance associated with the top and bottom electrodes are designated as R_1, L_1 and R_2, L_2, respectively, in the model. These are dependent on the electrode cross-sectional dimensions and length. The computed resistance and inductance are on the order of $R_1 = R_2 = 0.19\,\Omega$, $L_1 = 62$ nH, and $L_2 = 61$ nH [30]. The parasitic shunt capacitances at the two ends of the MIM capacitor are designated in the model as C_{p1} and C_{p2}, respectively. The C_{P1} and C_{P2} computed for the example above have values small compared to C_s and are on the order of 38 and 35 fF, respectively [30]. The numerically simulated and measured reflection and transmission coefficients [30] for the MIM capacitor in the example above are shown in Figure 9.29(*a*) and (*b*).

9.11.3 Parallel Metal-Insulator-Metal Capacitor

The circuit layout of the parallel MIM capacitor [30] is shown in Figure 9.30. The capacitor appears as a shunt element between the CPW center strip conductor and the ground planes. This element is very useful in circuits where a capacitor to ground is required. Moreover it results in small circuit size and consequently reduced parasitic effects. The lumped element equivalent circuit model for the capacitor is shown in Figure 9.31. As a typical example, consider a parallel MIM capacitor with the following parameters: $S = 100\,\mu$m, $W = 75\,\mu$m, $h = 450\,\mu$m, $\varepsilon_r = 12.9$, $Z_{0(CPW)} = 50\,\Omega$, $t_c = 0.2\,\mu$m $t_{top} = 3.0\,\mu$m, $t_{bot} = 0.48\,\mu$m, $\varepsilon_{rc} = 7.45$, and $L_c = 300\,\mu$m. The shunt capacitance C_p in the model can be calculated using Eq. (9.25) and is about 9.9 pF [30]. The computed parasitic resistance and inductance are small and on the order of $0.1\,\Omega$ and 0.27 pH, respectively [30].

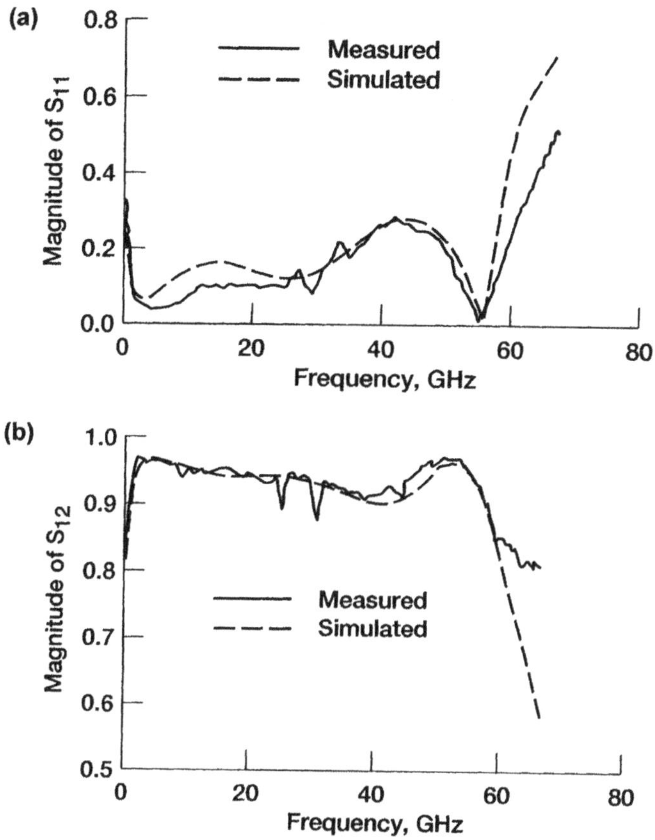

FIGURE 9.29 Calculated and measured reflection and transmission coefficient of coplanar waveguide series metal-insulation-metal capacitor, $S = 100\,\mu\text{m}$, $W = 75\,\mu\text{m}$, $Z_{0(\text{CPW})} = 50\,\Omega$, $h = 450\,\mu\text{m}$, $\varepsilon_r = 12.9\,\mu\text{m}$, $L_c = 300\,\mu\text{m}$, $t_c = 0.2\,\mu\text{m}$, $\varepsilon_{rc} = 7.45\,\mu\text{m}$, $t_{\text{bot}} = 0.48\,\mu\text{m}$, $t_{\text{top}} = 3.0\,\mu\text{m}$: (*a*) Reflection coefficient; (*b*) transmission coefficient. (Reference [30], ©Microwave Exhibitions and Publishers, 1994.)

9.11.4 Comparison between Coplanar Waveguide Interdigital and Metal-Insulator-Metal Capacitors

The capacitance range for CPW interdigital capacitors [30] and MIM capacitors [30] as a function of the geometry are given in Tables 9.6 and 9.7, respectively. These tables show that, the maximum attainable capacitance in the case of CPW interdigital and MIM capacitor is about 0.8 and 13.2 pF, respectively. Further, for a given capacitance value, the MIM capacitors are smaller in size. Last, the parasitic capacitances C_{p1} and C_{p2} are smaller in the case of MIM capacitors.

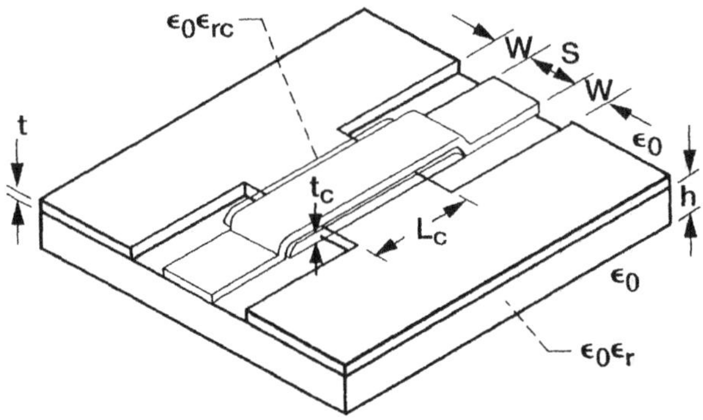

FIGURE 9.30 Coplanar waveguide parallel metal-insulator-metal capacitor.

TABLE 9.6 CPW Interdigital Capacitor Geometry and Values on a GaAs Wafer

n_f	W_f, μm	S_f, μm	L_f, μm	C_{p1}, (fF)	C_{p2}, (fF)	C_s (fF)
2	25	20	300	—	—	27
5	25	20	300	83	46	82
10	25	20	300	—	—	190
35	10	10	300	—	—	800

Note: $h = 450$ μm, $\varepsilon_r = 12.9$. Input/output CPW: $S = 100$ μm, $W = 75$ μm, $G = 200$ μm, $Z_{0(\mathrm{CPW})} = 50\ \Omega$, G = the ground plane width.

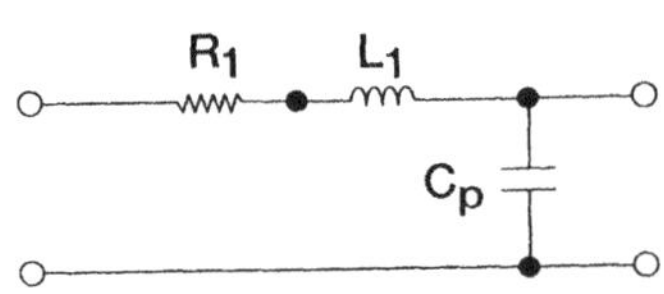

FIGURE 9.31 Lumped element equivalent circuit model for the parallel MIM capacitor.

9.12 COPLANAR WAVEGUIDE STUBS

Coplanar waveguide series as well as shunt stubs are essential for constructing filter circuits using capacitive and inductive elements [31]. The values of these capacitances and inductances are a function of the stub length and geometry. In this section several lumped element equivalent circuit models for CPW stubs are presented.

TABLE 9.7 CPW MIM Capacitor Geometry and Values on a GaAs Wafer

S, μm	W, μm	L_c, μm	C_{p1} (fF)	C_{p2} (fF)	C_s (pF)
100	75	300	38	35	9.9
100	75	50	—	—	1.65
		400	—	—	13.2
50	37	25	—	—	0.82
		200	—	—	6.6
25	20	12.5	—	—	0.41
		100	—	—	3.3
10	10	5	—	—	0.165
		40	—	—	1.32

Note: $h = 450\ \mu$m, $\varepsilon_r = 12.9$, $Z_{0(\mathrm{CPW})} = 50\ \Omega$, $\varepsilon_{rc} = 7.45$, $t_c = 0.2\ \mu$m, $t_{bot} = 0.48\ \mu$m, $t_{top} = 3.0\ \mu$m.

9.12.1 Open-End Coplanar Waveguide Series Stub

An open circuited stub formed by an interdigital structure in the center strip conductor of a coplanar waveguide is shown in Figure 9.32, gives rise to a series capacitance with negligible parasitics when the dimensions of the circuit are very small [32]. Hence, for the purpose of circuit analysis, the stub can be accurately modeled as a series capacitance C_s as shown in the inset in Figure 9.32. Typically the length L in Figure 9.32 is less than $0.1\lambda_{g(\mathrm{CPW})}$ when the circuit is used as a capacitor. As a design aid the experimentally determined capacitance as a function of the length L [32] for open circuited stub on a conductor-backed GaAs wafer of thickness h is presented in Figure 9.33. The figure shows that C_s increases linearly with L but decreases as the interdigital finger width S_1 increases. The capacitance values in this figure are obtained by computer optimization of the measured scattering parameters of the circuit across the frequency range of 1 to 26.5 GHz. Consequently the model above is valid over a wide frequency range.

The circuit in Figure 9.32 has been analyzed using space domain integral equation (SDIE) method in [33]. The computed as well as the measured reflection and transmission coefficients shows that when the transverse dimensions of the CPW as well as the finger length L_1 are very large, the simple capacitive series equivalent circuit model discussed above cannot accurately predict the circuit performance. In particular, the simple model cannot predict the bandpass resonance when the mean stub length approaches $0.25\lambda_{g(\mathrm{CPW})}$. Hence a new equivalent circuit model that is valid over the frequency range of 5 GHz to the first bandpass resonance is demonstrated in [33] and shown in Figure 9.34. In this model, as the finger length L_1 approaches zero, the inductances L_{s1}, L_{s2}, and L_{s3} reduce to zero and the capacitances C_{p1} becomes equal to C_{p2}, resulting in a symmetric capacitive π-equivalent circuit model that is expected of a series gap as explained in Section 9.5.

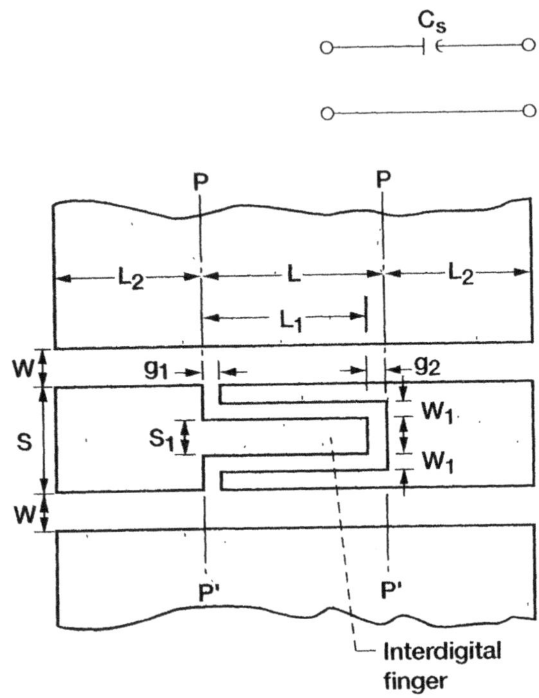

FIGURE 9.32 Open end coplanar waveguide series stub. Inset shows a simple equivalent circuit model for the stub.

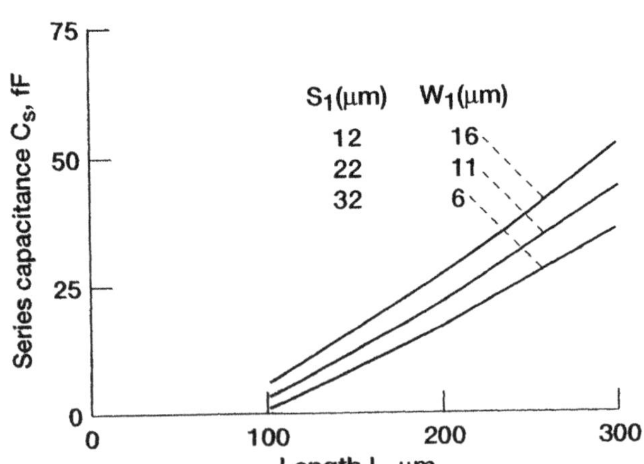

FIGURE 9.33 Series capacitance as a function of the length L ($S = 66\ \mu$m, $W = 40\ \mu$m, $h = 100\ \mu$m, $\varepsilon_r = 12.8$, $L_2 = 500\ \mu$m, $g_1 = g_2 = W_1$). (Reference [32], ©IEEE, 1992.)

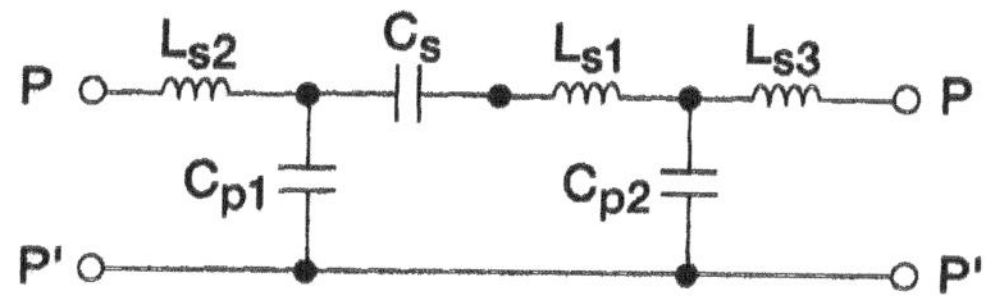

FIGURE 9.34 Lumped element equivalent circuit model for the open end coplanar waveguide series stub.

In [33] simple equations are given in terms of the finger length L_1 which predicts the equivalent circuit element values. These equations are strictly valid for a circuit with $S = 440\ \mu m$, $W = 240\ \mu m$, $S_1 = 140\ \mu m$, $W_1 = 90\ \mu m$, $g = 75\ \mu m$, and finger length L_1 through 2500 μm fabricated on an alumina substrate of thickness and dielectric constant equal to 546 μm and 9.9, respectively. In addition the equations take into account the presence of a supporting Duroid® substrate below of thickness and dielectric constant equal to 3175 μm and 2.2, respectively. The equations are as follows [33]:

$$C_s = 1.01 \times 10^{-4} L_1 + 1.642 \times 10^{-2}, \tag{9.26}$$

$$C_{p1} = 0.39 \times 10^{-4} L_1 + 1.765 \times 10^{-2}, \tag{9.27}$$

$$C_{p2} = 0.883 \times 10^{-4} L_1 + 1.765 \times 10^{-2}, \tag{9.28}$$

$$L_{s1} = 3.26 \times 10^{-4} L_1, \tag{9.29}$$

$$L_{s2} = 1.22 \times 10^{-4} L_1, \tag{9.30}$$

$$L_{s3} = 1.43 \times 10^{-4} L_1, \tag{9.31}$$

where the finger length L_1 is in μm, the inductances in nH, and the capacitances in pF.

The foregoing lumped element equivalent circuit model predicts the response up to the first resonant frequency with a 5 percent accuracy [33]. It is expected that similar linear relationships hold good for stubs with any dimensions. Thus it is enough to model two different stub lengths from which the characteristics of other lengths can be derived.

The open end series stub can also be implemented in the ground planes of a finite-width coplanar waveguide instead of the center strip conductor as demonstrated in [34]. The results indicate that the stubs in the ground planes can be modeled as two stubs in parallel to achieve twice the capacitance value than can be achieved for an identical length stub in the center strip conductor. However, the resonant frequencies in the two cases are identical.

9.12.2 Short-End Coplanar Waveguide Series Stub

Most of the discussion in the previous section pertaining to the open end coplanar waveguide series stub is also relevant to the short-end coplanar

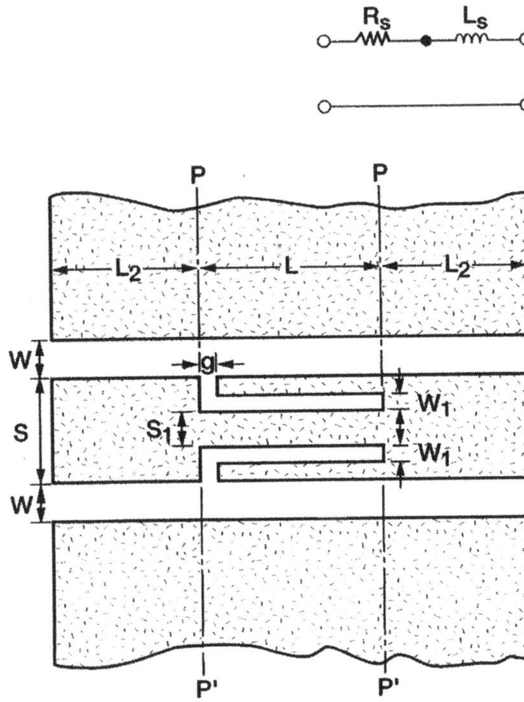

FIGURE 9.35 Short-end coplanar waveguide series stub. Inset shows a simple equivalent circuit model for the stub.

waveguide series stub, and hence many details are omitted here to avoid repetition. A short-end coplanar waveguide series stub is shown in Figure 9.35. The short-end stub can be modeled as a series combination of a resistor R_s and an inductor L_s [32] as shown in the inset in Figure 9.35. This model is valid when the dimensions of the circuit, including the finger length L, are small, typically less than $0.1\lambda_{g(CPW)}$. The experimentally determined R_s and L_s [32] are shown in Figure 9.36(*a*) and (*b*), as a function of L. The circuit is fabricated on a conductor backed GaAs wafer of thickness h and dielectric constant ε_r. The figures show that both R_s and L_s increase with L but decrease as S_1 increases.

Analogous to the open end coplanar waveguide stub, the simple model discussed above cannot predict the band-stop resonance when the mean stub length approaches $0.25\lambda_{g(CPW)}$. Hence a new equivalent circuit model that is valid over the frequency range of 5 GHz to the first band-stop resonance is demonstrated in [33] and shown in Figure 9.37. In this model, as the finger length L approaches g, L_{s2} will approach L_{s3} and C_{p1} will approach C_{p2}, resulting in a symmetric model as expected for a simple notch in the center strip conductor of the CPW. In [33] simple equations are given in terms of the finger length L, which predicts the equivalent circuit element values. These equations are strictly valid for a circuit with parameters same as the open end

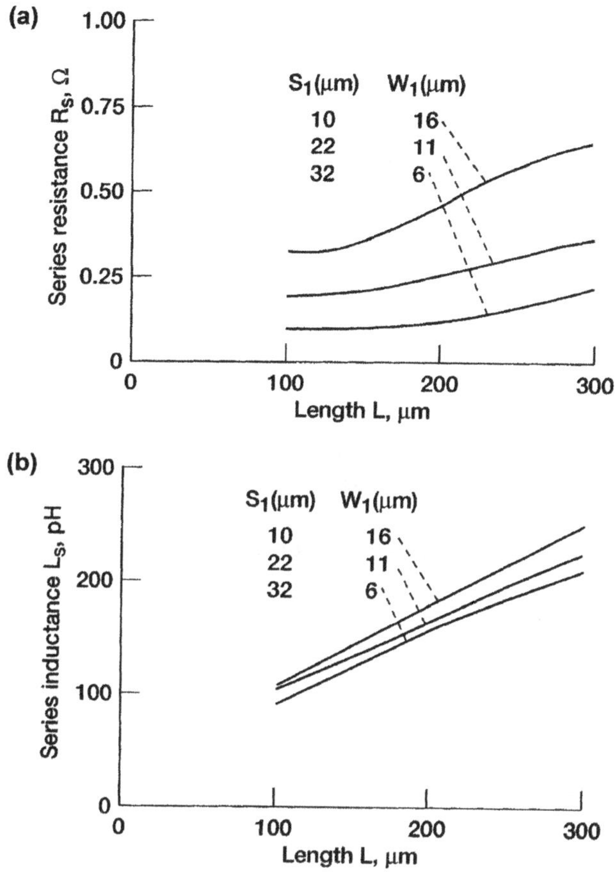

FIGURE 9.36 Equivalent circuit element values as a function of the length L ($S = 66\ \mu$m, $W = 40\ \mu$m, $h = 100\ \mu$m, $\varepsilon_r = 12.8$, $L_2 = 500\ \mu$m, $g = W_1$): (*a*) Series resistance; (*b*) series inductance. (Reference [32], ©IEEE, 1992.)

coplanar waveguide stub. The equations are as follows [33]:

$$C_s = 1.32 \times 10^{-4} L + 3.3515 \times 10^{-2}, \tag{9.32}$$

$$C_{p1} = 1.5959 \times 10^{-2}, \tag{9.33}$$

$$C_{p2} = 1.1249 \times 10^{-4} L + 7.522 \times 10^{-3}, \tag{9.34}$$

$$L_{s1} = 1.77 \times 10^{-4} L - 8.35 \times 10^{-4}, \tag{9.35}$$

$$L_{s2} = 2.6368 \times 10^{-4} L - 6.618 \times 10^{-3}, \tag{9.36}$$

$$L_{s3} = 1.8656 \times 10^{-4} L - 8.34 \times 10^{-4}, \tag{9.37}$$

where the finger length L is in μm, the inductance in nH, and the capacitance in pF.

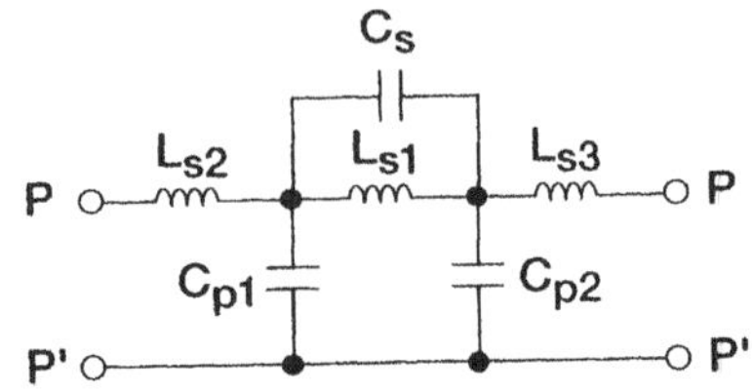

FIGURE 9.37 Lumped element equivalent circuit model for the short-end coplanar waveguide series stub.

As a concluding remark it may be mentioned that a short-end series stub can also be implemented in the ground planes of a finite width coplanar waveguide instead of the center strip conductor as demonstrated in [34]. The results indicate that the inductance and resistance is half the value of the same stub when located in the center strip conductor.

9.12.3 Combined Short and Open End Coplanar Waveguide Series Stubs

A combined short and open end series stubs in the center strip conductor of a coplanar waveguide is shown in Figure 9.38(*a*). When the stub lengths and parasitics are small, this circuit gives rise to a series combination of an inductance and a capacitance [35]. Consequently maximum power transfer takes place from the input to the output port at resonance. The stub layout shown in Figure 9.38(*b*) can, with some imagination, be easily recognized as a parallel combination of an inductance and a capacitance [35]. In this case the input and the output ports are isolated at resonance.

9.12.4 Coplanar Waveguide Shunt Stubs

An open end coplanar waveguide shunt stub printed in the CPW ground plane is shown in Figure 9.39, and the inset shows a similar shunt stub with a short end. When the length of these stubs are small, they give rise to a shunt capacitance and a shunt inductance, respectively. These circuits have been analyzed using the space domain integral equation (SDIE) method in [36] and [37]. The results show that the computed and measured scattering parameters are in good agreement over a wide frequency range. In addition in [36] it has been demonstrated that by bending the stubs as in the above figures, the loss of power due to radiation is significantly reduced. This is because the fields radiated by the two opposing stubs partially cancel. A lumped element equivalent circuit model for the cross junction in the shunt stubs can be found in [38]. Finally, the characteristics of shunt stubs printed within the center strip conductor of a CPW are presented in [39].

9.12.5 Coplanar Waveguide Radial Line Stub

Coplanar waveguide radial line single-stub [40] and double-stub [41] resonators are shown in Figures 9.40(*a*) and (*b*). These resonators have several

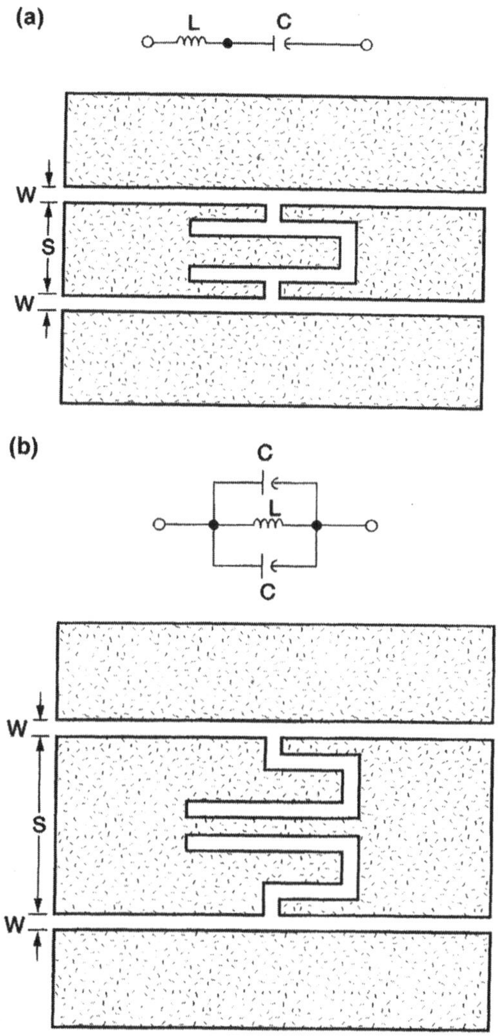

FIGURE 9.38 Combined short- and open end coplanar waveguide series stubs: (*a*) Series combination of an inductance and a capacitance; (*b*) parallel combination of an inductance and a capacitance.

applications, which include bias line filters requiring a point of virtual RF ground, mixer, and frequency multiplier circuits that require a reactance to terminate the diodes and harmonics. In general, the advantages of a radial line stub over a straight 50 Ω stub are smaller resonant length, wider bandwidth, and smaller discontinuity reactance at the junction with the main line. In this section, closed form equations to predict the resonant radius, for single-stub as well as double-stub resonators are given. These equations are obtained by curve-fitting the measured data.

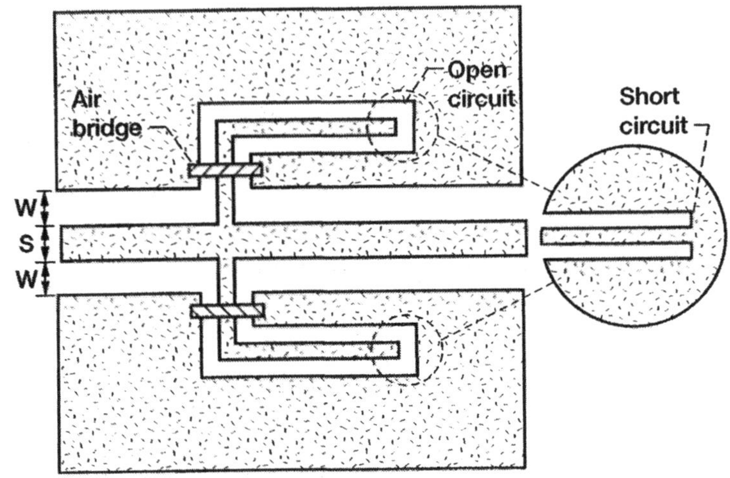

FIGURE 9.39 Open end coplanar waveguide shunt stub. Inset shows a short-end shunt stub.

The resonant outer radius for a coplanar waveguide and a grounded coplanar waveguide radial line single stub are, respectively, given by two equations [40]:

$$\log(R_2) = C_1 \log[\sqrt{\varepsilon_r} f_0] + C_2 \log(h) + C_3 \log(R_1) + C_4, \tag{9.38}$$

$$\log(R_2) = C_5 \log[\sqrt{\varepsilon_r}\, f_0] + C_6[\log(h) + \log(R_1)], \tag{9.39}$$

where f_0 is the resonant frequency in GHz, R_1 and R_2 are the inner and outer radius of the stub in meters, and h is the substrate thickness in meters. The coefficients are $C_1 = -1.0496$, $C_2 = 0.098375$, $C_3 = -0.1401$, $C_4 = -1.297$, $C_5 = -0.82462$, and $C_6 = 0.21201$. These equations are valid for the following range of parameters:

Coplanar waveguide,

$$\left.\begin{array}{l} 6 \leqslant \varepsilon_r \leqslant 11.7 \\ 0.254\,\text{mm} \leqslant h \leqslant 0.635\,\text{mm} \\ \theta = 60^\circ \\ 0.114\,\text{mm} \leqslant W \leqslant 0.234\,\text{mm} \\ 0.152\,\text{mm} \leqslant S \leqslant 0.508\,\text{mm} \\ 0.092\,\text{mm} \leqslant R_1 \leqslant 0.3\,\text{mm} \end{array}\right\}, \tag{9.40}$$

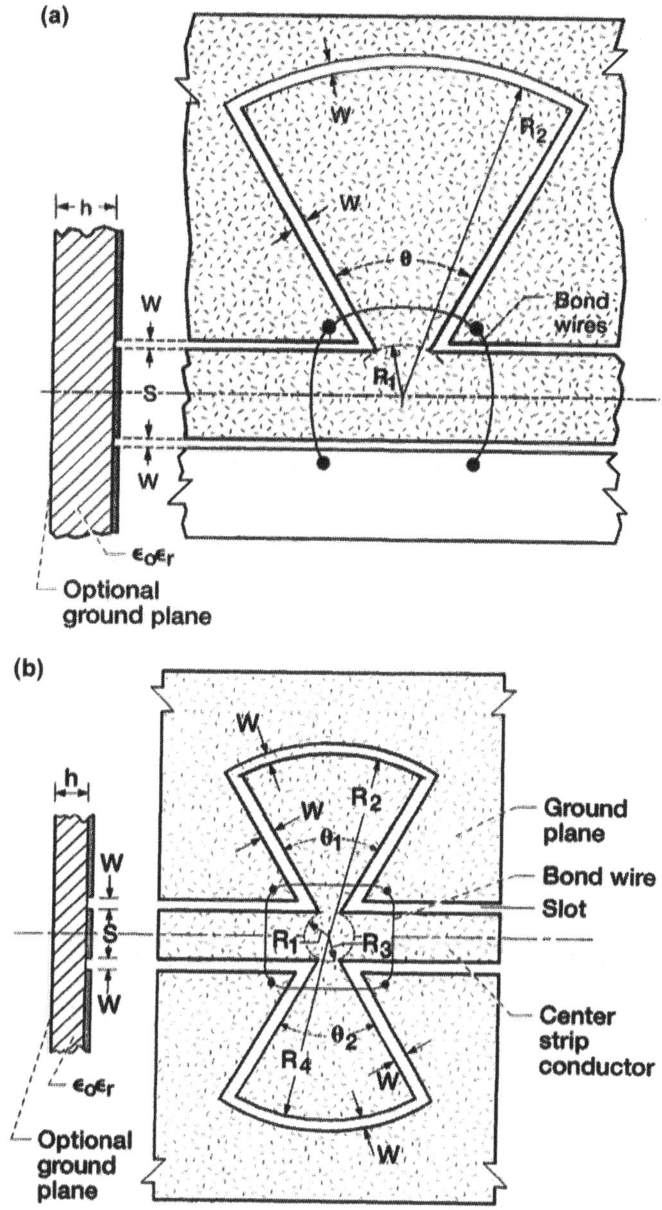

FIGURE 9.40 Coplanar waveguide radial line resonators: (*a*) Single stub; (*b*) double stub.

where ε_r is the substrate dielectric constant, θ is the sector angle W and S are the coplanar waveguide slot and strip width, respectively.

Grounded coplanar waveguide,

$$\varepsilon_r = 2.2,\ h = 0.254\ \text{mm},\ \theta = 60°,\ W = 0.127\ \text{mm},\ S = 0.518\ \text{mm, and}\ R_1 = 0.3\ \text{mm}.$$

The closed form Equations to predict the resonant outer radius for a coplanar waveguide and a grounded coplanar waveguide radial line symmetric double stub ($R_2 = R_4$ and $\theta_1 = \theta_4$) are, respectively, given by the following two equations [41]:

$$\log(R_2) = C_1 \log[\sqrt{\varepsilon_r} f_0] + C_2[\log(h) + \log(R_1)], \tag{9.41}$$

$$\log(R_2) = C_3 \log[\sqrt{\varepsilon_r} f_0] + C_4[\log(h) + \log(R_1)], \tag{9.42}$$

where f_0 is the resonant frequency in GHz, R_1 and R_2 are the inner and outer radius of the stub in meters, and h is the substrate thickness in meters. The coefficients are $C_1 = -0.89203$, $C_2 = 0.16733$, $C_3 = -0.79377$, and $C_4 = 0.20831$. The equations above are valid for the following parameters:

Coplanar waveguide,

$$\varepsilon_r = 10.5,\ h = 0.635\ \text{mm},\ \theta = 60°,\ W = 0.234\ \text{mm},$$
$$S = 0.33\ \text{mm, and}\ R_1 = 0.191\ \text{mm}.$$

Grounded coplanar waveguide,

$$\varepsilon_r = 2.2,\ h = 0.254\ \text{mm},\ \theta = 60°,\ W = 0.127\ \text{mm},$$
$$S = 0.518\ \text{mm, and}\ R_1 = 0.3\ \text{mm}.$$

The equations above are for a fixed sector angle $\theta = 60°$. If the sector angle changes, then the resonant frequency also changes. These changes can be understood by considering the characteristics of the CPW radial line single stub discussed earlier. For this stub the measured resonant frequency as a function of θ for a fixed R_2 [40] is shown in Figure 9.41. It is observed that, the resonant frequency decreases by 20 percent as θ increases from 30° to 90°. Note that R_1 does not remain constant but increases with θ.

9.13 COPLANAR WAVEGUIDE SHUNT INDUCTOR

Metal strips across the slots of a coplanar waveguide, as shown in Figure 9.42, effectively behave as shunt inductors. An approximate method to estimate the shunt inductance is to use the well-known formula for ribbon inductors, which

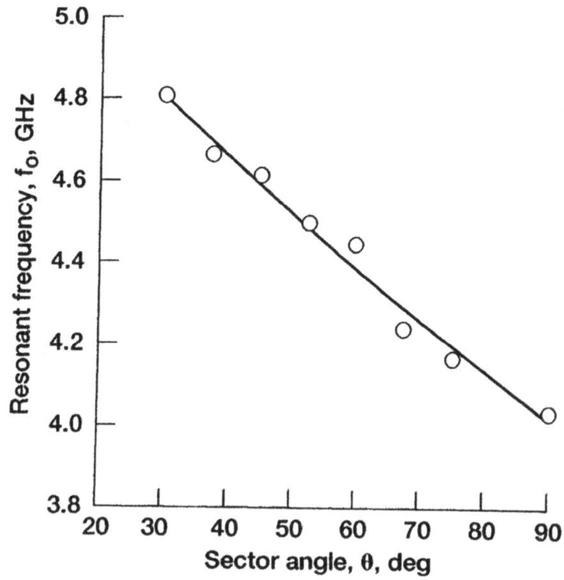

FIGURE 9.41 Measured resonant frequency of coplanar waveguide radial line single stub as a function of the sector angle θ ($S = 0.33$ mm, $W = 0.234$ mm, $h = 0.635$ mm, $\varepsilon_r = 10.5$, $R_2 = 5.08$ mm).

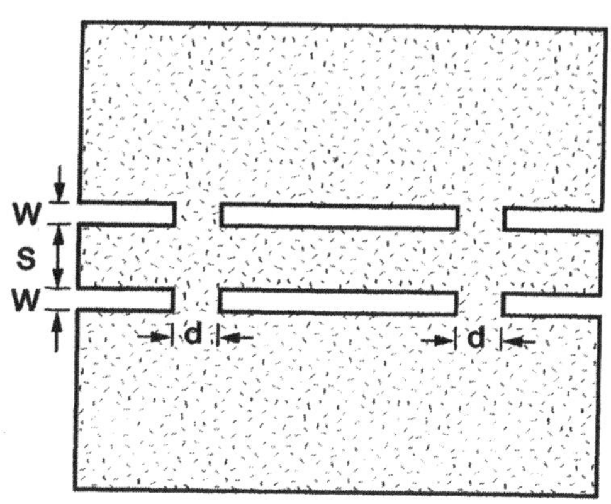

FIGURE 9.42 Coplanar waveguide shunt inductor.

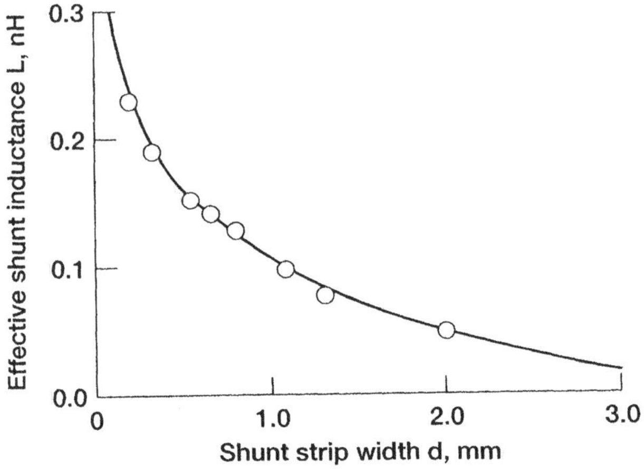

FIGURE 9.43 Measured effective inductance as a function of the strip width d ($\varepsilon_r = 2.2$, $h = 1.5748$ mm, $S = 2.5$ mm, $W = 1.0$ mm, $f_r = 4.5$ GHz). (Reference [42], ©IEEE, 1993.)

is as follows [42]:

$$L = 2W\left\{\ln\left(\frac{2\pi W}{d}\right) - 1 + \frac{d}{\pi W}\right\}\text{nH}, \tag{9.43}$$

where d and W, measured in centimeters, are the width and length of the inductor, respectively. However, the accuracy of this formula is quite poor, with error on the order of 10 percent or greater for small inductances resulting from the use of ribbons with wide width [42].

An alternative approach to obtain the effective inductance of the structure shown in Figure 9.42 is to conduct a series of experiments to measure the loaded quality factor Q_L and the transmission coefficient S_{21} of a single section resonator having variable shunt inductance values [42]. These resonators are made to resonate at roughly the same frequency f_r. The data gathered are then used to calculate the effective inductance L from the following relationship [42]:

$$Q_L = \frac{\pi}{4} z S_{21}\left\{\left(\frac{Z_s}{2\pi f_r L}\right)^2 + \left(\frac{z-1}{z}\right)^2\right\}, \tag{9.44}$$

where

$z = Z_0/Z_s$
Z_s = source and load impedance
Z_0 = coplanar waveguide characteristic impedance.

Figure 9.43 shows an example of the measured effective inductance as a function of the strip width d for a coplanar waveguide shunt inductor fabricated on a 1.5748 mm thick RT-5880 Duroid substrate ($\varepsilon_r = 2.2$).

REFERENCES

[1] R. N. Simons and G. E. Ponchak, "Modeling of Some Coplanar Waveguide Discontinuities," *IEEE Trans. Microwave Theory Tech.*, Vol. 36, No. 12, pp. 1796–1803, Dec. 1988.

[2] K. Beilenhoff, H. Klingbeil, W. Heinrich, and H. L. Hartnagel, "Open and Short Circuits in Coplanar MMIC's," *IEEE Trans. Microwave Theory Tech.*, Vol. 41, No. 9, pp. 1534–1537, Sept. 1993.

[3] K. Beilenhoff, W. Heinrich, and H. L. Hartnagel, "Finite-Difference Analysis of Open and Short Circuits in Coplanar MMIC's Including Finite Metallization Thickness and Mode Conversion," 1992 *IEEE MTT-S Int. Microwave Symp. Dig.*, Vol. 1, pp. 103–106, Albuquerque, NM, June 1–5, 1992.

[4] M.-H. Mao, R.-B. Wu, C.-H. Chen, and C.-H. Lin, "Characterization of Coplanar Waveguide Open End Capacitance-Theory and Experiment," *IEEE Trans. Microwave Theory Tech.*, Vol. 42, No. 6, pp. 1016–1024, June 1994.

[5] R. Bromme and R. H. Jansen, "Systematic Investigation of Coplanar Waveguide MIC/MMIC Structures Using a Unified Strip/Slot 3D Electromagnetic Simulator," *1991 IEEE MTT-S Int. Microwave Symp. Dig.*, Vol. 3, pp. 1081–1084, Boston, MA, June 10–14, 1991.

[6] N. I. Dib, W. P. Harokopus Jr., L. P. B. Katehi, C. C. Ling, and G. M. Rebeiz, "Study of a Novel Planar Transmission Line," *1991 IEEE MTT-S Int. Microwave Symp. Dig.*, Vol. 2, pp. 623–626, Boston, MA, June 10–14, 1991.

[7] N. I. Dib, W. P. Harokopus Jr., G. E. Ponchak, and L. P. B. Katehi, "A Comparative Study between Shielded and Open Coplanar Waveguide Discontinuities," *Int. J. Microwave Millimeter-Wave Computer-Aided Eng.*, Vol. 2, No. 4, pp. 331–341, Oct. 1992.

[8] F.-L. Lin and R.-B. Wu, "Analysis of Coplanar-Waveguide Discontinuties with Finite-Metallization Thickness and Nonrectangular Edge Profile," *IEEE Trans. Microwave Theory Tech.*, Vol. 45, No. 12, pp. 2131–2138, Dec. 1997.

[9] R. H. Jansen, "Hybrid Mode Analysis of End Effects of Planar Microwave and Millimeter Wave Transmission Lines," *IEE Proc.*, Vol. 128, Pt. H, No. 2, pp. 77–86, April 1981.

[10] W. J. Getsinger, "End-Effects in Quasi-TEM Transmission Lines," *IEEE Trans. Microwave Theory Tech.*, Vol. 41, No. 4, pp. 666–672, April 1993.

[11] M. Naghed and I. Wolff, "Equivalent Capacitance of Coplanar Waveguide Discontinuities and Interdigital Capacitors Using a Three-Dimensional Finite Difference Method," *IEEE Trans. Microwave Theory Tech.*, Vol. 38, No. 12, pp. 1808–1815, Dec. 1990.

[12] B. C. Wadell, *Transmission Line Design Handbook*, Norwood, MA: Artech House, 1991, pp. 278–280.

[13] E. Lan, S. M. El-Ghazaly, V. Nair, K. Eisenbeiser, and B. Ooms, "Wide Band CAD Model for Coplanar Waveguide Using FDTD Technique," *1997 IEEE-MTT-S Int. Microwave Symp. Dig.*, Vol. 3, pp. 1583–1586, Denver, CO, June 8–13, 1997.

[14] C.-W. Chiu and R.-B. Wu, "Capacitance Computation for CPW Discontinuities with Finite Metallization Thickness by Hybrid Finite-Element Method," *IEEE Trans. Microwave Theory Tech.*, Vol. 45, No. 4, pp. 498–504, April 1997.

[15] C.-W. Chiu, "Inductance Computation for Coplanar Waveguide Discontinuities with Finite Metallization Thickness," *IEE Proc., Microwave Antennas Propag.*, Vol. 145, No. 6, pp. 496–500, Dec. 1998.

[16] A. A. Omar, Y. L. Chow, L. Roy, and M. G. Stubbs, "Effect of Air-Bridges and Mitering on Coplanar Waveguide 90° Bends: Theory and Experiments," *1993 IEEE MTT-S Int. Microwave Symp. Dig.*, Vol. 2, pp. 823–826, June 14–18, 1993, Atlanta, GA.

[17] P. M. Watson and K. C. Gupta, "Design and Optimization of CPW Circuits Using EM-ANN Models for CPW Components," *IEEE Trans. Microwave Theory Tech.*, Vol. 45, No. 12, pp. 2515–2523, Dec. 1997.

[18] T. M. Weller, R. M. Henderson, S. V. Robertson, and L. P. B. Katehi, "Optimization of MM-Wave Distribution Networks Using Silicon-Based CPW," *1998 IEEE MTT-S Int. Microwave Symp. Dig.*, Vol. 2, pp. 537–540, June 7–12, 1998, Baltimore, MD.

[19] N. H. L. Koster, S. Kosslowski, R. Bertenburg, S. Heinen, and I. Wolff, "Investigations on Air Bridges Used for MMICs in CPW Technique," *Proc. 19th European Microwave Conf. Dig.*, pp. 666–671, Sept. 4–7, 1989, London.

[20] K. Beilenhoff, W. Heinrich, and H. L. Hartnagel, "The Scattering Behavior of Air Bridges in Coplanar MMIC's" *Proc. 21st European Microwave Conf. Dig.*, pp. 1131–1135, Sept. 9–12, 1991, Stuttgart, Germany.

[21] H. Jin and R. Vahldieck, "Calculation of Frequency-Dependent *S*-Parameters of CPW Air-Bridges Considering Finite Metallization Thickness and Conductivity," *1992 IEEE MTT-S Int. Microwave Symp., Dig.*, Vol. 1, pp. 207–210, June 1–5, 1992, Albuquerque, NM.

[22] H. Jin and R. Vahldieck, "Full-Wave Analysis of Coplanar Waveguide Discontinuities Using the Frequency Domain TLM Method," *IEEE Trans. Microwave Theory Tech.*, Vol. 41, No. 9, pp. 1538–1542, Sept. 1993.

[23] M. Rittweger, N. H. L. Koster, S. Kosslowski, R. Bertenburg, S. Heinen, and I. Wolff, "Full-Wave Analysis of a Modified Coplanar Air Bridge T-Junction," *Proc. 21st European Microwave Conf. Dig.*, pp. 993–998, Sept 9–12, 1991, Stuttgart, Germany.

[24] T. Becks and I. Wolff, "Full-Wave Analysis of Various Coplanar Bends and T-Junctions with Respect to Different Types of Air-Bridges," *1993 IEEE MTT-S Int. Microwave Symp. Dig.*, Vol. 2, pp. 697–700, June 14–18, 1993, Atlanta, GA.

[25] A. A. Omar and Y. L. Chow, "A Versatile Moment Method Solution of the Conventional and Modified Coplanar Waveguide T-Junctions," *IEEE Trans. Microwave Theory Tech.*, Vol. 41, No. 4, pp. 687–692, April 1993.

[26] D. M.-Syahkal, "Computation of Equivalent Circuits of CPW Discontinuities Using Quasi-static Spectral Domain Method," *IEEE Trans. Microwave Theory Tech.*, Vol. 44, No. 6, pp. 979–984, June 1996.

[27] M. Naghed, M. Rittweger, and I. Wolff, "A New Method for the Calculation of the Equivalent Inductances of Coplanar Waveguide Discontinuities," *1991 IEEE MTT-S Int. Microwave Symp. Dig.*, Vol. 2, pp. 747–750, June 10–14, 1991, Boston, Massachusetts.

[28] P. Pogatzki, D. Kother, R. Kulke, B. Hopf, T. Sporkmann, and I. Wolff, "Coplanar Hybrids Based on an Enhanced Inductor Model for Mixer Applications up to

MM-Wave Frequencies," *Proc. 24th European Microwave Conf. Dig.*, Vol. 1, pp. 254–257, Sept. 5–8, 1994, Cannes, France.

[29] M. Abdo-Tuko, M. Naghed, and I. Wolff, "Novel 18/36 GHz (M)MIC GaAs FET Frequency Doublers in CPW-Techniques under the Consideration of the Effects of Coplanar Discontinuities," *IEEE Trans. Microwave Theory Tech.*, Vol. 41, No. 8, pp. 1307–1315, Aug. 1993.

[30] R. Kulke, P. Pogatzki, D. Kother, T. Sporkmann, and I. Wolff, "Enhancement of Coplanar Capacitor Models and Verification up to 67 GHz for (M)MIC Circuit Design," *Proc. 24th European Microwave Conf. Dig.*, Vol. 1, pp. 258–262, Sept. 5–8, 1994, Cannes, France.

[31] M. Houdart, "Coplanar Lines: Application to Broadband Microwave Integrated Circuits," *Proc. 6th European Microwave Conf. Dig.*, pp. 49–53, Sept. 14–17, 1976, Rome.

[32] A. K. Sharma and H. Wang, "Experimental Models of Series and Shunt Elements in Coplanar MMICs," *1992 IEEE MTT-S Int. Microwave Symp. Dig.*, Vol. 3, pp. 1349–1352, June 1–5, 1992, Albuquerque, NM.

[33] N. I. Dib, L. P. B. Katehi, G. E. Ponchak, and R. N. Simons, "Theoretical and Experimental Characterization of Coplanar Waveguide Discontinuities for Filter Applications," *IEEE Trans. Microwave Theory Tech.*, Vol. 39, No. 5, pp. 873–882, May 1991.

[34] G. E. Ponchak and L. P. B. Katehi, "Open- and Short-Circuit Terminated Series Stubs in Finite-Width Coplanar Waveguide on Silicon," *IEEE Trans. Microwave Theory Tech.*, Vol. 45, No. 6, pp. 970–976, June 1997.

[35] K. Hettak, N. I. Dib, A.-F. Sheta, and S. Toutain, "A Class of Novel Uniplanar Series Resonators and Their Implementation in Original Applications," *IEEE Trans. Microwave Theory Tech.*, Vol. 46, No. 9, pp. 1270–1276, Sept. 1998.

[36] N. I. Dib, G. E. Ponchak, and L. P. B. Katehi, "A Theoretical and Experimental Study of Coplanar Waveguide Shunt Stubs," *IEEE Trans. Microwave Theory Tech.*, Vol. 41, No. 1, pp. 38–44, Jan. 1993.

[37] N. I. Dib, M. Gupta, G. E. Ponchak, and L. P. B. Katehi, "Characterization of Asymmetric Coplanar Waveguide Discontinuities," *IEEE Trans. Microwave Theory Tech.*, Vol. 41, No. 9, pp. 1549–1558, Sept. 1993.

[38] F. Mernyei, I. Aoki, and H. Matsuura, "New Cross T Junction for CPW Stub-Filters on MMIC's," *IEEE Microwave Guided Wave Lett.*, Vol. 5, No. 5, pp. 139–141, May 1995.

[39] K. Hettak, N. I. Dib, A. Omar, G.-Y. Delisle, M. Stubbs, and S. Toutain, "A Useful New Class of Miniature CPW Shunt Stubs and its Impact on Millimeter-Wave Integrated Circuits," *IEEE Trans. Microwave Theory Tech.*, Vol. 47, No. 12, pp. 2340–2349, Dec. 1999.

[40] R. N. Simons and S. R. Taub, "Coplanar Waveguide Radial Line Stub," *Electron. Lett.*, Vol. 29, No. 4, pp. 412–413, Feb. 1993.

[41] R. N. Simons and S. R. Taub, "Coplanar Waveguide Radial Line Double Stub and Application to Filter Circuits," *Electron. Lett.*, Vol. 29, No. 17, pp. 1584–1585, Aug. 1993.

[42] J. K. A. Everard and K. K. M. Cheng, "High Performance Direct Coupled Bandpass Filters on Coplanar Waveguide," *IEEE Trans. Microwave Theory Tech.*, Vol. 41, No. 9, pp. 1568–1573, Sept. 1993.

[43] B. P. Hopf, I. Wolff and M. Guglielmi, "Coplanar MMIC Active Bandpass Filters Using Negative Resistance Circuits," *IEEE Trans. Microwave Theory Tech.*, Vol. 42, No. 12, pp. 2598–2602, Dec. 1994.

CHAPTER 10

Coplanar Waveguide Transitions

10.1 INTRODUCTION

This chapter presents transitions between coplanar waveguide (CPW) to other microwave transmission media, such as microstrip, slotline, coplanar stripline (CPS), balanced stripline, and rectangular waveguide. Each section is devoted to a particular transmission media. Wherever possible the geometry and the measured performance characteristics of a transition are summarized in a table.

Section 10.2 describes several CPW-to-microstrip transitions. These transitions are broadly divided into two types, electromagnetically coupled and directly coupled. Section 10.3 describes two types of transitions that are compatible with CPW microwave probes. These transitions, when incorporated at the input/output ports of monolithic microwave integrated circuits (MMICs), facilitate their characterization using wafer probes. Section 10.4 describes several CPW-to-CPW transitions. These CPW lines are either on opposite sides of a dielectric substrate or on two substrates that are orthogonal to each other. This section also includes a transition from a grounded CPW to a CPW supported on a dielectric membrane. Such a transition is fabricated by micromachining a silicon wafer.

Section 10.5 presents several transitions between a CPW and a rectangular waveguide. These transitions resort to any one of the following techniques to couple power in and out: a tapered metal ridge attached to the broadwall inside the waveguide, an aperture in the waveguide wall or a metal post, or a probe that enters the waveguide through an opening.

Sections 10.6 and 10.7 present several different transitions between a CPW and a balanced transmission line, such as a slotline or a coplanar stripline (CPS). These transitions are ideally suited for applications involving printed dipole type antennas. Sections 10.8 and 10.9 present several different transitions between a CPS and a microstrip line or a slotline. This section also includes a transition between a micro-CPS and a microstrip line. The micro-CPS is a new variant of the CPS in which the two strip conductors are located at different heights and separated by a dielectric film. Last, Section 10.10 presents a transition between a CPW and a balanced stripline.

10.2 COPLANAR WAVEGUIDE-TO-MICROSTRIP TRANSITION

10.2.1 Coplanar Waveguide-to-Microstrip Transition Using Ribbon Bond

In this transition [1] the dielectric substrate with the microstrip line is adhesive bonded to the ground plane of the coplanar waveguide (CPW) as shown in Figure 10.1. This arrangement allows the two transmission lines to share a common ground plane. Electrical continuity between open ends of the CPW center strip conductor and the microstrip line is provided by a gold ribbon. The open end shunt capacitances of the two transmission lines, together with the series inductance of the gold ribbon, form a π-equivalent circuit. This type of circuit has a low-pass characteristic. Hence for broad bandwidth, the shunt capacitance as well as the series inductance must be kept small. In [1] the characteristics of an experimental transition on an alumina substrate are reported. Typical dimensions of the transition are given in the caption of Figure 10.1. The maximum VSWR of this transition is less than 1.4:1 over the frequency range of 2.0 to 18.0 GHz [1]. This includes the VSWR of the coaxial-to-microstrip and the CPW-to-coaxial transitions at the input and output ports.

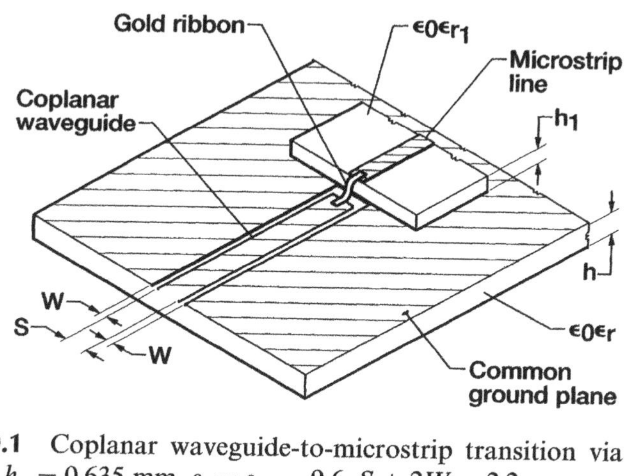

FIGURE 10.1 Coplanar waveguide-to-microstrip transition via bond wire/ribbon: $h = h_1 = 0.635$ mm, $\varepsilon_r = \varepsilon_{r1} = 9.6$, $S + 2W = 2.2$ mm.

10.2.2 Coplanar Waveguide-to-Microstrip Surface-to-Surface Transition via Electromagnetic Coupling

In this section the construction and characteristics of two types of transition are discussed. In the first type of transition [2], the microstrip line on a separate substrate is placed on top of the coplanar waveguide (CPW) such that the two transmission lines share a common ground plane and the two strip conductors overlap as shown in Figure 10.2(*a*). Coupling between the strip conductors is achieved electromagnetically and no wire or ribbon bonds are necessary. In [2] the characteristics of an experimental transition designed for 7.5 GHz operation are reported. Typical dimensions and the measured characteristics are summarized in Table 10.1. The characteristics of this transition are modeled using full-wave analysis and are shown to be in good agreement with the measured results [2]. The full-wave analysis is then used to predict the dimensions for two 35.0 GHz transitions with alumina and quartz as substrates [2], respectively. The dimensions of the 35.0 GHz transitions are also summarized in Table 10.1.

TABLE 10.1 Coplanar Waveguide-to-Microstrip Transition with Electromagnetic Coupling between Strip Conductors

Dimensions and Characteristics	Design 1	Design 2	Design 3
Center frequency (GHz)	7.5	35.0	35.0
	Coplanar waveguide		
SUBSTRATE MATERIAL	DUROID 6010.2	ALUMINA	QUARTZ
h (mm)	0.635	0.127	0.254
S (mm)	0.75	0.150	0.1
W (mm)	2.125	0.425	0.3
g (mm)	1.0	0.176	0.198
	Microstrip		
SUBSTRATE MATERIAL	DUROID 6010.2	GaAs	GaAs
h_1 (mm)	0.635	0.1	0.1
W_m (mm)	2.0	0.4	0.24
L (mm)	4.5	0.88	0.99
	Measured characteristics		
Return loss (dB)[a]	< -15.0	—	—
Bandwith (%)	47	—	—
Insertion loss (dB)[b]	1.7	—	—

[a]De-embedded for two back-to-back transitions.
[b]Includes two CPW to coaxial transitions at each end and 17.0 mm of 50 Ω microstrip in between.

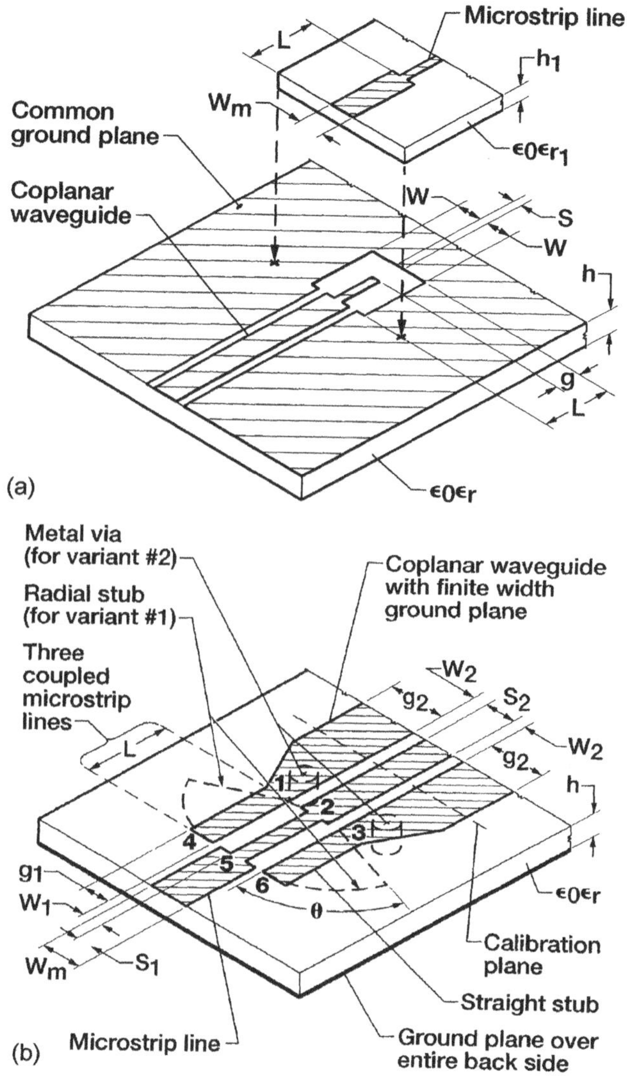

FIGURE 10.2 Coplanar waveguide-to-microstrip transition: (*a*) Electromagnetic coupling between strip conductors; (*b*) electromagnetic coupling between ground conductors.

In the second type of transition [1], [3], the microstrip line and the CPW are fabricated on a single substrate with a common continuous strip conductor as shown in Figure 10.2(*b*). In contrast to the previous design, the coupling between the respective ground planes is achieved electromagnetically. In addition the ground planes of the CPW are of finite width to suppress the excitation of the parasitic parallel plate mode.

This transition is modeled either as a bandpass filter whose center frequency is determined by the length of the coupled section [1] or as a six-port network formed by three coupled microstrip lines [3]. In this section the six-port network model will be briefly described. This model, as seen from Figure 10.2(*b*) has port 2 and port 5 as its input and output ports. Ports 4 and 6 are terminated in an open circuit. The length of the coupled section is one-quarter of the guide wavelength in the three coupled microstrip section [4] at the center frequency f_0 of operation. Therefore at f_0 the open circuits translate to a short circuit at ports 1 and 3. In order for this six-port network to behave as an efficient coupler, the insertion loss and the return loss have to be optimized. The three coupled microstrip lines that constitute the network support three modes of propagation, namely the odd mode, the even mode, and the odd and even mode. Therefore the insertion loss and the return loss are functions of the three characteristic impedances corresponding to the above modes [56].

There are two variants of this basic structure. In the first variant the lines 1–4 and 3–6 are replaced by wideband radial stubs to enhance the bandwidth [3]. The radial stubs are shown by dotted lines in Figure 10.2(*b*). In the second variant, the lines 1–4 and 3–6 are replaced by metal vias [5]. The vias are shown by dotted lines in Figure 10.2(*b*). The characteristics of the experimental transitions are reported in [1] and [3]. Typical dimensions as well as the measured characteristics of the transitions are summarized in Table 10.2.

10.2.3 Coplanar Waveguide-to-Microstrip Transition via a Phase-Shifting Network

This transition [6] is illustrated in Figure 10.3. At the input it consists of a 50 Ω microstrip line that branches into three orthogonal lines. The two outer line lengths are made longer than the inner line length such that the electric fields between points *a* and *b* and between *a* and *c* are equal in magnitude but 180° out of phase. Consequently at this plane a coplanar waveguide (CPW) mode is excited and then propagates along the three strip conductors. The longer path length is taken to be about $0.79\lambda_{gm}$, where λ_{gm} is the guide wavelength in the microstrip at the center frequency f_0 to compensate for the right angle bend parasitics.

The experimental transition [6] is fabricated on RT/Duroid 6010.5 of 0.254 mm thickness and ε_r of 10.5. The measured 3 dB insertion loss bandwidth is approximately 40 percent at f_0 of 9.2 GHz. The minimum insertion loss occurs at f_0 and is about 0.25 dB. The transition is numerically modeled using FDTD technique, and the measured and modeled characteristics are in good agreement [6].

10.2.4 Coplanar Waveguide-to-Microstrip Transition via a Metal Post

In this transition [7] power is coupled from a coplanar waveguide to a microstrip line located on the opposite sides of a common ground plane that

TABLE 10.2 Coplanar Waveguide-to-Microstrip Transition with Electromagnetic Coupling between Ground Conductors

Dimensions and Characteristics	Houdart and Aury Design		Gauthier *et al.* Design: With Straight Stubs	Gauthier *et al.* Design: With Radial Stubs
Center frequency (GHz)	8.0	14.5	94.0	94.0
Coplanar waveguide				
SUBSTRATE MATERIAL	ALUMINA, $\varepsilon_r = 9.6$		HIGH-RESISTIVITY SILICON, $\varepsilon_r = 11.7$	
h (μm)	635	635	120	120
S_1 (μm)	420	420	30	30
W_1 (μm)	310	310	55	55
g_1 (μm)	680	680	45	—
S_2 (μm)	—	—	50	50
W_2 (μm)	—	—	45	45
g_2 (μm)	—	—	145	145
Microstrip				
W_m (μm)	—	—	70	70
L (μm)	3900	2000	280	280
θ (deg)	—	—	—	60
Measured characteristics				
Return loss (dB)	[b]$\leqslant -15.0$	[b]$\leqslant -15.0$	$\leqslant -17.0$	$\leqslant -17.0$
Return loss bandwith (%)	75	41	—	—
Insertion loss (dB)	—	—	[a]0.4	[a]0.4
3-dB insertion loss bandwidth (%)	—	—	20	25

[a]For two back transitions with 860 μm long microstrip line in between.
[b]Includes two CPW to coaxial transitions.

has a circular aperture. The transition is formed by a metal post that enters the aperture and makes contact with the strip conductors of the coplanar waveguide and the microstrip line, respectively. The transition is shown in Figure 10.4. A pair of bond wires located adjacent to the post holds the elevated coplanar waveguide ground planes and the microstrip ground plane at a common potential. This arrangement of conductors results in a conductor-backed coplanar waveguide (CBCPW). In [7] the characteristics of an experimental transition are reported. Typical dimensions of the transition are given in the caption of Figure 10.4. The measured insertion loss of the transition is

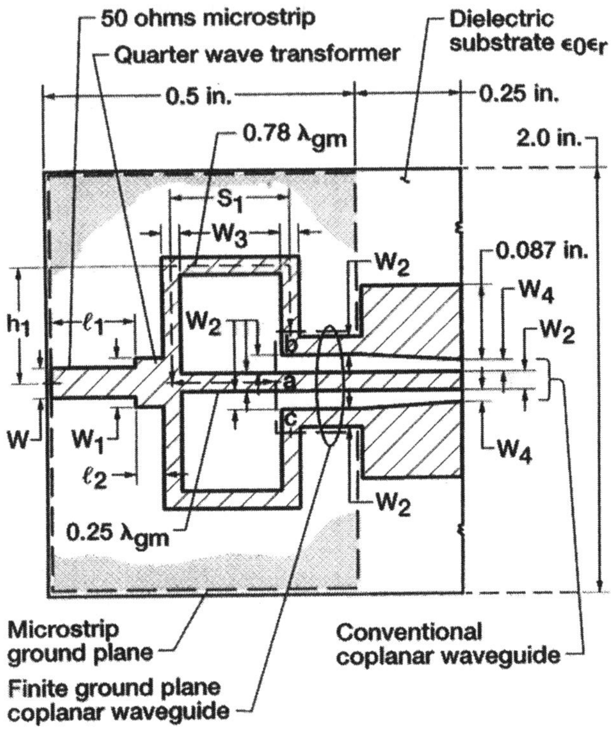

FIGURE 10.3 Coplanar waveguide-to-microstrip transition via a phase shifting network: $W = 0.23$ mm, $W_1 = 0.47$ mm, $W_2 = 0.254$ mm, $W_3 = 0.102$ mm, $l_2 = 2.72$ mm, $S_1 = 2.97$ mm, $h_1 = 3.3$ mm, $\varepsilon_r = 10.5$, substrate thickness = 0.254 mm.

less than 1.0 dB over the frequency range of 0.045 to 6.5 GHz. The insertion loss includes the losses occurring in the junction, the attenuation of the short length of CBCPW and microstrip line on either side of the junction, and the attenuation of the coaxial connectors of the test fixture. The measured return loss is less than −10.0 dB across the above frequency band.

10.2.5 Coplanar Waveguide-to-Microstrip Transition Using a Via-Hole Interconnect

In this transition which is shown in Figure 10.5, the coplanar waveguide (CPW) and the microstrip line are fabricated on opposite sides of a semiconductor wafer, such as semi-insulating GaAs or InP. The two lines are then interconnected by an etched and metallized via [8]. The first step in the fabrication process is the definition of the coplanar environment on the first side of the wafer with standard photolithography. The wafer is then electro-

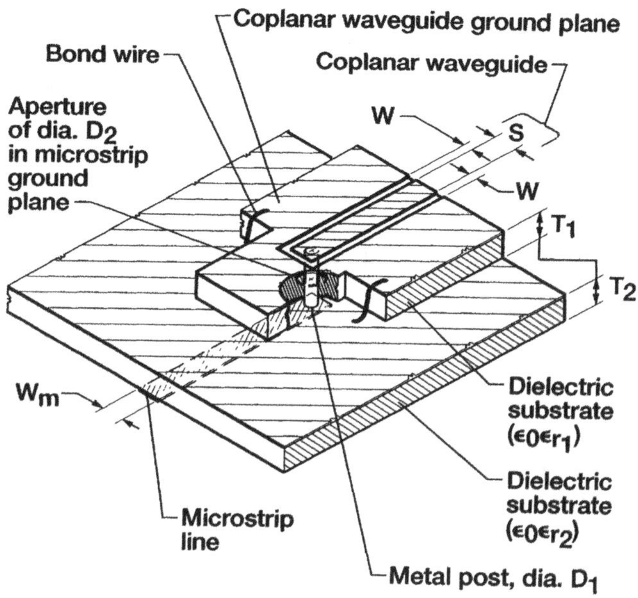

FIGURE 10.4 Coplanar waveguide-to-microstrip transition via a metal post: $S = 1.14$ mm, $W = 0.25$ mm, $W_m = 0.76$ mm, $D_1 = 1.02$ mm, $D_2 = 2.54$ mm, $T_1 = 3.18$ mm, $\varepsilon_{r1} = 2.2$, $T_2 = 0.25$ mm, $\varepsilon_{r2} = 2.2$.

plated with gold to a thickness at least three times the skin depth at the operating frequency; typically this is about 1.5 μm at Ka-band frequencies. In this design the gold thickness is 4 μm. At this point the wafer is thinned by lapping and polishing so that the final via interconnect has a small inductance. Typically for Ka-band and W-band applications, the wafers are thinned to

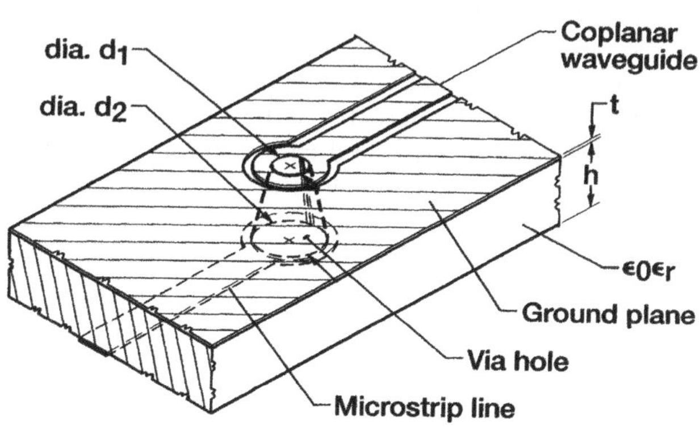

FIGURE 10.5 Coplanar waveguide-to-microstrip transition using a via-hole interconnect: $h = 100$ μm, $\varepsilon_r = 12.9$, $d_1 = 55$ μm, $d_2 = 140$ μm, $t = 4$ μm.

about 100 and 50 μm, respectively. Next the via hole is patterned and etched through the substrate with a reactive ion etch process. In the final step the microstrip line is defined on the backside and is then electroplated selectively. A single via interconnect between 50 Ω lines fabricated as above on a 100 μm thick semi-insulating GaAs wafer has a measured insertion loss of 0.15 dB at Ka-band frequencies [8].

10.2.6 Coplanar Waveguide-to-Microstrip Orthogonal Transition via Direct Connection

Two types of orthogonal transitions [9] are considered in this section. In these transitions a coplanar waveguide (CPW) on an orthogonal substrate makes a direct connection with a microstrip line on a horizontal substrate either from the bottom or from the top as shown in Figure 10.6(*a*) and (*b*). The transition with connection from the bottom will be described first. In this transition the ground plane of the microstrip line has a rectangular opening to avoid shorting the coplanar waveguide. The width W_1 of the opening is set equal to $S + 2W$, where S and W are the center strip conductor and slot width, respectively, of the CPW. The length L_1 of the opening for a given geometry and substrate material is experimentally determined to minimize insertion loss and provide good impedance match. Typically L_1 is about $0.025\lambda_{gm}$, where λ_{gm} is the guide wavelength in the microstrip line at the center frequency. The microstrip width W_m is chosen such that the characteristic impedance $Z_{0m} = 50\ \Omega$. The CPW center strip conductor width S is made equal to W_m and the slot width W is chosen such that the characteristic impedance $Z_{0(CPW)}$ is equal to 50 Ω. The CPW center strip conductor extends beyond the CPW ground plane by a distance L equal to the total thickness of the microstrip substrate and ground plane.

Next the transition with connection from the top is described. In this transition, before the two strip conductors meet, both the CPW and the microstrip line are tapered over a short length so that the characteristic impedance increases from 50 to 60 Ω. The CPW and the microstrip taper lengths T and T_1 are typically $0.045\lambda_{g(CPW)}$ and $0.055\lambda_{gm}$, respectively, where $\lambda_{g(CPW)}$ and λ_{gm} are the guide wavelength in the 50 Ω CPW and microstrip at the center frequency of operation. The width S_1 of the CPW center strip conductor is chosen to be equal to the strip width W_2 of the microstrip line. The slot width W_1 of the CPW is chosen so that the $Z_{0(CPW)}$ is 60 Ω. The CPW is terminated in an open circuit with an open end length L equal to the total thickness of the microstrip substrate and ground plane.

The experimental transitions [9] are fabricated on RT/Duroid 6006 substrate. Typical dimensions of the transitions are given in the captions of Figure 10.6(*a*) and (*b*). The de-embedded insertion loss and return loss are 0.3 dB and less than -17.0 dB across the frequency band of 2.0 to 14.0 GHz for the transition in Figure 10.6(*a*). The de-embedded insertion loss and return loss are 0.2 dB and less than -15.0 dB across the same frequency band for the transition in Figure 10.6(*b*).

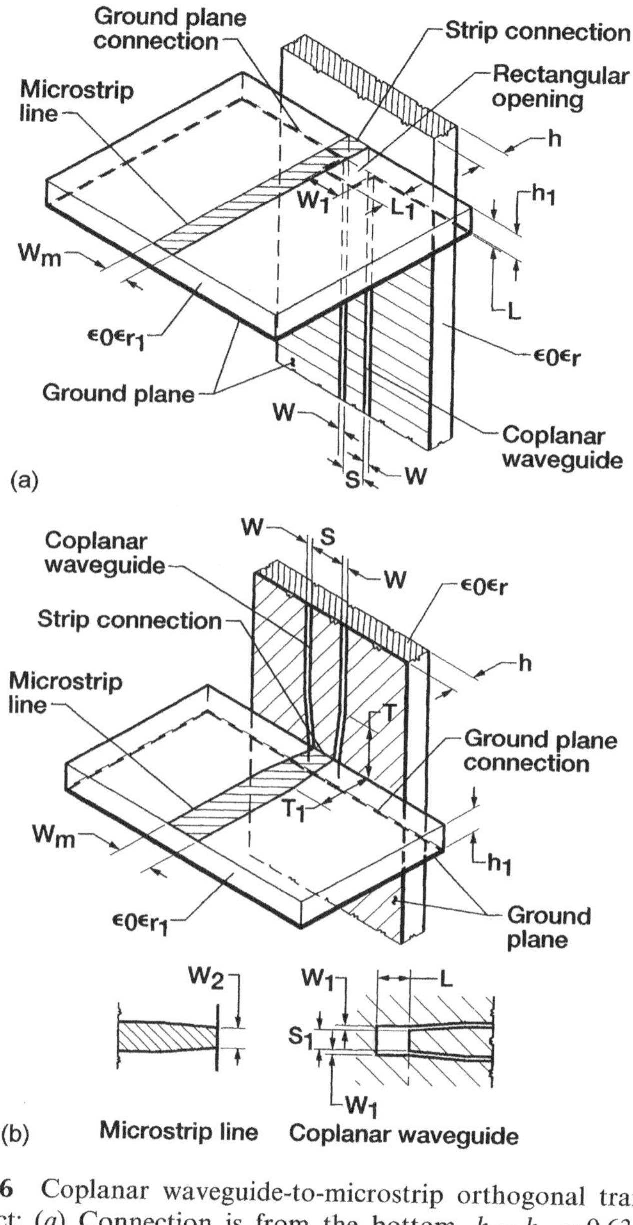

FIGURE 10.6 Coplanar waveguide-to-microstrip orthogonal transition via direct contact: (*a*) Connection is from the bottom, $h = h_1 = 0.63$ mm, $\varepsilon_r = \varepsilon_{r1} = 6.0$, $S = 0.93$ mm, $W = 0.20$ mm, $L = 0.65$ mm, $W_1 = 1.33$ mm, $L_1 = 0.42$ mm, $W_m = 0.93$ mm; (*b*) connection is from the top, $h = h_1 = 0.63$ mm, $\varepsilon_r = \varepsilon_{r1} = 6.0$, $S = 0.93$ mm, $W = 0.20$ mm, $T = 0.99$ mm, $L = 0.65$ mm, $S_1 = 0.65$ mm, $W_1 = 0.3$ mm, $W_m = 0.93$ mm, $T_1 = 0.99$ mm, $W_2 = 0.65$ mm.

10.3 TRANSITIONS FOR COPLANAR WAVEGUIDE WAFER PROBES

10.3.1 Coplanar Waveguide Wafer Probe-to-Microstrip Transition Using a Radial Stub

A coplanar waveguide wafer probe-to-microstrip transition [10] in which the ground continuity is established through the use of an electromagnetically coupled microstrip radial stub is shown in Figure 10.7. In this transition the center signal contact and the outer ground contacts of the wafer probe establish direct connection with the microstrip line and the microstrip radial stub on the substrate, respectively. The microstrip radial stub provides a virtual short circuit between the ground contacts of the wafer probe and the substrate ground plane.

For any given frequency of operation, the optimum stub length neglecting the fringing fields along the perimeter of the stub is derived in [11]. If the stub angle θ is large, and the stub inner radius r_1 is small compared to the wavelength of operation, the optimum stub length L is given by [11]

$$L \approx \frac{\lambda_0}{2\pi\sqrt{\varepsilon_r}}, \tag{10.1}$$

where λ_0 is the free space wavelength and ε_r is the substrate relative permittivity. The stub length is optimum for the following two reasons: First, for this length the stub impedance is the lowest hence the radial stub is close to being a virtual short circuit. Second, the stub impedance is also insensitive to the substrate thickness, which is a desirable feature when making measurements on substrates of different thicknesses.

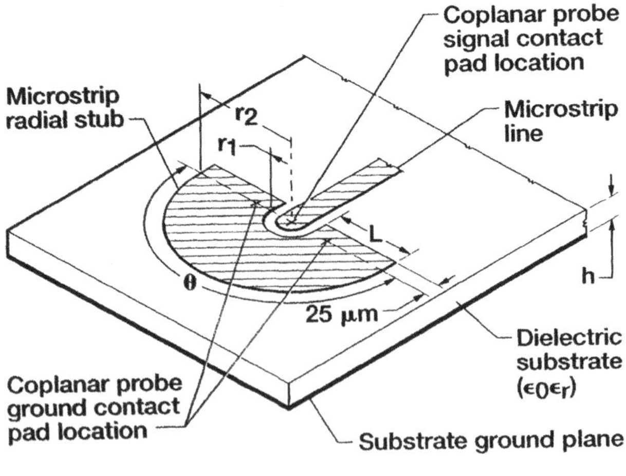

FIGURE 10.7 Coplanar probe-to-microstrip transition using a radial stub.

Quite often in practice measurements are made over a band of frequencies, in which case it is necessary to determine the maximum frequency of the stub operation. In general, as the frequency of operation increases, the stub electrical length increases and approaches approximately half a wavelength in the dielectric. At this frequency the radial stub and the substrate ground plane are decoupled, and the radial stub behaves as an open circuit with a large impedance. This condition therefore sets the upper limit of the operating frequency. The stub reactance becomes infinite when [10]

$$L \approx \frac{3.832\lambda_0}{2\pi\sqrt{\varepsilon_r}}. \tag{10.2}$$

For example, a coplanar waveguide wafer probe-to-microstrip transition for optimum operation at 10.6 GHz with a maximum operating frequency of 40.0 GHz has the following typical dimensions as determined from Eqs. (10.1) and (10.2): stub length $L = 1250\,\mu$m, inner radius $r_1 = 125\,\mu$m, outer radius $r_2 = 1375\,\mu$m, and stub angle $\theta = 180°$. The substrate is a semi-insulating GaAs of thickness $h = 92\,\mu$m and $\varepsilon_r = 12.9$.

10.3.2 Coplanar Waveguide Wafer Probe-to-Microstrip Transition Using Metal Vias

This transition [12] is shown in Figure 10.8. In this transition the center signal contact and the outer ground contacts of the wafer probe establish direct connection with the microstrip line and pads on the substrate, respectively. The pads are connected by metal vias to the substrate ground plane. The disadvantage of this approach is that the fabrication is more complicated than the transition discussed in Section 10.3.1. Typical dimensions for a transition which is good through 26.5 GHz are included in the caption of Figure 10.8.

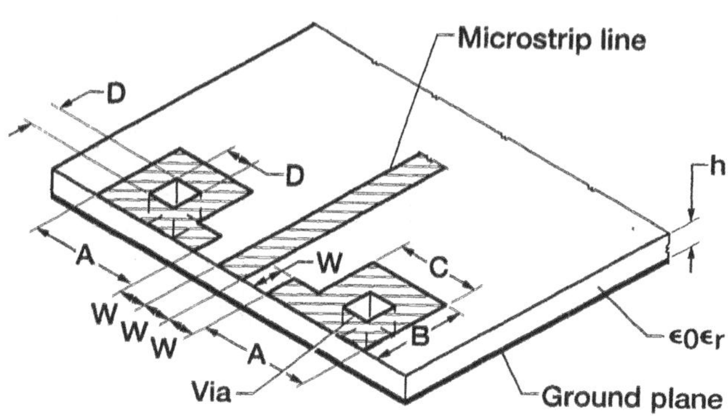

FIGURE 10.8 Coplanar probe-to-microstrip transition using metal vias: $h = 100\ \mu$m, $\varepsilon_r = 12.9$, $A = 250\ \mu$m, $B = 200\ \mu$m, C $= 200\ \mu$m, $D = 100\ \mu$m, $W = 75\ \mu$m.

10.4 TRANSITIONS BETWEEN COPLANAR WAVEGUIDES

10.4.1 Grounded Coplanar Waveguide-to-Microshield Coplanar Line

In a microshield coplanar line the center strip conductor and the upper ground planes are supported by a very thin dielectric membrane. This results in a line that is very fragile and therefore cannot withstand repeated probe contact pressure during characterization. To alleviate this problem, the probe contact pads are designed as a grounded coplanar waveguide (GCPW). Therefore a transition from GCPW-to-microshield coplanar line is necessary. One such transition design [13] is shown in Figure 10.9. The transition is fabricated on a high resistivity silicon wafer with resistivity greater than 2000 Ω-cm. The membrane is about 1.5 μm thick and is a tri-layer composite of $SiO_2/Si_3N_4/SiO_2$ with typical thickness of 7000 Å/3000Å/4000Å, respectively. The dielectric constant of the oxide is 3.9 and that of the nitride is 7.5. To minimize power leakage due to surface waves in the GCPW section of the transition, the length is kept small typically about $\lambda_g/10$ to $\lambda_g/8$, where λ_g is the guide wavelength. Further, to suppress leaky modes, metalized cavities are integrated in the back of the substrate and located on either side in close proximity to the probe contact pads. These metalized cavities short out the electric field setup between the top and bottom ground planes by the leaky

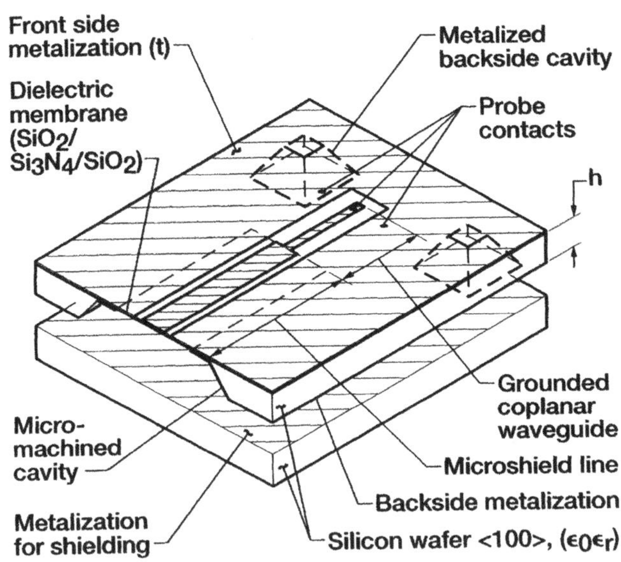

FIGURE 10.9 Grounded coplanar waveguide-to-microshield line: $h = 355$ μm, $t = 1.2$ μm, $\varepsilon_r = 11.7$. Grounded CPW: strip width = 50 μm, slot width = 125 μm, $Z_0 = 73$ Ω; microshield line: strip width = 250 μm, slot width = 25 μm, $Z_0 = 75$ Ω, cavity width = 1800 μm.

modes. Last to maintain CPW-like mode of propagation, the slot width to substrate thickness ratio is kept small. In [13] the characteristics of an experimental GCPW-to-microshield coplanar line transition are reported. Typical dimensions of the transition are summarized in the caption of Figure 10.9. The measured effective dielectric constant of the microshield line is 1.15. The measured return loss of two back-to-back transition over the frequency range of 10.0 to 40.0 GHz is less than -18.0 dB.

The transition above is fabricated on a high-resistivity silicon wafer. It is also possible to fabricate similar transition on a GaAs wafer. In [14] the microwave performance of a coplanar waveguide (CPW) transition supported on a Si_3N_4 or a polyimide film ($\varepsilon_r = 2.9$) that extends across a cavity micromachined from a semi-insulating GaAs substrate is demonstrated.

10.4.2 Vertical Fed-through Interconnect between Coplanar Waveguides with Finite-Width Ground Planes

A schematic illustrating two back-to-back vertical interconnects [15] between coplanar waveguides with finite-width ground planes (FW-CPW) on opposite sides of a dielectric substrate is shown in Figure 10.10. In this circuit the width of the CPW ground planes are twice the center strip conductor width. The diameters of the metal-filled vias are about the same as the center strip conductor width which is 0.254 mm. These features result in a very compact circuit. The circuit is fabricated using a low-temperature co-firing ceramic (LTCC) tape for the dielectric substrate and a screen-printing technique for the

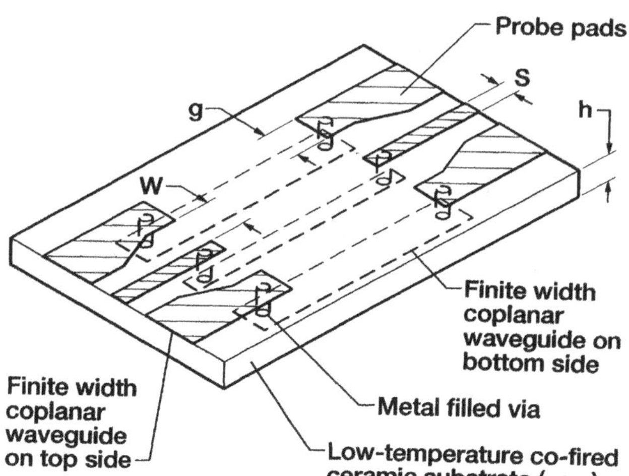

FIGURE 10.10 Vertical fed through interconnect between coplanar waveguides with finite width ground planes (FW-CPW): $h = 0.286$ mm, $\varepsilon_r = 5.9$, $S = 0.31$ mm, $W = 0.1$ mm, $g = 0.61$ mm.

deposition of gold conductor pattern. In [15] the characteristics of an experimental interconnect is reported. The interconnect is experimentally characterized by de-embedding the insertion loss and return loss using TRL standards and a wafer probe. The de-embedded insertion loss and return loss are about 0.3 and 30.0 dB, respectively, at 20.0 GHz. This vertical interconnect has potential applications in the construction of CPW patch array antennas [15].

10.4.3 Orthogonal Transition between Coplanar Waveguides

An orthogonal transition [16] between two coplanar waveguides (CPWs) is shown in Figure 10.11. In this transition continuity between the two CPWs is established by three metal pins. In the horizontal CPW three holes are provided to receive the pins. In addition to facilitate solder connection a circular island is provided in the center strip conductor. Further, for good impedance matching, the CPW beyond the island is terminated in a short-circuited stub. The length L_s of the short-circuited stub, the diameter D of the pin, and radius R_1 and R_2 of the island and the surrounding slot region, respectively, are experimentally optimized to obtain the best insertion loss and return loss characteristics. In [16] the characteristics of an experimental

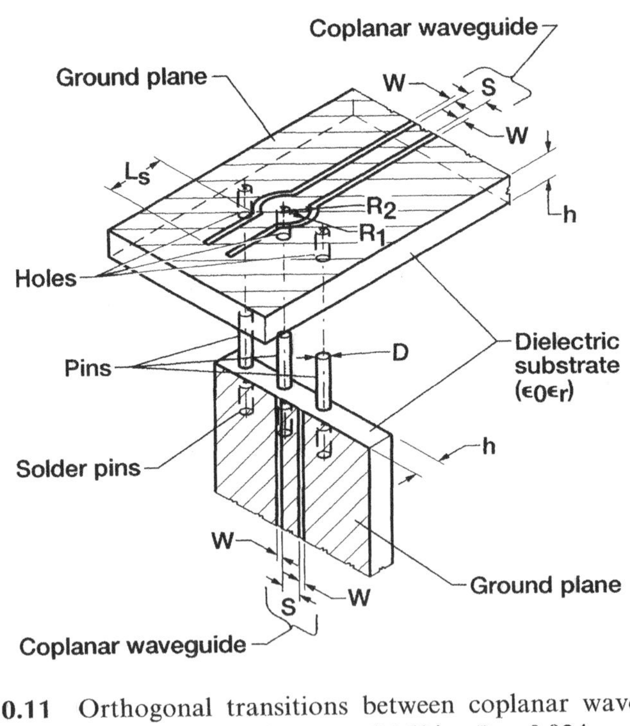

FIGURE 10.11 Orthogonal transitions between coplanar waveguides: $h = 0.0508$ cm, $\varepsilon_r = 2.22$, $R_1 = 0.043\lambda_g$, $R_2 = 0.063\lambda_g$, $D = 0.024$ cm, $L_s = 0.2\lambda_g$, where λ_g is the guide wavelength in the CPW at $f_0 = 18.0$ GHz.

transition are reported. The dimensions L_s, D, R_1, and R_2 are given in the caption of Figure 10.11.

The measured insertion loss and return loss of the transition is 1.7 and −20.0 dB, respectively, at the center frequency f_0 of 18.0 GHz. The measured insertion loss includes the losses occurring at the interconnection, the losses in the 25.4 mm length of CPW on either side of the interconnect and the losses in the two coaxial connectors used at the measurement ports. The bandwidth over which the return loss is less than −10.0 dB is about 8 percent at $f_0 = 18.0$ GHz. This transition has potential applications in the construction of linearly tapered slot antenna arrays [16].

10.4.4 Electromagnetically Coupled Transition between Stacked Coplanar Waveguides

In this transition coplanar waveguides on two separate substrates are stacked one on top of another and are proximity coupled [17]. In the arrangement shown in Figure 10.12, the feed line is located on the bottom side of the feed substrate and is terminated in a modified open circuit that also forms the coupling region. The line to which power is to be coupled is located on the top

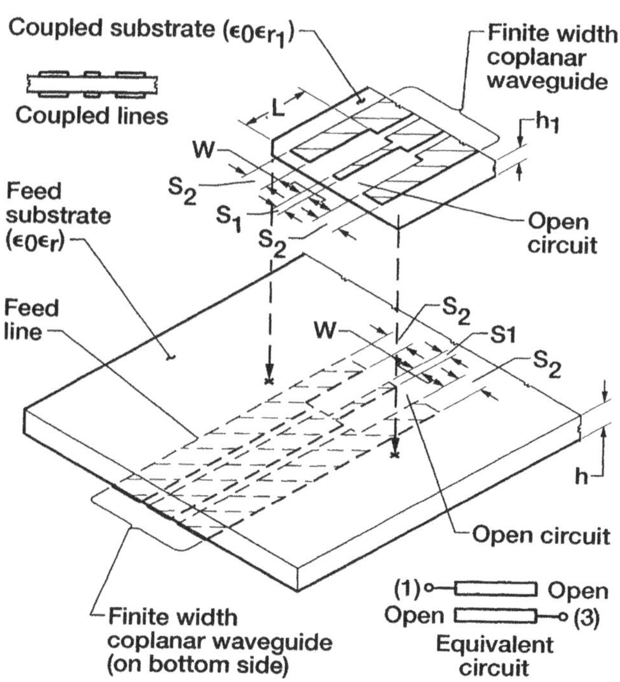

FIGURE 10.12 Electromagnetically coupled transition between stacked coplanar waveguides.

side of the coupled substrate and has an identical coupled region at the open end.

To analyze this transition, the coupled region is considered as a symmetric broadside coupled coplanar waveguide as shown in the inset on the left-hand side of Figure 10.12. This structure is then analyzed by considering the even and odd modes of propagation [17], as discussed in Chapter 7. The even- and odd-mode propagation parameters, such as the effective dielectric constant ($\varepsilon_{\mathrm{eff,e}}, \varepsilon_{\mathrm{eff,o}}$) and characteristic impedances ($Z_{0,\mathrm{e}}, Z_{0,\mathrm{o}}$), are related to the substrate parameters and the geometry of the coupled lines. Next the coupled region is represented by a two-port network as shown in the inset on the right-hand side of Figure 10.12. This two-port network is completely described in terms of its *ABCD*-parameters [18]. These *ABCD*-parameters are in turn expressed in terms of the propagation parameters as follows [18]:

$$A = \frac{Z_{0,\mathrm{e}} \cot \theta_\mathrm{e} + Z_{0,\mathrm{o}} \cot \theta_\mathrm{o}}{Z_{0,\mathrm{e}} \csc \theta_\mathrm{e} - Z_{0,\mathrm{o}} \csc \theta_\mathrm{o}} = D, \tag{10.3}$$

$$B = \frac{j}{2}\,\frac{Z_{0,\mathrm{e}}^2 + Z_{0,\mathrm{o}}^2 - 2Z_{0,\mathrm{e}}Z_{0,\mathrm{o}}(\cot \theta_\mathrm{e} \cot \theta_\mathrm{o} + \csc \theta_\mathrm{e} \csc \theta_\mathrm{o})}{Z_{0,\mathrm{e}} \csc \theta_\mathrm{e} - Z_{0,\mathrm{o}} \csc \theta_\mathrm{o}}, \tag{10.4}$$

$$C = \frac{2j}{Z_{0,\mathrm{e}} \csc \theta_\mathrm{e} - Z_{0,\mathrm{o}} \csc \theta_\mathrm{o}}, \tag{10.5}$$

where

$$\theta_\mathrm{e} = \beta_\mathrm{e} l = \frac{2\pi}{\lambda_0} \sqrt{\varepsilon_{\mathrm{eff,e}}}, \tag{10.6}$$

$$\theta_\mathrm{o} = \beta_\mathrm{o} l = \frac{2\pi}{\lambda_0} \sqrt{\varepsilon_{\mathrm{eff,o}}}. \tag{10.7}$$

By converting the *ABCD*-parameters to *S*-parameters, the reflection coefficient and the transmission coefficient can be determined. To accurately predict the center frequency, the capacitance associated with the open end of the CPW should be compensated by reducing the length L of the coupled region. In [17] the characteristics of an experimental transition are reported. Typical dimensions as well as the measured characteristics of two back-to-back transitions are summarized in Table 10.3. The insertion loss and return loss of the transitions in Table 10.3 are the de-embedded values.

10.4.5 Electromagnetically Coupled Transition between Orthogonal Coplanar Waveguides

This transition is formed when a coplanar waveguide on a separate substrate is placed perpendicular to a coplanar waveguide on a second substrate as shown in Figure 10.13. The preliminary experimental results are reported in [19].

TABLE 10.3 Stacked and Electromagnetically Coupled Coplanar Waveguides

Dimensions and Characteristics	
Center frequency, GHz	5.0
Feed substrate	
ε_r	10.2
h (mm)	1.27
S_1 (mm)	2.4
W (mm)	3.9
S_2 (mm)	4.8
L (mm)	6.3
Coupled substrate	
ε_{r1}	10.2
h_1 (mm)	0.635
Measured characteristics for two back-to-back transitions	
Return loss (dB)[a]	$\leqslant -20.0$
Bandwidth (%)[b]	25
Insertion loss (dB)	0.6

[a]With 13.0 mm of coplanar waveguide in between.
[b]Over which the return loss is $\leqslant -20.0$ dB.

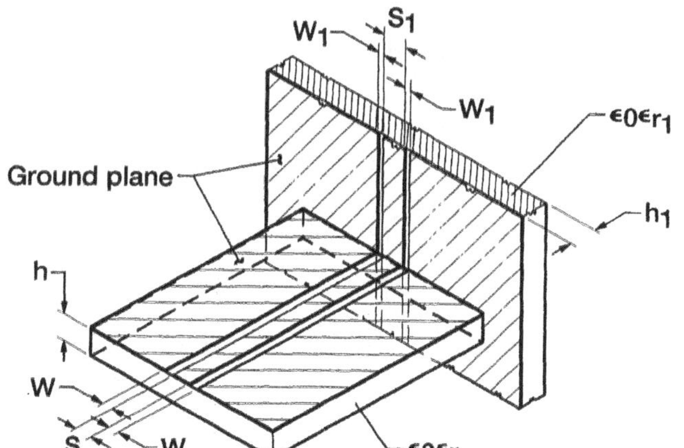

FIGURE 10.13 Electromagnetically coupled transition between orthogonal coplanar waveguides.

10.5 COPLANAR WAVEGUIDE-TO-RECTANGULAR WAVEGUIDE TRANSITION

10.5.1 Coplanar Waveguide-to-Ridge Waveguide In-line Transition

A coplanar waveguide (CPW)-to-ridge waveguide in-line transition [20] is shown in Figure 10.14. In this transition the center strip conductor of the CPW terminates in an abrupt open circuit, while the two ground planes on either sides extend further and meet at a short distance away from the open circuit forming an isosceles triangular slot region inbetween. The width of the triangle at the base is $S + 2W$, where S and W are the CPW center strip conductor and slot width, respectively. The ground plane beyond the apex of the triangle forms the lower broadwall of the rectangular waveguide. Since the slot is formed in the center of the waveguide broadwall, it does not loose power through radiation. Over the slot region a cosine tapered ridge extends from the top broadwall of the waveguide. When the waveguide and the circuit board are clamped together the ridge makes contact with the CPW center strip conductor at the open circuit. The width of the ridge is the same as the width of the CPW center strip conductor S to ensure small discontinuity reactance. The ridge acts as an impedance transformer between the standard rectangular waveguide and CPW. The characteristic impedance of the waveguide and the CPW are typically few hundreds of ohms and few tens of ohms, respectively.

The electric field distribution at cross sectional planes along the transition is shown in Figure 10.15. From this figure one can easily visualize that the ridge and the nonradiating slot gradually split the electromagnetic fields of the TE_{10} waveguide mode and rotate them through 90° to match the fields of the CPW.

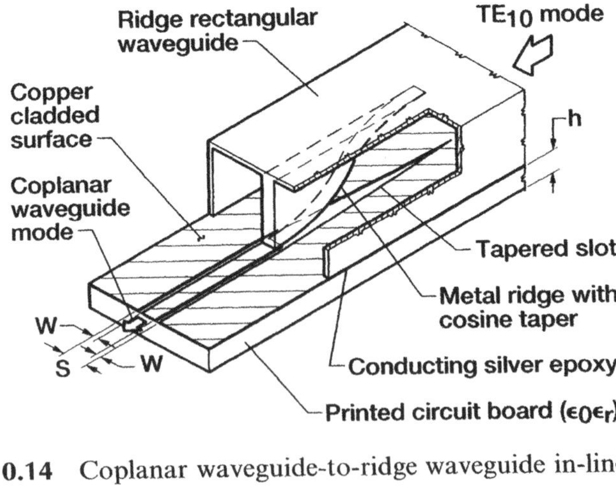

FIGURE 10.14 Coplanar waveguide-to-ridge waveguide in-line transition.

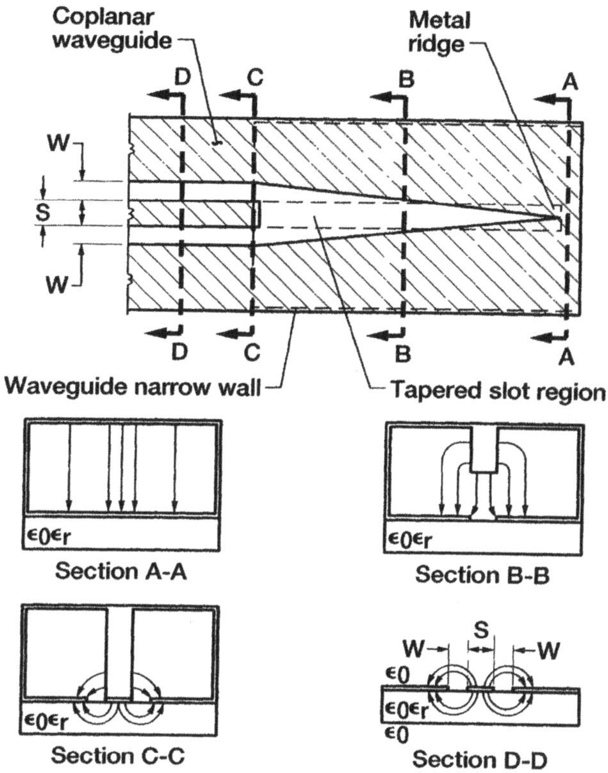

FIGURE 10.15 Electric-field distribution at various cross sections along the transition.

In [20] the characteristics of an experimental transition at K-band frequencies are reported. The CPW is fabricated on a $h = 3.175$ mm thick RT/Duroid 5880 ($\varepsilon_r = 2.2$) with single-sided copper cladding. The waveguide is a standard K-band waveguide to which a ridge with a cosine taper is attached. The cosine taper is approximately $1.5\lambda_g$ at the center frequency, where λ_g is the guide wavelength in the standard rectangular waveguide. The longitudinal edge formed by the substrate, and the waveguide is sealed by conductive silver epoxy. The strip conductor width S and the slot width W of the CPW are 0.813 and 0.203 mm, respectively, yielding a characteristic impedance of about 75 Ω. Meaurements made over the frequency range of 18.0 to 26.5 GHz show that for two back-to-back transitions with 20.32 mm length of CPW in between, the insertion loss is 1.75 ± 0.25 dB. The measured return loss is less than -11.0 dB. A scaled version of this transition at W-band on a GaAs substrate is reported in [21]. The insertion loss per transition is in the range of 1.0 to 3.0 dB and the return loss less than -10.0 dB across the frequency band of 75.0 to 110.0 GHz.

10.5.2 Coplanar Waveguide-to-Trough Waveguide Transition

The cross section of a transition [22] from a 50 Ω CPW to a ridge-trough waveguide is shown in Figure 10.16. The transition has five different waveguide sections. Each section has a characteristic impedance of 50 Ω and a length of quarter wavelength. The corresponding electric field distribution along the five cross sections are shown in Figure 10.17(*a*) to (*e*). The first cross section immediately behind the CPW resembles a ridge in a trough as shown in Figure 10.17(*a*). The electric fields in the left and the right half of the cross section are equal and opposite, and hence they match that of the CPW. At this cross section the waveguide has a width a_1 and a height b_1, the ridge has a width a_2 and a gap b_2, and last, the trough has a width a_3 and a depth b_3. The cutoff frequency f_c of the ridge-trough waveguide is expressed in terms of the inductance L and capacitance C as follows [22]:

$$f_c = \frac{1}{2\pi\sqrt{(L/2)\,C}}, \tag{10.8}$$

where

$$L = \frac{\mu(a_1 - a_2)}{2} b_1. \tag{10.9}$$

In this expression μ is permeability of the medium and

$$C = C_1 + 2(C_2 + C_3 + C_4). \tag{10.10}$$

The terms C_1 and C_2 are the electrostatic, parallel-plate capacitances of the ridge-trough waveguide and are expressed as [22]

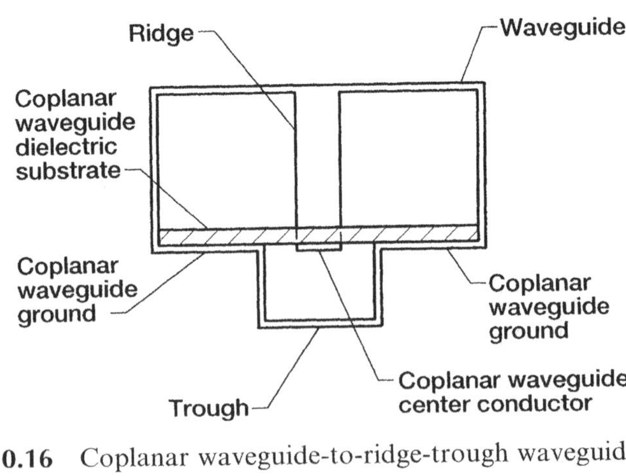

FIGURE 10.16 Coplanar waveguide-to-ridge-trough waveguide transition.

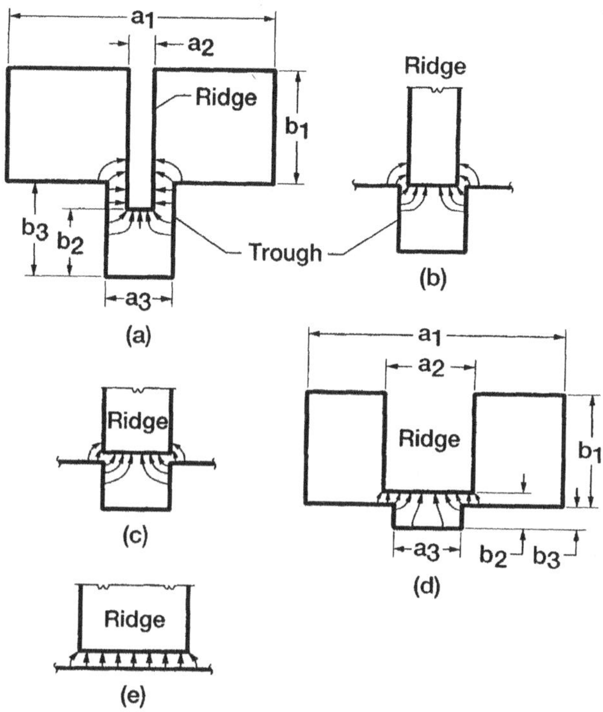

FIGURE 10.17 Five different waveguide cross sections of the transition: (*a*) Ridge in trough; (*b*) ridge even with trough; (*c*) ridge over trough; (*d*) ridge overlapping trough; (e) conventional ridge waveguide.

$$C_1 = \varepsilon \frac{a_2}{b_2}, \tag{10.11}$$

$$C_2 = \varepsilon \frac{2(b_3 - b_2)}{a_3 - a_2}, \tag{10.12}$$

and ε is the permittivity of the medium.

The terms C_3 and C_4 represent the fringing capacitances present at the bottom corners of the ridge and the upper corners of the trough respectively. The capacitance C_3 is expressed as [22]

$$C_3 = C_d(x), \tag{10.13}$$

where

$$C_d(x) = \frac{\varepsilon}{\pi}\left[\frac{x^2+1}{x}\cosh^{-1}\left(\frac{1+x^2}{1-x^2}\right) - 2\ln\left(\frac{4x}{1-x^2}\right)\right], \tag{10.14}$$

and

$$x = \frac{a_3 - a_2}{2a_3}. \qquad (10.15)$$

The capacitance C_4 is also expressed in terms of $C_d(x)$ as

$$C_4 = C_d(x); \qquad (10.16)$$

however, x is now given by

$$x = \frac{(a_3 - a_2)}{(a_1 - a_2)}. \qquad (10.17)$$

From the cutoff frequency, the cutoff wavelength of the ridge-trough waveguide can be determined as follows [22]

$$\lambda_c = \frac{c}{f_c}, \qquad (10.18)$$

where c is the velocity of light in free space.

The characteristic impedance Z_{0i} at infinite frequency for the ridge-trough waveguide shown in Figure 10.17(a) is expressed as [22]

$$Z_{0i} = \frac{120\pi}{A_1 + A_2 + A_3}, \qquad (10.19)$$

where

$$A_1 = \frac{\sin\theta_3}{b_2}, \qquad (10.20)$$

$$A_2 = \frac{\cos\theta_3}{\sin(\theta_1 + \theta_2)}\left[\left(\frac{2}{a_3 - a_2} - \frac{1}{b_1}\right)\cos\theta_1 - \frac{2\cos(\theta_1 + \theta_2)}{a_3 - a_2} + \frac{1}{b_1}\right], \qquad (10.21)$$

$$A_3 = \frac{2\cos\theta_3}{\varepsilon}\left[C_3 + C_4\frac{\sin\theta_1}{\sin(\theta_1 + \theta_2)}\right], \qquad (10.22)$$

$$\theta_1 = \frac{\pi(a_1 - a_2)}{\lambda_c}, \qquad (10.23)$$

$$\theta_2 = \frac{2\pi(b_3 - b_2)}{\lambda_c}, \qquad (10.24)$$

$$\theta_4 = \frac{\pi a_2}{\lambda_c}. \qquad (10.25)$$

Once the Z_{0i} at infinite frequency has been calculated, the characteristic impedance Z_0 at a finite frequency f, can be determined from the relation [22]

$$Z_0 = \frac{Z_{0i}}{\sqrt{1-(f_c/f)^2}}. \tag{10.26}$$

In the second cross section shown in Figure 10.17(*b*), the ridge is even with the trough. The Z_0 at this cross section is also determined using Eqs. (10.19) to (10.26) with $(b_3 - b_2)$ set equal to zero. In the third cross section shown in Figure 10.17(*c*), the ridge is over the trough. For this case also Eqs. (10.9) and (10.10) hold good. However, the constituent capacitances C_1 to C_4 take on a different form as follows [22]:

$$C_1 = \varepsilon \frac{a_3}{b_2}, \tag{10.27}$$

$$C_2 = \varepsilon \frac{a_2 - a_3}{2(b_2 - b_3)}, \tag{10.28}$$

$$C_3 = C_d(x) \quad \text{with} \quad x = \frac{(b_2 - b_3)}{b_1}, \tag{10.29}$$

$$C_4 = C_d(x) \quad \text{with} \quad x = \frac{(b_2 - b_3)}{b_2}, \tag{10.30}$$

where C_d is given by Eq. (10.14). The f_c is next determined from Eq. (10.8) with the new value of capacitances given by Eqs. (10.27) to (10.30).

The Z_{0i} at infinite frequency for the ridge-trough waveguide shown in Figure 10.17(*c*) is expressed as [22]

$$Z_{0i} = \frac{120\pi}{\dfrac{\lambda_c}{b_2\pi}\{B_1 + B_2\} + B_3}, \tag{10.31}$$

where

$$B_1 = \sin\phi_3, \tag{10.32}$$

$$B_2 = \frac{b_2\cos\phi_3}{\sin(\phi_1+\phi_2)}\left[\cos\phi_1\left(\frac{1}{b_2 - b_3} - \frac{1}{b_1}\right) - \frac{\cos(\phi_1+\phi_2)}{b_2 - b_3} + \frac{1}{b_1}\right], \tag{10.33}$$

$$B_3 = 2\cos\frac{\phi_3}{\varepsilon}\left(C_4 + \frac{C_3\sin\phi_1}{\sin(\phi_1+\phi_2)}\right), \tag{10.34}$$

$$\phi_1 = \frac{\pi(a_1 - a_2)}{\lambda_c}, \tag{10.35}$$

$$\phi_2 = \frac{\pi(a_2 - a_3)}{\lambda_c}, \tag{10.36}$$

$$\phi_3 = \frac{\pi a_3}{\lambda_c}, \tag{10.37}$$

and λ_c is given by Eq. (10.18).

The equations above simplify when the ridge width a_2 is the same as the trough width a_3. In this case $a_2 - a_3$ is zero, and hence ϕ_2 is zero. Last, the Z_0 is determined from Eq. (10.26). In the fourth cross section shown in Figure 10.17(*d*), the ridge over laps the tough. The Z_0 at this cross section is also determined using Eqs. (10.31) to (10.37) and (10.26). In the final cross section shown in Figure 10.17(*e*), there is no trough, and the cross section is that of a conventional ridge waveguide. The expression for L remains the same as in Eq. (10.9), but the capacitance C is expressed as [22]

$$C = \frac{\varepsilon a_2}{b_2} + 2C_d(x), \tag{10.38}$$

where $C_d(x)$ is given by Eq. (10.14) with $x = b_2/b_1$. The cutoff frequency is calculated from Eq. (10.8). The Z_{0i} at infinite frequency is given by [23]

$$Z_{0i} = \frac{120\pi}{D_1 + D_2}, \tag{10.39}$$

where

$$D_1 = 2\frac{C_d(x)}{\varepsilon} \cos\psi_2, \tag{10.40}$$

$$D_2 = \frac{1}{\pi}\frac{\lambda_c}{b_2}\left(\sin\psi_2 + \frac{b_2}{b_1}\cos\psi_2 \tan\frac{\psi_1}{2}\right), \tag{10.41}$$

$$\psi_1 = \frac{\pi(a_1 - a_2)}{\lambda_c}, \tag{10.42}$$

$$\psi_2 = \frac{\pi a_2}{\lambda_c}. \tag{10.43}$$

The Z_0 at any finite frequency is calculated from Eq. (10.26).

The conventional ridge waveguide is then matched to a standard rectangular waveguide using a seven section binomial transformer. The design of the binomial transformers are given in [24]. The characteristics of a W-band (75.0 to 110.0 GHz) CPW probe with this transition is reported in [25]. The insertion loss is less than 0.7 dB from 75.0 to 106.5 GHz, increasing to 0.8 dB at 110.0 GHz. The return loss is less than −15.0 dB over the entire W-band.

10.5.3 Coplanar Waveguide-to-Rectangular Waveguide Transition with a Tapered Ridge

The coplanar waveguide (CPW)-to-rectangular waveguide transition [26] with a tapered ridge is shown in Figure 10.18. The transition consists of a very thin metal ridge located longitudinally on the bottom broadwall of a rectangular waveguide. The shape of the ridge is approximately a cosine wave or a staircase. The ridge guides the RF power to a slot cut in the top broadwall. In the slot the RF power is split into two halves and guided in the two air regions on either side of the ridge. The magnitude and direction of the electric field lines in the two air regions are equal and opposite to each other as shown in the figure. Thus the ridge and the air regions form the center strip conductor and the slots respectively of a CPW with air as a dielectric substrate. In the figure the CPW center strip conductor and the slot widths are denoted as S and W, respectively. The thickness of the conductor is denoted as t. The length of the transition, which is a uniform CPW, is denoted as L. Typical values for these dimensions are summarized in Table 10.4.

The electric field lines at any cross section of the CPW is shown in the inset in Figure 10.18. From this inset it is easy to visualize the presence of two equipotential planes A–A and B–B located midway in the two slot regions. In addition a virtual magnetic wall exists midway along the center strip conductor

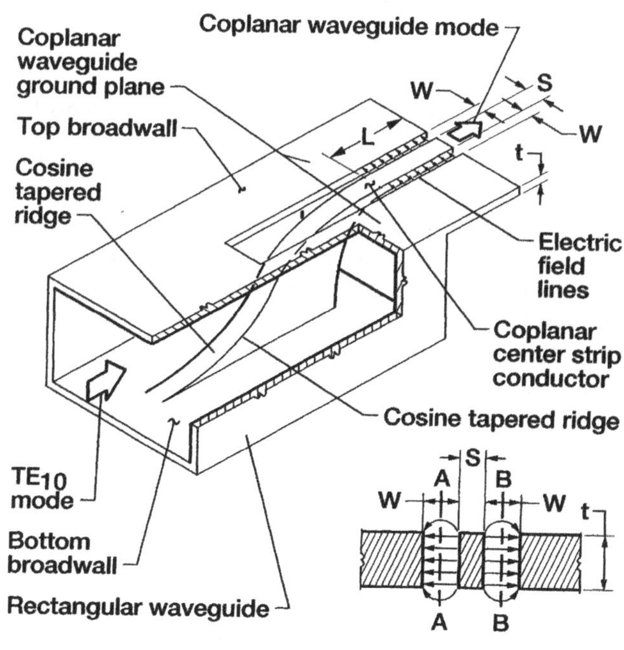

FIGURE 10.18 Coplanar waveguide-to-rectangular waveguide transition with a tapered ridge.

TABLE 10.4 Coplanar Waveguides-to-Rectangular Waveguide Transition with a Tapered Ridge

Dimensions and Characteristics	X-band	K_a-band	W-band
W (mm)	0.7	0.4	0.0635
S (mm)	1.6	1.2	0.127
t (mm)	4.0	1.0	1.0
L (mm)	1.25	0.65	—
Frequency range (GHz)	7.5–11.5	26.5–40.0	90.0–100.0
Measured:			
Return loss (dB)	−18.0	−14.0	−10.0
Insertion loss (dB)	1.0	1.5	1.0

of the CPW. Therefore the structure between the magnetic wall and the equipotential plane is isolated and modeled as a microstrip of width t and thickness $S/2$ supported on a air-dielectric substrate of thickness W/2 [26]. This partial structure has a characteristic impedance Z_0. The characteristic impedance of each slot which consists of two partial sections in series in therefore $2Z_0$. In a CPW the two slots are in parallel and therefore the characteristic impedance of the CPW in Z_0. In [26] the characteristics of several experimental back-to-back transitions at X-band, Ka-band, and W-band frequencies are reported. These characteristics are summarized in Table 10.4.

10.5.4 Coplanar Waveguide-to-Rectangular Waveguide End Launcher

A coplanar waveguide (CPW)-to-rectangular waveguide end launcher [27] is schematically illustrated in Figure 10.19. In the end launcher a dielectric substrate carrying the CPW enters the rectangular waveguide through an aperture in the center of the end wall. The CPW center strip conductor extends beyond the ground planes further into the waveguide and forms an L-shaped loop. The loop couples power from the quasi-TEM CPW mode to the TE_{10} waveguide mode. The center strip conductor width S and slot width W of the CPW for a given substrate thickness h are determined so as to provide a good impedance match to a 50 Ω coaxial connector. The waveguide height and width are standard for a given frequency band. The length, height, and width of the probe extending into the waveguide are denoted as L, H, W_1, and W_2, respectively. The length L is approximately $\lambda_{g(\text{waveguide})}/6$ at the center frequency f_0 of 30.75 GHz. The height H is half the substrate height. The width W_1 and W_2 are 0.4 and 0.508 mm, respectively. The end launcher operates with a return loss of better than −10.0 dB over approximately 20 percent bandwidth centered at f_0. The measured insertion loss of the end launcher is approximately 1.0 dB and includes the losses in the coaxial connector and in the coax-to-waveguide adapter of the test setup used in the characterization.

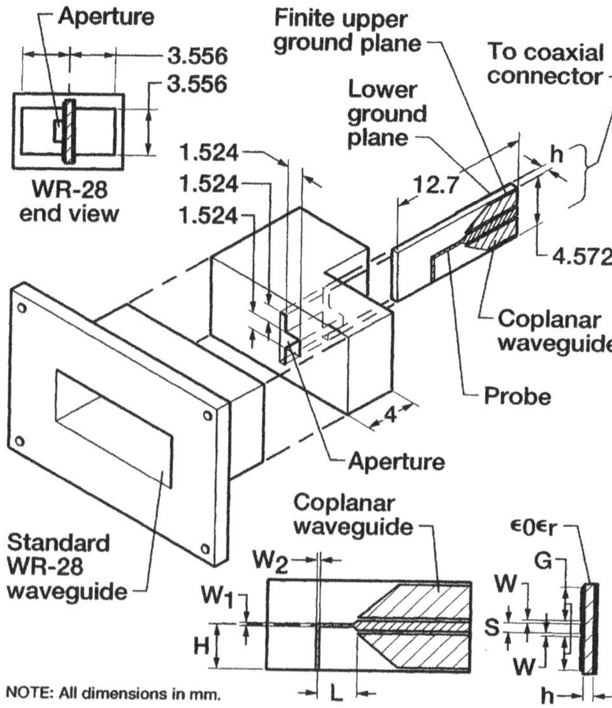

FIGURE 10.19 Coplanar waveguide-to-rectangular waveguide end launcher: $h = 0.254$ mm, $\varepsilon_r = 2.22$, $S = 0.554$ mm, $W = 0.254$ mm, $G = 1.25$ mm.

10.5.5 Coplanar Waveguide-to-Rectangular Waveguide Launcher with a Post

A channelized coplanar waveguide (CCPW)-to-rectangular waveguide post launcher [28] and the CCPW substrate with the coupling probe in place are schematically illustrated in Figures 10.20 and 10.21, respectively. The channel width C, center strip conductor width S, and slot width W of the CCPW for a given substrate thickness h are determined as explained in Chapter 2. The waveguide height, width, and broadwall thickness are denoted as a, b, and T, respectively, and are standard for a given frequency band. The axis of the two transmission lines are parallel; however, they can be oriented at any arbitrary angle. A metal post of diameter d_1 enters the waveguide through an aperture of diameter d_2 to form an electric probe. The length of the probe extending into the waveguide is L_p. The opposite end of the post is attached through the dielectric substrate to the center strip conductor of the CCPW. The waveguide and the CCPW are terminated in a fixed short circuit placed at a distance L_W and L_S from the center of the post, respectively.

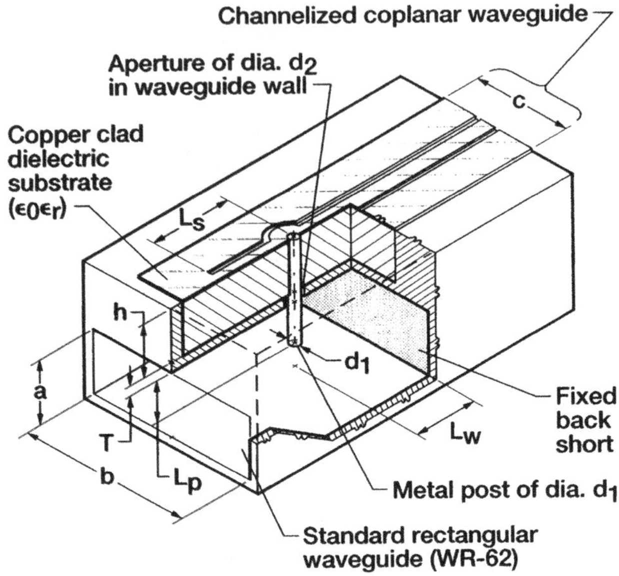

FIGURE 10.20 Cutaway drawing of channelized coplanar waveguide (CCPW)-to-rectangular waveguide post launcher: $a = 7.8994$ mm, $b = 15.7988$ mm, $C = 5.08$ mm, $h = 3.175$ mm, $d_1 = 1.3208$ mm, $d_2 = 5.08$ mm, $L_p = 5.715$ mm, $T = 1.016$, $L_W = 4.191$ mm, $\varepsilon_r = 2.2$.

In the experimental launcher [28], L_W and L_S are approximately $\lambda_{g(\text{waveguide})}/6$ and $\lambda_{g(\text{CCPW})}/6$, respectively, at the center frequency. The radii of curvature R_1 and R_2 of the circular inland in the CCPW center strip conductor and the ground plane at the location of the post are approximately $0.7d_1$ and $0.7d_1 + W$, respectively. Further the normalized diameter d_1/b and length L_p/a are equal

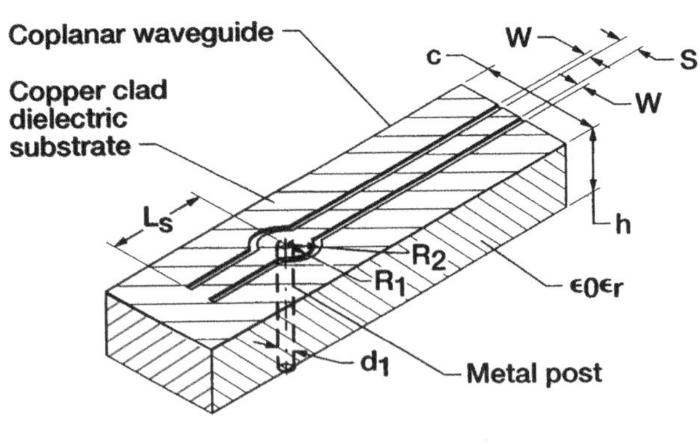

FIGURE 10.21 CCPW circuit board with a probe: $S = 1.143$ mm, $W = 0.254$ mm, $L_S = 2.54$ mm, $R_1 = 0.889$ mm, $R_2 = 1.143$ mm.

to 0.084 and 0.723, respectively. The post launcher operates over the entire Ku-band extending from 12.4 to 18.0 GHz. The measured return loss at the CCPW port is less than −10.0 dB, and the insertion loss between the CCPW and the waveguide ports is 1.0 dB. The insertion loss includes the losses in the coaxial connector and in the coax-to-waveguide adapter of the test setup.

10.5.6 Channelized Coplanar Waveguide-to-Rectangular Waveguide Launcher with an Aperture

A channelized coplanar waveguide (CCPW)-to-rectangular waveguide aperture launcher [28] and the CCPW substrate with an aperture in the lower ground plane are schematically illustrated in Figures 10.22 and 10.23, respectively. The lower ground plane is flush with the inner surface of the waveguide narrow wall and consequently the aperture interrupts the transverse surface currents. Thus power from the TE_{10} waveguide mode is coupled to the CCPW. The aperture width W_2 is chosen as 0.254 mm for ease of fabrication. The aperture length is equal to the channel width C. The waveguide and the CCPW are terminated in a short circuit placed at a distance L_W and L_S from the center

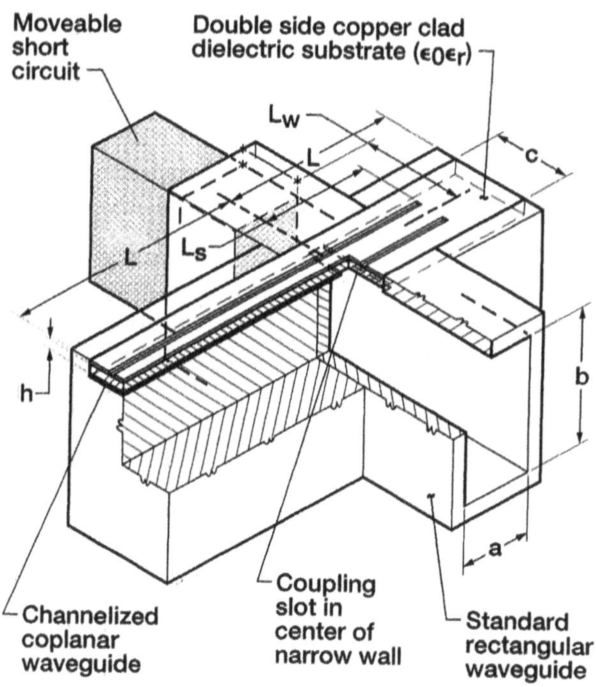

FIGURE 10.22 Cutaway drawing of channelized coplanar waveguide (CCPW)-to-rectangular waveguide aperture launcher: $a = 7.8994$ mm, $b = 15.7988$ mm, $C = 5.08$ mm, $d = 0.7874$ mm, $L = 25.4$ mm, $L_W = 11.43$ mm, $\varepsilon_r = 2.2$.

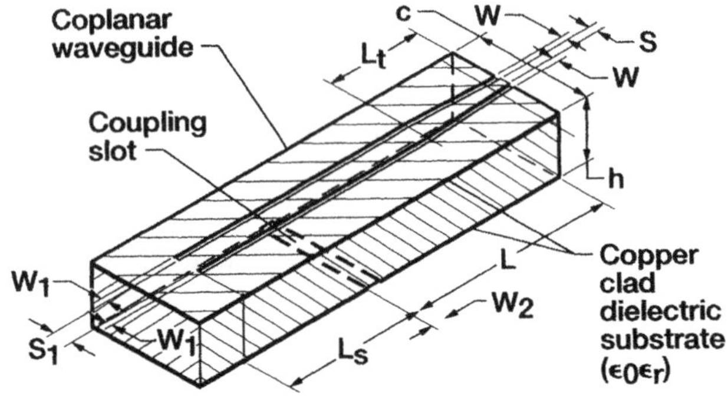

FIGURE 10.23 CCPW circuit board with an aperture: $S = 0.6858$ mm, $W = 0.3048$ mm, $S_1 = 1.397$ mm, $W_1 = 0.254$ mm, $L_t = 1.143$ mm, $W_2 = 0.254$ mm, $L_S = 11.0236$ mm.

of the aperture, respectively. The length of the impedance matching taper L_t is equal to $\lambda_{g(CCPW)}/2$.

In the experimental launcher [28], L_W and L_S are approximately $\lambda_{g(waveguide)}/4$ and $\lambda_{g(CCPW)}/2$, respectively, at the center frequency. The measured return loss at the CCPW port is less than -10.0 dB, and the insertion loss between the CCPW port and the waveguide port is about 1.5 dB across the frequency band of 11.0 to 12.0 GHz.

10.5.7 Coplanar Waveguide-to-Rectangular Waveguide Transition with a Printed Probe

A coplanar waveguide (CPW) to rectangular waveguide transition that makes use of a printed probe is shown in Figure 10.24. In this transition, the printed probe is formed by extending the CPW center strip conductor beyond the ground planes into the waveguide. The circuit board carrying the printed probe enters the waveguide through an aperture cut in the center of the broadwall parallel to the longitudinal axis. A tunable backshort is provided to maximize the power coupled from the waveguide to the CPW. In [29] the characteristics of an experimental transition fabricated on a 0.127 mm thick alumina substrate are reported. The measured insertion loss and return loss of two back-to-back transitions over the frequency range of 43.0 to 50.0 GHz is less than 2.0 dB and about -15.0 dB, respectively.

A CPW-to-rectangular waveguide transition can also be designed with the probe entering the waveguide through an aperture cut in the broadwall perpendicular to the waveguide longitudinal axis. However, in this kind of transition a probe in the shape of a rectangular patch is found to be very effective [30]. As in the previous example a tunable backshort is also necessary to maximize the coupling. The characteristics of an experimental transition at

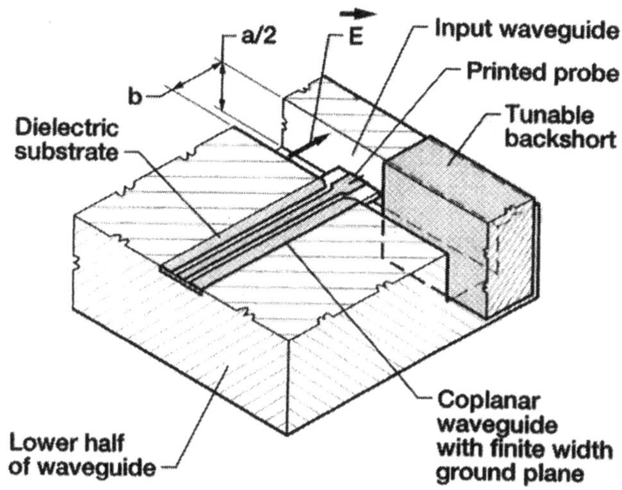

FIGURE 10.24 Coplanar waveguide-to-rectangular waveguide transition with a printed probe. The top half of the waveguide is removed. The inside dimensions of the waveguide are *a* and *b*, respectively.

X-band are reported in [30]. The CPW and the patch are fabricated on a 0.794 mm thick substrate with $\varepsilon_r = 2.35$. The CPW strip width S and slot width W are 3 mm and 0.1 mm, respectively. The CPW characteristic impedance $Z_{0(\mathrm{CPW})}$ is 50 Ω. The patch has an area of 10 × 6 mm and, the distance of the backshort from the patch is 5.6 mm. The measured return loss is about −15.0 dB over the frequency range of 9.8 to 12.4 GHz. Further it has been observed that the bandwidth of the transition is larger when the patch faces the backshort than when the patch faces the waveguide [30].

10.6 COPLANAR WAVEGUIDE-TO-SLOTLINE TRANSITION

In this section the design and characteristics of transitions between an unbalanced coplanar waveguide (CPW) and a balanced slotline will be presented. Transitions of this type are commonly referred to as baluns.

10.6.1 Coplanar Waveguide-to-Slotline Compensated Marchand Balun or Transition

The coplanar waveguide (CPW)-to-slotline transition and its lumped element equivalent circuit model first reported by Houdart [1] are shown in Figure 10.25(*a*) and (*b*). The advantage of this transition is its uniplanar construction. In an uniplanar circuit the conductors are all on the same side of the substrate, which simplifies fabrication. This transition for the purpose of design is

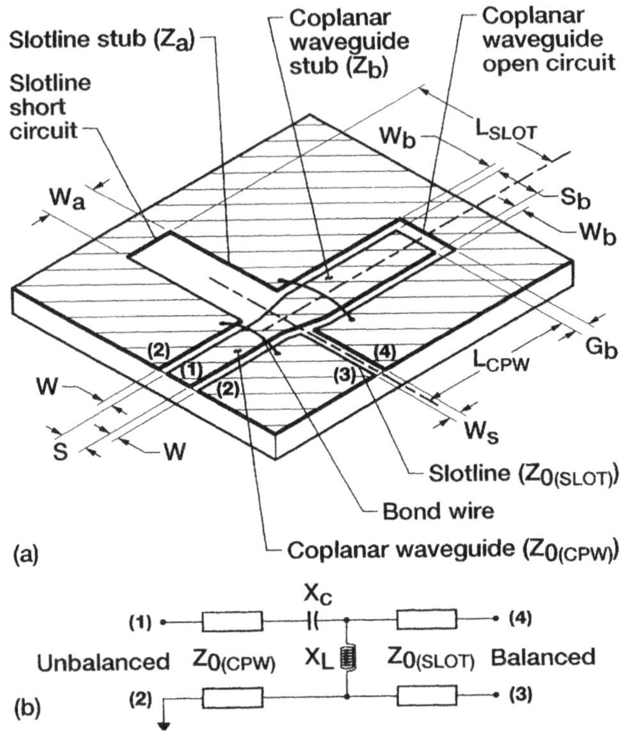

FIGURE 10.25 Coplanar waveguide-to-slotline compensated Marchand balun or transition: (*a*) Circuit layout; (*b*) lumped element equivalent circuit model.

considered as a planar version of the coaxial balun reported by Marchand [31] and Oltman [32]. According to [31], [32] the coaxial balun behaves as a bandpass network. In this section we will show that the CPW-to-slotline transition operates both as a balun and as an impedance transformer.

In Figure 10.25(*a*) the center strip conductor width, slot width, and characteristic impedance of the input CPW are denoted as S, W, and $Z_{0(\mathrm{CPW})}$, respectively. The center strip conductor width, slot width, open end gap width, length, and characteristic impedance of the open circuited CPW stub are denoted as S_b, W_b, G_b. L_{CPW}, and Z_b, respectively. The slot width and characteristic impedance of the output slotline are denoted as W_s and $Z_{0(\mathrm{slot})}$, respectively. The slot width, length, and characteristic impedance of the short circuited slotline stub are denoted as W_a, L_{slot}, and Z_a, respectively. To keep the design simple, the stub end effects, air-bridge, or wire bond reactances and the junction reactances are neglected. The design of the transition proceeds as follows: If f_1 and f_2 are the frequencies corresponding to the lower and upper band edges, the center frequency $f_0 = (f_1 + f_2)/2$ is first calculated. The second step involves the determination of Z_a. In [32] it is shown that the bandwidth of the balun is a maximum when Z_a is maximum. In order to maximize Z_a, the slot width W_a has to be large. However, W_a cannot be made arbitrarily large. The largest W_a has to be small compared to the guide wavelength in the slotline stub at f_2 to avoid multi-moding. Thus an appropriate Z_a is chosen. The third step involves, the determination of optimum Z_b. For $Z_{0(CPW)} = 50\,\Omega$ and for arbitrary $Z_{0(slot)}$, the optimum Z_b is given by [32]

$$Z_{b(opt)} = \left[\frac{Z_a \tan^2\left(\frac{\pi}{2}\frac{f_n}{f_0}\right)}{1 + \left(\frac{Z_a}{Z_{0(slot)}}\right)^2 \tan^2\left(\frac{\pi}{2}\frac{f_n}{f_0}\right)} \right] \tag{10.44}$$

where f_n is either f_1 or f_2. The length L_{slot} and L_{CPW} are quarter guide wavelength at f_0. Clearly, the balun also operates as an impedance transformer.

In Figure 10.25(*b*) the series compensating capacitive reactance X_c arises from the open circuited CPW stub and at any frequency f is given by the well-known relation

$$X_c = -jZ_b \cot\left(\frac{\pi}{2}\frac{f}{f_0}\right). \tag{10.45}$$

Similarly the shunt compensating reactance X_L arises from the short circuited slotline stub and is given by

$$X_L = jZ_a \tan\left(\frac{\pi}{2}\frac{f}{f_0}\right). \tag{10.46}$$

In [33], [34], [35], and [36] the characteristics of several experimental transitions are reported. Typical dimensions, circuit parameters and measured characteristics of these transitions are summarized in Table 10.5. This table shows that the compensated Marchand balun or transition has a wide bandwidth.

10.6.2 Coplanar Waveguide-to-Slotline Transition with Radial or Circular Stub Termination

A coplanar waveguide (CPW)-to-slotline uniplanar transition with radial stub termination [35], [36], and [37] is shown in Figure 10.26(*a*). In this transition the slotline radial open stub radius and angle are denoted as R_s and θ_s, respectively. The characteristic impedance of the slotline and the CPW are denoted as $Z_{0(slot)}$ and $Z_{0(CPW)}$, respectively. In [35], [36], and [37] the characteristics of several experimental transitions are reported. Typical dimensions, circuit parameters and the measured characteristics are summarized in

TABLE 10.5 Coplanar Waveguide-to-Slotline Marchand Balun or Transition

Dimensions and Characteristics	Reference [33]	Reference [34]	Reference [35]	Reference [36]
SUBSTRATE MATERIAL	ALUMINA	ALUMINA	HIGH-RESISTIVITY	RT/DUROID 6010.8
ε_r AND THICKNESS (mm)	9.9, 0.635	9.9, 0.635	SILICON, 11.9, 0.33	10.8, 1.27
		Coplanar waveguide		
S (mm)	0.145	—	0.1	0.5
W (mm)	0.347	—	0.07	0.25
$Z_{0(\mathrm{CPW})}$ (Ω)	50.0	50.0	53.0	50.0
S_b (mm)	0.145	—	0.2	0.5
W_b (mm)	0.347	—	0.025	0.25
G_b (mm)	—	—	0.2	—
Z_b (Ω)	50.0	25.0	30.9	50.0
L_{CPW} (mm)	6.82	—	1.1	10.41
		Slotline		
W_S (mm)	0.118	—	0.075	0.1
$Z_{0(\mathrm{SLOT})}$ (Ω)	—	50.0	70.0	50.0
W_a (mm)	0.118	—	0.205	0.1
Z_a (Ω)	—	100.0	100.0	50.0
L_{SLOT} (mm)	7.811	—	1.4	11.745
		Measured characteristics[a]		
Frequency range (GHz)	3.1–4.5	4.5–18.0	17.0–36.0	1.6–3.6
S_{11} (dB)	$\leqslant -20.0$	$\leqslant -10.0$	$\leqslant -10.0$	—
S_{21} (dB)	0.6–0.7	0.8–2.0	2.0	1.0

[a] For two back-to-back transitions.

Table 10.6. Since the slotline radial open stub has a wide bandwidth, the transition also has a wide bandwidth. In [36] the slotline radial stub is modeled by tandem-connected segments of piecewise uniform slotlines. Typically 10 line segments of equal length are used, but different characteristic impedance is shown to be sufficient. Based on this model the computed insertion loss agrees very well with the measured results.

Circuits containing active devices require a dc bias. For dc biasing a narrow isolating slot is etched in the ground plane near one end of the radial arc [35] as shown in Figure 10.26(*a*). Measurements made with and without the isolating slot indicate that the insertion loss is essentially identical in both cases indicating that the perturbation by the slot is minimum [35].

An alternate design [38], [39] for the transition is presented in Figure 10.26(*b*). In this design the slotline radial opens stub is replaced by a circular open stub. The characteristics of an experimental transition of this type are reported in [38]. These characteristics are summarized and compared with

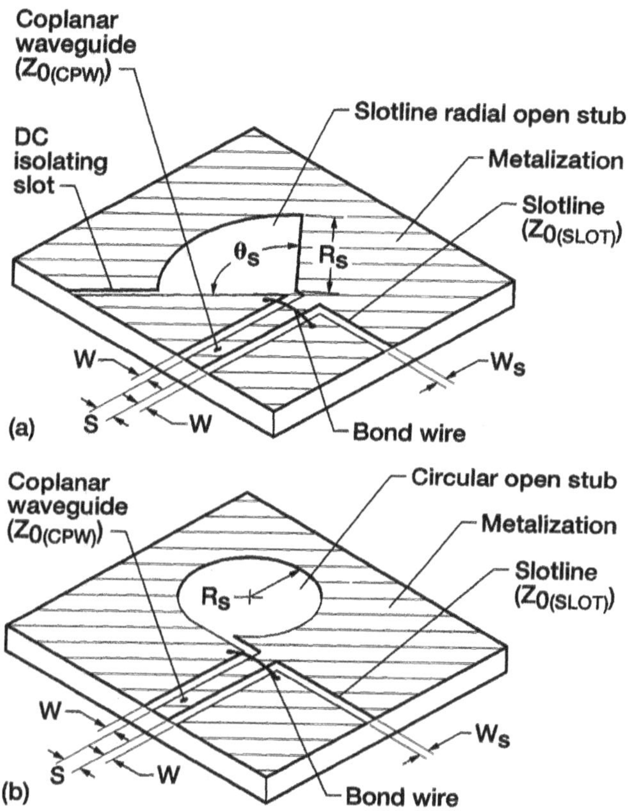

FIGURE 10.26 Coplanar waveguide-to-slotline transition: (*a*) Radial stub termination and dc isolation slot; (*b*) circular stub termination.

those of the previous design in Table 10.6. It is observed that this design also results in low insertion loss and wide bandwidth.

In [40] a transition with a straight stub inplace of a radial or circular stub is investigated. The transition is designed for operation over the frequency range of 0.24 to 60.0 GHz. The slotline stub length is chosen as $\lambda_{g(slot)}/4$ at the center frequency of 35.0 GHz. The characteristics of this transition are also summarized in Table 10.6 for comparison.

10.6.3 Coplanar Waveguide-to-Slotline Double-Y Balun or Transition

The double-Y uniplanar balun [35] is based on a radial six-port junction formed by placing alternatively three balanced and three unbalanced transmission lines. These lines have equal characteristic impedances. Further the length of adjacent pair of lines are made equal and terminated in a short circuit and an open circuit, respectively.

TABLE 10.6 Coplanar Waveguide-to-Slotline Transition

	Slotline Radial Open Stub		Circular Open Stub	Straight Stub
Dimensions and Measured Characteristics	References [36] and [37]	Reference [35]	Reference [38]	Reference [40]
SUBSTRATE MATERIAL ε_r AND THICKNESS (mm)	RT/DUROID 6010.8 10.8, 1.27	HIGH-RESITIVITY SILICON, 11.9, 0.33	— 10.0, 1.27	GaAs 13.1, 0.4
Coplanar waveguide				
S (mm)	0.5	0.025	0.127	0.03
W (mm)	0.25	0.025	0.051	0.025
$Z_{0(CPW)}$ (Ω)	50	50	50	50
Slotline				
W_S (mm)	0.1	0.025	0.053	0.025
$Z_{0(SLOT)}$ (Ω)	50	—	50	50
Radial stub radius, R_s (mm)	6.0	1.0	—	—
Radial stub angle, θ_s (deg)	60	90	—	—
Circular stub radius, R_s (mm)	—	—	1.59	—
Straight stub length (mm)	—	—	—	0.815
Measured characteristics				
Frequency range (GHz)	1.1–5.7	20.0–40.0	2.9–12.2	28.5–44.0
S_{21} (dB) for two back-to-back transitions	[a]$\leqslant 1.0$	[b]$\leqslant 3.5$	[c]$\leqslant 1.0$	[d]$\leqslant 2.0$
S_{21} (dB)	—	$\leqslant -10.0$	$\leqslant -10.0$	$\leqslant -10.0$

[a]With 8.0 mm of slotline in between and coaxial launchers.
[b]With 6.0 mm of slotline in between.
[c]With 8.9 mm of slotline in between and coaxial launchers.
[d]With slotline in between.

This type of baluns differs from the Marchand baluns in the following manner: First, the double-Y is unable to perform impedance transformation since all the lines have the same characteristic impedance. Second, the upper cutoff frequency in general is set when the line length becomes an odd multiple of $\lambda_g/8$, where λ_g is the wavelength in the transmission lines. Third, it is an all-pass network.

A double-Y balun constructed with slotline and coplanar waveguide (CPW) as transmission lines is shown in Figure 10.27(a). The bond wires across the CPW at the junction suppress the propagation of coupled slotline even mode.

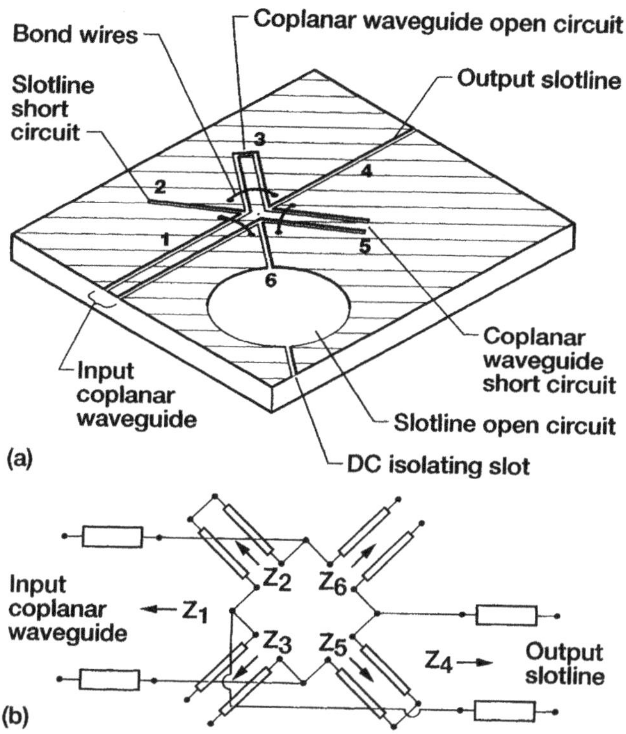

FIGURE 10.27 Coplanar waveguide-to-slotline double-Y balun or transition: (*a*) Circuit layout; (*b*) lumped element equivalent circuit model.

To understand the operation of this balun consider the equivalent circuit model shown in Figure 10.27(*b*). The circuit elements Z_1 to Z_6 refer to the input impedance of the stubs and the transmission lines as seen from the junction. As a simplification the parasitics associated with the junction are all neglected in the equivalent circuit model. Assuming $Z_2 = Z_5$ and $Z_3 = Z_6$ and that all impedances are normalized to Z_4, that is,

$$\frac{Z_n}{Z_4} = Z'_n, \qquad n = 1, 2, \ldots, 6, \tag{10.47}$$

it can be shown that the normalized input impedance at port 1 of the junction is [35]

$$Z'_{\text{in}} = \frac{1}{1 + Z'_2 + Z'_3}\left[Z'_2 Z'_3 + \frac{(1 + Z'_2)(1 + Z'_3)(Z'_2 + Z'_3)}{(2 + Z'_2 + Z'_3)}\right]. \tag{10.48}$$

For perfect transmission between the input CPW and the output slotline ports,

TABLE 10.7 Coplanar Waveguide-to-Slotline Double-Y Balun or Transition

Dimensions and Characteristics	Reference [35]	Reference [41]	
SUBSTRATE MATERIAL	HIGH-RESISTIVITY	—	—
ε_r AND THICKNESS (mm)	SILICON, 11.9, 0.33	9.8, 0.635	9.8, 0.635
	Coplanar waveguide		
S (mm)	—	0.25	0.25
W (mm)	—	0.10	0.10
$Z_{0(\mathrm{CPW})}$ (Ω)	50	50	50
Stub length (mm)	0.415	1.63	0.904
	Slotline		
W_s (mm)	—	0.1	0.1
Stub length (mm)	0.39	1.83	0.96
$Z_{0(\mathrm{SLOT})}$ (Ω)	50	50	50
Open circuit diameter (mm)	0.85	6.0	3.0
	Measured characteristics		
Frequency range (GHz)	20.0–35.0	1.45–8.7	2.3–12.8
S_{11} (dB)	[a]$\leqslant -15.0$	$\leqslant -17.0$	$\leqslant -17.0$
S_{21} (dB)	[a]3.0–4.0	1.0	1.0

[a]For two back-to-back transition.

Z'_{in} must equal to 1. This condition is equivalent to [35]

$$Z'_2 Z'_3 = 1. \tag{10.49}$$

The equation above can also be expressed in terms of the reflection coefficient as [35]

$$\Gamma_2 = -\Gamma_3, \tag{10.50}$$

where Γ_2 and Γ_3 are the reflection coefficients at ports 2 and 3, respectively.

In [35] and [41] the characteristics of experimental transitions fabricated on high-resistivity silicon and Duroid substrates, respectively, are reported. These characteristics are summarized in Table 10.7. The measured characteristics do not demonstrate all-pass characteristics. The reason for this is because the circular open circuit represents a good open circuit only over two to three octave frequency range.

Circuits with active devices require dc bias for operation. A provision for dc bias can be made by etching a narrow slot in the ground plane surrounding the circular open. This slot, when located diametrically opposite to the point of entry of the slotline from the junction, causes minimum perturbation to the balun characteristics [35].

10.6.4 Electromagnetically Coupled Finite-Width Coplanar Waveguide-to-Slotline Transition with Notches in the Ground Plane

An orthogonal transition [42] between a coplanar waveguide with finite-width ground planes (FW-CPW) and a slotline that are on opposite sides of a substrate is shown in Figure 10.28. The FW-CPW and the slotline are orthogonal to each other and are electromagnetically coupled. To enhance the coupling between the lines, two notches of length L_1 and L_2 and width W are located in the FW-CPW ground planes above the slotline.

The dimensions of the transition are experimentally optimized for the highest return loss bandwidth. The ratio of the lengths $L_1/L_2 \approx 0.33$, and $L_2 \approx 0.25\lambda_{g(slot)}$, where $\lambda_{g(slot)}$ is the guide wavelength in the slotline at the center frequency. The width $2G$ of the ground plane is about $0.65\lambda_{g(CPW)}$, where $\lambda_{g(CPW)}$ is the guide wavelength in the FW-CPW. The CPW is terminated in a short circuit at a distance of $L_c \approx 0.4\lambda_{g(CPW)}$ from the center of the notch. The slotline is terminated in a short circuit at a distance $L_s \approx 0.25\lambda_{g(slot)}$ from the notch. A pair of bond wires on either side of the notch suppresses the parasitic coupled slotline even mode on the FW-CPW.

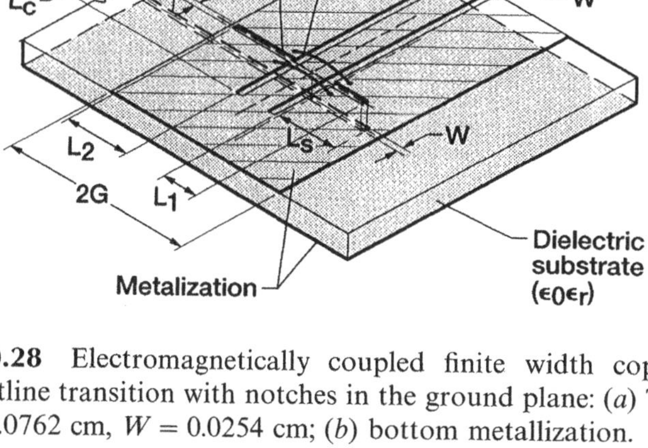

FIGURE 10.28 Electromagnetically coupled finite width coplanar waveguide-to-slotline transition with notches in the ground plane: (*a*) Top metallization, $S = 0.0762$ cm, $W = 0.0254$ cm; (*b*) bottom metallization.

In [42] the return loss characteristics of an experimental transition fabricated on a 0.508 mm thick RT/Duroid 5800 ($\varepsilon_r = 2.2$) substrate are reported. The return loss is measured at the FW-CPW port with the slotline port terminated in a wideband load, such as a tapered slot antenna. The transition has a return loss of less than -10.0 dB over 14 percent bandwidth centered at 21.0 GHz.

10.6.5 Electromagnetically Coupled Finite-Width Coplanar Waveguide-to-Slotline Transition with Extended Center Strip Conductor

An in-line transition [43] between a coplanar waveguide with finite-width ground planes (FW-CPW) and a slotline that are on opposite sides of a substrate is shown in Figure 10.29. The FW-CPW and the slot line are in line and are electromagnetically coupled. The in-line construction greatly reduces the overall circuit size.

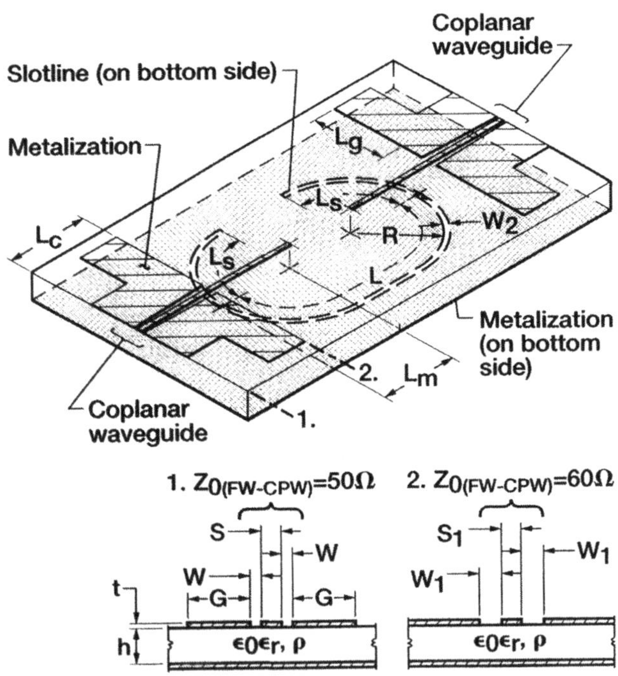

FIGURE 10.29 Electromagnetically coupled grounded finite width coplanar waveguide to slotline transition with extended center strip conductor. Top side circuit pattern: $S = S_1 = 0.1524$ mm, $W = 0.1067$ mm, $W_1 = 0.2515$ mm, $G = 2.54$ mm, $L_m = 4.6$ mm, $L_g = 3.98$ mm, $L_c = 5.08$ mm. Bottom side circuit pattern: $L = 18.7$ mm, $L_s = 6.83$ mm, $R = 4.35$ mm, $W_2 = 0.157$ mm. Substrate parameters: $h = 0.381$ mm, $\varepsilon_r = 11.7$, $\rho = 5000–10{,}000\ \Omega$ cm.

The design and construction of the transition is briefly described here. At the input port the characteristic impedance $Z_{0(\text{FW-CPW})}$ of the FW-CPW is 50 Ω, and at the open end the $Z_{0(\text{FW-CPW})}$ is 60 Ω. This change in impedance is realized by gradually flaring the FW-CPW slots. At the open end the top ground planes of the FW-CPW are terminated in two open circuited stubs of length L_g. These stubs are modeled as microstrip lines of very low characteristic impedance, typically about 10 Ω and length about $\lambda_{g(\text{microstrip})}/4$, where λ_g is the guide wavelength at the center frequency f_0. The stubs provide a virtual short circuit between the top ground planes of the FW-CPW and the slotline. The characteristic impedance of the slotline is 70 Ω. The center strip conductor of the FW-CPW, which extends beyond the terminated ground planes, forms the transition and is modeled as a microstrip line. The characteristic impedance of this microstrip line is 65 Ω. The distances L_s and L_m are approximately $\lambda_{g(\text{slotline})}/4$ and $\lambda_{g(\text{microstrip})}/4$, respectively, at f_0. The thickness t of the gold metallization is about three times the skin depth at f_0.

In [43] the characteristics of two back-to-back transitions fabricated on a high-resistivity silicon wafer for operation at $f_0 = 6.0$ GHz are reported. The measured insertion loss and return loss per transition are 1.5 dB and less than −10.0 dB, respectively, over 40 percent bandwidth centered at 4.55 GHz. This insertion loss includes, the insertion loss of a 5.08 mm long FW-CPW and half the length of the slotline between which is about 9.35 mm. The shift in the center frequency from 6.0 to 4.55 GHz is caused by discontinuity effects that are not adequately accounted for in the simplified design.

Figure 10.30 shows an alternate design [44] for constructing a transition of the above type. In this design the two microstrip stubs of length L_g are eliminated, and the top and bottom ground planes are directly connected by metal-filled vias. This technique works best on a soft substrate, such as RT/Duroid 5880. The advantage of this interconnect is its compactness; a feature much desired in an array antenna feed network. In [44] the return loss characteristics of an experimental transition is reported. The return loss is measured at the FW-CPW port with the slotline port terminated in a wideband load, such as a tapered slot antenna. The transition has a return loss of less than −10.0 dB over 90 percent bandwidth centered at 20.0 GHz.

10.6.6 Air-Bridge Coupled Coplanar Waveguide-to-Slotline Transition

A uniplanar transition [43] with air-bridge coupling between a coplanar waveguide (CPW) and a slotline is shown in Figure 10.31. The design and construction of the transition is briefly described here. At the input CPW port the characteristic impedance $Z_{0(\text{CPW})}$ is 50 Ω. The CPW transforms to a 60 Ω line that terminates in an open circuit. The characteristic impedance of the slotline is also 60 Ω. The circular bend at the feed end of the slotline provides a smooth transition. A 25 μm diameter bond wire between the open end of the CPW center strip conductor and the opposite edge of the slotline functions as an air bridge and couples RF energy. The length L_s is approximately $\lambda_{g(\text{slotline})}/4$,

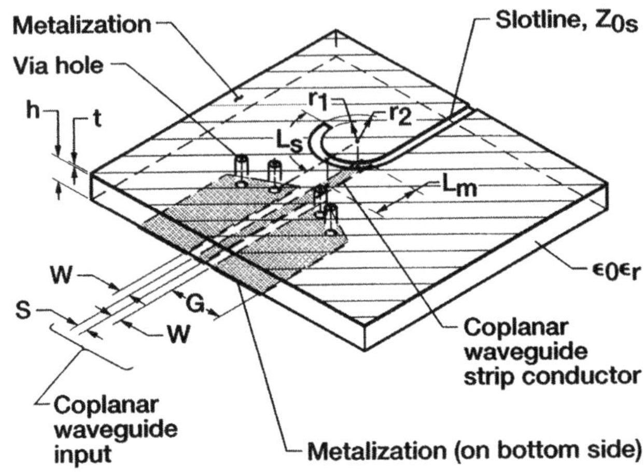

FIGURE 10.30 Electromagnetically coupled transition with ground planes connected together by metal filled vias. Top side circuit pattern: $S = 0.762$ mm, $W = 0.254$ mm, $G = 5.08$ mm, $L_m = 2.951$ mm. Bottom side circuit pattern: $L_s = 3.43$ mm, $r_1 = 2.171$ mm, $r_2 = 2.425$ mm. Substrate parameters: $h = 0.0508$ cm, $\varepsilon_r = 2.2$, $t = 17$ μm.

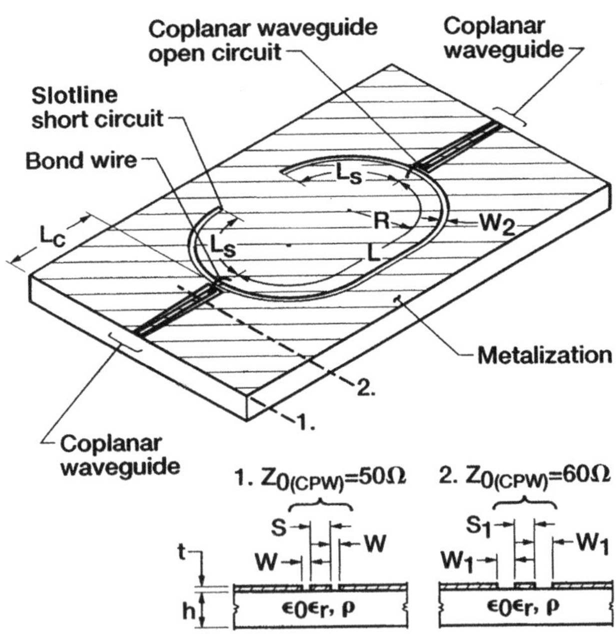

FIGURE 10.31 Air-bridge coupled coplanar waveguide-to-slotline transition. Circuit pattern: $S = S_1 = 0.1524$ mm, $W = 0.0838$ mm, $W_1 = 0.1524$ mm, $L = 19.77$ mm, $L_s = 6.83$ mm, $L_c = 5.08$ mm, $R = 4.35$ mm, $W_2 = 0.157$ mm. Substrate parameters: $h = 0.381$ cms, $\varepsilon_r = 11.7$, $\rho = 5000\text{–}10{,}000\ \Omega$ cm.

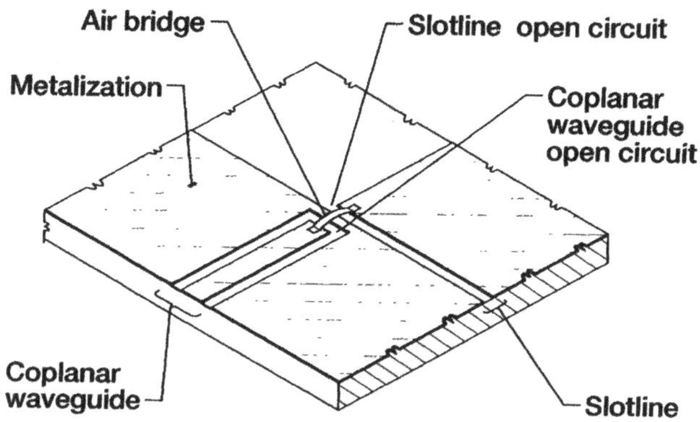

FIGURE 10.32 Air-bridge coupled coplanar waveguide-to-slotline transition.

where λ_g is the guide wavelength at the center frequency f_0. The thickness t of the metalization is three times the skin depth at f_0.

In [43] the characteristics of two back-to-back transitions fabricated on a high resistivity silicon wafer for operation at $f_0 = 6.0$ GHz are reported. The measured insertion loss and return loss per transition are 1.5 dB and less than -10.0 dB, respectively, over 30 percent bandwidth centered at 6.0 GHz. The above insertion loss includes, the insertion loss of a 5.08 mm long CPW, slotline stub of length L_s and half the length of the slotline in between which is about 9.9 mm.

Figure 10.32 shows an alternate design [45] for constructing a transition of the above type. In this design, the quarter wave short circuited stub of length L_s is eliminated and instead the slotline is terminated in abrupt open circuit. Further the CPW and the slotline are orthogonal to each other. In [45] the measured characteristics of an experimental transition are reported. The measured insertion loss and return loss for two back-to-back transitions are 1.6 dB and less than -12.5 dB, respectively, over the frequency band of 21.5 to 26.5 GHz. These losses include the insertion loss of 650 μm long slotline in between and two 1000 μm long CPWs at the input and output, respectively.

10.7 COPLANAR WAVEGUIDE-TO-COPLANAR STRIPLINE TRANSITION

In microwave circuits a transition or a balun is required to connect a balanced transmission line, such as, a coplanar stripline (CPS) to an unbalanced line, such as, a coplanar waveguide (CPW). In this section the design considerations for three types of uniplanar baluns are presented.

10.7.1 Coplanar Stripline-to-Coplanar Waveguide Balun

Figure 10.33 shows a CPS to a CPW balun [46]. In this balun at the balanced end currents of equal magnitude but opposite in direction flow along the two

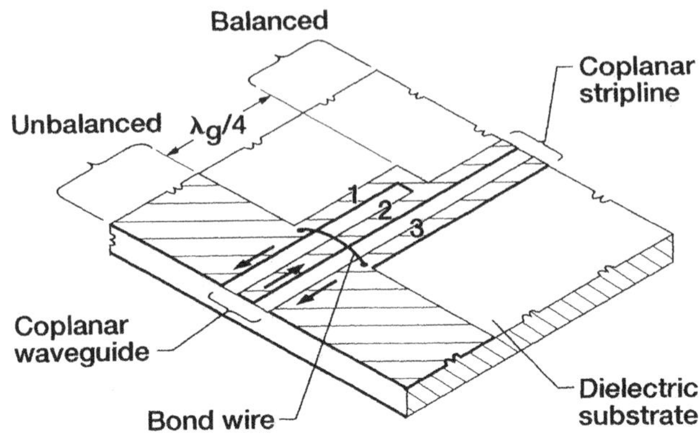

FIGURE 10.33 Coplanar stripline-to-coplanar waveguide transition with three strip transmission lines.

strip conductors. At the unbalanced end currents of equal magnitude but opposite in direction flow along the center strip conductor and the ground planes on either side. A short circuit placed between conductors 1 and 2 at the balanced end results in an open circuit quarter wavelength away at the unbalanced end, forcing all the currents to flow between conductors 2 and 3. Further conductors 1 and 3 are short circuited at the unbalanced end by a bond wire. This short circuit appears as an open circuit quarter wavelength away at the balanced end, and therefore conductor 1 is isolated from the balanced end.

This transition in addition to serving as a balun also provides impedance transformation. Since RF energy propagates between conductors 2 and 3, the characteristic impedance between these conductors determine the impedance transformation over the quarter wavelength section. In [47] the application of this transition to a CPW feed network for a dipole antenna array is demonstrated.

10.7.2 Coplanar Stripline-to-Coplanar Waveguide Balun with Slotline Radial Stub

Figure 10.34 shows a coplanar stripline (CPS)-to-coplanar waveguide (CPW) balun. In this balun one of the slots of the CPW is terminated in a broadband slotline radial open stub, while the other slot extends further to meet the CPS. In [37] the characteristics of an experimental balun are reported. The measured insertion loss and return loss of the balun are typically 1.0 dB and less than −13.0 dB, respectively, over the frequency range of 1.6 to 7.0 GHz. The balun has a bandwidth greater than two octaves.

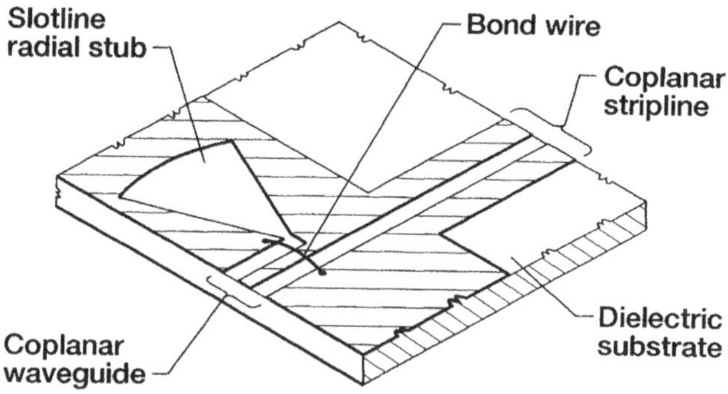

FIGURE 10.34 Coplanar stripline-to-coplanar waveguide transition with radial slotline stub.

10.7.3 Coplanar Stripline-to-Coplanar Waveguide Double-Y Balun

The double-Y junction in this balun [48] is formed by placing alternatively three coplanar striplines (CPS) and three coplanar waveguides with finite width ground planes (FW-CPW) as shown in Figure 10.35. The advantage of FW-CPW over CPW in this design is the greatly reduced parasitics. Because of small parasitics the FW-CPW terminations behave almost like ideal open and short circuits resulting in wider bandwidth. This balun is modeled as a six port network as explained in Section 10.6.3 and Eqs. (10.48) and (10.50) are also valid here. In [41] the characteristics of an experimental balun are

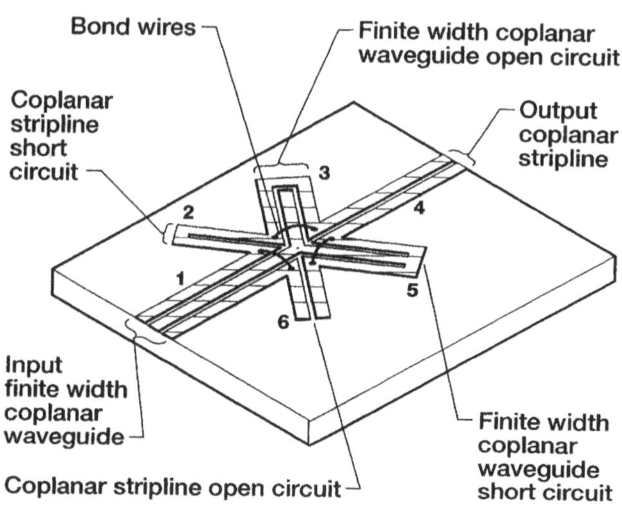

FIGURE 10.35 Coplanar stripline-to-coplanar waveguide double-Y balun or transition.

TABLE 10.8 Coplanar Waveguide with Finite-Width Ground Planes-to-Coplanar Stripline Double-Y Balun or Transition

Dimensions and Characteristics	Reference [41]
SUBSTRATE MATERIAL ε_r AND THICKNESS (mm)	ALUMINA 9.8, 0.635
Coplanar waveguide	
S (mm)	0.12
W (mm)	0.04
g (mm)	0.1
$Z_{0(CPW)}$ (Ω)	50.0
Stub length (mm)	0.996
Coplanar stripline	
W (mm)	0.2
S (mm)	0.02
$Z_{0(CPS)}$ (Ω)	50.0
Stub length (mm)	1.0
Measured characteristics	
Frequency range (GHz)	dc – 12.0
S_{11} (dB)	$\leqslant -13.0$
S_{21} (dB)	[a]2.0

[a]For two back-to-back transition.

reported. The geometry as well as typical measured characteristics are summarized in Table 10.8. These characteristics show that the balun has a bandwidth of several decades.

10.8 COPLANAR STRIPLINE-TO-MICROSTRIP TRANSITION

10.8.1 Coplanar Stripline-to-Microstrip Transition with an Electromagnetically Coupled Radial Stub

Coplanar stripline (CPS)-to-microstrip transitions [49] in a back-to-back arrangement are shown in Figure 10.36. In each transition one of the CPS strip conductor is terminated in a microstrip radial stub of radius R and angle ϕ, while the other strip conductor is extended forward to form the microstrip line. The CPS strip conductor width and separation are W and S, respectively. The width of the microstrip line is W_m. The resonance frequency of the radial stub

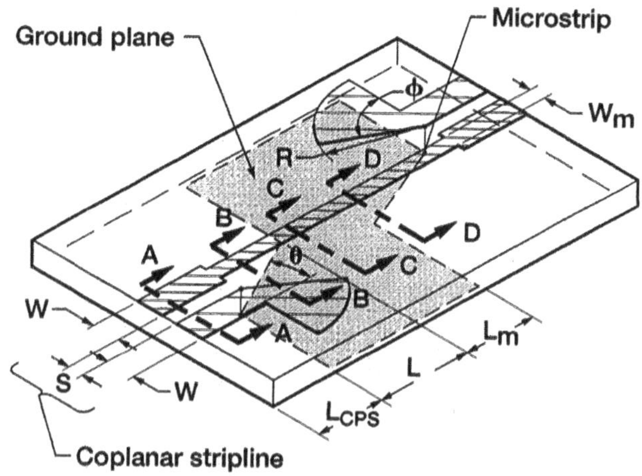

FIGURE 10.36 Two back-to-back coplanar stripline-to-microstrip transition. Top side circuit pattern: $W = 0.69$ mm, $S = 0.19$ mm, $R = 5.08$ mm, $\phi = 60°$, $W_m = 0.19$ mm, $L_{CPS} = 6.9$ mm. Bottom side circuit pattern: $L = 7.1$ mm, $L_m = 12.7$ mm, $\theta \approx 7°$. Substrate parameters: $h = 0.254$ mm, $\varepsilon_r = 10.5$.

depends on the radius R and angle ϕ. At resonance, the microstrip radial stub provides a virtual RF short circuit between the CPS strip conductor and the microstrip ground plane on the opposite side of the substrate. In the CPS the electric-field lines are across the strip conductors and parallel to the substrate. In the microstrip line, the electric-field lines are normal to the substrate. Hence, to gradually rotate the electric field lines by 90°, the microstrip ground plane edge is tapered at an angle θ. The electric-field lines at various cross sections along the transition are illustrated in Figure 10.37.

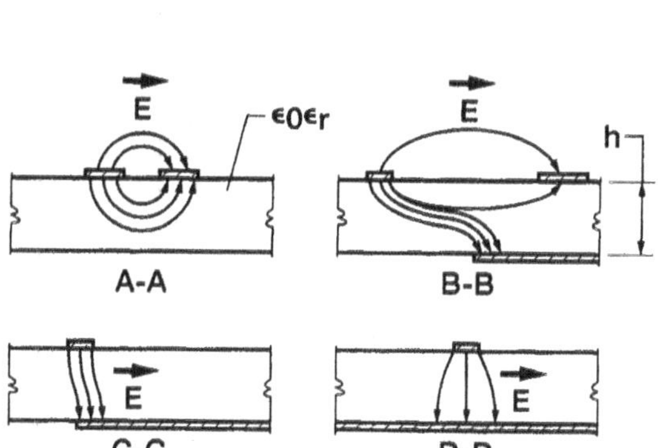

FIGURE 10.37 Electric-field lines at various cross-sections along the transition.

In [49] the characteristics of an experimental transition fabricated on a RT/Duroid 6010 substrate are reported. The measured insertion loss and return loss for two back-to-back transitions over the frequency range of 5.1 to 6.1 GHz is 2.4 dB (max) and less than −10.0 dB, respectively. The −10.0 dB return loss bandwidth of both transitions is about 18 percent centered at 5.55 GHz. The numerical simulation of the transition carried out using FDTD technique shows good agreement between measured and modeled results [49]. In [5] and in [50] a transition from an asymmetric CPS (ACPS)-to-microstrip with radial stubs and with metal vias, respectively, is demonstrated.

10.8.2 Uniplanar Coplanar Stripline-to-Microstrip Transition

A uniplanar coplanar stripline (CPS)-to-microstrip back-to-back transition [51] in a balanced arrangement is shown in Figure 10.38. In this transition, a microstrip line of characteristic impedance 50 Ω is coupled to two orthogonal microstrip lines of characteristic impedance 70 Ω through a quarter-wave stepped impedance-matching transformer. The characteristic impedance of 70 Ω is chosen for ease of fabrication. In an ideal transition, the mean path length of the folded loop "*a*-through-*b*" is equal to $0.5\lambda_{g(\text{microstrip})}$, where $\lambda_{g(\text{microstrip})}$ is the guide wavelength in the microstrip of characteristic impedance 70 Ω at the center frequency f_0. In a practical transition for wideband

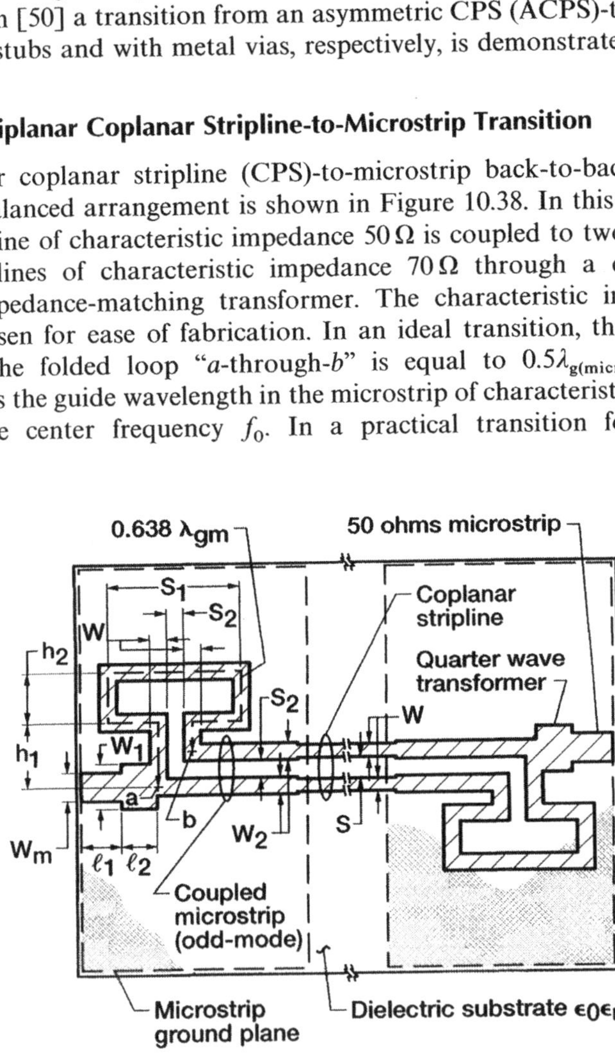

FIGURE 10.38 Two back-to-back balanced uniplanar coplanar stripline-to-microstrip transition: $W_m = 0.23$ mm, $W_1 = 0.33$ mm, $W_2 = W = S_2 = S = 0.1$ mm, $l_1 = 4.9$ mm, $l_2 = 2.8$ mm, $S_1 = 2.1$ mm, $h_1 = 0.54$ mm, $h_2 = 1.34$ mm. Substrate thickness: $h = 0.254$ mm, $\varepsilon_r = 10.5$.

operation, the right angle bend parasitic elements have to be compensated. A simple solution to this problem is to increase the mean path length of the microstrip. Hence the mean path length is chosen as $0.638\lambda_{g(\text{microstrip})}$. This design ensures that the phase of the signals at the input locations a and b to the coupled microstrip line are 180° out of phase. The coupled microstrip lines are excited in the odd mode with the electric-field lines predominantly across the gap S_2. The dimensions of the gap S_2 is chosen such that the odd-mode characteristic impedance is 50 Ω. The horizontal orientation of the electric-field lines in the coupled microstrip makes the transition to a CPS simple. The width W and separation S of the CPS in the experimental transition is the same as the coupled microstrip width and separation.

In [51] the measured characteristics of the transition shown in Figure 10.38 are reported. The transition is designed for operation at a center frequency $f_0 = 10.0$ GHz. However, the measurements show that f_0 has shifted to 9.6 GHz. The measured insertion loss and return loss is about 2.0 dB and less than −10.0 dB, respectively, over 42 percent bandwidth centered at 9.6 GHz. The numerical simulation of the transition carried out using FDTD technique shows that the measured and modeled results are in good agreement [51].

10.8.3 Coplanar Stripline-to-Microstrip Transition

Figure 10.39 shows a coplanar stripline (CPS)-to-microstrip transition [52] in a stacked configuration. In this transition the microstrip ground plane is attached using electrically conducting epoxy to one of the strip conductors of the CPS. The connection between the remaining strip conductors is by a gold ribbon. In [52] the characteristics of an experimental transition are reported. The geometry as well as the measured characteristics of the transition are summarized in Table 10.9. The transition has a wide bandwidth.

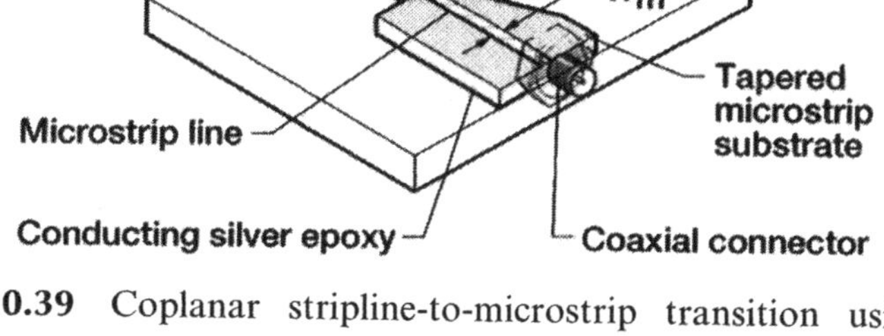

FIGURE 10.39 Coplanar stripline-to-microstrip transition using ribbon bond.

10.8.4 Micro-coplanar Stripline-to-Microstrip Transition

Figure 10.40 shows a micro-coplanar stripline (MCPS)-to-microstrip transition [53] in a back-to-back arrangement. In this transition the width of the microstrip ground plane is abruptly reduced to form a symmetric parallel-plate line. The parallel-plate conductors are next flared out to form the MCPS. The cross section at three locations along the transition are shown in the inset in Figure 10.40. Section $A-A$, $B-B$, and $C-C$ show the microstrip, the parallel-plate, and the MCPS regions, respectively. The relevant parameters of the transition are as follows: $h_1 = 254\ \mu\text{m}$, $\varepsilon_{r1} = 10.5$, $\varepsilon_{r2} \approx 1.0$, $W = 889\ \mu\text{m}$, $W_m = 254\ \mu\text{m}$, and the characteristic impedance $Z_{0(\text{microstrip})} = 50\ \Omega$. In [53] the characteristics of an experimental transition are reported. The measured insertion loss and return loss are 1.0 dB and less than -16.0 dB, respectively, for two back-to-back transitions over the frequency range of 0.045 to 14.0 GHz. These losses also include the loss of the test fixture. The measurements show that the transition has a broad bandwidth.

TABLE 10.9 Coplanar Stripline-to-Microstrip Transition

Dimensions and Characteristics	Reference [52]
Coplanar stripline	
SUBSTRATE THICKNESS (μm)	3000
ε_r	10.0
W (μm)	490
S (μm)	58
$Z_{0(\text{CPS})}$ (Ω)	50
Microstrip line	
SUBSTRATE THICKNESS (μm)	127
ε_r	2.0
W_m (μm)	385
$Z_{0(\text{m})}$ (Ω)	50
Gold ribbon	
Width (μm)	350
Length (μm)	250
Measured characteristics	
Frequency range (GHz)	dc $-$ 18.0
S_{11} (dB)	[a] $\leqslant -13.0$

[a] For two back-to-back transitions.

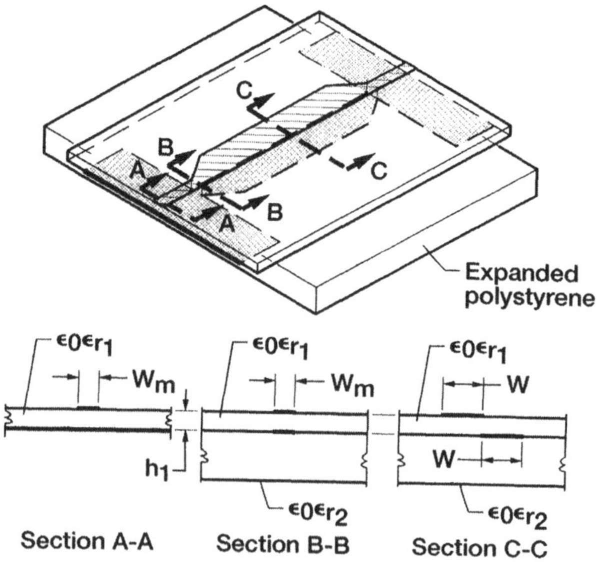

FIGURE 10.40 Two back-to-back micro-coplanar stripline-to-microstrip transition.

10.9 COPLANAR STRIPLINE-TO-SLOTLINE TRANSITION

An electromagnetically coupled coplanar stripline (CPS)-to-slotline transitions [54] in a back-to-back arrangement is shown in Figure 10.41. The CPS and the slotline are on opposite sides of a dielectric substrate. To elucidate the design of the transition, the cross sections at four locations along the length are shown in Figure 10.42. The input port at section A–A is a CPS with strip

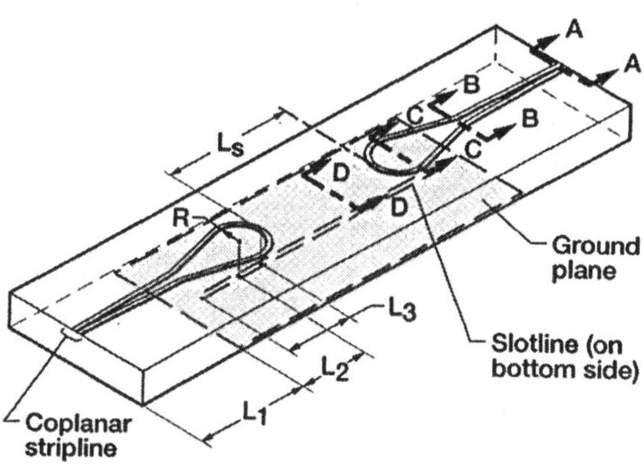

FIGURE 10.41 Two back-to-back electromagnetically coupled coplanar stripline-to-slotline transition.

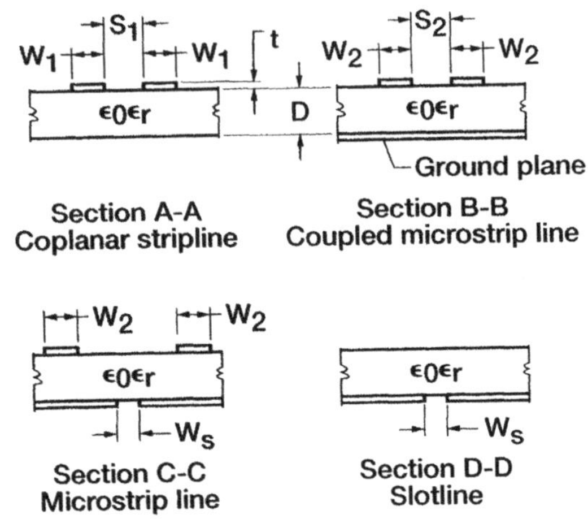

FIGURE 10.42 Cross-sectional view of the transition.

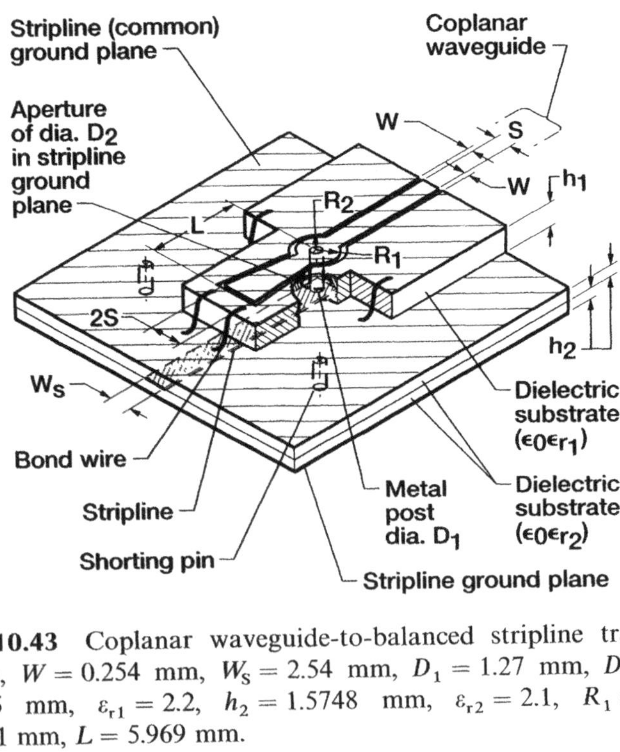

FIGURE 10.43 Coplanar waveguide-to-balanced stripline transition: $S = 1.397$ mm, $W = 0.254$ mm, $W_S = 2.54$ mm, $D_1 = 1.27$ mm, $D_2 = 2.54$ mm, $h_1 = 3.175$ mm, $\varepsilon_{r1} = 2.2$, $h_2 = 1.5748$ mm, $\varepsilon_{r2} = 2.1$, $R_1 = 1.397$ mm, $R_2 = 1.651$ mm, $L = 5.969$ mm.

TABLE 10.10 Coplanar Stripline-to-Slotline Transition

Dimensions and Characteristics	Reference [54]
SUBSTRATE	HIGH-RESISTIVITY SILICON ($\rho \approx 2500$–$3300\,\Omega \cdot \mathrm{cm}$)
Thickness D (mm)	0.295
ε_r	11.7
t (μm)	$\approx$2.5
Coplanar stripline	
W_1 (mm)	0.28
S_1 (mm)	0.254
W_2 (mm)	0.24
S_2 (mm)	0.254
L_1 (mm)	6.35
L_2 (mm)	3.94
R (mm)	1.27
Slot line	
W_s (mm)	0.1143
L_s (mm)	7.13
L_3 (mm)	3.8
Measured characteristics per transition	
f_0 (GHz)	10.0
S_{11} (dB)	$\leqslant -10.0$
S_{21} (dB)	1.45
Bandwidth (%) at f_0	30.0

conductor width W_1 and separation S_1. At section B–B the CPS overlaps the ground plane of the slotline and transforms to coupled microstrip lines of width W_2 with odd-mode excitation. The separation S_2 between the coupled microstrip lines gradually diverges beyond B–B to C–C, but the strip width remains fixed at W_2. At section C–C the coupled microstrip lines are totally decoupled and behave as two independent lines. Beyond this plane the CPS is terminated in a short circuit arc of radius R. Coupling between the CPS and the slotline takes place at the location where the short circuit arc crosses the slotline at right angles. The slotline has a slot of width W_s. Further the slotline is terminated in a short circuit at a distance L_3, which is approximately $\lambda_{g(\mathrm{slotline})}/4$ where $\lambda_{g(\mathrm{slotline})}$ is the guide wavelength in the slotline at the design frequency f_0. In [54] the characteristics of an experimental transition are reported. Table 10.10 summarizes the transition performance.

10.10 COPLANAR WAVEGUIDE-TO-BALANCED STRIPLINE TRANSITION

A coplanar waveguide (CPW)-to-balanced stripline transition [55] is shown in Figure 10.43. In this transition the center strip conduction of the CPW and the stripline are electrically connected by a metal port through an aperture in the ground plane. The diameters of the aperture and the metal post are experimentally optimized to obtain the best insertion loss characteristics. The ground planes of the stripline are connected together by shorting pins to ensure both surfaces are at equipotential. Electrical continuity between the CPW and the stripline ground planes is maintained by bond wires. In [55] the characteristics of an experimental transition are reported. The measured insertion loss and return loss of the transition are less than 0.5 dB and −10.0 dB, respectively, over the frequency range of 0.045 to 3.0 GHz. This transition has applications in the feed network of patch antennas.

REFERENCES

[1] M. Houdart and C. Aury, "Various Excitation of Coplanar Waveguide," *1979 IEEE MTT-S Int. Microwave Symp. Dig.*, Orlando, FL, pp. 116–118, April 30–May 2, 1979.

[2] J. J. Burke and R. W. Jackson, "Surface-to-Surface Transition via Electromagnetic Coupling of Microstrip and Coplanar Waveguide," *IEEE Trans. Microwave Theory Tech.*, Vol. 37, No. 3, pp. 519–524, March 1989.

[3] G. P. Gauthier, L. P. B. Katehi, and G. M. Rebeiz, "W-Band Finite Ground Coplanar Waveguide (FGCPW) to Microstrip Line Transition," *1998 IEEE MTT-S Int. Microwave Symp. Dig.*, Baltimore, MD, Vol. 1, pp. 107–109, June 7–12, 1998.

[4] D. Pavlidis and H. L. Hartnagel, "The Design and Performance of Three-Line Microstrip Couplers," *IEEE Trans. Microwave Theory Tech.*, Vol. 24, No. 10, pp. 631–640, Oct. 1976.

[5] D. Jaisson, "A Single-Balanced Mixer with a Coplanar Balun," *Microwave J.*, Vol. 35, No. 7, pp. 87–96, July 1992. (Corrections: p. 201, Sept. 1992.)

[6] N. I. Dib, R. N. Simons, and L. P. B. Katehi, "New Uniplanar Transitions for Circuit and Antenna Applications," *IEEE Trans. Microwave Theory Tech.*, Vol. 43, No. 12, pp. 2868–2873, Dec. 1995.

[7] R. N. Simons and R. Q. Lee, "Coplanar-Waveguide/Microstrip Probe Coupler and Applications to Antennas," *Electron. Lett.*, Vol. 26, No. 24, pp. 1998–1999, Nov. 1990.

[8] B. Golja, H. B. Sequeira, S. Duncan, G. Mendenilla, and N. E. Byer, "A Coplanar-to-Microstrip Transition for W-Band Circuit Fabrication with 100-μm-thick GaAs Wafers," *IEEE Microwave Guided Wave Lett.*, Vol. 3, No. 2, pp. 29–31, Feb. 1993.

[9] L. Hyden, S. Hagelin, P. Starski, and K. Yhland, "Analysis and Design of a Vertical CPW Transition Between Microstrip Planes," *1997 IEEE MTT-S Int. Microwave Symp. Dig.*, Denver, CO., Vol. 2, pp. 727–730, June 8–13, 1997.

[10] D. F. Williams and T. H. Miers, "A Coplanar Probe to Microstrip Transition," *IEEE Trans. Microwave Theory Tech.*, Vol. 36, No. 7, pp. 1219–1223, July 1988.

[11] H. A. Atwater, "Microstrip Reactive Circuit Elements," *IEEE Trans. Microwave Theory Tech.*, Vol. 31, No. 6, pp. 488–491, June 1983.

[12] D. Harvey, "A Lumped Coplanar to Microstrip Transition Model for De-Embedding *S*-Parameters Measured on GaAs Wafers," *29th Automatic RF Tech. Group Conf. Proc.*, Las Vegas, NV, pp. 204–217, June 1987.

[13] T. M. Weller, L. P. B. Katehi, and G. M. Rebeiz, "High Performance Microshield Line Components," *IEEE Trans. Microwave Theory Tech.*, Vol. 43, No. 3, pp. 534–543, March 1995.

[14] P. Salzenstein, O. Dupuis, M. Helal, E. Lheurette, O. Vanbesien, P. Mounaix, and D. Lippens, "Coplanar Waveguides on Dielectric Membranes Micromachined on a GaAs Substrate," *Electron. Lett.*, Vol. 32, No. 9, pp. 821–822, April 1996.

[15] R. N. Simons, R. Q. Lee, K. A. Shalkhauser, J. Owens, J. Demarco, J. Leen, and D. Sturzebecher, "Finite Width Coplanar Waveguide Patch Antenna with Vertical Fed through Interconnect," *1996 IEEE AP-S Int. Symp. and URSI Radio Science Meeting Dig.*, Baltimore, MD, Vol. 2, pp. 1338–1341, July 21–26, 1996.

[16] R. N. Simons, T. D. Perl, and R. Q. Lee, "New Coplanar Waveguide Feed Network for 2 × 2 Linearly Tapered Slot Antenna Subarray," *Microwave Optical Tech. Lett.*, Vol. 5, No. 9, pp. 420–423, Aug. 1992.

[17] R. W. Jackson and D. W. Matolak, "Surface-to-Surface Transition via Electromagnetic Coupling of Coplanar Waveguides," *IEEE Trans. Microwave Theory Tech.*, Vol. 35, No. 11, pp. 1027–1032, Nov. 1987.

[18] G. I. Zysman and A. K. Johnson, "Coupled Transmission Line Networks in an Inhomogeneous Dielectric Medium," *IEEE Trans. Microwave Theory Tech.*, Vol. 17, No. 10, pp. 753–759, Oct. 1969.

[19] R. Q. Lee and R. N. Simons, "Electromagnetically Coupled Feed Network for an Array Module of Four Microstrip Elements," *1988 IEEE Int. Antennas Propag. Symp. Dig.*, Syracuse, New York, Vol. 3, pp. 1018–1021, June 1988.

[20] G. E. Ponchak and R. N. Simons, "A New Rectangular Waveguide to Coplanar Waveguide Transition," *1990 IEEE MTT-S Int. Microwave Symp. Dig.*, Dallas, TX, Vol. 1, pp. 491–492, May 8–10, 1990.

[21] J. R. Kessler and P. D. Coleman, "Impedance and Complex Propagation Constant Measurements at W-Band on Planar Circuits," *Int. J. Infrared Millimeter Waves*, Vol. 13, No. 4, pp. 397–424, April 1992.

[22] E. M. Godshalk, "A V-Band Wafer Probe Using Ridge-Trough Waveguide," *IEEE Trans. Microwave Theory Tech.*, Vol. 39, No. 12, pp. 2218–2228, Dec. 1991.

[23] T.-S. Chen, "Calculation of the Parameters of Ridge Waveguides," *IRE Trans. Microwave Theory Tech.*, Vol. 5, No. 1, pp. 12–17, Jan. 1957.

[24] G. L. Matthaei, L. Young, and E. M. T. Jones, *Microwave Filters, Impedance-Matching Networks, and Coupling Structures*, Norwood, MA: Artech House, 1980, ch. 6.

[25] E. M. Godshalk and J. Pence, "Low-Cost Wafer Probe Scales 110-GHz Summit," *Microwaves RF*, Vol. 32, No. 3, pp. 162–167, March 1993.

[26] G. C. Dalman, "A Simple mm-Wave Transition from Waveguide to Coplanar Waveguide," *Microwave J.*, Vol. 35, No. 10, pp. 109–112, Oct. 1992.

[27] R. N. Simons and S. R. Taub, "New Coplanar Waveguide to Rectangular Waveguide End Launcher," *Electron. Lett.*, Vol. 28, No. 12, pp. 1138–1139, June 1992.

[28] R. N. Simons, "New Channelized Coplanar Waveguide to Rectangular Waveguide Post and Slot Couplers," *Electron. Lett.*, Vol. 27, No. 10, pp. 856–857, May 1991.

[29] G. S. Dow, T. N. Ton, and K. Nakano, "Q-Band Coplanar Waveguide Amplifiers," *1989 IEEE MTT-S Int. Microwave Symp. Dig.*, Long Beach, CA, Vol. 2, pp. 809–812, June 13–15, 1989.

[30] J. Machac and W. Menzel, "On the Design of Waveguide-to-Microstrip and Waveguide-to-Coplanar Line Transitions," *23rd European Microwave Conf. Proc.*, Vol. 1, pp. 615–616, Madrid, Spain, Sept. 6–9, 1993.

[31] N. Marchand, "Transmission-Line Conversion Transformers," *Electronics*, Vol. 17, pp. 142–145, Dec. 1944.

[32] G. Oltman, "The Compensated Balun," *IEEE Trans. Microwave Theory Tech.*, Vol. 14, No. 3, pp. 112–119, March 1966.

[33] V. F. Hanna and L. Ramboz, "Broadband Planar Coplanar Waveguide-Slotline Transition," *12th European Microwave Conf. Proc.*, pp. 628–631, Helsinki, Finland, Sept. 13–17, 1982.

[34] D. Cahana, "A New, Single Plane, Double-Balanced Mixer," *Appl. Microwave*, Vol. 1, No. 2, pp. 78–83, Aug–Sept. 1989.

[35] W. Grammer and K. S. Yngvesson, "Coplanar Waveguide Transitions to Slotline: Design and Microprobe Characterization," *IEEE Trans. Microwave Theory Tech.*, Vol. 41, No. 9, pp. 1653–1658, Sept. 1993.

[36] C.-H. Ho, L. Fan, and K. Chang, "New Uniplanar Coplanar Waveguide Hybrid-Ring Couplers and Magic-T's," *IEEE Trans. Microwave Theory Tech.*, Vol. 42, No. 12, pp. 2440–2448, Dec. 1994.

[37] C.-H. Ho, L. Fan, and K. Chang, "Broad-Band Uniplanar Hybrid-Ring and Branch-line Couplers," *IEEE Trans. Microwave Theory Tech.*, Vol. 41, No. 12, pp. 2116–2125, Dec. 1993.

[38] T. Q. Ho and S. M. Hart, "A Broad-Band Coplanar Waveguide to Slotline Transition," *IEEE Microwave Guided Wave Lett.*, Vol. 2, No. 10, pp. 415–416, Oct. 1992.

[39] D. F. Filipovic, R. F. Bradley, and G. M. Rebeiz, "A Planar Broadband MIC Balanced Varactor Doubler Using a Novel Grounded-CPW to Slotline Transition," *1994 IEEE MTT-S Int. Microwave Symp.*, San Diego, CA, Vol. 3, pp. 1633–1636, 1994.

[40] K. Elgaid, D. Edgar, S. Ferguson, S. Broadfoot, M. Taylor, and S. Beaumont, "Compact Low Loss Coplanar Waveguide To Slotline Transition for MMMIC Applications," *Electron. Lett.*, Vol. 32, No. 18, pp. 1677–1678, Aug. 1996.

[41] B. Jokanovic and V. Trifunovic, "Double-Y Baluns for MMICs and Wireless Applications," *Microwave J.*, Vol. 41, No. 1, pp. 70–92, Jan. 1998.

[42] R. N. Simons, R. Q. Lee, and T. D. Perl, "New Techniques for Exciting Linearly Tapered Slot Antennas With Coplanar Waveguide," *Electron. Lett.*, Vol. 28, No. 7, pp. 620–621, March 1992.

[43] R. N. Simons, S. R. Taub, and P. G. Young, "Novel Coplanar Waveguide to Slotline Transition on High Resistivity Silicon," *Electron. Lett.*, Vol. 28, No. 24, pp. 2209–2210, Nov. 1992.

[44] R. N. Simons, R. Q. Lee, T. D. Perl, and J. Silvestro, "Effect of Dielectric Overlay on a Linearly Tapered Slot Antenna Excited by a Coplanar Waveguide," *Microwave Optical Tech. Lett.*, Vol. 6, No. 4, pp. 225–228, March 1993.

[45] T. Hirota, Y. Tarusawa, and H. Ogawa, "Uniplanar MMIC Hybrids—A Proposed New MMIC Structure," *IEEE Trans. Microwave Theory Tech.*, Vol. 35, No. 6, pp. 576–581, June 1987.

[46] R. E. DeBrecht, "Coplanar Balun Circuits for GaAs FET High-Power Push-Pull Amplifiers," *1973 IEEE G-MTT Int. Microwave Symp. Dig.*, Boulder, CO, pp. 309–311, June 4–6, 1973.

[47] R. N. Simons, G. E. Ponchak, R. Q. Lee, and N. S. Fernandez, "Coplanar Waveguide Fed Phased Array Antenna," *1990 IEEE Antennas Propag. Symp. Dig.*, Vol. 4, Dallas, TX, pp. 1778–1781, May 1990.

[48] V. Trifunovic and B. Jokanovic, "Four Decade Bandwidth Uniplanar Balun," *Electron. Lett.*, Vol. 28, No. 6, pp. 534–535, March 1992.

[49] R. N. Simons, N. I. Dib, and L. P. B. Katehi, "Coplanar Stripline to Microstrip Transition," *Electron. Lett.*, Vol. 31, No. 20, pp. 1725–1726, Sept. 1995.

[50] D. Jaisson, "Transition from a Microstrip Line to an Asymmetrical Coplanar Waveguide," *Microwave J.*, Vol. 37, No. 6, pp. 112–115, June 1994.

[51] R. N. Simons, N. I. Dib, R. Q. Lee, and L. P. B. Katehi, "Integrated Uniplanar Transition for Linearly Tapered Slot Antenna," *IEEE Trans. Antennas Propag.*, Vol. 43, No. 9, pp. 998–1002, Sept. 1995.

[52] H. Y. Lee and T. Itoh, "Wideband and Low Return Loss Coplanar Strip Feed Using Intermediate Microstrip," *Electron. Lett.*, Vol. 24, No. 9, pp. 1207–1208, Sept. 1988.

[53] K. Goverdhanam, R. N. Simons, and L. P. B. Katehi, "Micro-Coplanar Stripline—New Transmission Media for Microwave Applications," *1998 IEEE MTT-S Int. Microwave Symp. Dig.*, Baltimore, MD, Vol. 2, pp. 1035–1038, June 7–12, 1998.

[54] R. N. Simons, "Novel Coplanar Stripline to Slotline Transition on High Resistivity Silicon," *Electron. Lett.*, Vol. 30, No. 8, pp. 654–655, April 1994.

[55] R. N. Simons, R. Q. Lee, and G. R. Lindamood, "New Coplanar Waveguide/Stripline Feed Network for Seven Patch Hexagonal CP Subarray," *Electron. Lett.*, Vol. 27, No. 6, pp. 533–535, March 1991.

[56] V. K. Tripathi, "On the Analysis of Symmetrical Three-Line Microstrip Circuits," *IEEE Trans. Microwave Theory Tech.*, Vol. 25, No. 9, pp. 726–729, Sept. 1977.

CHAPTER 11

Directional Couplers, Hybrids, and Magic-Ts

11.1 INTRODUCTION

The design and construction of coupled line directional coupler is presented in Section 11.2 of this chapter. These couplers are realized using edge coupled coplanar waveguides (CPWs) or broadside coupled CPWs. The conventional edge coupled CPW coupler is suited for weak coupling, typically 10 or 20 dB. The broadside coupled CPW coupler is suited for tight coupling, typically, 3 dB.

The design and construction of CPW quadrature (90°) hybrid or branch-line directional coupler is presented in Section 11.3. Both the standard as well as the reduced size branch-line coupler design are discussed. The advantage of the reduced size design is that it facilitates lowering the cost of monolithic integrated circuit (MMIC) based subsystems. In addition applications involving multistage balanced MMIC amplifiers require impedance transforming branch-line couplers. The design of these couplers are also discussed.

The design and construction of a CPW 180° ring hybrid or rat race is presented in Section 11.4. The 180° ring hybrid has application in balanced mixers. The design and construction of CPW passive and active magic-T circuits are described in Sections 11.5 and 11.6, respectively. The magic-T circuits are commonly used in monopulse receivers.

11.2 COUPLED-LINE DIRECTIONAL COUPLERS

When two CPWs are brought in close proximity their electromagnetic fields interact and power is coupled from one CPW to the other. Figure 11.1 shows three examples of coupled CPW structures, the edge coupled CPW, the edge coupled grounded CPW, and the broadside coupled CPW. The propagation

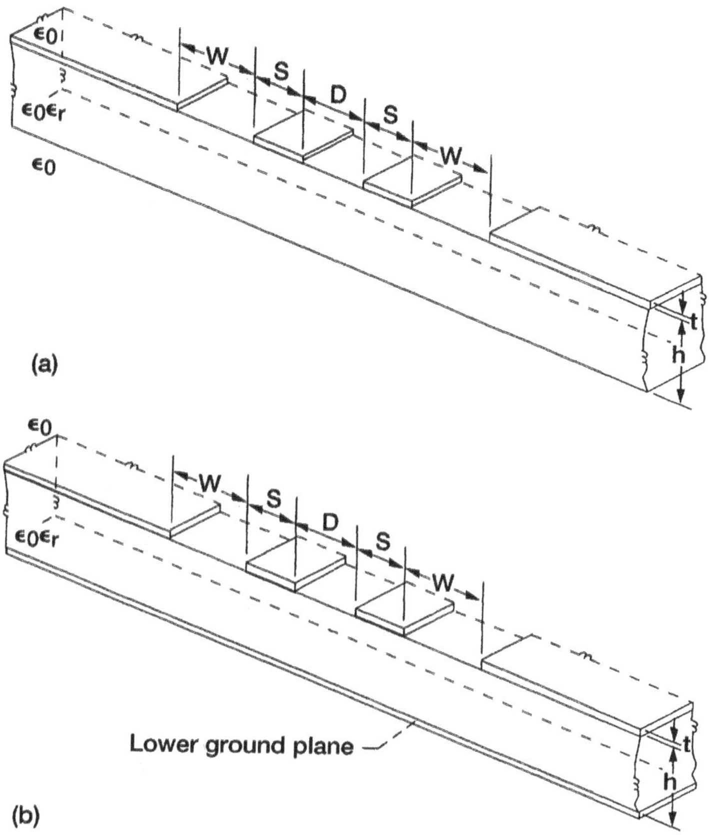

FIGURE 11.1 Coupled-line configurations: (*a*) Edge coupled CPW; (*b*) edge coupled grounded CPW.

along these CPW structures is described by two normal modes, namely the even and the odd modes. In Chapter 7 the normal modes of coupled CPW were explained in detail, and whereever possible in this book analytical expressions were presented to compute the even-mode and odd-mode effective dielectric constants and characteristic impedances. These characteristics are essential for designing directional couplers.

In weakly coupled lines, the distance of separation between the two lines is large, and hence the wave propagates with a velocity approximately the same as that on a single line. In this situation the even-mode and odd-mode phase velocities are assumed to be equal [1] and expressed as

$$v_{\text{ph(even)}} = v_{\text{ph(odd)}} = v_{\text{ph}} = \frac{c}{\sqrt{\varepsilon_{\text{eff}}}}, \tag{11.1}$$

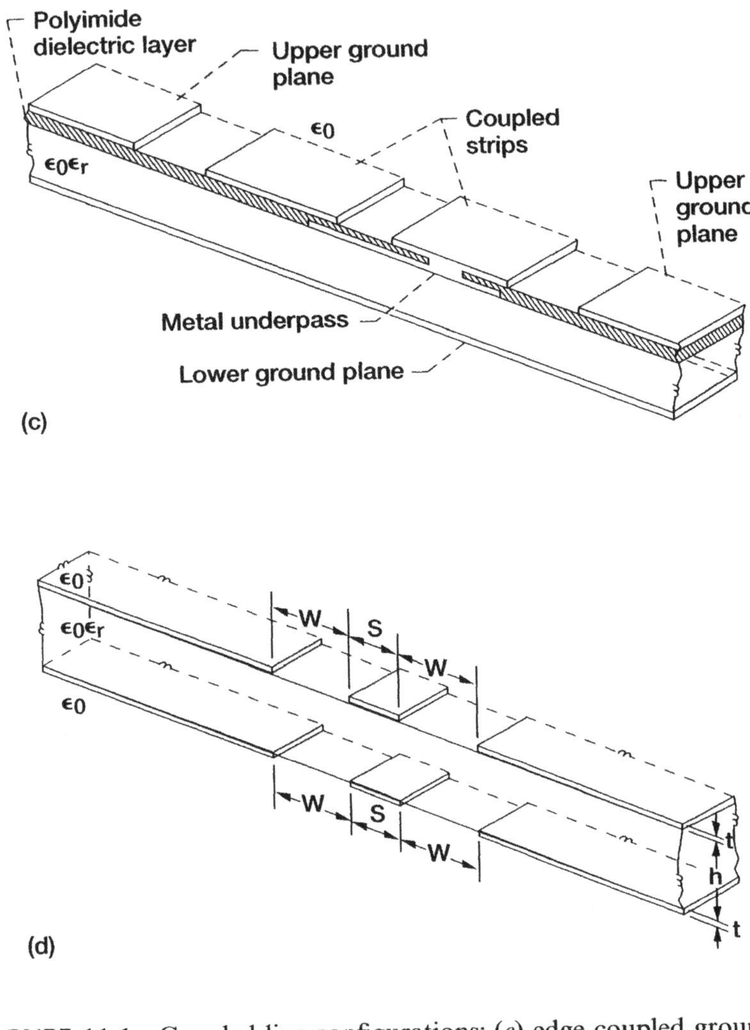

FIGURE 11.1 Coupled-line configurations: (*c*) edge coupled grounded CPW with metal underpass; (*d*) broadside coupled CPW.

and

$$\lambda_g = \frac{v_{ph}}{f}, \tag{11.2}$$

where

c = velocity of light in free space

ε_{eff} = effective dielectric constant

λ_g = guide wavelength

f = frequency.

In addition, for maximum coupling the length L of the coupled section is $\lambda_g/4$ [2]. The following two relations [2] also hold good:

$$Z_0^2 = Z_{0,e} Z_{0,o} \tag{11.3}$$

and

$$K = \frac{Z_{0,e} - Z_{0,o}}{Z_{0,e} + Z_{0,o}}, \tag{11.4}$$

where

Z_0 = characteristic impedance of the feed line

$Z_{0,e}$ and $Z_{0,o}$ = even-mode and odd-mode characteristic impedance

K = voltage coupling coefficient.

Last, it is worth mentioning that under the preceding conditions there will be no reflection at the input port of the coupler and also the coupler will have infinite directivity [2].

In the case of tightly coupled lines, the even-mode and odd-mode phase velocities are unequal [3], and so the length L of the coupled section is determined from the expression

$$L = \frac{\pi}{\beta_e + \beta_o}, \tag{11.5}$$

where β_e and β_o are the even-mode and odd-mode propagation constants. For this case the coupler will not have infinite directivity.

11.2.1 Edge Coupled CPW Directional Couplers

A 10 dB edge coupled CPW directional coupler for operation at 3.0 GHz is demonstrated in [1]. The schematic of the coupler is shown in Figure 11.2.

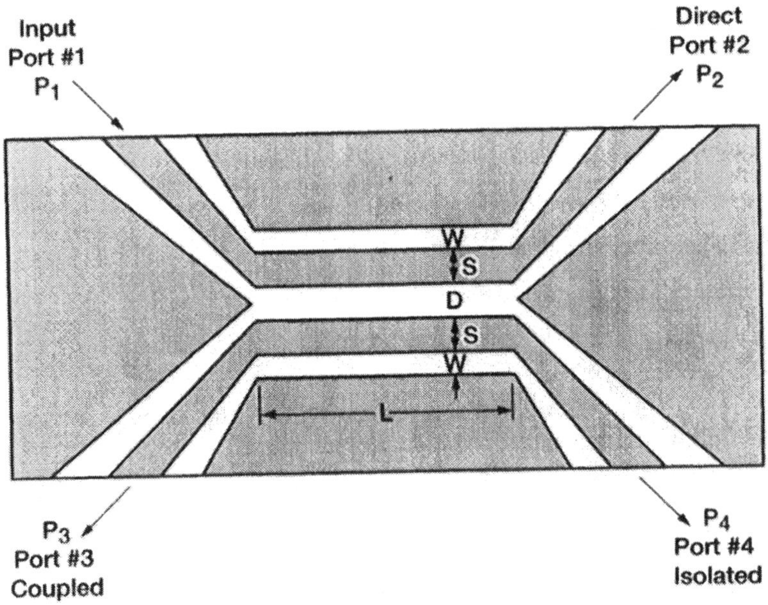

FIGURE 11.2 Edge coupled CPW directional coupler. The characteristic impedance of the CPW at the four ports in 50 Ω. The dielectric substrate thickness h is 0.075 in. and relative permittivity $\varepsilon_r = 16.0$.

Briefly, the design of the coupler is as follows: First, from the coupling requirement of 10 dB the even-mode and odd-mode characteristic impedances are determined from Eqs. (11.3) and (11.4). These impedances turn out to be $Z_{0,e} = 69.5\,\Omega$ and $Z_{0,o} = 36.0\,\Omega$ [1]. The graph of the computed even-mode and odd-mode characteristic impedances as a function of the geometry suggest the following choice: the strip width $S = 0.014$ in., the slot width $W = 0.009$ in., and the distance of separation $D = 0.013$ in. [1]. The computed v_{ph} for this weakly coupled lines is about 0.34 times c. Hence λ_g is 1.3385 in., and the length L of the coupled section is 0.334 in. [1].

The measured coupling (P_3/P_1) is on the order of 10.3 ± 0.7 dB over the frequency range of 1.7 to 2.5 GHz [1]. It is observed that the center frequency has shifter from 3.0 to 2.1 GHz due to discontinuity capacitance that cause a significant increase in the effective coupler length L. The measured return loss at any port and directivity (P_4/P_3) are -28.0 and 17.0 dB, respectively, over this frequency range. The departure from zero reflection and infinite directivity is because of the difference in the even-mode and odd-mode phase velocities.

11.2.2 Edge Coupled Grounded CPW Directional Couplers

In this section the design and performance of directional couplers with weak and tight coupling are examined. An edge coupled grounded CPW directional coupler for a coupling of about 8.0 dB at a center frequency of 3.1 GHz is

TABLE 11.1 Edge Coupled Grounded CPW Directional Coupler

Circuit parameters and measured characteristics	
Substrate (alumina), ε_r	9.9
Thickness, mm	0.635
Metalization (gold) thickness, t, μm	4.0
Coupled section	
GCPW center strip width, S, mm	0.76
GCPW slot width, W, mm	0.345
GCPW distance of separation, D, mm	0.05
Feed line	
GCPW center strip width, mm	0.52
GCPW slot width, mm	0.22
Measured characteristics	
Center frequency, GHz	3.1
Bandwidth, GHz	2.8–3.5
Coupling, dB	8.25–8.65
Directivity, dB	$>$19.0

demonstrated in [4]. The design makes use of Eqs. (11.1) to (11.4) to address the requirements. In Table 11.1 the coupler geometrical parameters as well as the measured performance are summarized. It is observed that the measured coupling of 8.45 ± 0.2 dB is in good agreement with the design value of 8.2 dB.

To achieve tight coupling on the order of 3.0 dB, the separation D between the strip conductors has to be very small. A small gap concentrates the electric fields and increases the conductor loss. Hence a better approach is to use a metal underpass as shown in the cross-sectional view in Figure 11.1(*c*). A 3.0 dB coupler with an underpass designed for operation at 24.0 GHz is reported in [5]. The geometry of the coupler is rather complicated, and hence simple design equations are not available. Therefore the coupler is designed with the help of computer simulation tools, such as em [6]. To facilitate characterization using on-wafer RF probing techniques the four ports are terminated in 50 Ω grounded CPW. The measured performance of the coupler shows that the coupling is about 4.0 ± 1.0 dB, the isolation greater than 15.0 dB and the input return loss less than -12.0 dB.

11.2.3 Broadside Coupled CPW Directional Coupler

A schematic of a broadside coupled CPW directional coupler is shown in Figure 11.3. The coupler is designed for a special application which requires a

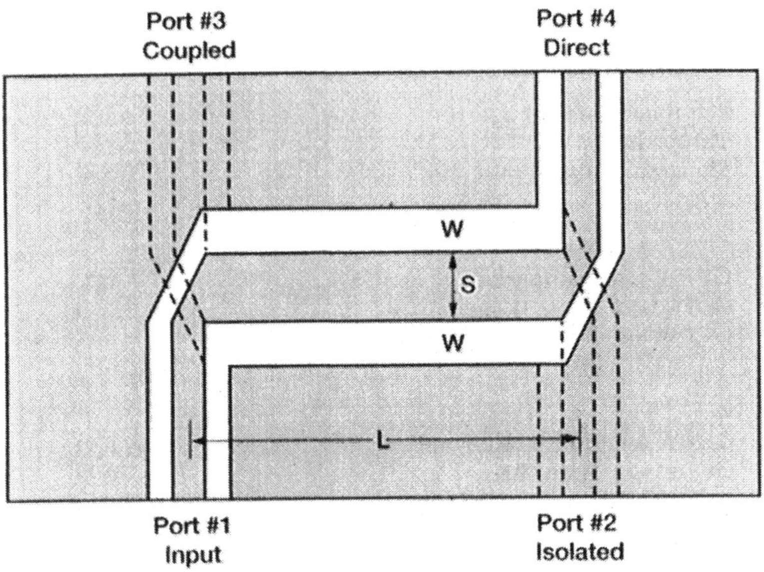

FIGURE 11.3 Broadside coupled CPW directional coupler. The characteristic impedance at the four ports is 50 Ω. (From Reference [3], © 1996 IEEE.)

coupling of 3.0 dB at closely spaced frequencies of 0.9 and 1.5 GHz [3]. To meet this requirement, the coupler is designed for a more tighter coupling of 2.65 dB at an in between frequency of 1.2 GHz. The rational behind this design is that the coupling tends to become weaker and approach the target value of 3.0 dB at frequencies on either side of 1.2 GHz. The dimensions of the coupler are determined from (11.3) to (11.5). Table 11.2 summarizes the coupler geometrical parameters and the measured performance over the frequency range of 0.7 to 1.7 GHz. The design and measured coupling values are in good agreement.

11.3 QUADRATURE (90°) HYBRID

A typical quadrature 90° hybrid constructed with CPW and slotline [7] is shown in Figure 11.4. This hybrid is also known as a branch-line hybrid or coupler. The branch-line coupler shown in Figure 11.4 has a rectangular shape, but the coupler may be reformed into a ring shape. To understand the operating principle of the branch-line coupler, consider the equivalent circuit model shown in Figure 11.5. In this model the input, direct, coupled, and isolated ports are designated as ports 1, 2, 3, and 4, respectively. All of these ports are terminated in feed lines of characteristic impedance Z_0 which normally is 50 Ω. The characteristic impedance of the branch- and

TABLE 11.2 Broadside Coupled CPW Directional Coupler

Circuit parameters and measured characteristics	
Substrate, ε_r	3.2
Thickness, mm	1.6
Coupled section	
CPW center strip width, S, mm	6.5
CPW slot width, W, mm	3.5
Length of coupled section, L, mm	38.5
Computed data	
$Z_{0,e}$, Ω	128.96
$Z_{0,o}$, Ω	19.54
$\varepsilon_{eff,e}$	2.28
$\varepsilon_{eff,o}$	3.06
Measured characteristics	
Coupling, S_{31}, S_{41}, dB	
at 0.88 GHz	−3.2
at 1.55 GHz	−3.3
Input return loss, S_{11}, dB	< -15.0
Isolation, S_{21}, dB	< -15.0

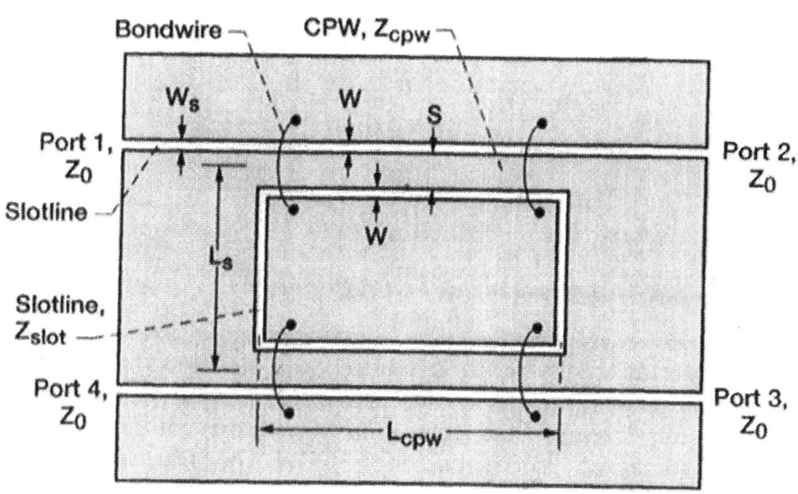

FIGURE 11.4 Schematic of a quadrature 90° hybrid or a branch-line coupler realized using a combination of coplanar waveguide and slotline. (From Reference [7], © 1993 IEEE.)

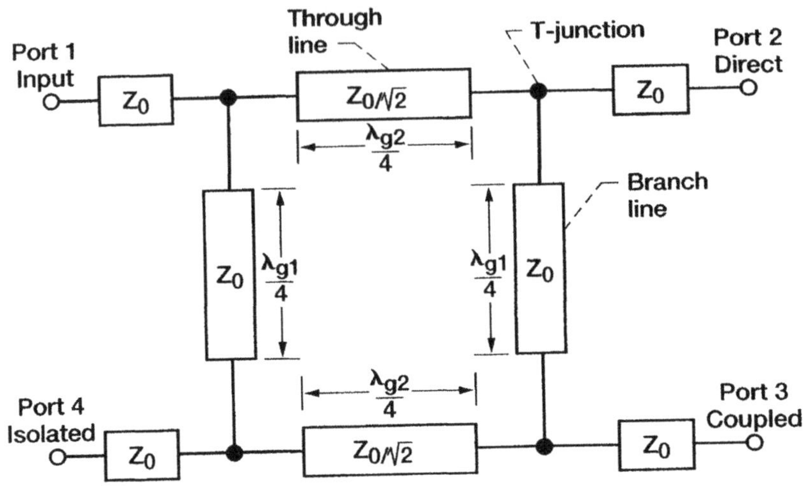

FIGURE 11.5 Equivalent circuit model of a branch-line coupler. The characteristic impedance and the guide wavelength are denoted as Z_0 and λ_g.

through-lines are Z_0 and $Z_0/\sqrt{2}$, respectively. The length of both the branch- and through-lines are quarter guide wavelength at the center frequency of operation. A signal entering port 1 is equally divided between the direct port 2 and the coupled port 3, with 90° phase difference. Port 1 is perfectly matched. Port 4 is the isolated port, and hence no power is coupled to this port. The scattering matrix for an ideal 3-dB quadrature hybrid or branch-line hybrid or coupler has the following form [8]:

$$[S] = \frac{-1}{\sqrt{2}}\begin{bmatrix} 0 & j & 1 & 0 \\ j & 0 & 0 & 1 \\ 1 & 0 & 0 & j \\ 0 & 1 & j & 0 \end{bmatrix}. \tag{11.6}$$

11.3.1 Standard 3-dB Branch-Line Hybrid

The coupler shown in Figure 11.4 is a good example of a standard 3-dB branch-line hybrid. The hybrid is fabricated on a 1.27 mm thick RT/Duroid 6010.8 ($\varepsilon_r = 10.8$) substrate [7]. In this circuit as a precaution bond wires are placed at the four corners of the rectangular loop to prevent the even mode of the coupled slotline from propagating on the CPW. The measured performance characteristics of the hybrid is summarized in Table 11.3. The hybrid has a bandwidth of 40 percent. A branch-line hybrid can be realized using a combination of CPW and asymmetric coplanar stripline (ACPS) also with similar characteristics as demonstrated in [9].

TABLE 11.3 Standard 3-dB Branch-Line Hybrid

Circuit parameters and measured characteristics	
CPW center strip width, S, mm	0.71
CPW slot width, W, mm	0.1
Length of CPW through-line, L_{CPW}, mm	10.39
Slotline slot width, W_s, mm	0.1
Length of slot branch-line, L_s, mm	7.77
Measured characteristics	
Center frequency, GHz	3.0
Insertion loss (includes fixture losses), dB	-3.75 ± 0.25
Phase quadrature, deg	83 ± 3
Input and output return loss, dB	< -19.0
Isolation, S_{41}, dB	< -20.0
Bandwidth, %	40

11.3.2 Size Reduction Procedure for Branch-Line Hybrid

Consider the quarter wavelength transmission line shown in Figure 11.6(a). This line constitutes the through-line and the branch-line of a standard branch-line hybrid. The line itself consists of distributed series inductance and shunt capacitance. The goal here is to replace this quarter wavelength long line by a physically shorter but electrically equivalent line. Normally, if a line is arbitrarily shortened, the inductance and capacitance reduce. However, the reductions in the inductance and the capacitance can be overcome by increasing the characteristic impedance and by adding lumped capacitors as shown in Figure 11.6(b).

Let $[Y_a]$ and $[Y_b]$ denote the admittance matrices of the lines in Figures 11.6(a) and (b). Then [10]

$$[Y_a] = \frac{1}{jZ_0}\begin{bmatrix} 0 & 1 \\ 1 & 0 \end{bmatrix} \tag{11.7}$$

$$[Y_b] = \begin{bmatrix} \dfrac{\cos\theta}{jZ\sin\theta} + j\omega C & \dfrac{1}{jZ\sin\theta} \\ \dfrac{1}{jZ\sin\theta} & \dfrac{\cos\theta}{jZ\sin\theta} + j\omega C \end{bmatrix}, \tag{11.8}$$

where Z_0, Z, θ, ω, and C are the characteristic impedance of the quarter wavelength line, the characteristic impedance of the shortened line, the electrical length of the shortened line, the angular frequency, and the lumped

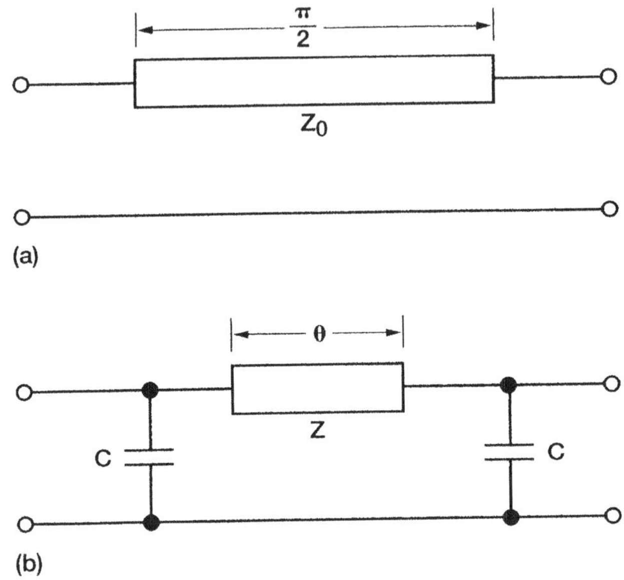

FIGURE 11.6 (*a*) Quarter wavelength transmission line; (*b*) equivalent shortened transmission line.

capacitance, respectively. Since these two lines are electrically equivalent, their admittance matrices should be identical. Hence equating (11.7) and (11.8), we obtain [10]

$$Z = \frac{Z_0}{\sin\theta}, \tag{11.9}$$

$$\omega C = \frac{\cos\theta}{Z_0}. \tag{11.10}$$

These equations suggest that the line length can be made shorter, for example, from 1/4 to 1/8 of the wavelength, by increasing the characteristic impedance of the line by a factor of $\sqrt{2}$. The lumped capacitance C can be easily incorporated in a CPW, since all the conductors are on the top side of the substrate.

11.3.3 Reduced Size 3-dB Branch-Line Hybrid

An equivalent circuit of standard and reduced size branch-line hybrid are shown in Figures 11.7(*a*) and (*b*). In a standard hybrid the characteristic impedance of the branch-line and the through-line are Z_0 and $Z_0/\sqrt{2}$, respectively. If in the reduced size hybrid the characteristic impedance of the both the branch-line and the through-line are Z and Z is greater than Z_0, then

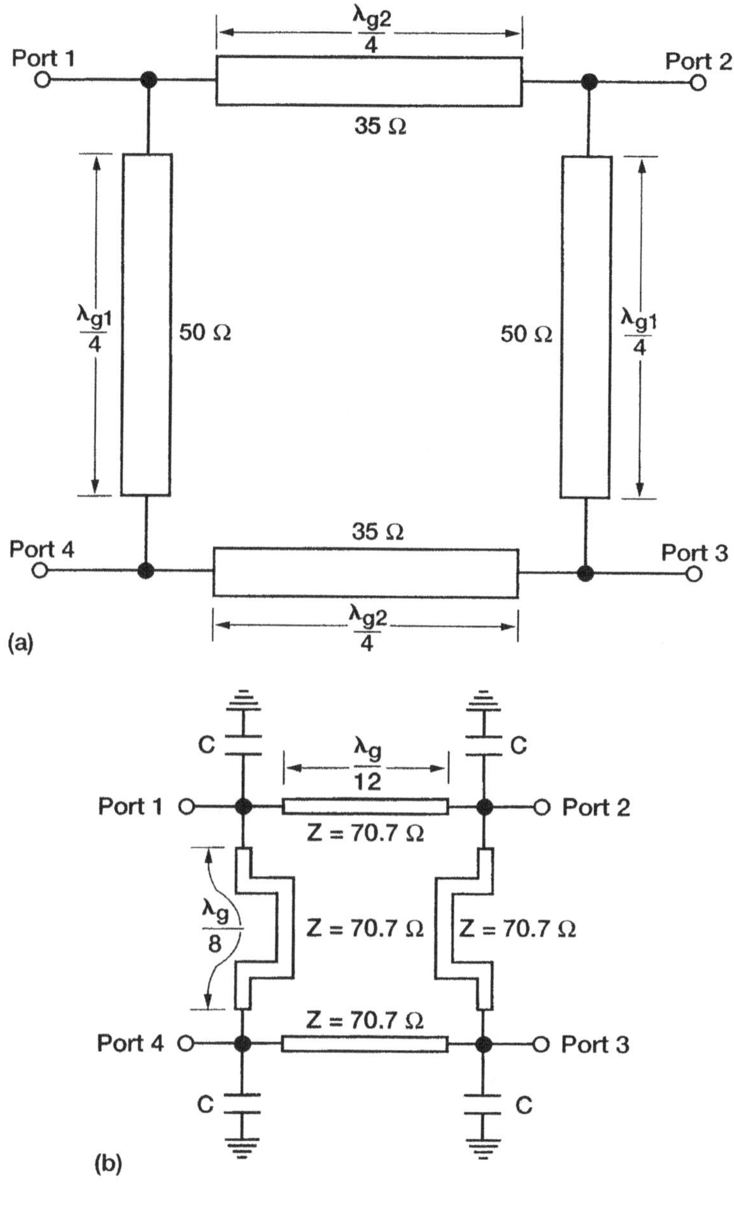

FIGURE 11.7 (*a*) Standard branch-line hybrid; (*b*) reduced size branch-line hybrid. The guide wavelength on the line is denoted as λ_g.

the electrical length of the branch-line θ_1, the electrical length of the through-line θ_2, and the lumped capacitance C are given by [10]

$$\theta_1 = \arcsin\left(\frac{Z_0}{Z}\right), \tag{11.11}$$

$$\theta_2 = \arcsin\left(\frac{Z_0}{\sqrt{2}Z}\right), \tag{11.12}$$

$$\omega C Z_0 = \sqrt{1 - \left(\frac{Z_0}{Z}\right)^2} + \sqrt{2 - \left(\frac{Z_0}{Z}\right)^2}. \tag{11.13}$$

For example, when $Z_0 = 50\ \Omega$ and $Z = 70.7\,\Omega$, the equations above give $\theta_1 = 45°$ and $\theta_2 = 30°$. This implies that the branch-line and the through-line lengths θ_1 and θ_2 are $\lambda_g/8$ and $\lambda_g/12$, where λ_g is the guide wavelength.

The phase difference between the power to the direct port and the power to the coupled port, that is, the phase (S_{21}/S_{31}), is computed in [10] for the reduced size hybrid and the standard quarter wavelength hybrid. The results show that the bandwidth over which the phase difference is 90° is narrower in the case of the reduced size hybrid. In a CPW, if the center strip conductor width is held fixed and the slot width is increased, the characteristic impedance increases and the conductor loss reduce. Hence the insertion loss of the reduced size hybrid that has a higher impedance than the standard hybrid is not degraded by size reduction.

An experimental reduced size hybrid fabricated on a semi-insulating GaAs is reported in [10]. The constant impedance transmission line is a CPW with a center strip width of 10 μm and characteristic impedance of 70.0 Ω. The length of the branch-line and the through-line are $\lambda_{g(\mathrm{CPW})}/8$ and $\lambda_{g(\mathrm{CPW})}/12$, respectively, at 25 GHz. Air-bridges are used a the CPW junctions to tie the ground planes at equal potential and suppress the coupled slotline even mode. Metal-insulator-metal (MIM) shunt capacitors ($\approx 40 \times 40\,\mu$m) are located at the corners of inner ground metal. The insulator film is Si_3N_4. The capacitance $C \approx 0.25$ pF. The size of the fabricated hybrid is 500 × 500 μm. This is more than 80 percent smaller than a standard branch-line hybrid.

The measured performance of the hybrid [10] over the frequency band of 24.0 to 26.0 GHz shows that the power coupled from port 1 to ports 2 and 3 are within 5.0 ± 0.5 dB, return loss at port 1 better than −10.0 dB, and the isolation between ports 1 and 4 is better than −12.0 dB. As a concluding remark, it may be noted that in [23] a small 90° hybrid coupler based on the Wheatstone bridge principle is demonstrated at a center frequency of 4.0 GHz.

11.3.4 Reduced Size Impedance Transforming Branch-Line Hybrid

The standard impedance transforming hybrid using quarter wavelength long lines with different characteristic impedances is shown in Figure 11.8(a). The

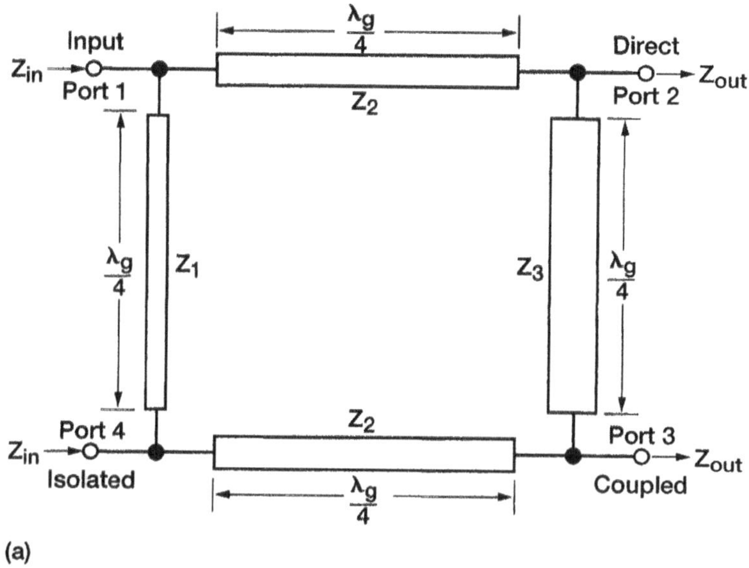

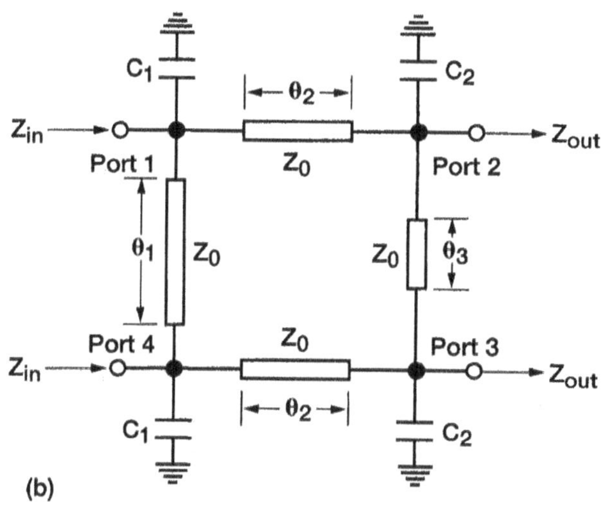

FIGURE 11.8 (*a*) Standard impedance transforming branch-line hybrid: (*b*) reduced size impedance transforming branch-line hybrid.

coupling constant C_0 is defined as [11]

$$C_0 = \frac{\text{coupling coefficient}}{\text{transmission coefficient}} = \left|\frac{S_{31}}{S_{21}}\right|. \tag{11.14}$$

The corresponding line characteristic impedances, in terms of C_0 and the termination impedances of the hybrid ports, Z_{in} and Z_{out}, can be expressed as [11]

$$Z_1 = \frac{Z_{in}}{C_0}, \tag{11.15}$$

$$Z_2 = \sqrt{\frac{Z_{in}Z_{out}}{1 + C_0^2}}, \tag{11.16}$$

$$Z_3 = \frac{Z_{out}}{C_0}. \tag{11.17}$$

Notice that for 3 dB, coupling $C_0 = 1$. Further, for $Z_{in} = Z_{out} = 50\,\Omega$, the impedance transforming hybrid reduces to a standard hybrid as discussed in Section 11.3.

The reduced size impedance transforming hybrid using lines of single characteristic impedance Z_0 but different lengths is shown in Figure 11.8(*b*). For an equivalent performance, the electrical lengths of the transmission lines are given by [11]

$$\theta_1 = \arcsin\left(\frac{Z_{in}}{Z_0 C_0}\right), \tag{11.18}$$

$$\theta_2 = \arcsin\left(\frac{\sqrt{Z_{in}Z_{out}/(1 + C_0^2)}}{Z_0}\right), \tag{11.19}$$

$$\theta_3 = \arcsin\left(\frac{Z_{out}}{Z_0 C_0}\right). \tag{11.20}$$

The shunt capacitor values are obtained from the expression [11]

$$\omega C_1 = \sqrt{\frac{C_0^2}{Z_{in}^2} - \frac{1}{Z_0^2}} + \sqrt{\frac{1 + C_0^2}{Z_{in}Z_{out}} - \frac{1}{Z_0^2}}, \tag{11.21}$$

$$\omega C_2 = \sqrt{\frac{C_0^2}{Z_{out}^2} - \frac{1}{Z_0^2}} + \sqrt{\frac{1 + C_0^2}{Z_{in}Z_{out}} - \frac{1}{Z_0^2}}. \tag{11.22}$$

For a 3 dB hybrid, $C_0 = 1$, and the equations above reduce to those in [12].

An experimental reduced size impedance matching hybrid using constant impedance CPW transmission lines and fabricated on a semi-insulating GaAs substrate is reported in [12]. Usually, the center frequency f_0, transforming impedances Z_{in} and Z_{out}, and CPW characteristic impedance Z_0 are known. Therefore the design involves determining θ_1, θ_2, θ_3, and the lumped capacitances C_1 and C_2 using Eqs. (11.18) to (11.22). The CPW circuit parameters, chip size, and the measured characteristics are summarized in Table 11.4.

TABLE 11.4 Reduced Size Impedance Transforming 3-dB Branch-Line Hybrid

Circuit parameters, measured characteristics and chip size	
Substrate, GaAs, ε_r	12.9
Thickness, μm	200
Given circuit parameters	
Z_{in}, Ω	50.0
Z_{out}, Ω	15.0
Z_0, Ω	55.0
CPW center strip width, μm	20.0
Computed parameters	
θ_1, deg	65.4
θ_2, deg	20.6
θ_3, deg	15.8
C_1, pF	0.6
C_2, pF	1.19
Measured characteristics	
Center frequency, f_0, GHz	14.8
Power coupled to the output ports, dB (coupled port and direct port)	−5.0
Return loss, dB	
at input port	< −12.0
at output port	< −15.0
Isolation, S_{41}, dB	< −20.0
Frequency Range, GHz	14 to 16
Size	
Chip, μm^2	800 × 800
Circuit size compared to standard impedance transforming hybrid	1/6

11.4 180° HYBRID

The 180° hybrid is in general a four-port network as represented in Figure 11.9(a). A signal incident at port 1 is evenly divided in-phase between the two output ports 2 and 3. Port 4 is the isolated port and hence no power is coupled to this port. On the other hand, if the signal is incident at port 4, it is evenly divided between the two output ports 2 and 3, and in addition the output signals are 180° out of phase as represented in Figure 11.9(b). In this case port 1 is the isolated port.

The 180° hybrid is also a signal combiner as represented in Figure 11.9(c). When signals are incident at ports 2 and 3, their sum and difference emerge from port 1 and port 4, respectively. Hence ports 1 and 4 are also referred to

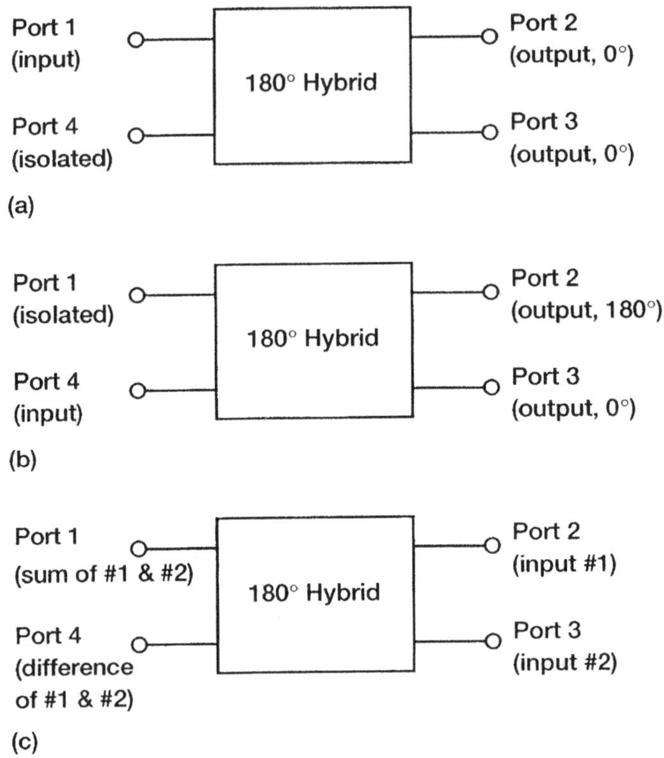

FIGURE 11.9 Representation of a 180° hybrid: (*a*) As a power divider with port 1 as input; (*b*) as a power divider with port 4 as input; (*c*) as a power combiner with ports 2 and 3 as input.

as the sum and difference ports, respectively. The scattering matrix for an ideal 3-dB 180° hybrid has the following form [8]:

$$[S] = \frac{-j}{\sqrt{2}} \begin{bmatrix} 0 & 1 & 1 & 0 \\ 1 & 0 & 0 & -1 \\ 1 & 0 & 0 & 1 \\ 0 & -1 & 1 & 0 \end{bmatrix}. \tag{11.23}$$

The reason for the $1/\sqrt{2}$ factor is that with all impedances equal an equal power split reduces voltages (and currents) by $1/\sqrt{2}$. The factor $\mp j$ depends on whether the path length from the input port to the output ports are $\lambda_g/4$ or $3\lambda_g/4$, respectively. This will be evident when a practical circuit is considered in the next section. The 180° hybrid is constructed either as a ring hybrid or as a magic-T. The ring hybrid is also known as the rat race.

11.4.1 Standard 180° Ring Hybrid

A standard 180° ring hybrid realized using a combination of CPW and asymmetric coplanar stripline (ACPS) [9] is shown in Figure 11.10. The circuit consists of four CPW-to-ACPS T-junctions and an ACPS ring. The circumference of the ACPS ring is $1.5\lambda_{g(ACPS)}$, where $\lambda_{g(ACPS)}$ is the guide wavelength in the ACPS. The ACPS ring is divided into three $\lambda_{g(ACPS)}/4$ sections and one $3\lambda_{g(ACPS)}/4$ section. The characteristic impedance of the ACPS and CPW are

$$Z_{ACPS} = \sqrt{2}\,Z_{CPW} \tag{11.24}$$

and

$$Z_{CPW} = Z_0, \tag{11.25}$$

where Z_0 is the terminating impedance or the source impedance.

An experimental hybrid built on a 0.635 mm thick RT/Duroid 6010.8 (ε_r = 10.8) substrate is reported in [9]. To prevent the coupled slotline mode from

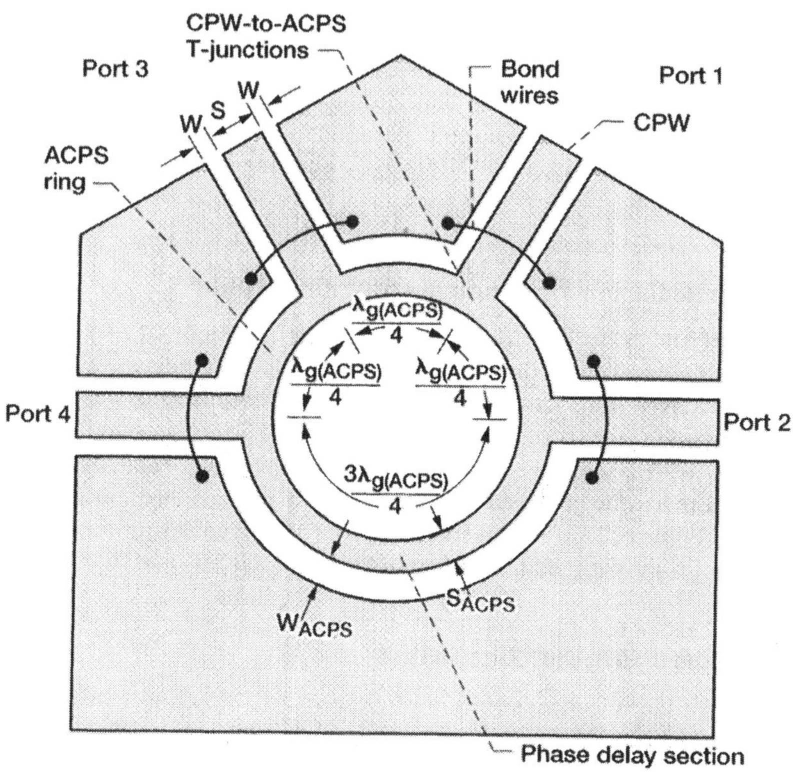

FIGURE 11.10 Standard 180° ring hybrid with coplanar waveguide terminations and asymmetric coplanar stripline ring. (From Reference [9], © 1997 IEEE.)

TABLE 11.5 Standard 180° Ring Hybrid

Circuit parameters and measured characteristics	
CIRCUIT PARAMETERS	
CPW center strip width, S, mm	0.60
CPW slot width, W, mm	0.29
Z_{CPW}, Ω	50
ACPS strip width, S_{ACPS}, mm	0.40
ACPS slot width, W_{ACPS}, mm	0.27
Z_{ACPS}, Ω	71
$\lambda_{g(ACPS)}/4$, mm	10.73
MEASURED CHARACTERISTICS	
Center frequency, GHz	30
Coupling (includes fixture losses) ($\lvert S_{34}\rvert$ or $\lvert S_{24}\rvert$), dB	3.5 ± 0.4
Input return loss ($\lvert S_{44}\rvert$), dB	< -23.0
Isolation ($\lvert S_{14}\rvert$), dB	< -21.0
Bandwidth, MHz, %	750 (25%)

interfering with the circuit performance, bond wires or air-bridges are located at the CPW-to-ACPS T-junctions. This circuit operates as explained in Section 11.4. The CPW and ACPS circuit parameters as well as the measured characteristics are summarized in Table 11.5. The ring hybrid has a bandwidth of about 25 percent.

11.4.2 Size Reduction Procedure for 180° Ring Hybrid

The standard 180° ring hybrid or rat race shown in Figure 11.10 consists of three $\lambda_g/4$ and a $3\lambda_g/4$ transmission lines, where λ_g is the guide wavelength. The quarter wavelength lines can be shortened to one-eighth of a wavelength by increasing their characteristic impedance by a factor of $\sqrt{2}$ and adding lumped capacitances at the ends as discussed in Section 11.3.2. The three-quarter wavelength line can be replaced by an equivalent lumped element Pi-network as shown in Figure 11.11. The elements of the Pi-network are determined by equating the Pi-network *ABCD* matrix to the line *ABCD* matrix.

11.4.3 Reduced Size 180° Ring Hybrid

The equivalent circuit representation of a standard 180° ring hybrid is shown in Figure 11.12(*a*). To reduce the size of the hybrid, first, the quarter wavelength lines of 70 Ω characteristic impedance are replaced by one-eighth wavelength lines of 100 Ω characteristic impedance with lumped capacitances at the ends. Second, the three-quarter wavelength line of 70 Ω characteristic

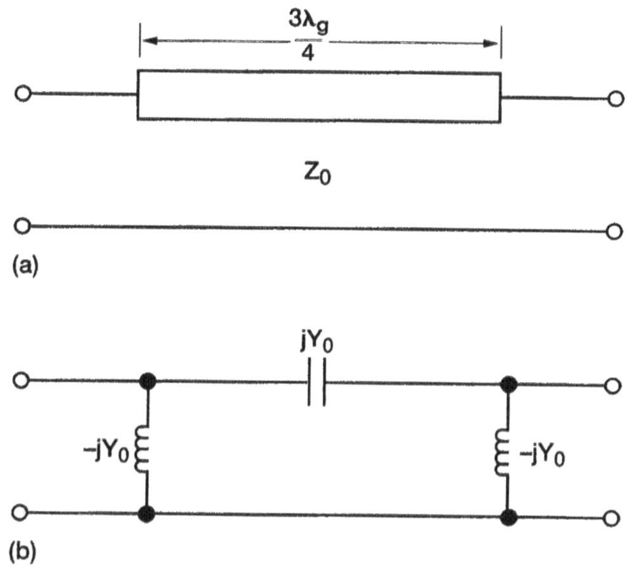

FIGURE 11.11 (*a*) Three-quarter wavelength transmission line. The wavelength on the line is denoted as λ_g; (*b*) equivalent lumped element Pi-network.

impedance is replaced by a Pi-network consisting of lumped series capacitance and shunt inductances. The lumped shunt inductances and capacitances almost cancel each other, resulting in the equivalent circuit [10] shown in Figure 11.12(*b*).

Besides size reduction the reduced size ring hybrid has the advantage of being able to interchange ports at the location of the series metal-insulator-metal (MIM) capacitor [10] as illustrated in Figure 11.13. In Figure 11.13(*a*) the output ports 2 and 3 are located on opposite sides of port 4. By interchanging the location of port 2 and 4, the output ports 2 and 3 are located side by side as shown in Figure 11.13(*b*). Further, if port 4 is used as the input port then, the output at ports 2 and 3 are 180° out of phase as discussed in Section 11.4. This feature greatly simplifies balanced mixer construction using a ring hybrid [24].

A experimental reduced size ring hybrid fabricated on a semi-insulating GaAs substrate with elevated CPW as the transmission line is reported in [13]. The elevated CPW has the advantage of reducing both the insertion loss as well as the chip area as discussed in Chapter 2. The hybrid consists of three one-eighth wavelength 100 Ω elevated CPW, a series and two shunt capacitors in place of the three $\lambda_g/4$, and a $3\lambda_g/4$ 70 Ω transmission lines as in a standard ring hybrid. In the experimental hybrid the elevated CPW is meandered to further reduce the overall size. The ring hybrid is schematically shown in Figure 11.14.

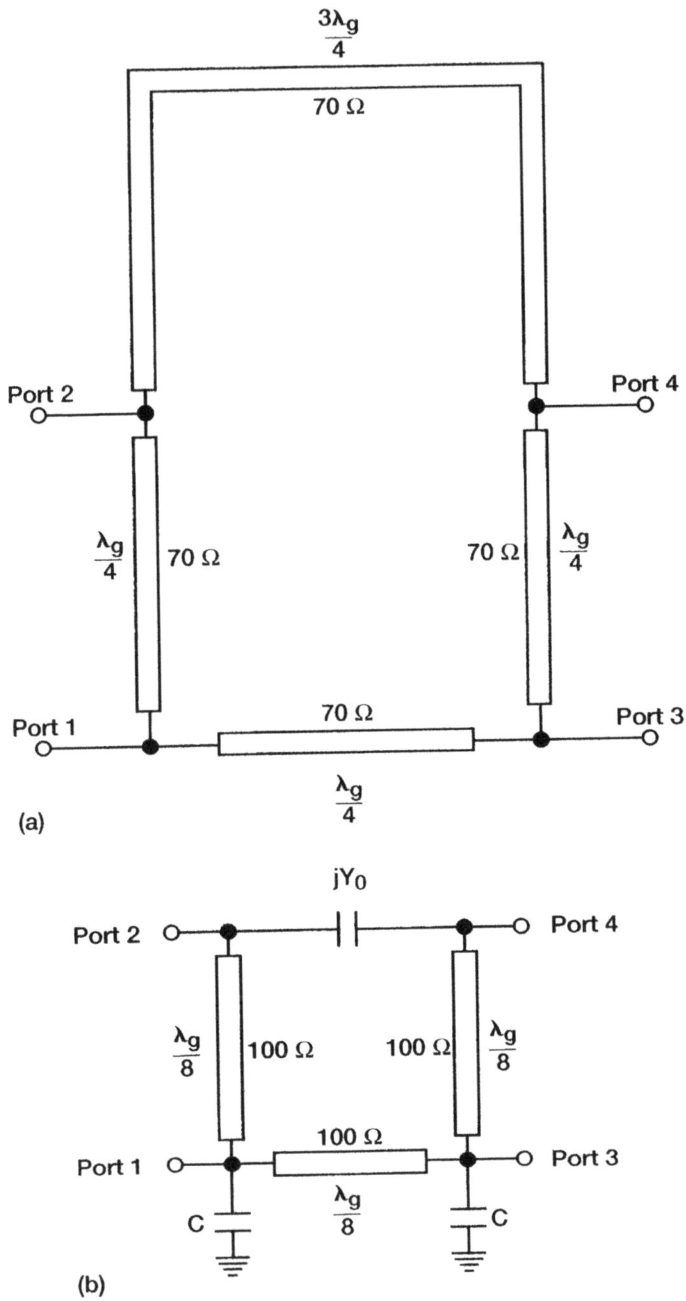

FIGURE 11.12 Representation of (*a*) standard 180° ring hybrid or rate race with transmission lines and (*b*) equivalent reduced size 180° ring hybrid or rat race with transmission lines and lumped elements. The wavelength on the line is denoted as λ_g.

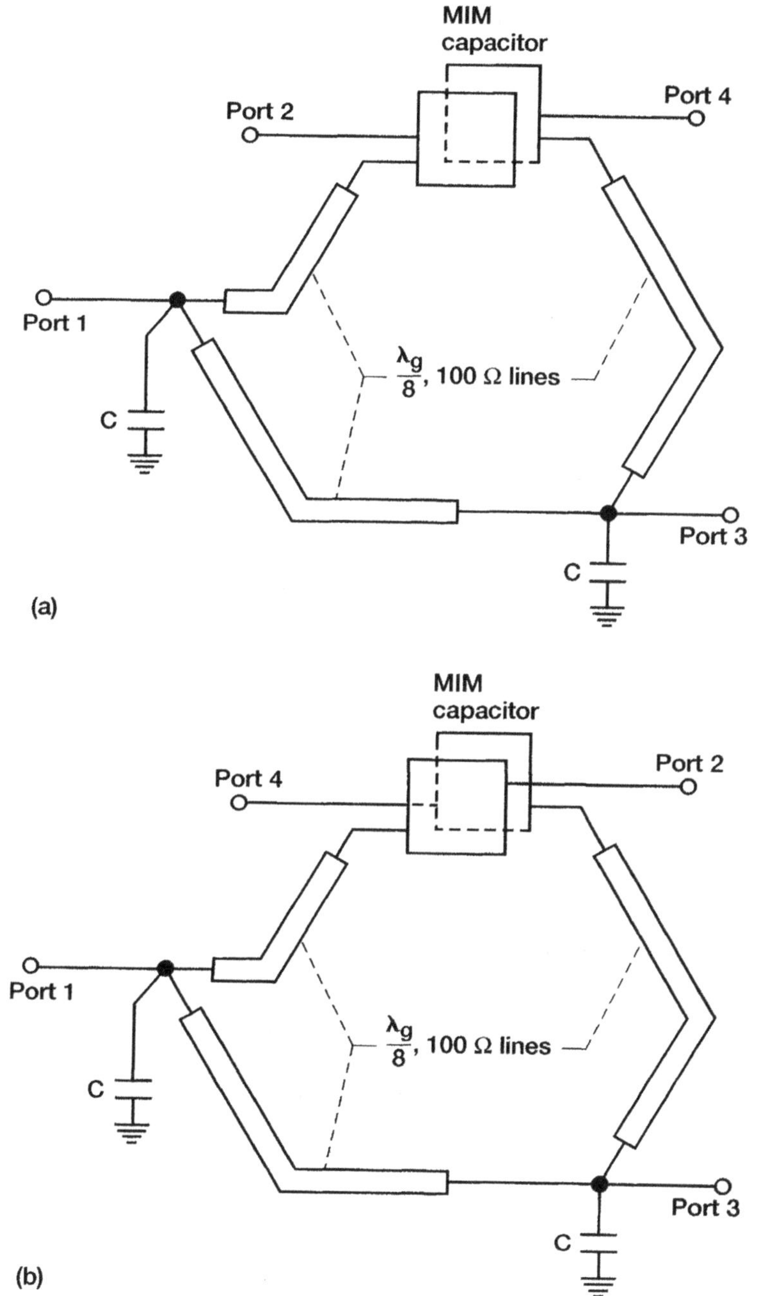

FIGURE 11.13 (*a*) Normal port layout; (*b*) port layout convenient for balanced mixer applications.

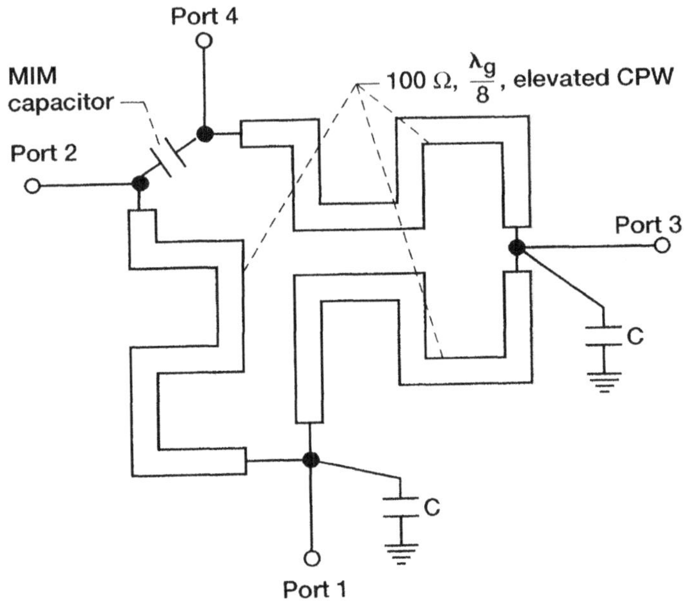

FIGURE 11.14 Reduced size 180° ring hybrid with meandered elevated CPW.

The measurements show that the coupling loss ($|S_{12}|, |S_{42}|$) of the hybrid is on the order of 4.2 ± 0.8 dB, return loss ($|S_{22}|$) better than 17.0 dB and isolation ($|S_{32}|$) better than 18.0 dB over the frequency range 15.0 ± 2.0 GHz. The chip size is on the order of 0.50×0.55 mm.

11.4.4 Reverse-Phase 180° Ring Hybrid

In the hybrid circuits shown in Figure 11.15, the three-quarter wavelength long phase delay section of the standard 180° ring hybrid is replaced by a quarter wavelength long line and a 180° phase shifter. The quarter wavelength line and the 180° phase shifter are implemented either by using a CPW-to-slotline T-junction [14] as shown in Figure 11.15(*a*) or by a phase-reversing asymmetric coplanar stripline (ACPS) section [9] as shown in Figure 11.15(*b*). The advantage gained in using either of the two approaches is that the phase shift is independent of frequency and hence the reverse-phase hybrid has a wider bandwidth. In addition, the reverse-phase hybrid is smaller in size since the circumference of the ring is one wavelength. Figure 11.15(*c*) shows the transmission line model for the circuit in Figure 11.15(*b*). The circuit parameters as well as the measured performance of the ACPS reverse-phase 180° ring hybrid shown in Figure 11.15(*b*) are summarized in Table 11.6. The reverse-phase 180° ring hybrid has a bandwidth of one octave.

TABLE 11.6 ACPS Reverse-Phase 180° Ring Hybrid

Circuit Parameters and Measured Characteristics	Standard Circuit	Reduced Size Circuit
RT/Duroid 6010.8 ($\varepsilon_r = 10.8$) thickness, mm	0.635	0.635
CPW feed line		
Strip width, S, mm	0.6	0.6
Slot width, W, mm	0.29	0.29
Characteristic impedance, Z_{CPW}, Ω	50	50
Slotline stub		
Radial stub radius, mm	6.0	—
Radial stub angle, deg	90	—
ACPS ring		
Strip width, S_{ACPS}, mm	0.4	—
Slot width, W_{ACPS}, mm	0.27	—
Length of each arm, deg	90	7.2
Characteristic impedance, Z_{ACPS}, Ω	70.7	66.9
Measured characteristics		
Center frequency, GHz	3.0	3.0
Bandwidth, GHz	2.0–4.0	2.0–4.0
Coupling (S_{12} or S_{42}), dB (includes fixture losses)	-3.95 ± 0.45	-3.5 ± 0.4
Amplitude balance, dB	± 0.4	± 0.3
Phase balance, deg	± 4.0	± 3.1
Input return loss, S_{22}, dB	< -13.5	< -12.0
Isolation, S_{32}, dB	< -23.0	< -29.8

11.4.5 Reduced Size Reverse-Phase 180° Ring Hybrid

The reverse-phase 180° ring hybrid discussed in the previous section has a circumference that is one wavelength. This circumference can be further reduced and the overall size of the circuit made smaller if the characteristics impedance of the feed line and the ring line are chosen according to the following relation [15]:

$$Z_{\text{ring}} = Z_{\text{feed}}\sqrt{2(1 - \cot^2\theta)}, \tag{11.26}$$

where Z_{ring} and Z_{feed} are the characteristic impedance of the ring line and feed line, respectively and θ is the electrical length of one quarter of the ring

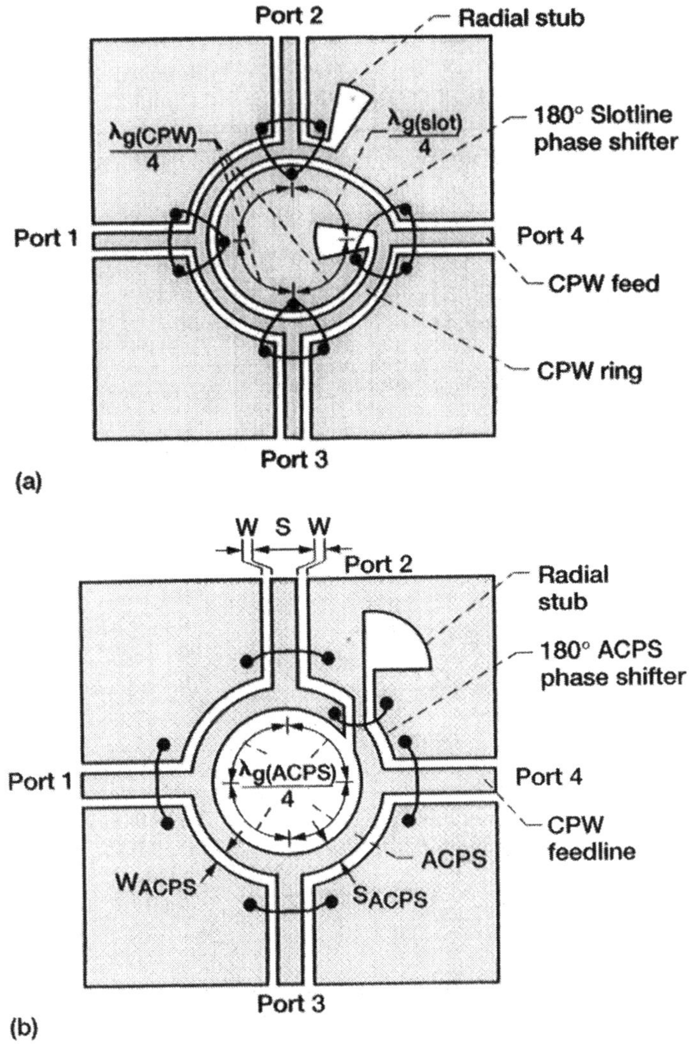

FIGURE 11.15 Reverse-phase 180° ring hybrid: (*a*) With CPW-to-slotline 180° reverse-phase section (from Reference [14], © 1994 IEEE); (*b*) with ACPS 180° reverse-phase section (from Reference [9], © 1997 IEEE); (*c*) equivalent transmission line model for the ACPS ring hybrid.

circumference. In this equation, θ can take any value between 45° and 90° [15]. By choosing θ equal to 72°, the four arm lengths reduce from one-quarter of a wavelength to one-fifth of a wavelength, and therefore the circumference reduces from one wavelength to 0.8 wavelength [9], [16]. The Z_{feed} is normally 50 Ω, and hence the characteristic impedance of the ring line also reduces from 70.7 to 66.9 Ω. The circuit layouts of reduced size reverse-phase 180° ring

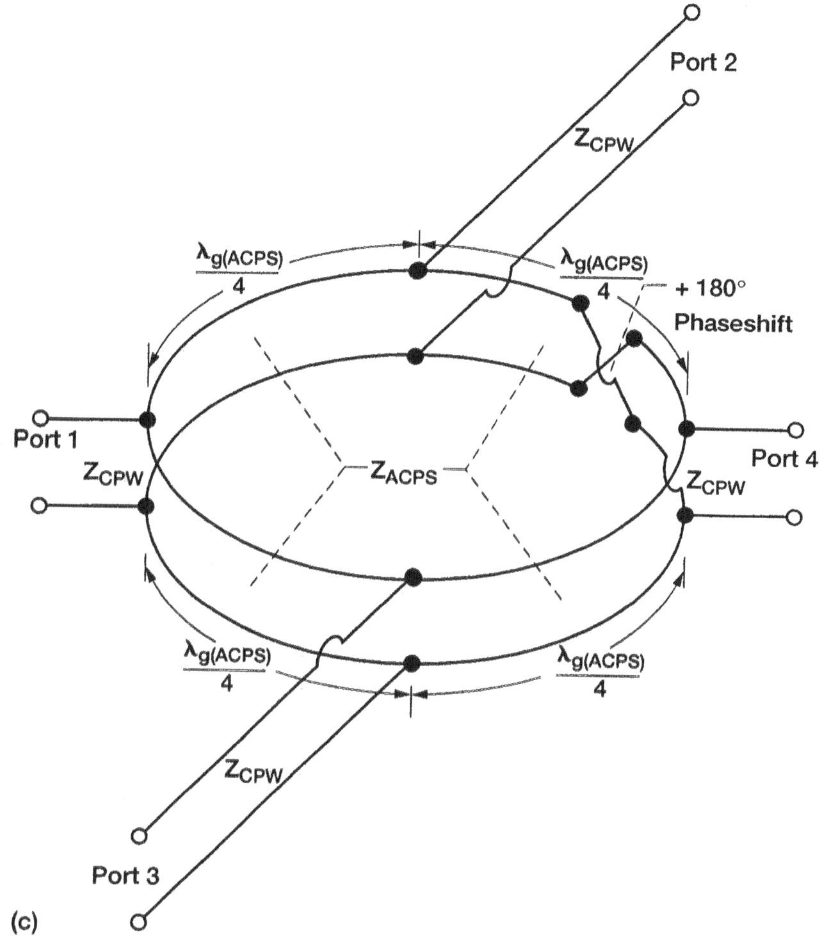

FIGURE 11.15 Continued

hybrid are identical to those of the standard reverse phase 180° ring hybrid shown in Figures 11.15(*a*) and (*b*). Except that the reduced size ring hybrid has arm lengths and characteristic impedances which are shorter and smaller, respectively. The measured performance of a CPS reduced size reverse-phase 180° ring hybrid [9] is also summarized in Table 11.6.

11.5 STANDARD 3-dB MAGIC-T

As explained in Section 11.4, the magic-T is also a 180° hybrid and can be completely described by the scattering matrix given in (11.23). In Figure 11.16(*a*) a magic-T [14] formed with three CPW T-junctions and a 180°

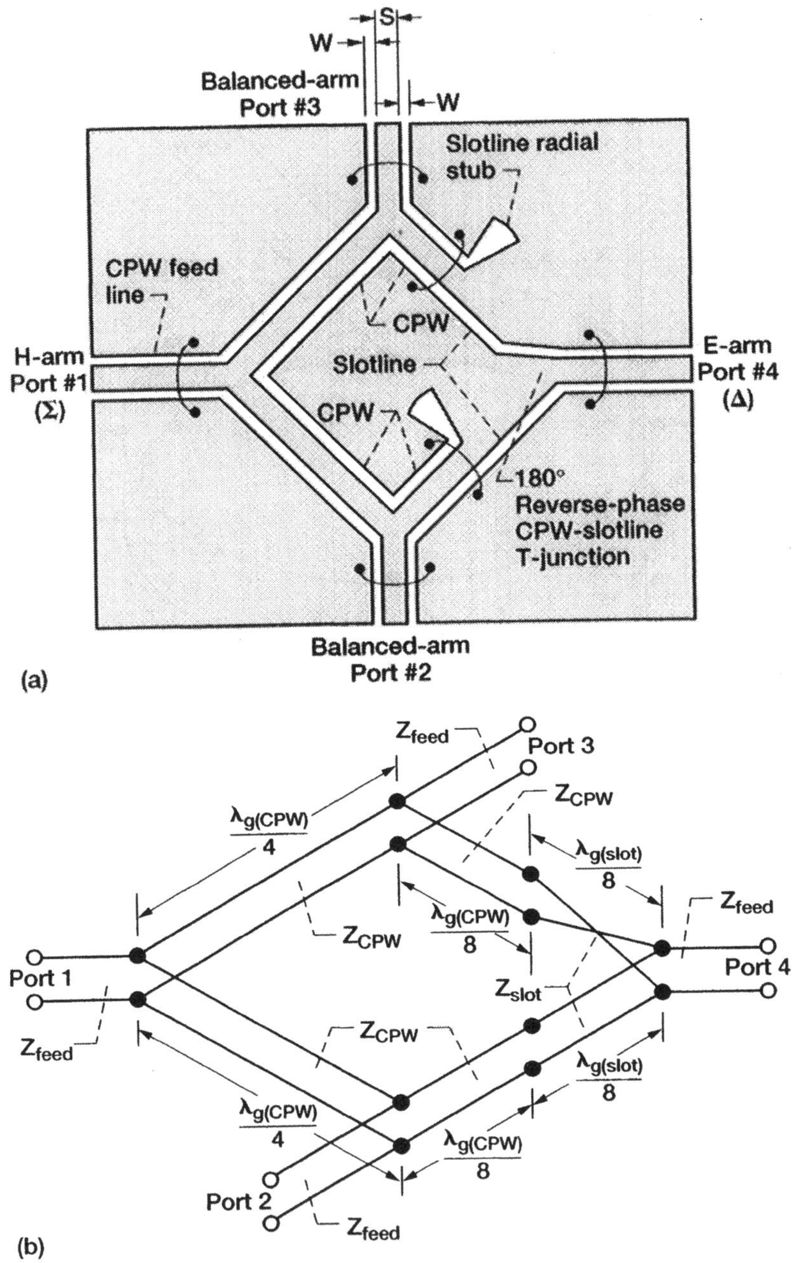

FIGURE 11.16 Standard 3-dB magic-T: (*a*) circuit configuration (from Reference [14], © 1994 IEEE); (*b*) equivalent transmission line model. The wavelength on the line is denoted as λ_g.

reverse-phase CPW-slotline T-junction is shown. In Figure 11.16(*a*) port 1 and port 4 correspond to the H-arm and E-arm of a conventional waveguide magic-T. In addition the ports 1 and 4 are known as the sum (Σ) and difference (Δ) ports, respectively, as explained in Section 11.4. Ports 2 and 3 are the power dividing balanced-arms. Figure 11.16(*b*) shows the transmission line model of the above magic-T. In this figure the twisted transmission line represents the 180° phase reversal of the CPW-slotline T-junction. The characteristic impedance of the feed line, the CPW line and the slotline are indicated as Z_{feed}, Z_{CPW}, and Z_{slot}, respectively. The guide wavelength on these lines are indicated as λ_g.

To obtain a relation between the various characteristic impedances consider a signal feed to port 1 (H-arm) of the magic-T as shown in Figure 11.17(*a*). The arrows indicate the direction of the electric field lines in the CPW and the slotline. The signal divides into two components and arrive in phase at ports 2 and 3. The two components arrive at port 4 (E-arm) 180° out of phase and cancel each other. In this case the plane of symmetry at the H-arm and the E-arm correspond to an open circuit (magnetic wall) and a short circuit (electric wall), respectively. The equivalent circuit of the magic-T for this excitation is shown in Figure 11.17(*b*).

Next consider a signal fed to port 4 (E-arm) of the magic-T as shown in Figure 11.18(*a*). Once again, the signal divides into two components and arrive at ports 2 and 3 but with a 180° phase difference. The phase difference is because of the 180° reverse-phase CPW-slotline T-junction. Two components arrive at port 1 (H-arm) 180° out of phase and cancel each other. In this case the plane of symmetry at the H-arm and the E-arm correspond to a short circuit (electric wall) and an open circuit (magnetic wall), respectively. The equivalent circuit of the magic-T for this excitation is shown in Figure 11.18(*b*).

By superimposing the above two cases, the total reflection coefficient at the balanced-arm, port 2 is obtained as follows [14]:

$$S_{22} = \tfrac{1}{2}\Gamma_{++} + \tfrac{1}{2}\Gamma_{+-}, \tag{11.27}$$

where Γ_{++} and Γ_{+-} are the voltage reflection coefficients at port 2. To achieve impedance matching at port 2, $|S_{22}|$ must be equal to zero. For this to happen, the following conditions must be satisfied:

$$Z_{\text{slot}} = Z_{\text{CPW}} \tag{11.28}$$

and

$$Z_{\text{CPW}} = \sqrt{2}\, Z_{\text{feed}}. \tag{11.29}$$

A similar reasoning holds true for port 3 impedance match. The isolation between ports 1 (H-arm) and 4(E-arm) is perfect as long as the 180° reverse-phase CWP-slotline T-junction is ideal.

A magic-T [7] can also be constructed using a three CPW T-junctions and a slotline T-junction as shown in Figure 11.19. In this circuit, the slotline

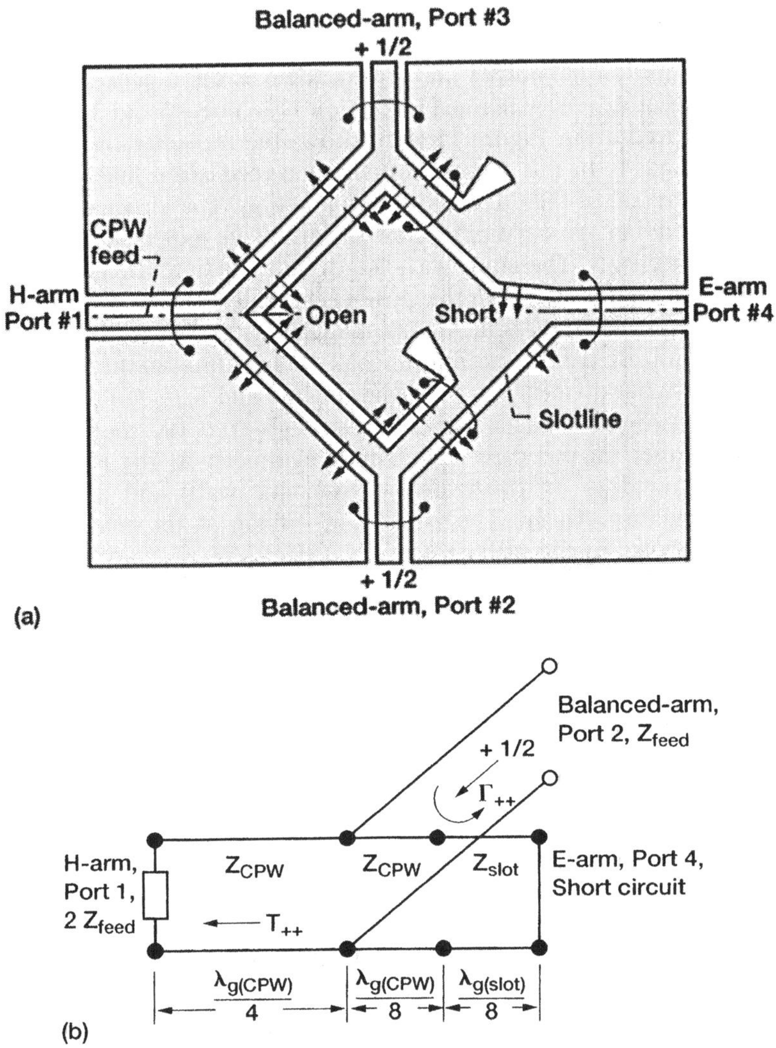

FIGURE 11.17 In-phase excitation at ports 2 and 3 of the magic-T: (*a*) electric-field distribution. The arrows indicate the direction of the electric field. (*b*) Simplified equivalent circuit.

T-junction behaves as a phase inverter or as a 180° reverse-phase network. To achieve impedance matching at ports 2 and 3, Z_{slot} is equal to $\sqrt{2}\,Z_{\text{feed}}$.

An experimental ring magic-T (Figure 11.19) fabricated on a 1.27 mm thicken RT/Duroid 6010.8 ($\varepsilon_r = 10.8$) substrate is reported in [7]. The CPW and the slotline circuit parameters as well as the measured characteristics are summarized in Table 11.7. The ring magic-T has an octave bandwidth.

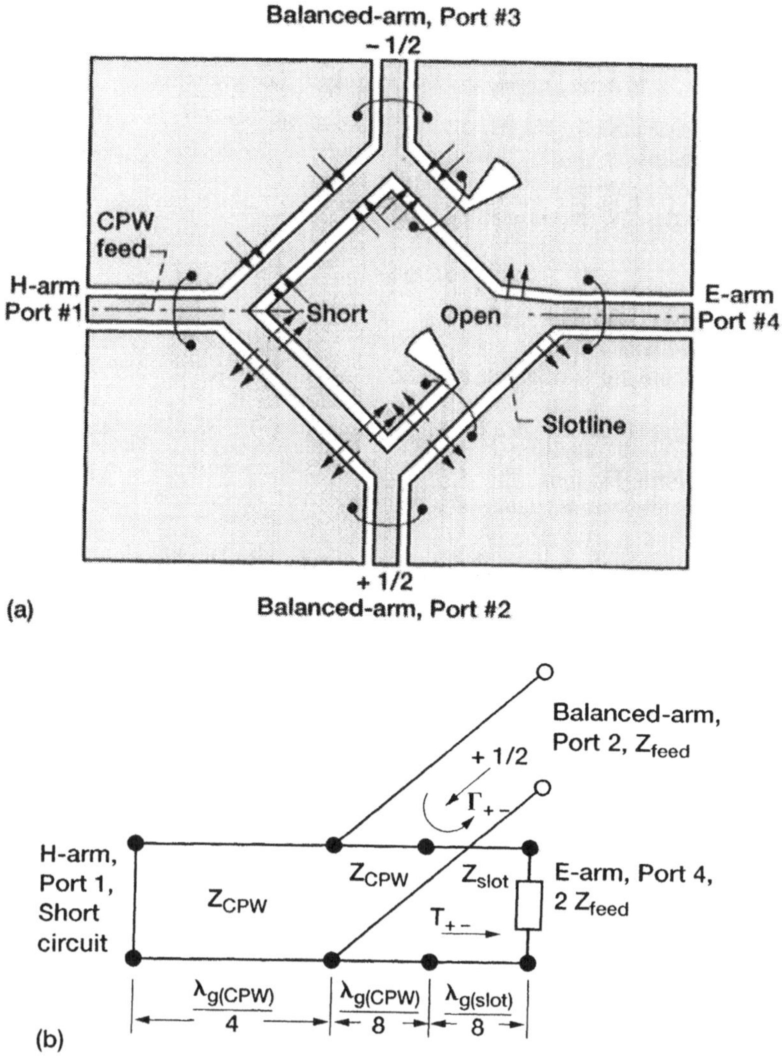

FIGURE 11.18 180° out-of-phase excitation at ports 2 and 3 of the magic-T: (*a*) Electric-field distribution; (*b*) simplified equivalent circuit.

11.5.1 Reduced Size 3-dB Magic-T

The circumference of the magic-T circuits discussed in the previous section is one guide wavelength. To reduce the circumference and thereby the overall size consider the hybrid ring topology [17] shown in Figure 11.20. In this circuit it is assumed that the transmission lines in opposite ring arms are identical and that the hybrid ring is matched at all ports. The phase shifter may be located anywhere in any of the four arms. In this hybrid a signal fed to port 1 (H-arm)

TABLE 11.7 Standard 3-dB Ring Magic-T

Circuit parameters and measured characteristics	
CPW FEED LINE	
Strip width, S, mm	0.51
Slot width, W, mm	0.25
Characteristic impedance, Z_{feed}, Ω	50
Slotline ring	
Slot width, W_s, mm	0.43
Radius, mm	7.77
Characteristic impedance, Z_{slot}, Ω	70.7
Slotline feed	
Slot width, W_f, mm	0.1
Characteristic impedance, $Z_{\text{slot feed}}$, Ω	50
Slotline radial stub	
Stub radius, mm	6.0
Stub angle, deg	90
MEASURED CHARACTERISTICS	
Center frequency, f_0, GHz	3.0
Bandwidth, GHz	2.0–4.0
E-arm power division	
Amplitude, dB	± 0.4
Phase, deg	181.0 ± 1.0
Insertion loss at f_0, dB	< -0.6
H-arm power divisions	
Amplitude, dB	± 0.3
Phase, deg	± 0.75
Insertion loss at f_0, dB	< -0.5
Isolation between E- and H-arms, dB	< -36.0
Isolation between the balanced arms 1 and 2, dB	< -12.0

will be split equally and in-phase between the two output ports 2 and 3 (balanced-arms). Similarly a signal feed to port 4 (E-arm) will be split equally and out-of-phase between the two output ports 2 and 3. The following design equations hold good [17]:

$$Z_{c1}^2 = Z_0^2 \frac{\sin^2\theta_1 + \sin^2\theta_2 - 2\cos\theta_1\cos\theta_2}{\sin^2\theta_1}, \tag{11.30}$$

$$Z_{c2}^2 = Z_0^2 \frac{\sin^2\theta_1 + \sin^2\theta_2 - 2\cos\theta_1\cos\theta_2}{\sin^2\theta_2}, \tag{11.31}$$

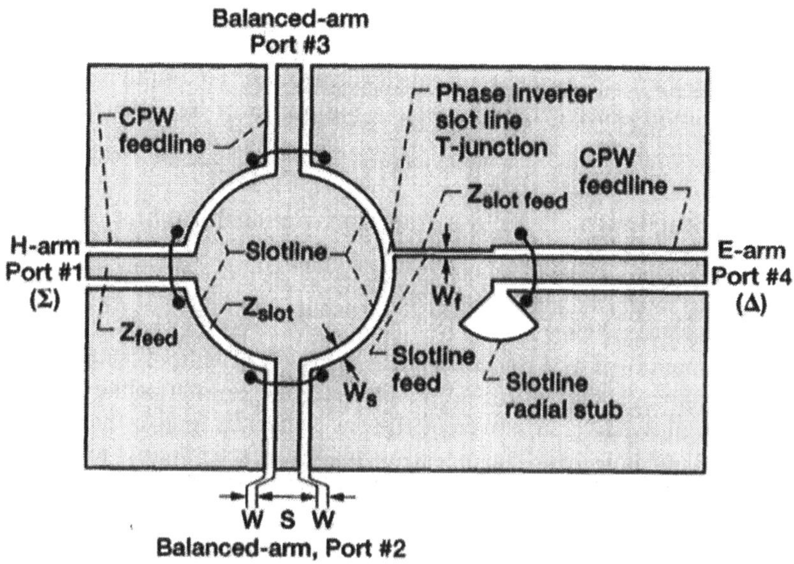

FIGURE 11.19 Standard 3-dB ring magic-T. (From Reference [7], © 1993 IEEE.)

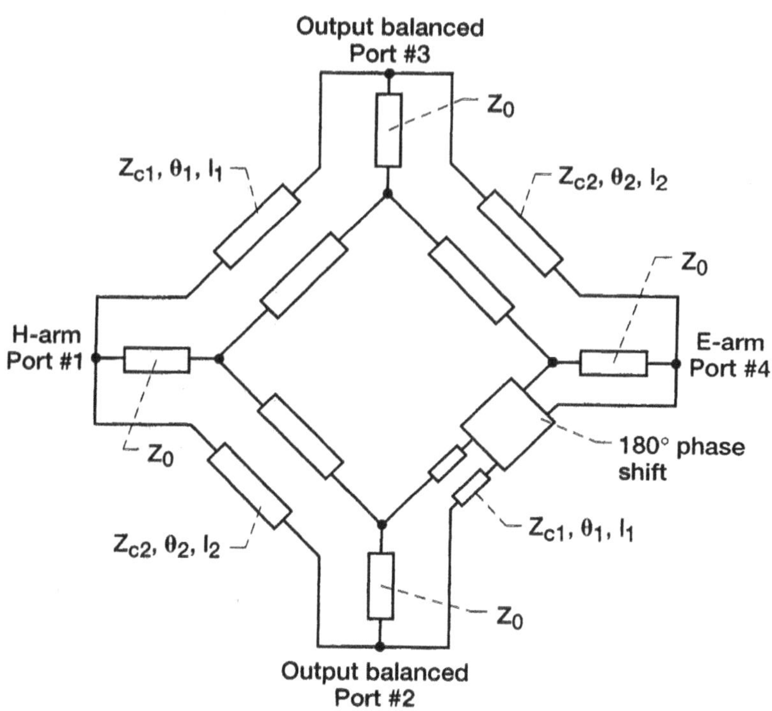

FIGURE 11.20 Generalized ring magic-T.

where θ_i ($i = 1, 2$) is the electrical length of arm i at the center frequency f_0. The impedances Z_{c1} and Z_{c2} are the characteristic impedances of the lines in the ring arms. The impedance Z_0 is the terminating impedance. To obtain real values for characteristic impedances, the following condition must be satisfied [17]:

$$\cos\theta_1 + \cos\theta_2 < \sqrt{2}. \tag{11.32}$$

In a symmetrical design all the four ring arms are identical. In such a case the equation above reduces to [15]

$$Z_c = Z_0\sqrt{2}\sqrt{1 - \cot^2\theta}. \tag{11.33}$$

An experimental reduced size magic-T [15], [17] realized with symmetric coplanar striplines is shown in Figure 11.21(*a*). Coplanar striplines are chosen because they allow easy crossover of the two strips to realize the 180° phase shifter needed at the E-arm. The crossover may be placed at any position along the arm or at the T-junction as shown in Figure 11.21(*b*).

In this design [15] θ is equal to 60° and for Z_0 equal to 50 Ω, Z_c from Eq. (11.33) is equal to 57.6 Ω. This design results in a ring circumference of 0.67 guide wavelength. The magic-T is fabricated on a 254 μm thick alumina substrate ($\varepsilon_r = 9.9$) and the conductor thickness is 4 μm. The geometrical parameters and the measured performance of the magic-T are summarized in Table 11.8. The magic-T has a bandwidth slightly larger than one octave.

As a concluding remark, it may be noted that there are other designs for magic-Ts that can be found in [16], [18], [19], [20].

11.6 ACTIVE MAGIC-T

The magic-T circuits discussed in earlier sections are based on short sections of transmission lines, which are interconnected, and hence are passive circuits. In this section an active magic-T [21] composed of a GaAs FET with two gate fingers, and very short sections of CPW and slotline is described. The schematic and equivalent circuit of the magic-T are shown in Figure 11.22(*a*) and (*b*). The GaAs FET source contact and gate fingers are extended to form the center strip conductor and the ground conductors of a CPW, respectively. The CPW is the H-arm as well as port 1 of the magic-T. The extended gate fingers and the two drain contacts form two slotlines that together with the CPW form an in-phase power divider. The two slotlines or the two output ports of the power divider are the two balanced-arms as well as ports 2 and 3 of the magic-T. In this arrangement the gate serves as the common or the ground conductor. The two drain contacts and the gate contact form a slotline series T-junction which is equivalent to an out-of-phase power divider. The input port of this power divider is the E-arm or port 4 of the magic-T. Having discussed the construction of the magic-T, expressions which relate the reflection coefficient at the four ports, the power division ratio between the input

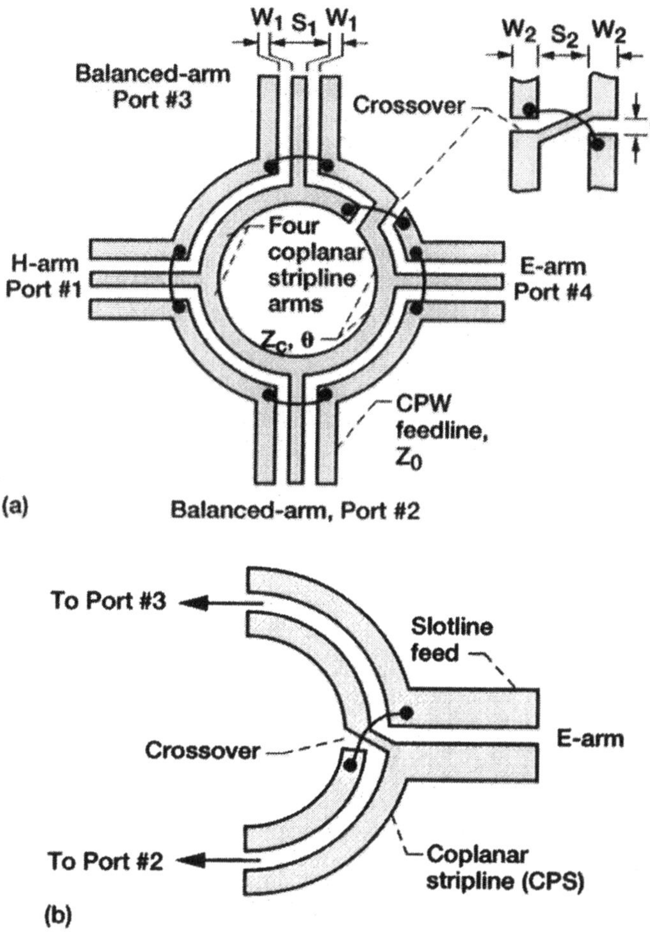

FIGURE 11.21 (*a*) Reduced size ring magic-T (from Reference [15], © IEE); (*b*) alternate implementation with slotline feed for E-arm (from Reference [17], © IEE).

H-arm and E-arm and output ports 2 and 3, and the isolation between the input and output ports are presented next. The reflection coefficient at the four ports are given by [21]

$$S_{11} = \frac{1 - 2g_m Z_H}{1 + 2g_m Z_H}, \tag{11.34}$$

$$|S_{22}| = |S_{33}| = \frac{|\Gamma_{+-} + \Gamma_{++}|}{2}, \tag{11.35}$$

$$|S_{44}| = |\Gamma_{+-}|, \tag{11.36}$$

TABLE 11.8 Reduced Size 3-dB Ring Magic-T

Circuit parameters and measured characteristics	
Substrate (alumina), ε_r	9.9
Thickness, μm	254
Metalization thickness, μm	4.0
Coplanar waveguide	
Strip width, S_1, μm	50
Slot width, W_1, μm	30
Z_0, Ω	50
Coplanar stripline	
Strip width, W_2, μm	300
Slot width, S_2, μm	60
Z_c, Ω,	57.6
θ, deg	60
Circumference, λ_g	0.67
Ring radius, μm	3478
Crossover gap, g, μm	100
Measured performance	
Center frequency, f_0, GHz	6
Frequency range, GHz	4.5–9.2
Transmission when fed at H-arm mag (S_{21}), mag (S_{31}) (dB)	4.3 ± 0.3
Transmission when fed at E-arm mag (S_{24}), mag (S_{34}) (dB)	4.3 ± 0.3
Output magnitude balance when fed at H-arm mag (S_{21}/S_{31}) (dB)	0.3 ± 0.05
Output phase balance when fed at H-arm phase (S_{21}/S_{31}) (deg)	0 ± 4
Output magnitude balance when fed at E-arm mag (S_{24}/S_{34}) (dB)	0.4 ± 0.25
Output phase balance when fed at E-arm phase (S_{24}/S_{34}) (deg)	180 ± 4
Return loss, mag (S_{22}) and mag (S_{33}), (dB)	< -10.0
Isolation, mag (S_{EH}), mag (S_{32}), (dB)	< -23.0

where g_m is the transconductance of the FET, Z_H is the characteristic impedance of the CPW or the H-arm, and

$$\Gamma_{+-} = \frac{\left(\dfrac{2Z_E R_{dg}}{Z_E + 2R_{dg}}\right) - 2Z_0}{\left(\dfrac{2Z_E R_{dg}}{Z_E + 2R_{dg}}\right) + 2Z_0}, \tag{11.37}$$

$$\Gamma_{++} = \frac{R_{dg} - Z_0}{R_{dg} + Z_0}, \tag{11.38}$$

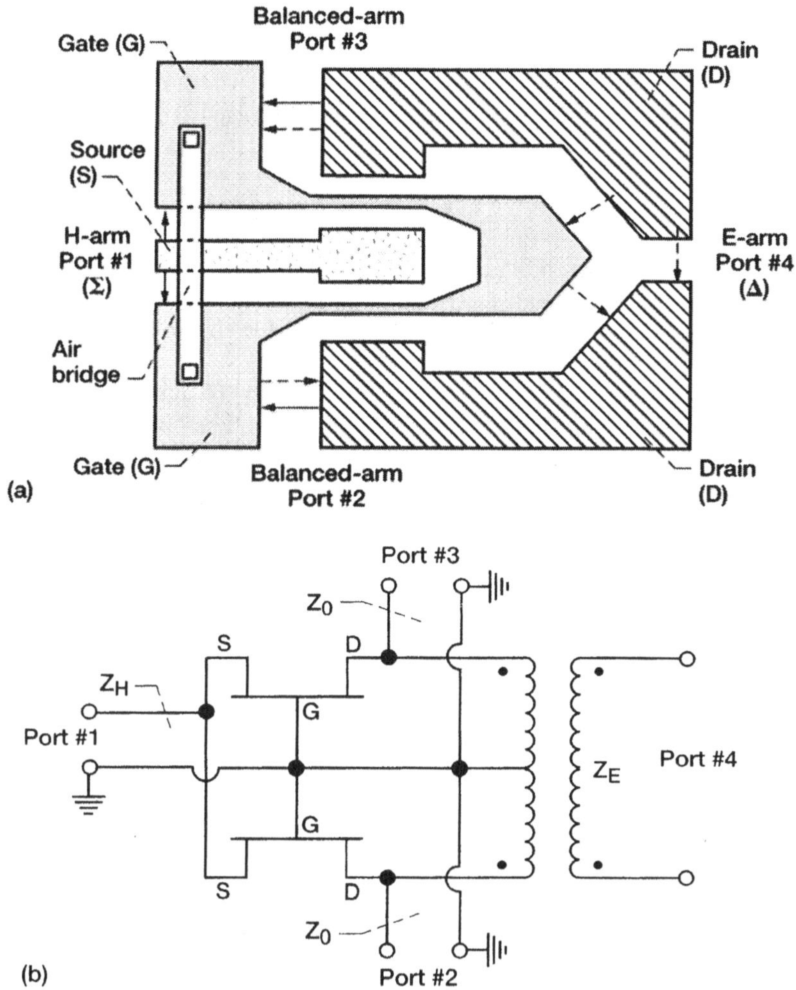

FIGURE 11.22 (*a*) Active magic-T; (*b*) equivalent circuit model. (From Reference [21], © 1989 IEEE.)

where Z_E is the characteristic impedance of the input slotline or E-arm, R_{dg} is the drain-to-gate resistance and Z_0 is the characteristic impedance of the output slot lines at ports 2 and 3.

The power division ratio between the input H-arm and E-arm and output ports 2 and 3 are [21]

$$|S_{21}| = |S_{31}| = \frac{2g_m\sqrt{Z_H Z_0}}{1 + 2g_m Z_H} \cdot \frac{R_{dg}}{R_{dg} + Z_0}, \qquad (11.39)$$

$$|S_{24}| = |S_{34}| = \sqrt{\frac{1 - |\Gamma_{+-}|^2}{2} \cdot \frac{2R_{dg}}{2R_{dg} + Z_0}}. \tag{11.40}$$

The isolation between the two output ports is given by [21]

$$|S_{23}| = |S_{32}| = \frac{|\Gamma_{+-} - \Gamma_{++}|}{2}. \tag{11.41}$$

The isolation between the output ports 2 and 3 and the input H-arm or port 1 is due to the unilateral characteristics of the FET. The isolation between the two input ports 1 and 4 or H-arm and E-arm is due to the orientation of the fields in the CPW and slotline.

An experimental active magic-T fabricated using 0.5 μm gate GaAs FET is reported in [21]. Each gate finger is 75 μm wide. The chip size is $1 \times 1 \times 0.6$ mm. The measured coupling loss from the input ports 1 (H-arm) or 4 (E-arm) and the output ports 2 and 3 is within 5 ± 1 dB. The measured return loss is better than 10 dB and isolation greater than 20 dB. These measurements are performed over the frequency range of 2 to 18 GHz. The active magic-T has a very wide bandwidth. The noise performance of active balun circuits is discussed in [22].

REFERENCES

[1] C. P. Wen, "Coplanar-Waveguide Directional Couplers," *IEEE Trans. Microwave Theory Tech.*, Vol. 18, No. 6, pp. 318–322, June 1970.

[2] R. E. Collin, *Foundations for Microwave Engineering*, 2nd ed., New York: McGraw-Hill, 1992, sec. 6.4.

[3] F. Tefiku, E. Yamashita, and J. Funada, "Novel Directional Couplers Using Broadside-Coupled Coplanar Waveguides for Double-Sided Printed Antennas," *IEEE Trans. Microwave Theory Tech.*, Vol. 44, No. 2, pp. 275–282, Feb. 1996.

[4] V. F. Hanna, "Parameters of Coplanar Directional Couplers with Lower Ground Plane," *Proc. 15th European Microwave Conf. Dig.*, pp. 820–825, Paris, France, Sept. 9–13, 1985.

[5] M. Gillick, I. D. Robertson, and J. S. Joshi, "A 12-36 GHz MMIC 3dB Coplanar Waveguide Directional Coupler," *Proc. 22nd European Microwave Conf. Dig.*, pp. 724–728, Helsinki, Finland, Aug. 24–27, 1992.

[6] em is a registered trademark of Sonnet Software, Inc.

[7] C.-H. Ho, L. Fan, and K. Chang, "Broad-Band Uniplanar Hybrid-Ring and Branch-Line Couplers," *IEEE Trans. Microwave Theory Tech.*, Vol. 41, No. 12, pp. 2116–2125, Dec. 1993.

[8] D. M. Pozar, *Microwave Engineering*, Reading, MA: Addison-Wesley 1990, ch. 8.

[9] B. R. Heimer, L. Fan, and K. Chang, "Uniplanar Hybrid Couplers Using Asymmetrical Coplanar Striplines," *IEEE Trans. Microwave Theory Tech.*, Vol. 45, No. 12, pp. 2234–2240, Dec. 1997.

[10] T. Hirota, A. Minakawa, and M. Muraguchi, "Reduced-Size Branch-Line and Rat-Race Hybrids for Uniplanar MMIC's," *IEEE Trans. Microwave Theory Tech.*, Vol. 38, No. 3, pp. 270–275, March 1990.

[11] M. Gillick and I. D. Robertson, "Design Analysis and Fabrication of Lumped-Distributed Branch-Line Couplers for Coplanar Waveguide MMICs," *Microwave Optical Tech. Lett.*, Vol. 6, No. 12, pp. 720–723, Sept. 1993.

[12] M. Gillick, I. D. Robertson, and J. S. Joshi, "Coplanar Waveguide Two-Stage Balanced MMIC Amplifier Using Impedance-Transforming Lumped-Distributed Branchline Couplers," *IEE Proc. Microwave Antennas Propag.*, Vol. 141, No. 4, pp. 241–245, Aug. 1994.

[13] H. Kamitsuna, "A Very Small, Low-Loss MMIC Rat-Race Hybrid Using Elevated Coplanar Waveguides," *IEEE Microwave Guided Wave Lett.*, Vol. 2, No. 8, pp. 337–339, Aug. 1992.

[14] C.-H. Ho, L. Fan, and K. Chang, "New Uniplanar Coplanar Waveguide Hybrid-Ring Couplers and Magic-T's," *IEEE Trans. Microwave Theory Tech.*, Vol. 42, No. 12, pp. 2440–2448, Dec. 1994.

[15] M.-H. Murgulescu, E. Moisan, P. Legaud, E. Penard, and I. Zaquine, "New Wideband, 0. $67\lambda_g$ Circumference 180° Hybrid Ring Coupler," *Electron. Lett.*, Vol. 30, No. 4, pp. 299–300, Feb. 1994.

[16] L. Fan, C.-H. Ho, S. Kanamaluru and K. Chang, "Wide-Band Reduced-Size Uniplanar Magic-T, Hybrid-Ring, and deRonde's CPW-Slot Couplers," *IEEE Trans. Microwave Theory Tech.*, Vol. 43, No. 12, pp. 2749–2758, Dec. 1995.

[17] M.-H. Murgulescu, E. Penard, and I. Zaquine, "Design Formulas for Generalized 180° hybrid Ring Couplers," *Electron. Lett.*, Vol. 30, No. 7, pp. 573–574, March 1994.

[18] M. Aikawa and H. Ogawa, "A New MIC Magic-T Using Coupled Slot Lines," *IEEE Trans. Microwave Theory Tech.*, Vol. 28, No. 6, pp. 523–528, June 1980.

[19] T. Hirota, Y. Tarusawa, and H. Ogawa, "Uniplanar MMIC Hybrids—A proposed New MMIC Structure," *IEEE Trans. Microwave Theory Tech.*, Vol. 35, No. 6, pp. 576–581, June 1987.

[20] M. Aikawa and H. Ogawa, "Double-Sided MIC's and Their Applications," *IEEE Trans. Microwave Theory Tech.*, Vol. 37, No. 2, pp. 406–413, Feb. 1989.

[21] T. Tokumitsu, S. Hara, and M. Aikawa, "Very Small Ultra-Wide-Band MMIC Magic-T and Applications to Combiners and Dividers," *IEEE Trans. Microwave Theory Tech.*, Vol. 37, No. 12, pp. 1985–1990, Dec. 1989.

[22] C.-S. Lee, M.-G. Kim, J.-J. Lee, K.-E. Pyun, and H.-M. Park, "A Low Noise Amplifier for a Multi-band and Multi-mode Handset," *1998 IEEE Radio Frequency Integrated Circuits (RFIC) Symp.*, Baltimore, MD, pp. 47–50, June 7–11, 1998.

[23] M.-H. Murgulescu, P. Legaud, E. Penard, and I. Zaquine, "New Small 90° Hybrid Coupler," *Electron. Lett.*, Vol. 30, No. 16, pp. 1289–1290, Aug. 1994.

[24] T. Hirota and M. Muraguchi, "K-Band Frequency Up-Converters Using Reduced-Size Couplers and Dividers," *1991 IEEE GaAs IC Symp.*, Monterey, CA, pp. 53–56, Oct. 20–23, 1991.

CHAPTER 12

Coplanar Waveguide Applications

12.1 INTRODUCTION

This chapter discusses several emerging applications of coplanar waveguides. These applications include microelectromechanical systems (MEMS) based switches and phase shifters, high-T_c superconducting transmission lines, tunable devices using ferroelectric materials, photonic bandgap structures, and printed circuit antennas. These components and systems make possible several new applications involving the transmission, display, and processing of information at a very high data rate through satellites and wireless networks.

12.2 MEMS COPLANAR WAVEGUIDE CAPACITIVE METAL MEMBRANE SHUNT SWITCH

A coplanar waveguide (CPW) capacitive membrane shunt switch [1] is schematically illustrated in Figure 12.1. The switch consists of a thin metallic membrane, suspended across the CPW center strip conductor. The membrane is supported at either ends by metal posts attached to the CPW ground planes. The thickness, width, and length of the membrane are denoted as t_m, W_m, and L_m, respectively. In such an arrangement the membrane and the CPW center strip conductor behave as the top and bottom electrodes, respectively, of a parallel-plate capacitor. Since the membrane is free to move, the parallel-plate capacitance can be increased by decreasing the distance of separation. This is done by applying a dc voltage between the membrane and the center strip. The dc potential sets up an electrostatic force of attraction that pulls the membrane toward the center strip, thereby decreasing the distance of separation and increasing the capacitance. A dielectric film deposited over the bottom electrode prevents stiction when the metal surfaces come in contact.

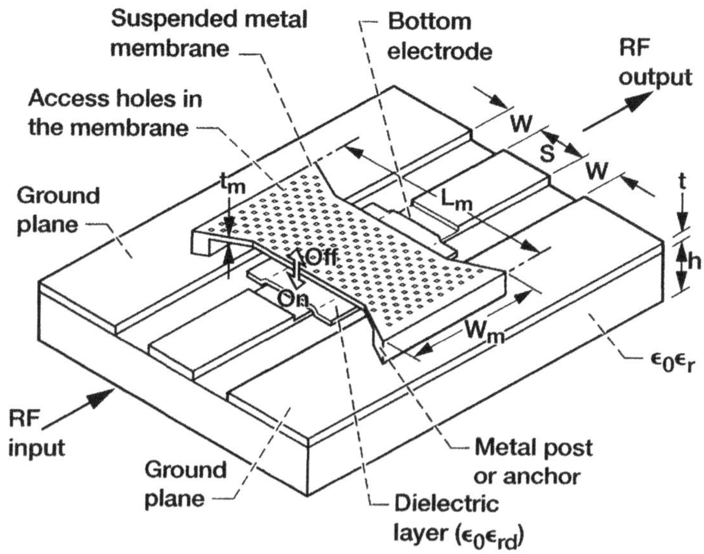

FIGURE 12.1 Schematic of a shunt microelectromechanical systems (MEMS) capacitive switch using a bowtie shaped metal membrane in a coplanar waveguide (CPW) environment.

12.2.1 OFF and ON Capacitances

The cross section of the switch when it is OFF, that is, when the membrane is in the up position, is shown in Figure 12.2(*a*). The OFF capacitance of the switch is given by [2].

$$C_{\mathrm{OFF}} = \frac{1}{(t_{\mathrm{d}}/A\varepsilon_0\varepsilon_{\mathrm{rd}}) + (h_{\mathrm{a}}/A\varepsilon_0)}, \tag{12.1}$$

where

A = overlap area between the membrane and the bottom electrode

h_{a} = height of the air gap between the membrane and the dielectric layer

t_{d} = thickness of the dielectric film

$\varepsilon_{\mathrm{rd}}$ = dielectric constant of the film

When the switch is ON, the membrane is in the down position as shown in Figure 12.2(*b*). The ON capacitance of the switch is given by [2]

$$C_{\mathrm{ON}} = \frac{A\varepsilon_0\varepsilon_{\mathrm{rd}}}{t_{\mathrm{d}}}. \tag{12.2}$$

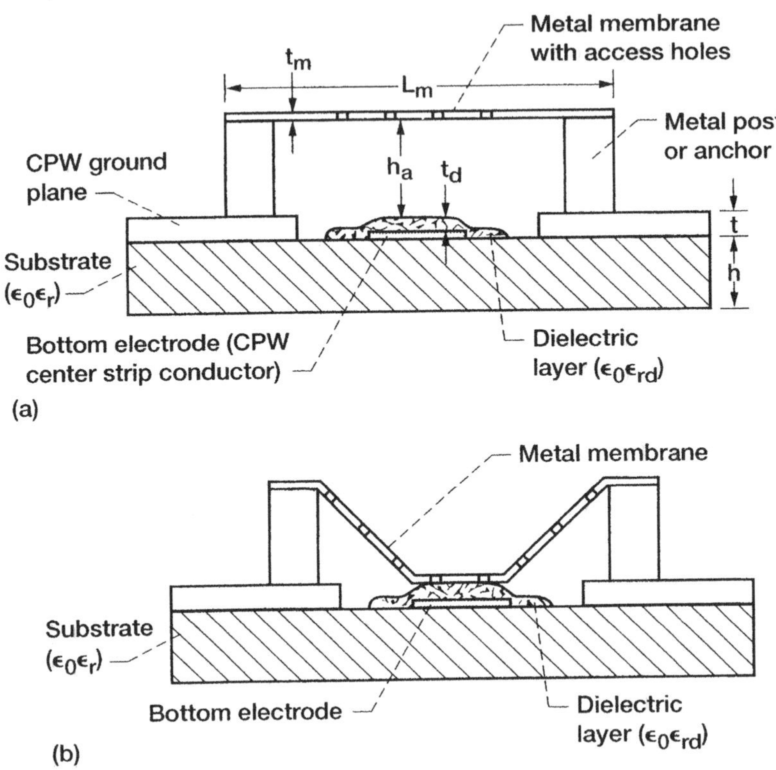

FIGURE 12.2 Cross section of the switch: (*a*) Switch is OFF when the membrane is up; (*b*) switch is ON when the membrane is down.

For example, if t_d is 2000 Å, ε_{rd} , is 6.7 (PECVD silicon nitride), contact area A is 120 μm by 120 μm, and h_a is 3.6 μm, then from Eqs. (12.1) and (12.2), the C_{OFF} and C_{ON} are 34 fF and 4.3 pF, respectively. This example clearly shows that C_{OFF} is very small compared to C_{ON} and the ratio C_{ON}/C_{OFF} is about 123. For most applications it is sufficient if the ratio C_{ON}/C_{OFF} is about 100.

12.2.2 Figure of Merit

The figure of merit of the switch in defined in terms of a cutoff frequency as follows [2]:

$$f_c = \frac{1}{2\pi R_{ON} C_{OFF}}, \tag{12.3}$$

where

R_{ON} = effective ON resistance or the parasitic resistance of the switch, which is on the order of 0.35 Ω to 0.4 Ω [2].

For the switch in the example in Section 12.2.1, the calculated f_c using Eq. (12.3) is 9000 GHz. This value contrasts with the f_c of 280 GHz and 730 GHz for a GaAs MESFET and a PIN diode, respectively, which operate at 20 GHz [2]. Thus the f_c of a MEMS switch is greater than 12 times that of the best available electronic switches. The higher value of f_c for MEMS based switches implies that these switches will operate with significantly lower loss and lower parasites than switches based on MESFETs and PIN diodes.

12.2.3 Pull Down Voltage

To obtain an expression for the pull down voltage, the MEMS switch discussed in Section 12.2.1 is represented by a simple lumped model. The model [3] consists of a parallel-plate capacitor with a fixed lower plate and a movable upper plate, which is attached to a linear spring as shown in Figure 12.3. The stiffness constant of the spring is K, the dc potential across the capacitor is V, and the gap between the capacitor plates is h_a. This model has single degree of freedom, that is, the gap between the movable and the fixed plates of the capacitor. The electrostatic force between the parallel plates of the capacitor when a voltage V is applied is given by [4]

$$F_{\text{electrostatic}} = \frac{\varepsilon_0 S W_m V^2}{2(h_a - \Delta)^2} \text{ Newtons,} \tag{12.4}$$

where

S = CPW center strip conductor width

W_m = membrane width

Δ = deflection of the plate

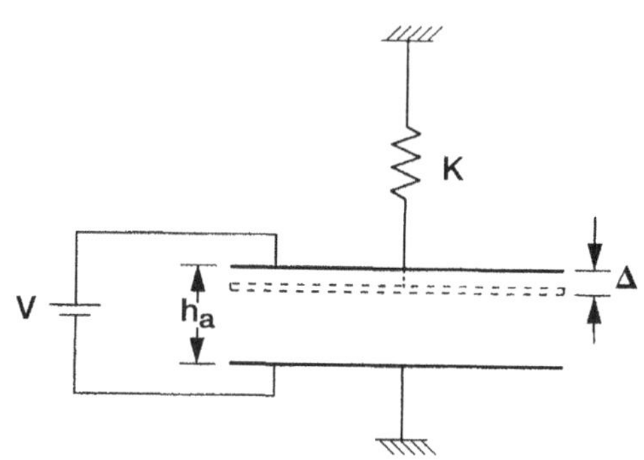

FIGURE 12.3 A simple lumped capacitor-spring model for the membrane switch.

The restoring spring force of the beam is given by [4]

$$F_{\text{spring}} = K\Delta \text{ Newtons.} \tag{12.5}$$

Since the system is in equilibrium, we can equate the electrostatic force between the parallel plates to the restoring spring force of the membrane:

$$\frac{\varepsilon_0 S W_m V^2}{2(h_a - \Delta)^2} = K\Delta. \tag{12.6}$$

Solving for the voltage, we have

$$V = (h_a - \Delta)\sqrt{\frac{2K\Delta}{\varepsilon_0 S W_m}} \text{ Volts.} \tag{12.7}$$

A generic plot of the voltage V as a function of the deflection [4] is shown in Figure 12.4. The figure shows that for deflection from zero to one-third of the initial gap height, $h_a/3$, the voltage required to hold the membrane in its deflected position increases monotonically with the deflection Δ. Further increases in deflection of the membrane result in the deflected position requiring a monotonically decreasing voltage to hold the beam. Therefore the system is unstable, and the membrane snaps down to turn ON the switch. This critical voltage is the pull down voltage. Substituting for $\Delta = h_a/3$ in Eq. (12.7) and simplifying obtains

$$V_p = \left[\frac{8K}{27\varepsilon_0 S W_m} h_a^3\right]^{1/2} \text{ Volts.} \tag{12.8}$$

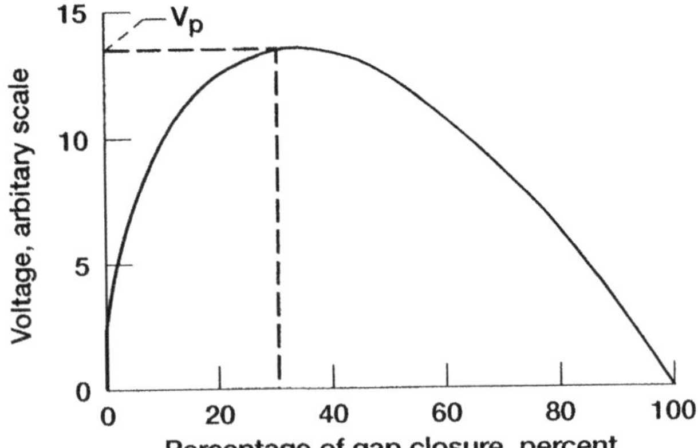

FIGURE 12.4 The characteristic relationship between the applied voltage and the percentage of gap closure. (From Reference [4], © IEEE 1997.)

The spring constant K is given by [3]

$$K = \frac{32Et_{\mathrm{m}}^3 W_{\mathrm{m}}}{L_{\mathrm{m}}^3} + \frac{8\sigma(1-\nu)t_{\mathrm{m}} W_{\mathrm{m}}}{L_{\mathrm{m}}} \text{ Newtons/meter,} \tag{12.9}$$

where

E = Young's modulus of the membrane material

σ = internal residual tensile stress in the membrane

ν = Poison's ratio for the membrane material

t_{m} = membrane thickness

L_{m} = membrane length.

For the switch in the example of Section 12.2.1, if $W_{\mathrm{m}} = 120\ \mu\mathrm{m}$, $L_{\mathrm{m}} = 280\ \mu\mathrm{m}$, $E = 70$ MPa, $\sigma = 120$ MPa, and $\nu = 0.33$, then $K = 138$ MPa·m. Further, if $S = 120\ \mu\mathrm{m}$ and $h_{\mathrm{a}} = 3.6\ \mu\mathrm{m}$, then the calculated V_{p} from Eq. (12.8) is $V_{\mathrm{p}} \approx$ 122 V. This valve is about 2.5 times higher than the actuation voltage of 50 V used in the practical switch [2]. Nevertheless, the model serves the purpose of providing a first-cut analysis to gain physical insight and understand of the actuation mechanism. To accurately predict V_{p}, a three-dimensional numerical simulation must be carried out.

12.2.4 Fabrication Process

Surface micromachining techniques are utilized to fabricate the switches. Typically the process requires about five mask layers. In [1] and [2] high-resistivity silicon ($\rho > 5\ \mathrm{K}\Omega\cdot\mathrm{cm}$) and aluminum are chosen as the materials for the substrate and the conductors, respectively. In addition silicon nitride ($\varepsilon_{\mathrm{rd}} = 6.7$) is used as a insulator between the conductors. These materials allow the possibility of integrating CMOS devices with MEMS devices to construct multifunction assemblies [2]. The fabrication process steps are schematically illustrated in Figure 12.5, and briefly described below:

1. A 1 μm thick layer of silicon dioxide is thermally grown over the silicon wafer and serves as a buffer layer.
2. A 0.5 μm thick layer of refractory metal such as tungsten is sputtered and patterned to define the bottom electrode of the switch. Tungsten gives a smooth surface finish to the bottom electrode. The good surface finish minimizes the air gaps and ensures good electrical contact when the membrane is in the down position.
3. A 2000 Å thick layer of PECVD silicon nitride ($\varepsilon_{\mathrm{rd}} = 6.7$) is deposited and patterned to insulate the bottom electrode. This film prevents the dc

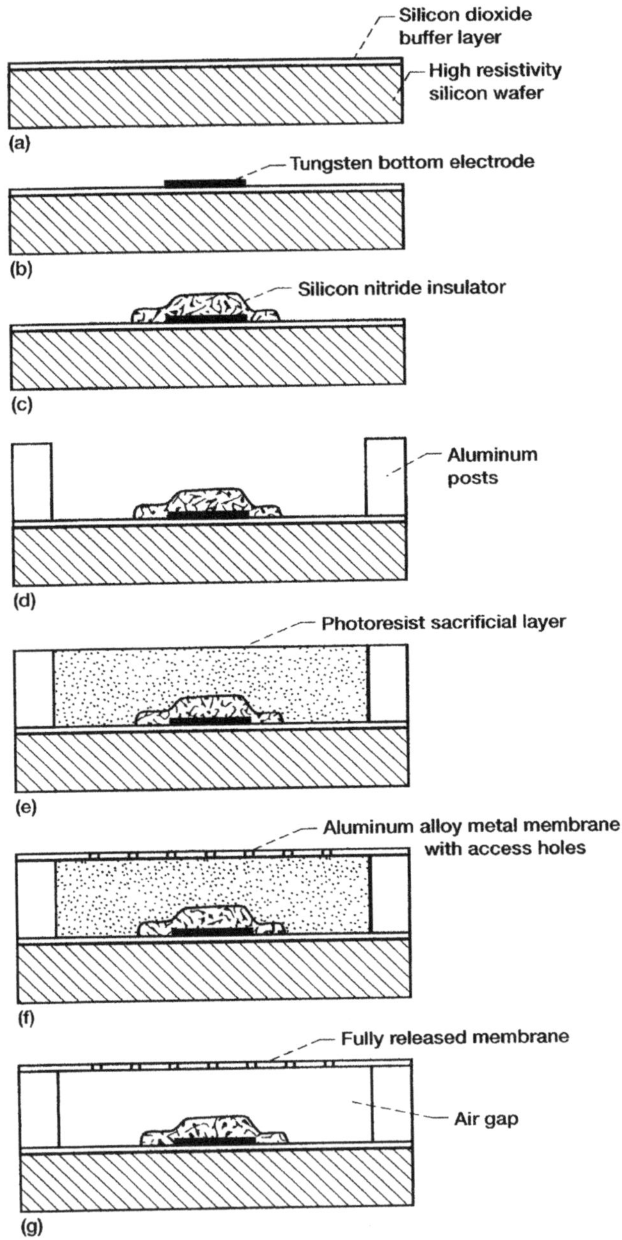

FIGURE 12.5 Schematic illustration of process flow: (*a*) Silicon dioxide deposition; (*b*) tungstun bottom electrode deposition; (*c*) silicon nitride insulator deposition and patterning; (*d*) aluminium metal posts deposition and patterning; (*e*) photoresist sacrificial spacer layer spin coating and patterning; (*f*) Aluminum alloy metal membrane deposition and patterning; (*g*) sacrificial spacer layer removal by oxygen plasma etch.

control voltage, applied between the membrane and the bottom electrode, from shorting out when the membrane is pulled down. This film thickness is sufficient thin to allow RF signals to capacitively couple from the membrane to the bottom electrode.

4. A 4 μm thick aluminium layer is evaporated and wet-etched using a commercial etchant to define the CPW conductors and the post for the membrane. The attenuation of RF signals on aluminium CPWs is about 0.06 dB/mm at 10 GHz [1], which is considered to be small.
5. A photoresist sacrificial spacer layer is spin-coated and patterned.
6. An aluminum alloy membrane layer is sputtered and wet-etched. The thickness of the membrane is less than 0.5 μm.
7. The sacrificial spacer layer is removed by oxygen plasma etch to release the membrane. To accelerate the release rate 2 μm square openings are provided in the aluminum alloy membrane.

12.2.5 Switching Time and Switching Energy

The measured switching speed for the switch in the example in Section 12.2.1 shows that, the time taken for the switch to turn on and off is on the order of 5.3 and 3.5 μs, respectively [2], when the RF power is in the few mW range. The corresponding pull down voltage and the energy needed to turn on the switch are 50 V and about 10 nJ, respectively [2]. The switch is turned off by simply turning off the supply voltage. If we assume that this cycle can take place at a rate of 5 KHz, then the drive power that is expended is 50 μW. Finally, the switch lifetime test results reported in [2] shows that, the membranes structural integrity is maintained, even after subjecting it to 500 million switching cycles, as evident from the absence of mechanical failures.

In general, the MEMS switches operate at a slower speed than electronic switches. For example, the above switching time of few microseconds contrasts with the switching time of GaAs MESFET and PIN diodes which is about 150 ps and 1 ns, respectively [5], when the RF power is in the mW range. The drive power required in the case of a MESFET switch is higher than a MEMS switch. For example, the driver circuit power expended in turning on the above MESFET switch in 150 ps at a switching frequency of 5 MHz is about 114 mW [5].

12.2.6 Insertion Loss and Isolation

The measured insertion loss of the switch in the example in Section 12.2.1 when it is OFF, that is, when the membrane is in the up position, is shown as a function of the frequency in Figure 12.6 [2]. The insertion loss is observed to increase by a few tenths of a dB as the frequency is swept from 0 to 40 GHz. For example, at 20 GHz and 35 GHz the insertion loss is about 0.14 dB and 0.25 dB, respectively, which is considered to be very small. The measured

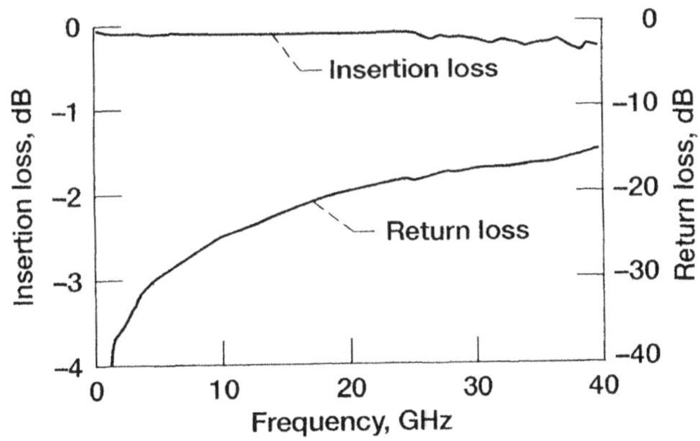

FIGURE 12.6 Measured insertion loss and return loss when the switch is OFF or the membrane is in the up position. (From Reference [2], © IEEE 1999.)

return loss is better than 15 dB over the entire frequency range. This shows that the switch is well matched to the characteristic impedance of the CPW. Good impedance match is possible because of the low C_{OFF}, which is on the order of few tens of fF.

The measured isolation of the switch when it is ON, that is, when the membrane is in the down position, is shown as a function of the frequency in Figure 12.7 [2]. The isolation is observed to monotonically improve as the frequency is swept from 0 to 40 GHz. For example, at 10 GHz and 35 GHz

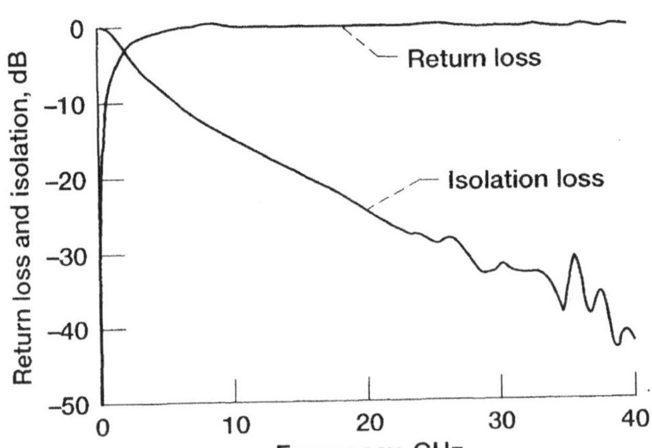

FIGURE 12.7 Measured isolation and return loss when the switch is ON or the membrane is in the down position. (From Reference [2], © IEEE 1999.)

the isolation is about 15 dB and 35 dB, respectively, which is considered to be acceptable for switching RF signals. The isolation below 10 GHz is mainly limited by C_{ON}, which is on the order of few pF. Higher isolation at lower frequencies can be obtained by increasing the size of the bottom electrode or cascading series and shunt versions of the switch.

There are several publications which report the insertion loss and isolation of GaAs MESFET [6] and PIN diode [7] switches. In [6] it is shown that the insertion loss and isolation of recessed-back GaAs MMIC FET switch is on the order of 0.5 dB and 70 dB, respectively over the frequency range of 8.4 to 12.0 GHz. In [7] it is shown that the insertion loss and isolation of vertical GaAs PIN diode switch is on the order of 1.0 dB and 45 dB, respectively, over the frequency range of 100 MHz to 20.0 GHz.

There are several other designs available in the literature for the construction of MEMS switches. For example, in [8] switches using serpentine and cantilever springs for low actuation voltage are reported. In [9] and [10] a cantilevered arm with electrical contacts is electrostatically actuated to close the gap in the signal line. These switches have an isolation $\geqslant 50$ dB and insertion loss $\leqslant 0.2$ dB at few GHz frequencies.

12.3 MEMS COPLANAR WAVEGUIDE DISTRIBUTED PHASE SHIFTER

A coplanar waveguide distributed true-time delay phase shifter [11], [12] is schematically shown in Figure 12.8. In this phase shifter, the CPW is periodically loaded with a MEMS air bridge to provide capacitance control. The capacitance control is brought about by the application of a dc voltage between the CPW center strip conductor and the air-bridge. The control voltage sets up an electrostatic force of attraction that lowers the height of the air-bridge and increases the parallel-plate capacitance, C_{MEMS}, formed by the air-bridge and the center strip conductor. If the distance of separation between the air-bridges is d. Then the distributed MEMS capacitance per unit length is C_{MEMS}/d. Assuming that the circuit is lossless and a TEM mode propagates on the CPW along the z-direction, the lumped element equivalent circuit model for an incremental length of the loaded CPW is as shown in Figure 12.9.

The phase velocity of the electromagnetic wave on the loaded CPW is given by the well-known relation

$$v_l = \frac{1}{\sqrt{L_u[C_u + (C_{MEMS}/d)]}}, \tag{12.10}$$

where L_u and C_u are the inductance and capacitance per unit length, respectively, of the unloaded CPW. These quantities are related to the effective dielectric constant ε_{eff}, the unloaded characteristic impedance of the CPW Z_0, and the velocity of light in free space c, through the following well-known

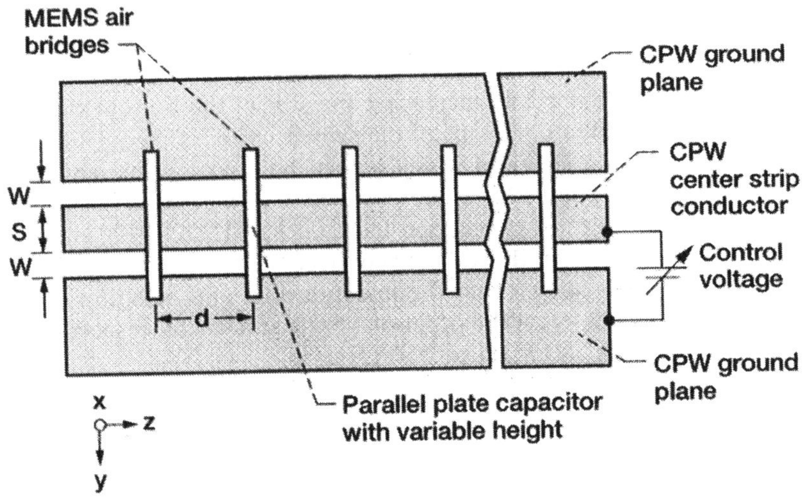

FIGURE 12.8 Schematic of a coplanar waveguide periodically loaded with MEMS air-bridge variable capacitors.

relations

$$L_u = C_u Z_0^2, \tag{12.11}$$

$$C_u = \frac{\sqrt{\varepsilon_{\text{eff}}}}{cZ_0}. \tag{12.12}$$

The characteristic impedance of the loaded CPW is given by the well-known relation

$$Z_l = \sqrt{\frac{L_u}{[C_u + (C_{\text{MEMS}}/d)]}}. \tag{12.13}$$

Thus from Eq. (12.10) it is evident that by increasing C_{MEMS}/d, the phase velocity v_l can be reducted. If v_{l1} and v_{l2} are the phase velocities for two

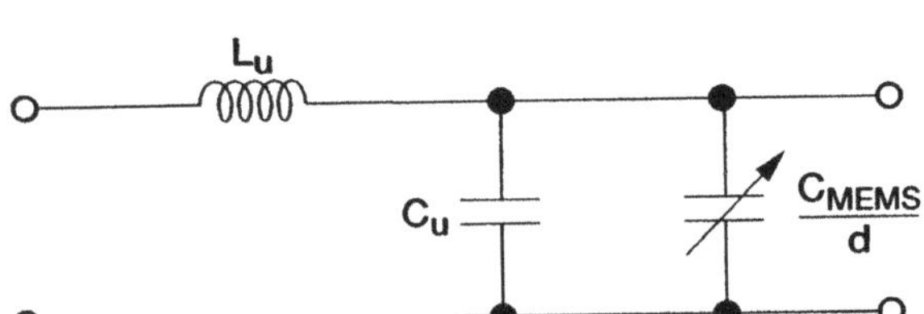

FIGURE 12.9 Lumped element equivalent circuit model for an incremental section of the loaded coplanar waveguide.

different settings of the control voltage, then the differential phase shift in degrees per meter assuming $v_{l1} > v_{l2}$ is

$$\Delta\phi = 360f\left(\frac{1}{v_{l2}} - \frac{1}{v_{l1}}\right)\text{deg/meter}, \tag{12.14}$$

where f is the frequency of operation.

The periodic structure has an upper frequency limit due to the Bragg reflection occuring at [11]

$$f_{\text{Bragg}} = \frac{1}{\pi d\sqrt{L_u[C_u + (C_{\text{MEMS}}/d)]}}. \tag{12.15}$$

12.3.1 MEMS Air-Bridge Capacitance

The capacitance C_{MEMS} can be controlled by either lowering the height of or by snapping down the air-bridge [11], [12]. Each of these options will be discussed below and expressions for the capacitances will be presented.

An isolated MEMS air-bridge capacitor, whose height can be varied through electrostatic actuation, is schematically illustrated in Figure 12.10. The capacitance of the air-bridge for a height h_a is given by

$$C_{\text{MEMS}} = \frac{W_m S \varepsilon_0}{h_a}, \tag{12.16}$$

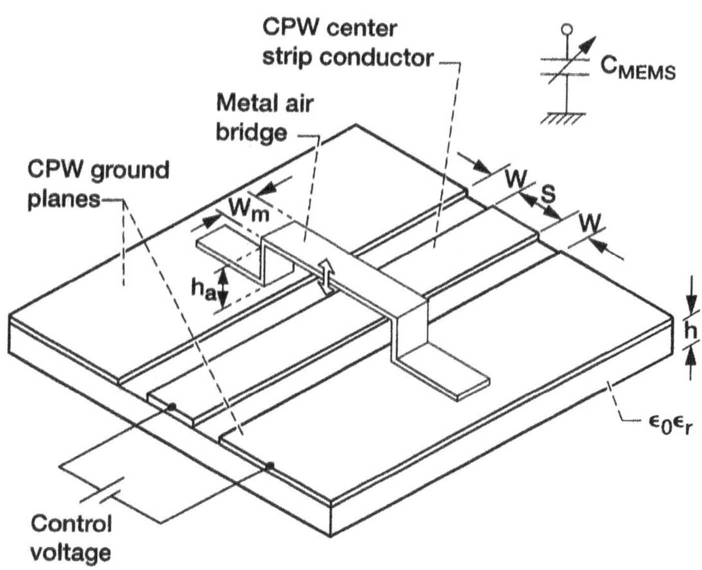

FIGURE 12.10 Schematic of a MEMS air-bridge capacitor with variable height. Inset shows the equivalent capacitance.

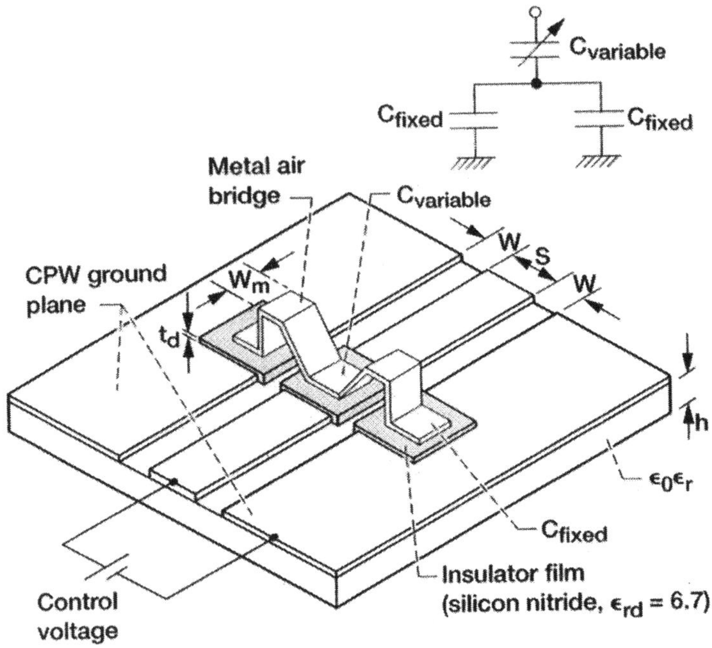

FIGURE 12.11 Schematic of a MEMS air-bridge capacitor in the snap down position and the bridge post loaded with fixed thin film capacitors. Inset shows the equivalent series capacitance.

where W_m and S are the air-bridge and the CPW center strip conductor widths, respectively.

An isolated MEMS air bridge in the snap down position is schematically illustrated in Figure 12.11. In this circuit besides the variable air-bridge capacitance, C_{variable}, fixed capacitances, C_{fixed}, are added to the air-bridge posts to enhance the capacitive loading.

These two capacitances appear in series as shown in the inset in Figure 12.11, and therefore the total capacitance can be expressed as

$$C_{\text{MEMS}} = 2\frac{C_{\text{variable}}C_{\text{fixed}}}{C_{\text{variable}} + 2C_{\text{fixed}}}, \tag{12.17}$$

where

$$C_{\text{variable}} = \frac{W_m S \varepsilon_0 \varepsilon_{rd}}{t_d}, \tag{12.18}$$

and ε_{rd} and t_d are the dielectric constant and thickness of the insulating film.

12.3.2 Fabrication and Measured Performance

The fabrication of the MEMS air bridge capacitors is similar to the MEMS metal membrane capacitors discussed in Section 12.2.4. The measured performance of two phase shifters with MEMS capacitors shown in Figures 12.10 and 12.11, respectively, are compared in Table 12.1.

TABLE 12.1 Phase Shifter Parameters and Measured Characteristics

	From Reference [11]	From Reference [12]
	CPW parameters	
Substrate	Quartz, $\varepsilon_r = 3.78$ $\tan\delta = 0.0001$ $h = 500\ \mu m$	Glass, $\varepsilon_r = 5.7$ $\tan\delta = 0.001$ —
Center conductor width, μm	100	100
Slot width, μm	100	60
Characteristic impedance (unloaded), Ω	—	69
	MEMS capacitors	
Bridge width, μm	35	40
Bridge length or span, μm	300	300
Spacing between bridges, μm	197	780
Total number of bridges	38	11
Bridge height above the center strip conductor, μm	1.2	3.0
Capacitance in UP state or zero bias (fF)	34	10
Capacitance in DOWN state or with bias (fF)	—	75
C_{max}/C_{min}	1.17	7.5
Bias voltage, V	13	75
	Conductor and dielectric thickness	
Center conductor, μm	Au: 0.8	Ti/Au/Ti = 0.02/0.7/0.02
Ground planes, μm	Au: 2.5	Ti/Au/Ti = 0.02/0.7/0.02
Bridges, μm	Au: 2.5	Ti/Au = 0.02/1
SiN layer, Å	—	5000
Total length, mm	7.6	8.58
Measured insertion loss for 360° differential phase shift, dB	5.1 @ 40 GHz 4.0 @ 60 GHz	2.34 @ 25 GHz 2.25 @ 35 GHz
Measured return loss, dB	Better than 10 dB over 0–43 GHz	Better than 11 dB over 0–35 GHz

12.4 HIGH-TEMPERATURE SUPERCONDUCTING COPLANAR WAVEGUIDE CIRCUITS

The Nobel Prize winning discovery in 1986, of high-transition-temperature superconductivity in the Ba-La-Cu oxide system [13] renewed the interest in superconducting circuits for high-frequency applications. Since that time several other oxide superconductors such as $YBa_2Cu_3O_{7-x}$ with transition temperature above the boiling point of liquid nitrogen, 77 K, have been discovered [13] and their application in high-frequency circuits has been demonstrated [14], [15]. These new oxide superconductors have been called high-temperature superconductors (HTS) in order to distinguish this class of superconductors from those studied prior to 1986 which have a critical temperature, T_c, below 77 K.

Most high-frequency applications require epitaxial c-axis oriented thin films for good electrical properties. Such films are usually obtained either by laser ablation or by sputtering from stoichiometric targets. In this section we focus mainly on the characteristics and applications of $YBa_2Cu_3O_{7-x}$ thin films. The substrate materials that are most widely used with $YBa_2Cu_3O_{7-x}$ are MgO ($\varepsilon_r = 10$ at 77 K) and $LaAlO_3$ ($\varepsilon_r = 24$ at 77 K).

12.4.1 High-Frequency Electrical Properties of Normal Metal Films

Consider a plane wave incident normal to the surface of a good conductor. As the wave propagates inward it suffers rapid attenuation. The distance at which the wave is attenuated to $1/e$ (36.8 percent) of its initial value is called the skin depth. The skin depth is commonly denoted as δ, and it is expressed as [16]

$$\delta = \sqrt{\frac{2}{\omega\mu\sigma(T)}}, \tag{12.19}$$

where

ω = angular frequency = $2\pi f$
$\mu = \mu_o = 4\pi \times 10^{-7}$ H/m
$\sigma(T)$ = the conductivity of the metal in Siemens/meter at a temperature T

One can easily observe from Eq. (12.19) that the skin depth varies inversely as the square root of the frequency.

The surface resistance of a sheet of conductor with thickness δ and conductivity $\sigma(T)$ is given by [16]

$$R_s = \frac{1}{\sigma(T)\delta}\ \Omega/\text{square}. \tag{12.20}$$

Equation (12.20) is valid as long as the conductor thickness is much greater than the skin depth. The equation shows that the surface resistance varies as the square root of the frequency. Last, since σ is a function of temperature, both δ and R_s are temperature dependent. As an example, the conductivity of copper is 5.8×10^7 S/m at room temperature, and hence at 10 GHz, δ and R_s are 0.66 μm and 26 mΩ/square, respectively. As the frequency is increased to 20 GHz, δ decreases to 0.47 μm while R_s increases to 36.9 mΩ/square. On the other hand, as the temperature is reduced to 77 K, the conductivity of copper increases to 4.5×10^8 S/m, and at 10 GHz both δ and R_s decrease to 0.33 μm and 13 mΩ/square, respectively. It is important to note that the surface resistance R_s is a measure of the time-average power dissipated per unit area of surface [16]. Hence the attenuation constant due to conductor losses α_c in any transmission line is proportional to R_s [16]. In the next section, and in Section 12.4.6, the reduction in α_c through the use of superconductors is discussed.

12.4.2 High-Frequency Electrical Properties of Epitaxial High-T_c Superconducting Films

The main advantage of using superconductors instead of normal conductors in high-frequency circuits, is the reduction of conductor losses. Therefore the most important electrical parameter is the surface resistance of the superconductor. The surface resistance is the real part of the surface impedance, Z_s, which is generally expressed as

$$Z_s = R_s + jX_s. \tag{12.21}$$

The surface impedance, Z_s, of $YBa_2Cu_3O_{7-x}$ in the local London limit, that is, when the electronic mean free path is larger than the superconducting coherence length, can be described using the two-fluid model [17]. According to this model

$$Z_s = \frac{\omega^2 \mu_o^2 \lambda_L^3(T)\sigma_1}{2} + j\omega\mu_o\lambda_L(T), \tag{12.22}$$

where $\lambda_L(T)$ is the London or the magnetic penetration depth which is a function of the temperature T, σ_1 is the real part of the complex conductivity.

Comparing Eqs. (12.21) and (12.22), we obtain the surface resistance and surface inductance

$$R_s = \frac{\omega^2 \mu_o^2 \lambda_L^3(T)\sigma_1}{2} \tag{12.23}$$

and

$$L_s = \omega\mu_0\lambda_L(T), \tag{12.24}$$

respectively. Notice that the surface resistance for a superconductor varies as the square of the frequency and as the cube of the penetration depth. Also the surface inductance is directly proportional to the penetration depth. Later on, when we discuss the quality factor and the resonant frequency of coplanar waveguide resonators, typical values for R_s will be presented.

The penetration depth λ_L is dependent on the temperature. A mathematical relation for the exact temperature dependance for high-T_c superconductors has not yet been established. However, based on experimental observations, the temperature dependence can be described by [17]

$$\lambda_L(T) = \frac{\lambda_L(0)}{\sqrt{1 - 0.1(T/T_c) - 0.9(T/T_c)^2}}, \tag{12.25}$$

where

T = temperature
T_c = critical temperature
$\lambda_L(0)$ = penetration depth at the reference temperature $T = 0\,\text{K}$

In the equation above the material constant $\lambda_L(0)$ is experimentally determined for several high-T_c $YBa_2Cu_3O_{7-x}$ thin film samples [17] and are listed in Table 12.2. Notice that $\lambda_L(0)$ increases with the transition width ΔT_c. Where as there is no significant dependence of $\lambda_L(0)$ on the critical current density J_c. Thus, for high-quality films, $\lambda_L(0)$ are in the range of 160 to 200 nm. Last, unlike the skin depth δ for normal conductors, the penetration depth $\lambda_L(0)$ for supercon-

TABLE 12.2 Measured Magnetic Penetration Depth[a] for Epitaxial c-Axis Oriented $YBa_2Cu_3O_{7-x}$ Thin Film Samples Deposited by dc-Magnetron Sputtering

Sample Number	Circuit Number	J_c (10^6 A/cm^2)	T_c (K)	ΔT_c (K)	$\lambda_L(0)$ (nm)
1	1	1.7	90.0	0.5	160
	2				175
2	1	3.0	90.5	0.8	200
	2				185
3	1	4.2	91.0	0.5	155
	2				165
	3				170
4	1	1.0	86.5	2.5	240
	2				270
	3				250

[a] Accuracy ±30 nm.
Note: Film thickness = 200 nm:
Source: From Reference [17], © 1993 American Institute of Physics.

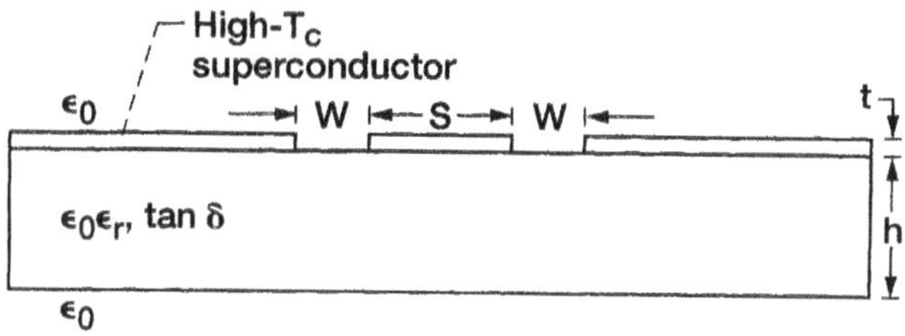

FIGURE 12.12 A high-T_c superconducting coplanar waveguide.

ductors is an intrinsic property of the material, and it is independent of the frequency.

12.4.3 Kinetic and External Inductances of a Superconducting Coplanar Waveguide

A high-T_c superconducting CPW and its lumped element equivalent circuit model for an incremental section are shown in Figures 12.12 and 12.13. In this model the total series inductance per unit length, L, consists of two parts, namely the internal inductance or the kinetic inductance per unit length, which accounts for magnetic energy storage within the conductor, and the external inductance per unit length, which accounts for magnetic energy storage external to the conductor [17]. That is,

$$L = L_{\text{kinetic}} + L_{\text{external}}. \tag{12.26}$$

The kinetic inductance originates from the inertia of the paired carriers in the superconductor [17], while the external inductance arises from flux linkages [16].

The kinetic inductance per unit length, L_{kinetic}, is given by [17]

$$L_{\text{kinetic}} = \mu_0 \lambda_L(T) \frac{C}{4ADK(k)} \left\{ \frac{1.7}{\sinh(t/2\lambda_L)} + \frac{0.4}{\sqrt{[(B/A)^2 - 1][1-(B/D)^2]}} \right\} \tag{12.27}$$

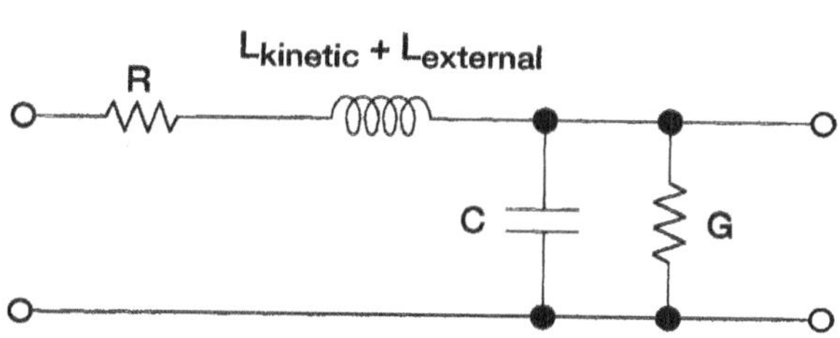

FIGURE 12.13 Lumped element equivalent circuit model for an incremental section of the superconducting coplanar waveguide.

where

$$A = -\frac{t}{\pi} + \frac{1}{2}\sqrt{\left(\frac{2t}{\pi}\right)^2 + S^2}, \tag{12.28}$$

$$B = \frac{S^2}{4A}, \tag{12.29}$$

$$C = B - \frac{t}{\pi} + \sqrt{\left(\frac{t}{\pi}\right)^2 + W^2}, \tag{12.30}$$

$$D = \frac{2t}{\pi} + C, \tag{12.31}$$

$$\left.\begin{aligned} t &= \text{superconductor thickness} \\ S &= \text{CPW center strip conductor width} \\ W &= \text{CPW slot width} \\ \mu_0 &= 4\pi \times 10^{-7}\text{H/m} \end{aligned}\right\}. \tag{12.32}$$

Notice that L_{kinetic} is a function of the penetration depth $\lambda_L(T)$.

The external inductance per unit length, L_{external}, is given by [17]

$$L_{\text{external}} = \frac{\mu_0}{4}\frac{K(k')}{K(k)}, \tag{12.33}$$

where $K(k)$ and $K(k')$ are the complete elliptic integrals of the first kind with their modules given by

$$k = \frac{S}{S + 2W} \tag{12.34}$$

$$k' = \sqrt{1 - k^2}. \tag{12.35}$$

As examples, the calculated external and kinetic inductances per unit length for three CPW structures [18] are tabulated in Table 12.3 and graphically illustrated in Figure 12.14. The results in Table 12.3 and Figure 12.14 show that the external and kinetics inductances per unit length increases as the aspect ratio $S/(S + 2W)$ of the CPW decreases. In addition for a fixed aspect ratio, the kinetic inductance increases as the penetration depth increases.

12.4.4 Resonant Frequency and Unloaded Quality Factor

Typical geometry of a linear and meander high-T_c superconducting coplanar waveguide series capacitively coupled resonators [19] are shown in Figure 12.15. A schematic showing the flip-chip assembly of the resonator within a microwave package for characterization [19] is shown in Figure 12.16. The measured normalized resonant frequency as a function of the normalized temperature for a CPW resonator is shown in Figure 12.17. In this figure T_0

TABLE 12.3 Calculated External Inductance per Unit Length for CPW Lines on a 0.5 mm Thick $LaAlO_3$ Substrate ($\varepsilon_r = 24$)

CPW Structure	Center Strip Conductor Width, S (μm)	Slot Width, W (μm)	Aspect Ratio $S/(S+2W)$	Conductor Thickness (μm)	External Inductance per Unit Length (nH/m)
1	74	46	0.446	0.3	427.6
2	34	26	0.395	0.3	454.0
3	8	13	0.235	0.3	562.7

Source: From Reference [18], © 1993 Trans Tech Publications.

is a fixed low temperature that is normally, the lowest attainable temperature of the measuring system. Notice that the resonant frequency decreases as the temperature increases. The decrease in resonance frequency occurs because the penetration depth increases with temperature, as given by Eq. (12.25).

The unloaded quality factor of a series resonator, assuming that G in the equivalent circuit model of Figure 12.13 is negligible, is given by [20]

$$Q_0 = \frac{\omega_0 L}{R}, \tag{12.36}$$

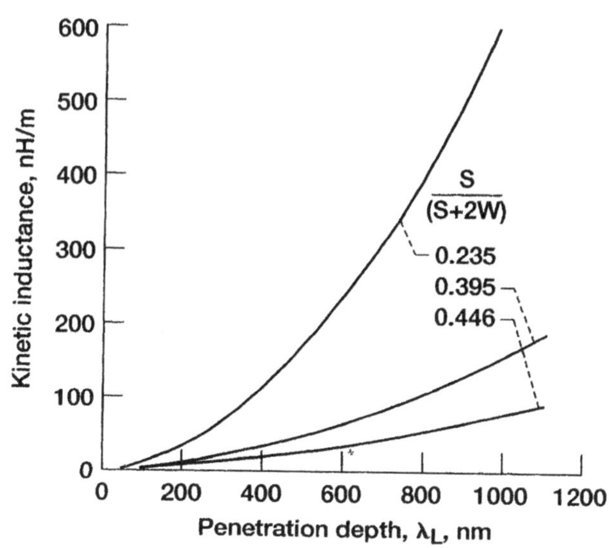

FIGURE 12.14 Computed kinetic inductance per unit length as a function of the London penetration depth for coplanar waveguide with aspect ratio as parameter. (From Reference [18], © 1993 Trans Tech Publications.)

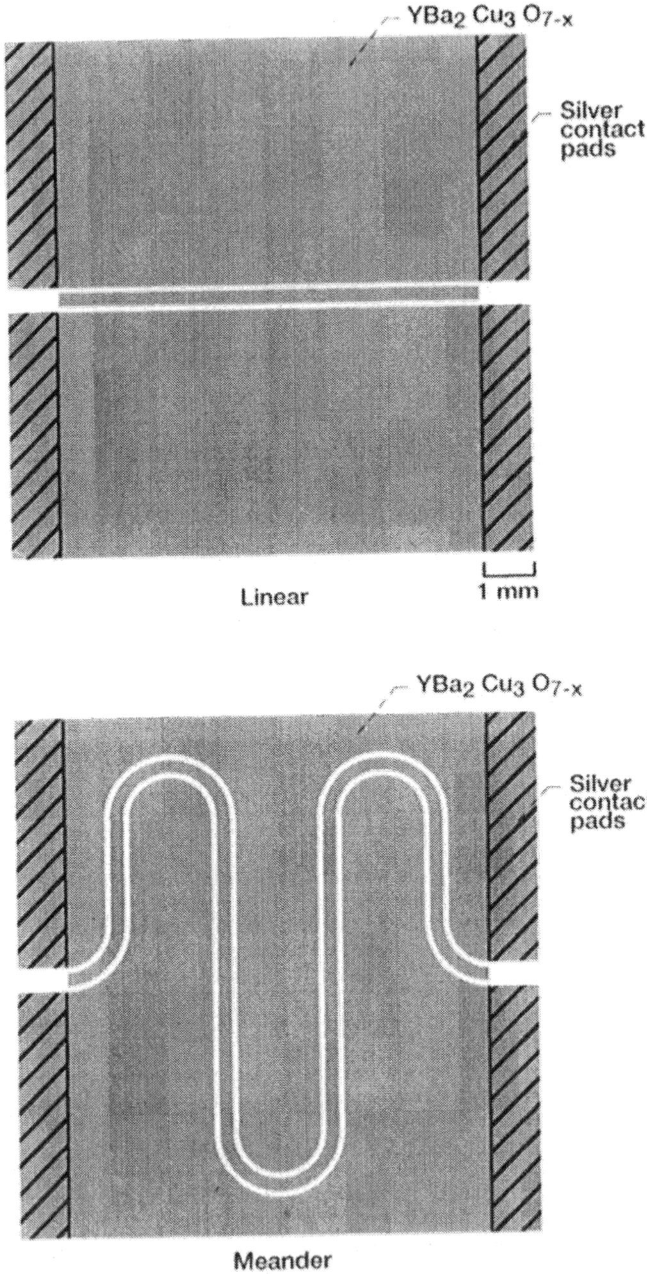

FIGURE 12.15 Typical geometry of a linear and meander high-T_c superconducting coplanar waveguide series capacitively coupled resonators. The shaded areas at the end of each ground plane represents the silver contact pads. (From Reference [19], © 1995 IEEE.)

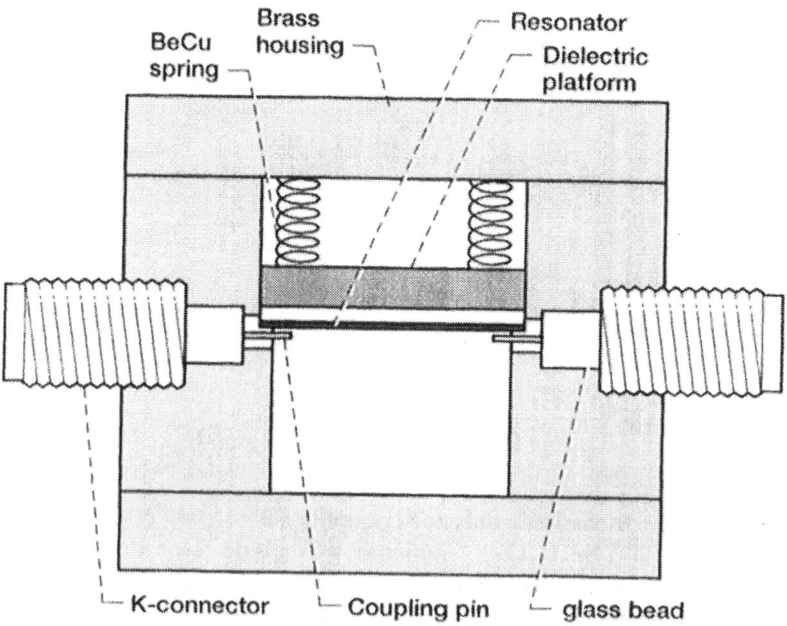

FIGURE 12.16 Schematic showing the flip chip assembly of the superconducting resonator within a microwave package mounted inside a closed-cycle cryo-cooler. (From Reference [19], © 1995 IEEE.)

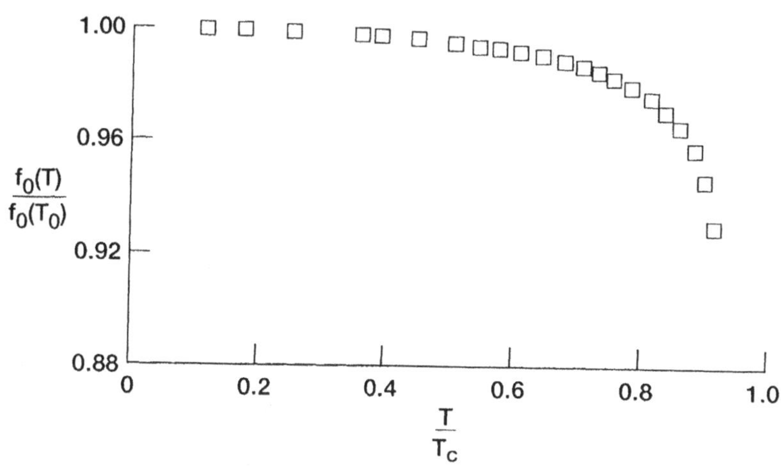

FIGURE 12.17 Measured normalized resonance frequency as a function of the temperature for a $YBa_2C_3O_{7-x}$ coplanar waveguide resonator. For material characteristics and resonator dimensions refer to data for sample 3/circuit 1 in Tables 12.2 and 12.4. (From Reference [17], © 1993 American Institute of Physics.)

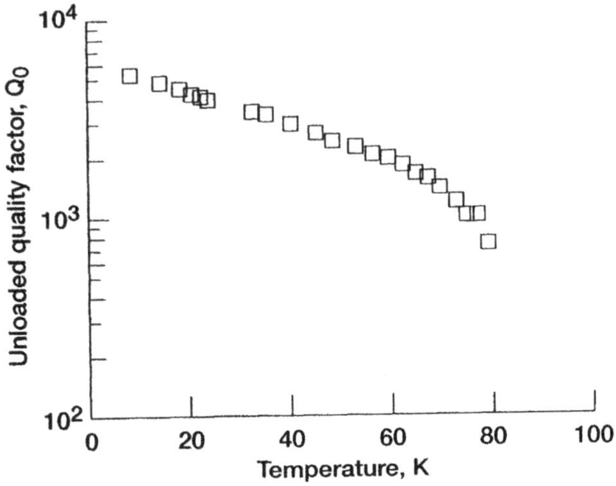

FIGURE 12.18 Measured unloaded quality factor as a function of the temperature for a $YBa_2C_3O_{7-x}$ coplanar waveguide resonator on a $LaAlO_3$ substrate. For material characteristics and resonator dimensions refer to data for sample 3/circuit 3 in Tables 12.2 and 12.4. The f_0 is 6.2 GHz. (From Reference [17], © 1993 American Institute of Physics.)

where

L = total series inductance per unit length
R = series resistance per unit length
ω_0 = angular frequency at resonance equal to $2\pi f_0$.

The quality factor as determined from a practical CPW resonator, is usually the loaded quality factor Q_L. At resonance the unloaded quality factor Q_0 is related to the loaded quality factor Q_L and the insertion loss L_0 in dB through the relation [21]

$$Q_0 = \frac{Q_L}{1-10^{-(L_0/20)}}. \tag{12.37}$$

The measured unloaded quality factor, Q_0, for a CPW resonator as a function of the temperature is shown in Figure 12.18. Notice that Q_0 decreases as the temperature increases. The reason for the decrease will be evident from the equations in the next section.

12.4.5 Surface resistance of High-T_c Superconducting Coplanar Waveguide

Temperature Dependence The two-fluid model is used in explaining the current flow in high-T_c superconductors. The model assumes that the total conduction current density $\vec{J}$ consists of the superconducting current density $\vec{J}_s$ and the normal electron current density $\vec{J}_n$. That is,

$$\vec{J} = \vec{J}_s + \vec{J}_n. \tag{12.38}$$

The current density $\vec{J}$ is related to the electric field $\vec{E}$ through the relation

$$\vec{J} = \sigma \vec{E}, \tag{12.39}$$

where σ is the complex conductivity expressed as

$$\sigma = \sigma_1 - j\sigma_2. \tag{12.40}$$

The real part of the complex conductivity σ_1 is the normal state conductivity of the high-T_c superconductor determined at $T = T_c$. The imaginary part of the complex conductivity σ_2 is equal to [17].

$$\sigma_2 = \frac{1}{\omega \mu_0 \lambda_L^2(T)}, \tag{12.41}$$

where $\lambda_L(T)$ is the temperature dependent penetration depth given by Eq. (12.25).

From the two-fluid model, the series resistance per unit length, R, of the high-T_c superconducting CPW is given by [22]

$$R = \frac{\sigma_1 \omega L_{\text{kinetic}}}{\sigma_2}, \tag{12.42}$$

where L_{kinetic} is the penetration depth dependent kinetic inductance given by Eq. (12.27).

By substituting for R and σ_2 from Eqs. (12.36) and (12.41) into Eq. (12.42) and rearranging the terms, the real part of the complex conductivity σ_1 at $\omega = \omega_0$ is obtained:

$$\sigma_1 = \frac{L}{\omega_0 \mu_0 Q_0 L_{\text{kinetic}} \lambda_L^2(T)}. \tag{12.43}$$

TABLE 12.4 Surface Resistance Determined from Measured Quality Factor and Resonance Frequency of CPW Resonators Fabricated on $YBa_2Cu_3O_{7-x}$ Samples of Table 12.2

Sample Number	Circuit Number	CPW Resonant S (μm)	W (μm)	Aspect Ratio $S/(S+2W)$	Surface Resistance R_s ($\mu\Omega$/square) $T = 10$ K	$T = 77$ K
1	1	80	40	0.5	150	356
	2	25	12.5	0.5	38	215
2	1	40	20	0.5	50	203
	2	25	12.5	0.5	84	400
3	1	80	40	0.5	170	280
	2	40	20	0.5	47	150
	3	25	12.5	0.5	65	105
4	1	80	40	0.5	270	>1 mΩ
	2	40	20	0.5	310	—
	3	25	12.5	0.5	390	—

Note: Resonator length = 6.5 mm; Resonance frequency = 6.2 GHz.
Source: From Reference [17], © 1993 American Institute of Physics.

Finally, by substituting for σ_1 from Eq. (12.43) into Eq. (12.23), the surface resistance R_s at $\omega = \omega_0$ is obtained:

$$R_s = \frac{\omega_0 \mu_0 L \lambda_L(T)}{2Q_0 L_{\text{kinetic}}}. \tag{12.44}$$

For our example, the surface resistance obtained from the measurements of the quality factor and the resonance frequency of CPW resonators [17], fabricated on $YBa_2Cu_3O_{7-x}$ samples of Table 12.2, is summarized in Table 12.4. The results in Table 12.4 show that the lowest achieved value of R_s is about 38 $\mu\Omega$/square and 105 $\mu\Omega$/square at the temperatures of 10 K and 77 K, respectively, when the frequency is 6.2 GHz. When the R_s values in Table 12.4 are compared with the J_c and ΔT_c values in Table 12.2, it is observed that R_s increases significantly when J_c is less than 10^6 A/cm^2 at 77 K and ΔT_c is greater than 2 K.

Frequency Dependence The experimentally determined surface resistance R_s for $YBa_2Cu_3O_{7-x}$ at 77 K over the frequency range of 5 to 20 GHz [17] is shown in Figure 12.19. For comparison the surface resistance R_s for copper at 77 K is also included. It is observed that at 77 K and 10 GHz, the R_s of $YBa_2Cu_3O_{7-x}$ is about 60 times lower than that of cooled copper. Furthermore the experimental data points appear to lie on a line that has a quadratic frequency dependance, consistent with Eq. (12.23).

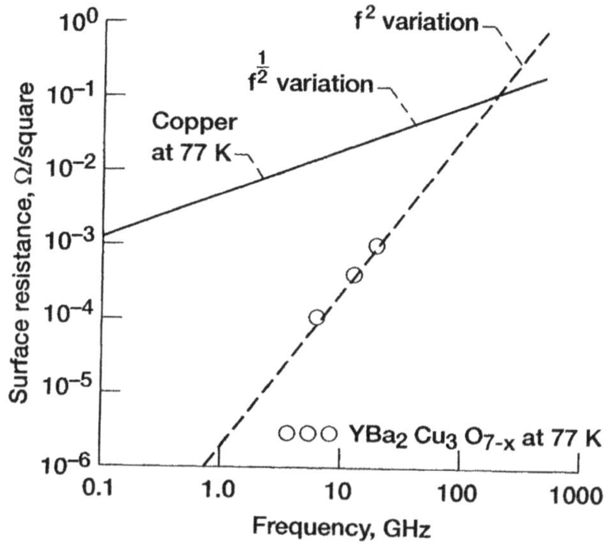

FIGURE 12.19 Measured surface resistance of $YBa_2Cu_3O_{7-x}$ as a function of frequency at 77 K. (From Reference [17], © 1993 American Institute of Physics.)

12.4.6 Attenuation Constant

The unloaded quality factor Q_0 for any transmission line resonator can be expressed as [23]

$$\frac{1}{Q_0} = \frac{1}{Q_c} + \frac{1}{Q_d} + \frac{1}{Q_r}, \tag{12.45}$$

where

Q_c = quality factor due to conductor losses

Q_d = quality factor due to dielectric losses

Q_r = quality factor due to radiation losses

The radiation losses can be minimized by providing a metal shield all round the CPW resonator. The quality factor due to dielectric losses is

$$\frac{1}{Q_d} \approx \tan\delta, \tag{12.46}$$

where tan δ is the loss tangent of the substrate material. The loss tangent for both MgO and $LaAlO_3$ at 77 K is on the order of 10^{-4}. Hence the dielectric losses are negligible. Therefore the unloaded quality factor is

$$Q_0 \approx Q_c. \tag{12.47}$$

The quality factor due to conductor losses is related to the attenuation constant due to conductor losses, α_c, through the relation [23]

$$Q_c = \frac{\pi}{\alpha_c \lambda_g}, \tag{12.48}$$

where λ_g is the guide wavelength and is equal to twice the resonator length for the fundamental mode. For example, the experimentally determined unloaded quality factor of the CPW resonator, indicated as sample 3/circuit 3 in Table 12.4, is about 1200 at 77 K [24]. The corresponding attenuation constant α_c from Eq. (12.48) is about 1.8 dB/m at 6.2 GHz. This film has a surface resistance of about 105 $\mu\Omega$/square. Closed-form equations to compute the attenuation constant of superconducting CPWs are given in Section 8.6.

12.5 FERROELECTRIC COPLANAR WAVEGUIDE CIRCUITS

Recently there has been renewed interest in tunable microwave components based on thin film ferroelectric materials for applications at room as well as cryogenic temperatures. The materials that are commonly used at these temperatures are barium strontium titanate and strontium titanate, respectively. In this section we focus on the dielectric characteristics of these materials and their applications in microwave phase shifters for phased array antennas.

12.5.1 Characteristics of Barium Strontium Titanate Thin Films

Dielectric Characteristics as a Function of Temperature and Voltage Thin films of barium strontium titanate, $Ba_{1-x}Sr_xTiO_3$, are experimentally observed to have complex permittivity, $\hat{\varepsilon}$, which is a function of both temperature and voltage [25]. The complex permittivity can be expressed as [26]

$$\hat{\varepsilon} = \varepsilon' - j\varepsilon'', \tag{12.49}$$

where ε' and ε'' are the real and imaginary parts, respectively. The loss tangent of the material is defined as [26]

$$\tan \delta = \frac{\omega\varepsilon'' + \sigma}{\omega\varepsilon'}, \tag{12.50}$$

where ω and σ are the angular frequency and conductivity, respectively. The conductivity is usually considered to be negligibly small. Hence tan δ can be approximated as

$$\tan\delta \approx \frac{\varepsilon''}{\varepsilon'}. \tag{12.51}$$

It is a common practice to specify, at microwave frequencies, the dielectric property of materials by the real part of the permittivity, $\varepsilon' = \varepsilon_r \varepsilon_0$, where ε_r is the relative permittivity or dielectric constant, and by the loss tangent, tan δ. Often it is preferable to choose a material with low tan δ in order to minimize power dissipation.

The ε_r and tan δ values of $Ba_{1-x}Sr_xTiO_3$ are extracted from capacitance measurements. For example, in [25] an interdigital capacitor fabricated over a 0.5 μm thick *c*-axis oriented $Ba_{0.5}Sr_{0.5}TiO_3$ film on a single crystal (001) $LaAlO_3$ substrate is used to estimate these values. Room temperature (300 K) measurements resulted in ε_r and tan δ values of 1430 and 0.007 with no externally applied electric field, and 960 and 0.001 with an externally applied electric field intensity of 2.33 V/μm, respectively. The single crystal (001) $LaAlO_3$ ($\varepsilon_r = 24$) is selected as the substrate material, first, to ensure good lattice match with the $Ba_{0.5}Sr_{0.5}TiO_3$ thin film and thus minimize interfacial strain, and second, for its low loss tangent, tan δ, which is typically $<10^{-5}$. The ε_r and tan δ characteristics of the this film are illustrated in Figure 12.20*a* and *b* over a wide range of temperature and voltage values.

Dependance of Dielectric Constant on Film Thickness The ε_r value of $Ba_{1-x}Sr_xTiO_3$ thin films usually depends on the film thickness, *t*. For $t < 0.2$ μm, the thinner the film, the smaller is the ε_r value. However, the ε_r value becomes less thickness dependent when $t > 0.5$ μm [25].

Dielectric Tunability The dielectric tunability (%) at a fixed temperature is defined as [27]

$$\text{Tunability} = \frac{(\varepsilon_{r,0V} - \varepsilon_{r,\text{V-applied}})}{\varepsilon_{r,0V}} \times 100, \tag{12.52}$$

where $\varepsilon_{r,0V}$ and $\varepsilon_{r,\text{V-applied}}$ are the dielectric constant at zero-bias voltage and at the applied bias voltage, respectively. For the $Ba_{0.5}Sr_{0.5}TiO_3$ thin film with characteristics as discussed in the previous section, the calculated tunability at room temperature is 33 percent when an electric field intensity of 2.33 V/μm is applied across the sample [25].

Curie Temperature The Curie temperature, T_C, that is, the temperature below which a material is ferroelectric and above which the material is paraelectric,

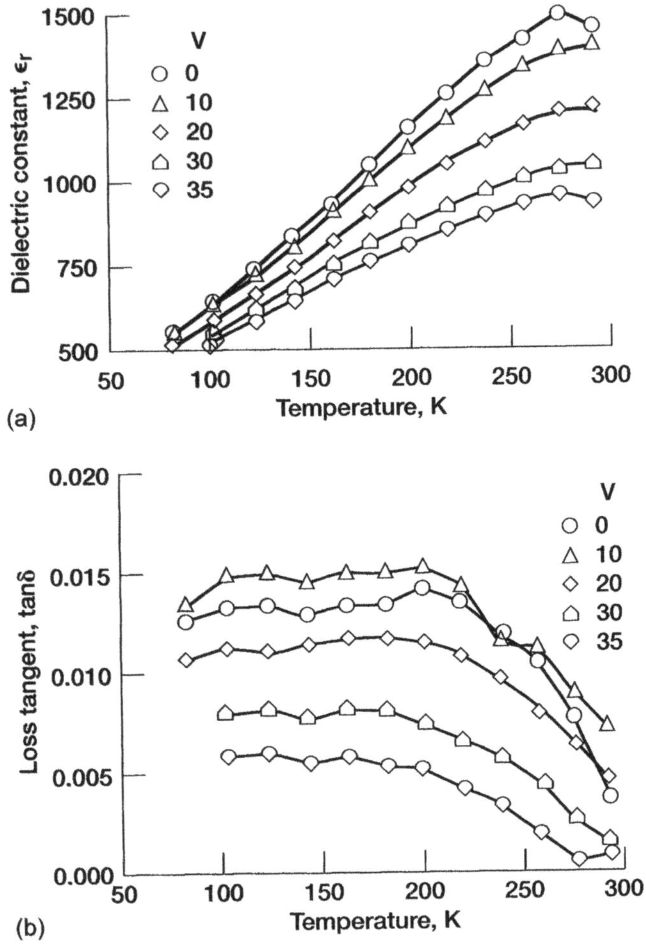

FIGURE 12.20 Measured dielectric characteristics of 0.5 μm thick *c*-axis oriented $Ba_{0.5}Sr_{0.5}TiO_3$ film on a single crystal (001) $LaAlO_3$ substrate at 1 MHz as a function of temperature and voltage: (*a*) Dielectric constant; (*b*) loss tangent. (From reference [25], © 1999 American Institute of Physics.)

is an important parameter. The T_C for $Ba_{1-x}Sr_xTiO_3$ thin films with $x = 0.4$ is about 275 K for a typical film thickness of 0.5 μm deposited on a single-crystal $LaAlO_3$ [27]. Often tunable microwave components that utilize this material for tunability operate above the Curie temperature, typically at room temperature. It is therefore worthwhile to mention that although at room temperature, the $Ba_{0.6}Sr_{0.4}TiO_3$ material operates in the paraelectric regime, it is customary for researchers in this field to refer to the material as ferroelectric.

12.5.2 Characteristics of Strontium Titanate Thin Films

Dielectric Characteristics as a Function of Temperature and Voltage The ε_r and tan δ values of strontium titanate ($SrTiO_3$) thin film are extracted from capacitance measurements [28]. The capacitor used in these experiments consists of a $SrTiO_3$ thin film sandwiched between two parallel metal plates. The capacitor is fabricated on a single-crystal $LaAlO_3$. The top and bottom plates of the capacitor are made from gold and strontium ruthenium oxide ($SrRuO_3$), respectively. It has been experimentally observed that the interfacial effects between the film and the electrodes become small when the film thickness is large [28]. Hence the thickness of the $SrTiO_3$ film is chosen as 2.5 μm [28]. Further, if the lattice constants of the film and the bottom plate are matched, the lattice mismatch-induced stress is reduced. Hence, the bottom plate is fabricated from $SrRuO_3$ [28], since the lattice constant of $SrRuO_3$ and $SrTiO_3$ are very well matched. The small interfacial effects and the reduced stress contribute to enhancing the dielectric properties of the film so that they approach those of a bulk single crystal material.

The ε_r and tan δ extracted from the measured capacitance of the parallel plate structure above [28] are presented as a function of the temperature in Figure 12.21*a* and *b*. These results corresponds to the case where the externally applied dc electric field intensity in zero. It is observed from Figure 12.21*a* that ε_r increases from about 300 at room temperature to about 930 at 4 K. On the other hand, it is observed from Figure 12.21*b* that tan δ is on the order of 1×10^{-3} and lower across the temperature range of 150 to 300 K. Below 150 K, tan δ increases and approaches 6×10^{-3} at 4 K. Further, two relaxation peaks occur in the tan δ characteristics at about 25 and 65 K, respectively.

The ε_r and tan δ as a function of the externally applied dc electric field intensity, with the temperature held constant, is shown in Figure 12.22. The figure indicates the presence of hystereses in both ε_r and tan δ, which is due to either ferroelectric ordering or interfacial space charge [28].

Dielectric Tunability and Figure of Merit The dielectric tunability of the film is given by Eq. (12.52). The measured tunability of $SrTiO_3$ film of thickness 2.5 μm over the temperature range of 4 to 100 K is about 70 percent when the applied electric field intensity is in the range of 40 to 50 V/μm.

The figure of merit is the ratio of the tunability to the loss and is expressed as [28]

$$K = \frac{(\varepsilon_{r,0V} - \varepsilon_{r,V\text{-applied}})}{\varepsilon_{r,0V}} \cdot \frac{1}{(\tan\,\delta)_{max}}, \tag{12.53}$$

where $(\tan\,\delta)_{max}$ is the maximum loss under all the applied electric field intensities.

The measured figure-of-merit for the $SrTiO_3$ film of thickness 2.5 μm is close to 200 up to a temperature of 70 K and then increases to 400 at a temperature of 120 K. The figure-of-merit remains at this value until the temperature

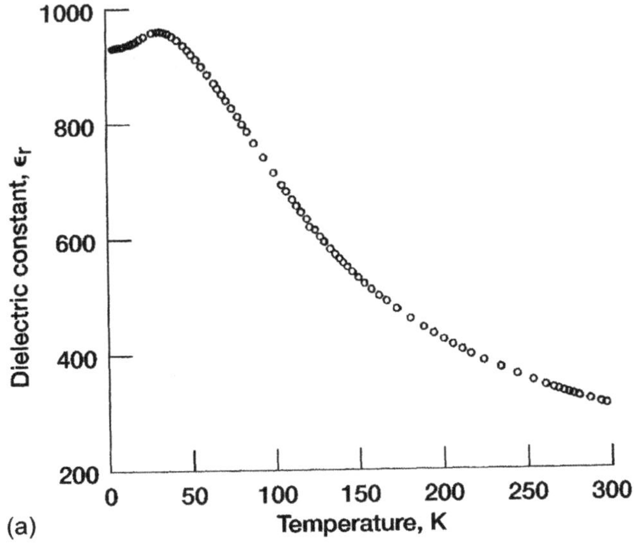

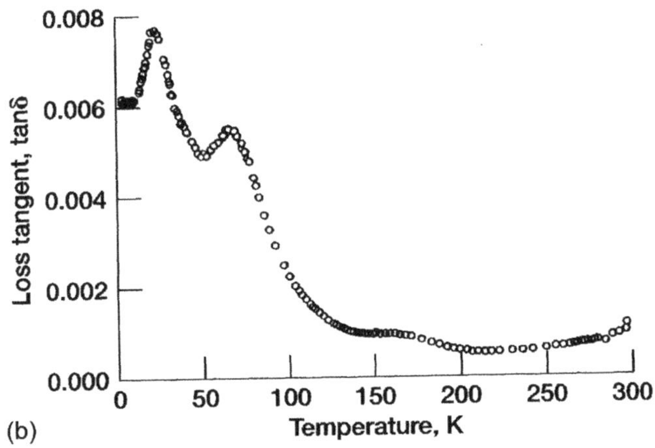

FIGURE 12.21 Measured dielectric characteristics of 2.5 μm thick $SrTiO_3$ film on a single crystal (001) $LaAlO_3$ substrate at 1 KHz as a function of temperature: (*a*) Dielectric constant; (*b*) loss tangent. (From reference [28], © 1998 American Institute of Physics.)

exceeds 200 K and then rapidly decreases. Hence $SrTiO_3$ thin films are useful for tunable microwave devices operating at temperatures above 70 K.

12.5.3 Grounded Coplanar Waveguide Phase Shifter

The layout of a $(Ba,Sr)TiO_3$ coplanar waveguide Ka-band phase shifter [29] is shown in Figure 12.23. The phase shifter consists of a uniform grounded

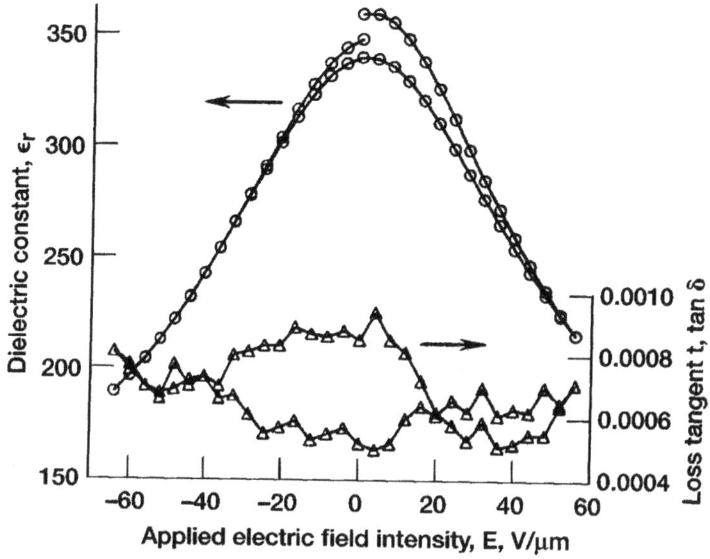

FIGURE 12.22 Measured dielectric characteristics of 2.5 μm thick $SrTiO_3$ film as a function of the applied dc electric field intensity at a fixed temperature of 190 K. (From reference [28], © 1998 American Institute of Physics.)

coplanar waveguide of length L with tapered impedance matching transformers of length L_1 at either ends. The circuit is fabricated on a multilayer dielectric that consists of a (Ba,Sr)TiO_3 thin film deposited over a MgO substrate. The cross section of the phase shifter is shown in Figure 12.24.

The principle of operation of the phase shifter is based on a change in the propagation constant of the grounded coplanar waveguide because of a change

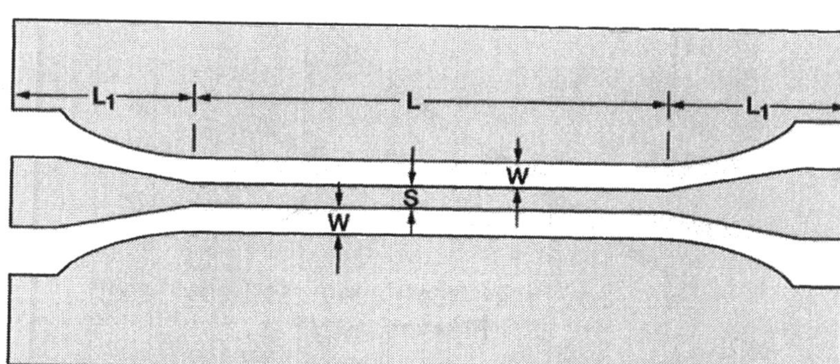

FIGURE 12.23 Schematic showing the layout of the Ka-band phase-shifter: $L = 6$ mm, $L_1 = 3$ mm.

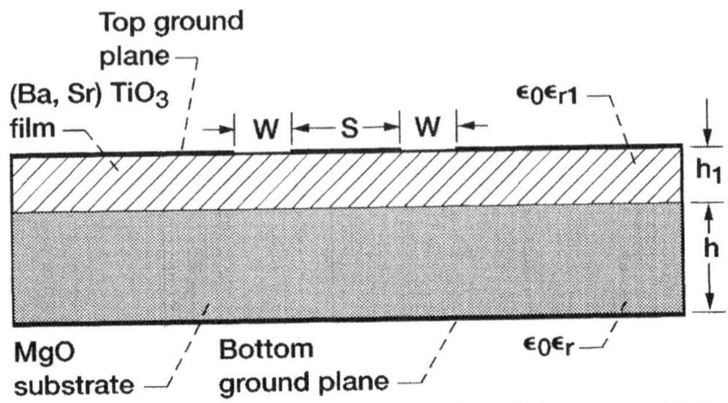

FIGURE 12.24 Schematic showing the cross-section of the phase-shifter: $h = 500\ \mu$m, $h_1 = 8\ \mu$m, $S = 60\ \mu$m, $W = 40\ \mu$m, $\varepsilon_r = 9.7$, ε_{r1} is bias dependent.

in the dielectric constant of the $(Ba,Sr)TiO_3$ film with an externally applied dc bias voltage. The measured phase characteristics, at a fixed frequency of 31.34 GHz, as a function of the bias voltage [29] is shown in Figure 12.25. It is observed that the phase shift increases almost linearly from 0 to 370° as the bias voltage is varied from 0 to 350 Volts. The measured insertion loss is in the range of 4.0 to 6.0 dB over the whole range of bias voltages.

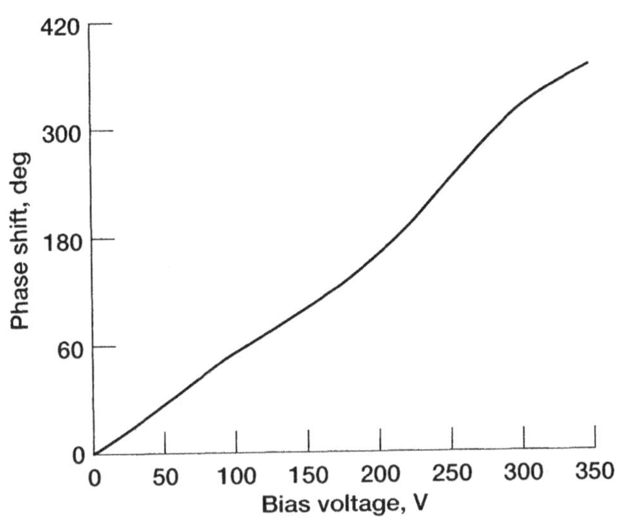

FIGURE 12.25 Measured phase characteristics as a function of the bias voltage at a fixed frequency of 31.34 GHz. (From reference [29], © 2000 Gordon and Breach Science Publishers.)

12.6 COPLANAR PHOTONIC-BANDGAP STRUCTURE

Photonic-bandgap structures are artificial electromagnetic crystals having a spatially periodic dielectric constant with a lattice parameter comparable to the wavelength of the electromagnetic wave. When suitably designed, a photonic-bandgap structure is capable of suppressing the propagation of electromagnetic wave along certain directions over a band of frequencies. The first experimental demonstration of a three-dimensional photonic-bandgap was by Yablonovitch et al. [30]. Since that time several other photonic-bandgap structures have been investigated [31, 32], and applications such as high-impedance ground plane structures for antennas have been demonstrated [31]. This section is devoted to a two-dimensional photonic-bandgap structure that is compatible with coplanar waveguides.

12.6.1 Nonleaky Conductor-Backed Coplanar Waveguide

A conductor-backed coplanar waveguide (CBCPW) can support, besides the dominant CPW mode, two other parasitic modes, namely the parallel plate mode and the microstrip mode as discussed in Sections 3.4 and 4.3, respectively. These parasitic modes, when excited, couple power from the dominant coplanar waveguide mode to a leaky mode. When compared to a conventional coplanar waveguide (CPW), the loss of power through leakage is responsible for the higher transmission loss in a CBCPW.

A nonleaky conductor backed coplanar waveguide [33] is schematically illustrated in Figure 12.26. In this transmission line the top ground planes on either side of the center strip conductor are replaced by two-dimensional photonic-bandgap lattice. The lattice consists of a periodic array of square metal pads with recess and interconnected by narrow lines. An unit cell of the lattice is shown in the inset in Figure 12.26. Over a predetermined frequency band, the photonic-bandgap lattice behaves as a perfect magnetic conductor (PMC), and has a band-stop-type characteristic, as will be shown in the sections that follow.

Effective Surface Impedance Model The unit cell is modeled [34] along the x-direction as a transmission line of length equal to the substrate thickness h. The propagation constant on this line is denoted as β. Furthermore, to take into account the presence of a lower ground plane, the line is terminated at the bottom by a perfect electric conductor (PEC). In the lattice the gaps between adjacent pads and the interconnecting lines between the pads give rise to a capacitance C and an inductance L [34]. Hence the transmission line is terminated at the top by a parallel equivalent circuit consisting of a capacitor C and an inductor L. The equivalent circuit model is shown in Figure 12.27. Based on this model the surface of a photonic-bandgap lattice is assigned a sheet impedance that is equal to the impedance of the parallel equivalent

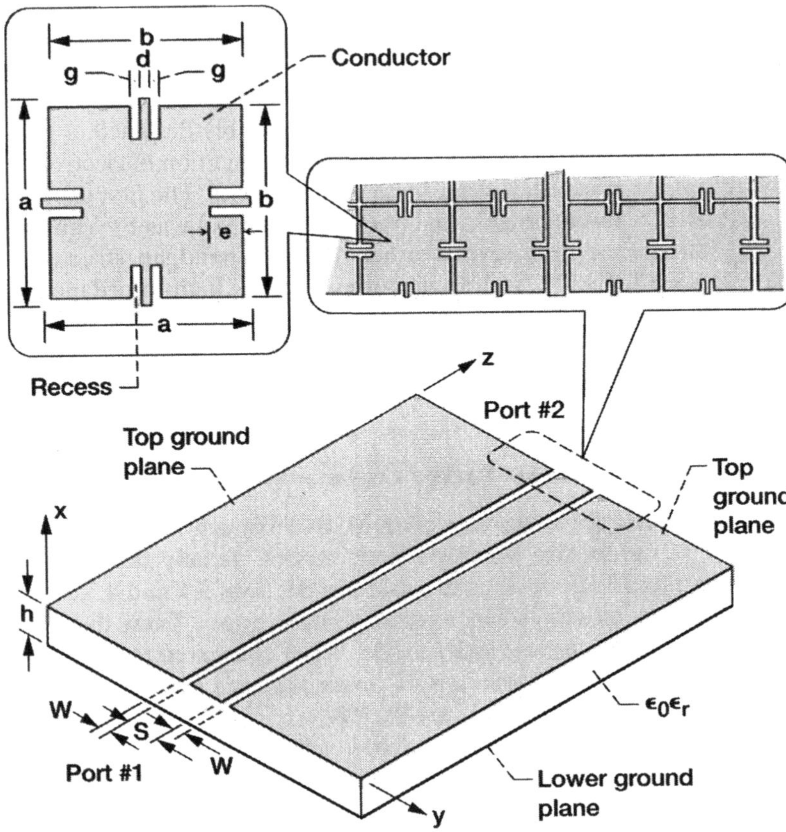

FIGURE 12.26 Schematic of nonleaky conductor backed coplanar waveguide with two-dimensonal photonic bandgap structure in the top ground planes: $h = 0.635$ mm, $\varepsilon_r = 10.2$, $S = 2.032$ mm, $W = 0.508$ mm, $a = 3.048$ mm, $b = 2.7432$ mm, $d = 0.3048$ mm, $g = 0.3048$ mm, $e = 0.762$ mm.

circuit, and is expressed as

$$Z = \frac{j\omega L}{1 - \omega^2 LC} \tag{12.54}$$

This equation suggests that the surface impedance is very high at the resonant frequency ω_0 given by

$$\omega_0 = \frac{1}{\sqrt{LC}} \tag{12.55}$$

Consequently at ω_0 the surface behaves as a perfect magnetic conductor (PMC). This type of impedance characteristic makes the photonic-bandgap

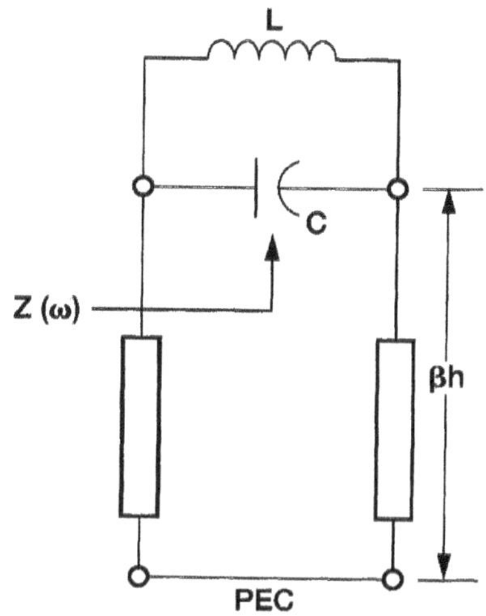

FIGURE 12.27 Equivalent circuit model for the unit cell.

structure reject a predetermined band of frequencies. Over this high-impedance surface or ground plane, the tangential electric and magnetic fields have an antinode and a node, respectively. Furthermore, from Eq. (12.54), the surface impedance below and above ω_0 is inductive and capacitive, respectively. This feature is exploited in the experimental characterization of a photonic-bandgap structure as explaned in the next section.

Photonic-Bandgap Structure Phase Characteristics The perfect magnetic conductor (PMC) behavior of the photonic bandgap structure can be verified, by illuminating the surface with an uniform plane wave and observing the phase of the reflected wave [34]. When the frequency of the incident wave is far below or above ω_0, the surface according to Eq. (12.54) has either an inductive or a capacitive impedance. Consequently the phase of the reflected wave from this surface, depending on frequency being greater than or less than ω_0, is either π or $-\pi$. At ω_0 the surface has a very high impedance, and therefore the phase of the reflected wave is zero. The phase falls within $+\pi/2$ and $-\pi/2$ when the magnitude of the surface impedance exceeds the impedance of free space.

Measured Phase Characteristics of Photonic-Bandgap Structure The experimental setup shown in Figure 12.28 is used to measure the phase of the reflected signal from the surface of the photonic-bandgap structure. The phase of the reflection coefficient is initially measured with a perfect electric conductor (PEC) in place of the photonic-bandgap structure. Next the measurement

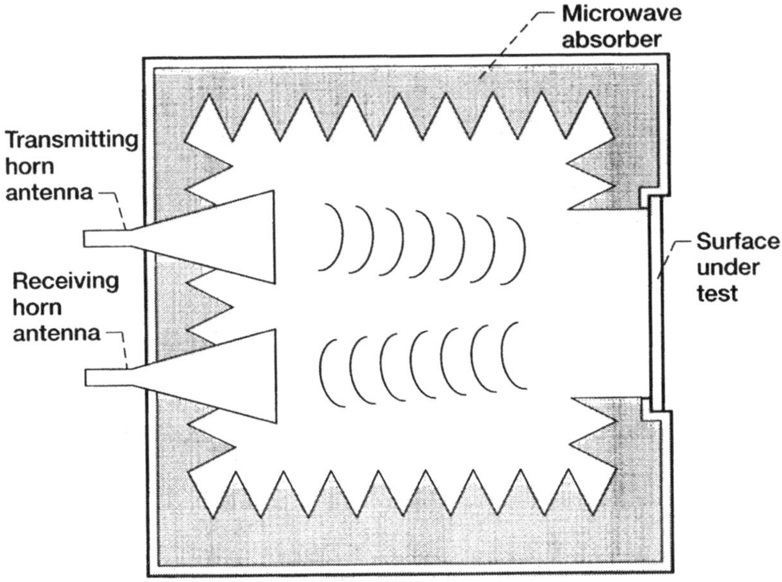

FIGURE 12.28 Experimental setup for measuring the phase of the reflection coefficient of a surface.

is repeated with the photonic-bandgap structure in place of the PEC. The difference in the phase for the two settings is now calculated as a function of the frequency. A typical phase characteristic is graphically illustrated in Figure 12.29 for the photonic-bandgap structure shown in Figure 12.26. The photonic-bandgap structure in this experiment [34] is a 40 × 60 array with a period a equal to 3.048 mm. The resonant frequency of the photonic-bandgap structure is about 14.5 GHz [34].

Measured Transmission Loss of Nonleaky Conductor-Backed Coplanar Waveguide The measured transmission losses [33] as a function of the frequency for nonleaky CBCPW, conventional coplanar waveguide (CPW), and CBCPW are presented in Figure 12.30. It is observed that because of the absence of parasitic parallel plate modes, the transmission loss in the case of CPW is the lowest. In the case of CBCPW, significant amount of power is lost in the form of leakage, and hence the transmission loss is very high. On the other hand, the photonic-bandgap structure from which the nonleaky CBCPW ground planes are constructed has a band-stop characteristic over the frequency range of 9 to 14 GHz [33]. Consequently over this frequency range the propagation of the parasitic parallel plate mode is suppressed, and the transmission loss of the nonleaky CBCPW approaches that of the CPW.

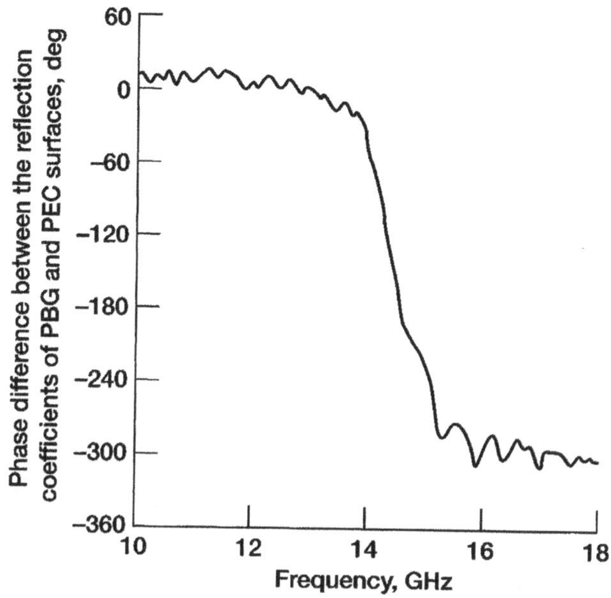

FIGURE 12.29 Measured phase difference between the reflection coefficients of a photonic-bandgap surface and a perfect electrically conducting surface. (From reference [34], © IEEE 1999.)

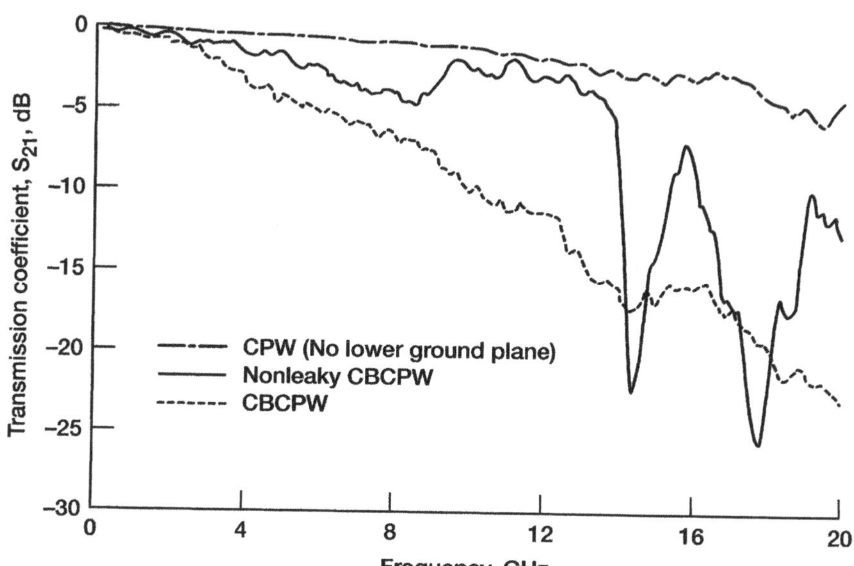

FIGURE 12.30 Comparison of measured transmission coefficient S_{21} of nonleaky conductor-backed coplanar waveguide with conventional CPW and CBCPW. (From reference [33], © IEEE 1999.)

12.7 COPLANAR WAVEGUIDE PATCH ANTENNAS

In this section the characteristics of patch antennas with coplanar waveguide feed are presented. The feed system in these antennas is directly coupled, electromagnetically coupled, or aperture coupled to the patch. The advantages of using coplanar waveguide in the feed are as follows: First, it allows series as well as shunt mounting of components without the use of via holes. Second, it allows mounting of MMIC circuits such as phase shifters using solder bumps and flip chip techniques.

12.7.1 Grounded Coplanar Waveguide Patch Antenna

A grounded coplanar waveguide (GCPW) patch antenna and feed [35] are shown in Figure 12.31. The patch is formed by widening the center strip conductor of the GCPW. The feed is a short length of GCPW. The GCPW is excited at the end opposite the patch by a surface launch connector. The center conductor of the connector passes through the dielectric substrate and is attached to the GCPW center strip conductor. The outer conductor of the connector is attached to the GCPW lower ground plane. This type of launcher efficiently transitions the coaxial TEM mode to the GCPW quasi-TEM mode. The upper and lower ground planes of the GCPW are electrically connected by a series of shorting pins. These pins suppress the excitation of the parasitic parallel plate mode between the upper and lower ground planes. The upper ground planes are electrically connected by a bond wire to suppress the excitation of the parasitic coupled slotline mode.

The patch operates in the fundamental mode, when the periphery is approximately one wavelength, that is, when the distance *A-B-C-D* in Figure 12.31 is half a wavelength or 180°. The fundamental mode produces a linearly polarized radiation pattern directed normal to the upper ground planes. The polarization is parallel to the line *AD*. The electric field lines in the slot region between the patch and the ground planes are also shown in the figure. The radiation from the patch takes place mainly from sides *CE* and *FB*. Sides *BC* and *EF* contribute very little to the radiation, since the fields go to zero and reverse direction at the midpoints of these sides.

The measured characteristics [35] for a linearly polarized single patch show that the *E*- and *H*-plane 3 dB beamwidths are typically about 110° and 80°, respectively. The gain of the antenna is about 3.5 dBi, and the 3:1 VSWR bandwidth is about 6.5 percent at 2.3 GHz.

When the periphery of the patch is two wavelengths, the next higher-order mode is excited. This mode gives rise to a linearly polarized pattern with a null normal to the patch [35]. This mode can be useful in generating high-gain patterns with tilted beams [35]. A circular polarization is produced by feeding a square patch on two adjacent sides through a 90° hybrid as explained in [35] and [36].

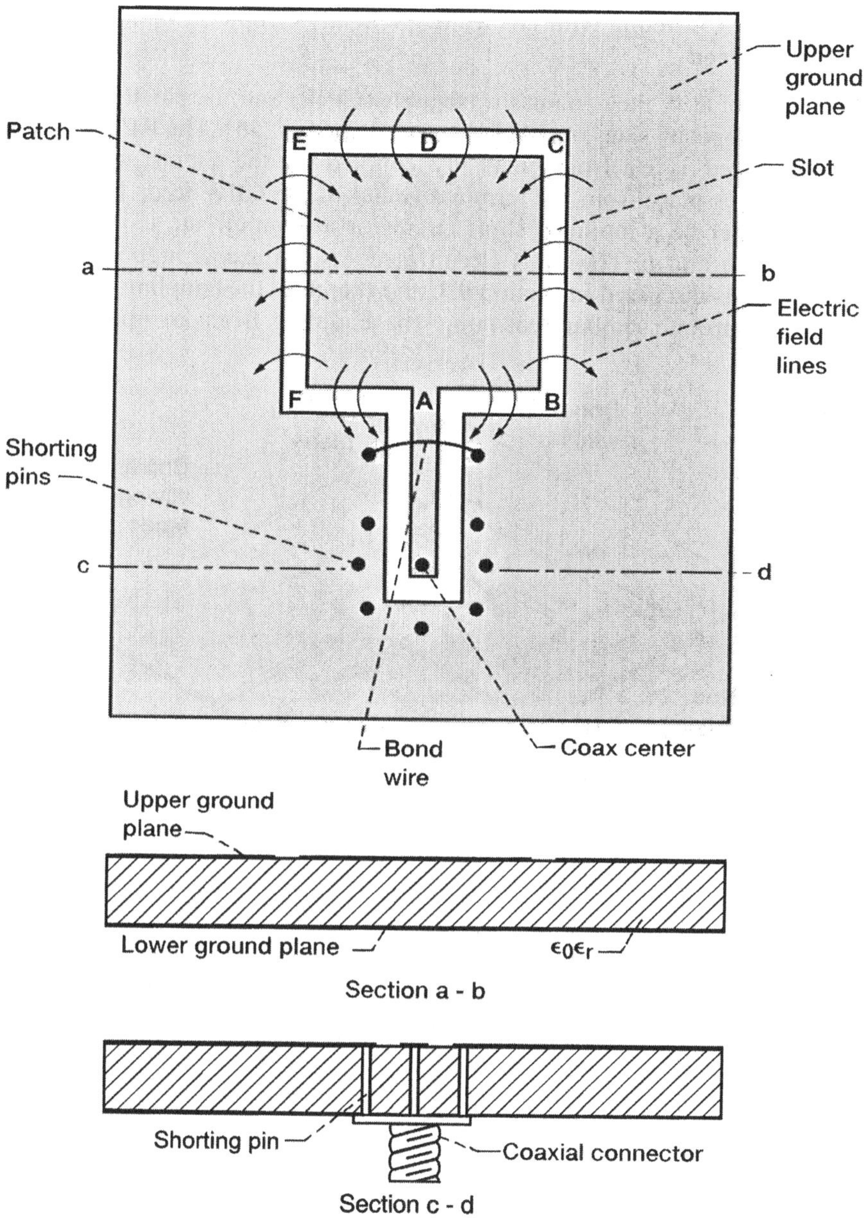

FIGURE 12.31 Grounded coplanar waveguide patch antenna operating in the fundamental mode.

12.7.2 Patch Antenna with Electromagnetically Coupled Coplanar Waveguide Feed

In this type of feed, the coplanar waveguide (CPW) and the patch antenna are located on opposite sides of a dielectric substrate [37, 38]. The RF power from the CPW feed is electromagnetically coupled to the patch antenna. The coupling takes place from the terminated end of the CPW feed. The termination can either be a modified short circuit or a open circuit as illustrated in Figure 12.32*a* and *b*. The short circuit gives rise to excess inductance at the termination as discussed in Section 9.3, and therefore the coupling is termed as inductive. Through similar reasoning, the coupling from an open circuit is

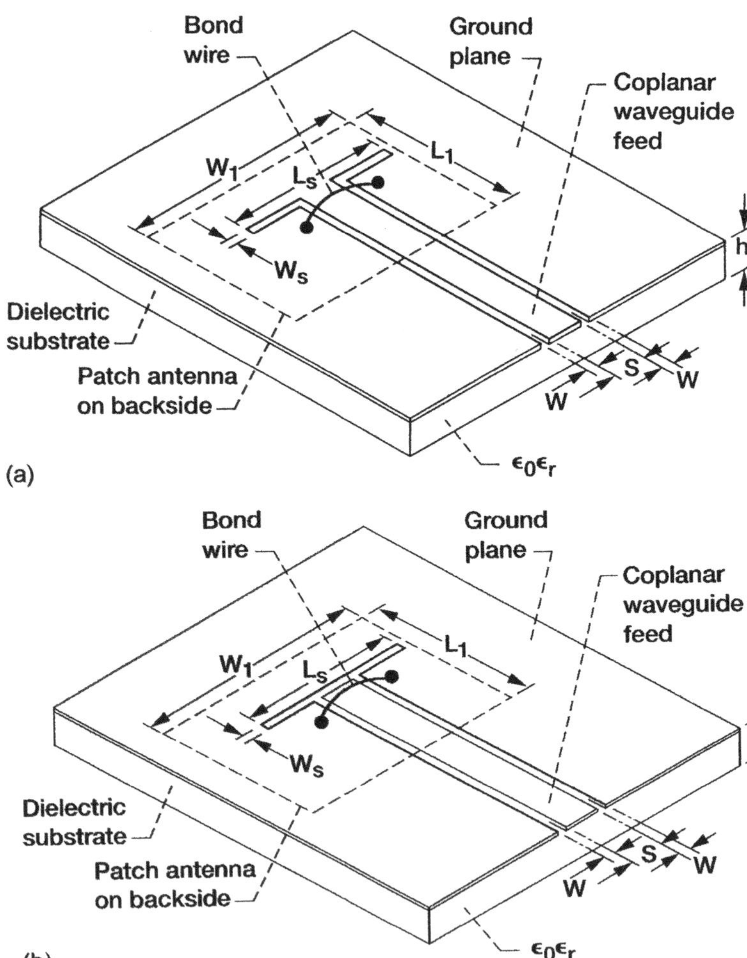

FIGURE 12.32 Patch antenna with electromagnetically coupled coplanar waveguide feed: (*a*) Inductively coupled; (*b*) capacitively coupled.

TABLE 12.5 Measured Performance of Patch Antenna with Electromagnetically Coupled Coplanar Waveguide Feeds

Parameters	Inductively Coupled	Capacitively Coupled
Center frequency, GHz	4.67	4.3
Slot length, L_s (mm)	16.0	14.0
−10.0 dB return loss bandwidth, percent	3.5	2.8
3 dB beamwidth, degrees		
H-Plane	70	80
E-Plane	67	—
Front-to-Back Ratio, dB	10	9.5

Note: $S = 3.0$ mm, $W = 0.06$ mm, $h = 1.58$ mm, $\varepsilon_r = 2.22$, $L_1 = 18.3$ mm, $W_1 = 22.0$ mm, $W_s = 1.0$ mm.

termed capacitive. In the antennas discussed in [37], the short-circuit and the open-circuit terminations are located symmetrically above the center of the patch. The mode that is excited in the cavity formed by the patch is the TM_{10} mode. The circuit dimensions and the measured performance [37] of the patch antenna with inductive and capacitive coupling are presented in Table 12.5. The radiation from the patch is linearly polarized and parallel to side of length L_1.

12.7.3 Coplanar Waveguide Aperture-Coupled Patch Antenna

Aperture-coupled feeding is attractive because of certain advantages, such as no physical contact between the feed and radiator, wider bandwidth, and better isolation between antennas and the feed network. Furthermore aperture-coupled feeding allows independent optimization of antennas and feed networks by using substrates of different thickness or permittivity. In this section aperture-coupled patch antennas with coplanar waveguide feed are described.

Grounded Coplanar Waveguide Aperture-Coupled Patch Antenna A grounded coplanar waveguide (GCPW) aperture-coupled patch antenna is schematically illustrated in Figure 12.33. In the experimental antenna [39], the patch and the GCPW feed structure, with a series gap L_1 in the center conductor, are fabricated on separate substrates, and the aperture is etched in the common ground plane. The aperature is located directly above the series gap. Thus microwave power is coupled from the GCPW feed line to the patch through the aperture. In the experimental antenna, the patch is displaced by about 0.32 cm from the center of the aperture and also, the GCPW stub length L is adjusted to provide the best impedance match to the feed line. The radiation from the antenna is linearly polarized with the plane of polarization parallel to the patch side of dimension a.

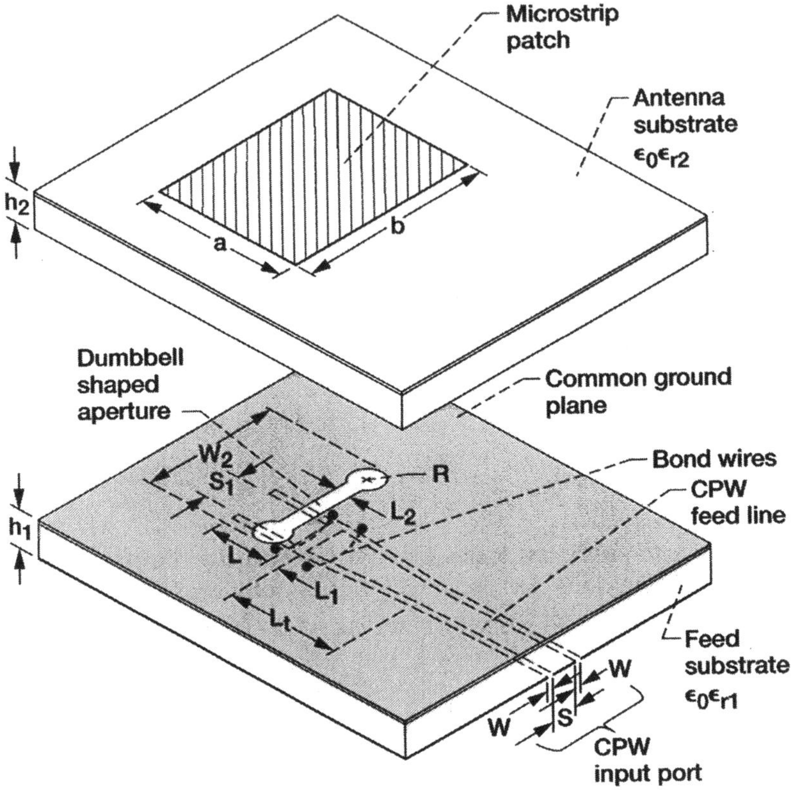

FIGURE 12.33 Grounded coplanar waveguide aperture-coupled patch antenna: $S = 0.076$ cm, $W = 0.025$ cm, $L = 0.711$ cm, $L_1 = 0.025$ cm, $L_2 = 0.205$ cm, $W_2 = 0.69$ cm, $a = 0.76$ cm, $b = 1.14$ cm, $h_1 = 0.051$ cm, $\varepsilon_{r1} = 2.2$, $h_2 = 0.025$ cm, $\varepsilon_{r2} = 2.2$, $L_t = 0.711$ cm, $S_1 = 0.355$ cm, $R = 0.0843$ cm.

The measured return loss [39] is shown in Figure 12.34. The return loss is about -16.9 dB at 12.65 GHz. Typical measured *E*- and *H*-plane radiation patterns [39] for the GCPW aperture-coupled patch antenna is shown in Figure 12.35*a* and *b*. The patterns look fairly symmetrical and exhibit a 3-dB beamwidth of about 60° and 50° in the *E*- and *H*-plane, respectively. The measured front-to-back ratio is about 14.0 dB, which is typical for an aperture coupled patch antenna. Last, in [40] an aperture-coupled patch antenna with series slotline stubs in the top ground planes, instead of a series gap in the center strip conductor, is demonstrated. The characteristics of this antenna is similar to that presented above.

Finite Width Coplanar Waveguide Aperture-Coupled Micromachined Patch Antenna Printed antennas fabricated on high dielectric constant substrates suffer from excess surface wave losses and lower radiation efficiency [41]. The

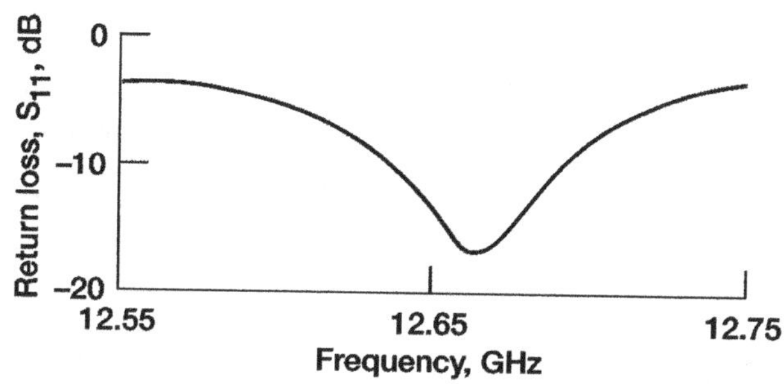

FIGURE 12.34 Measured return loss as a function of the frequency for feed structure with series gap and a dumbell-shaped aperture.

surface wave losses can be reduced by decreasing the substrate thickness, typically the substrate thickness should be less than $\lambda_d/10$, where λ_d is the wavelength in the dielectric. For example, in silicon ($\varepsilon_r = 11.7$) at 94.0 GHz this thickness corresponds to about 100 μm. The radiation efficiency can be improved by locally removing the substrate using micromachining technique and synthezing in a low dielectric constant region around the antenna.

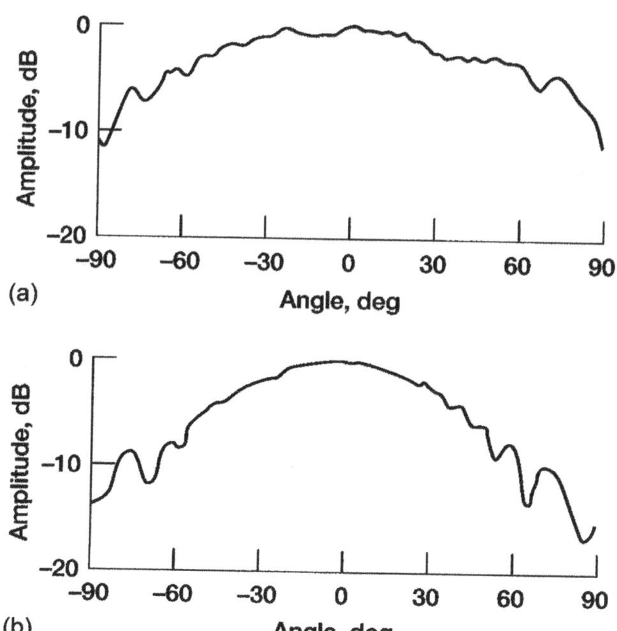

FIGURE 12.35 Measured radiation patterns for the grounded coplanar waveguide aperture-coupled patch antenna: (*a*) *E*-plane; (*b*) *H*-plane.

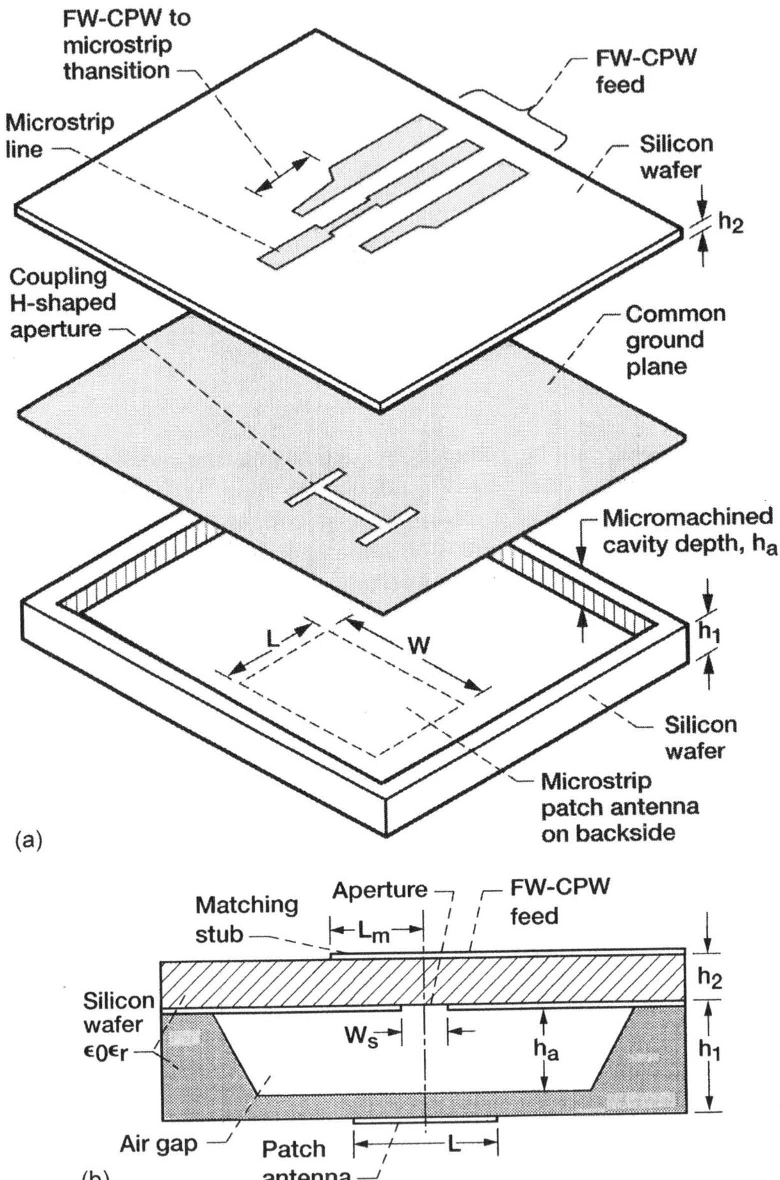

FIGURE 12.36 A W-band finite width coplanar waveguide aperture-coupled micromachined patch antenna: (*a*) Schematic $h_1 = 200$ μm, $h_2 = 100$ μm, $h_a = 150$ μm, $L = 800$ μm, $W = 800$ μm; (*b*) cross-section, $L_m = 160$ μm, $W_s = 50$ μm, $\varepsilon_r = 11.7$.

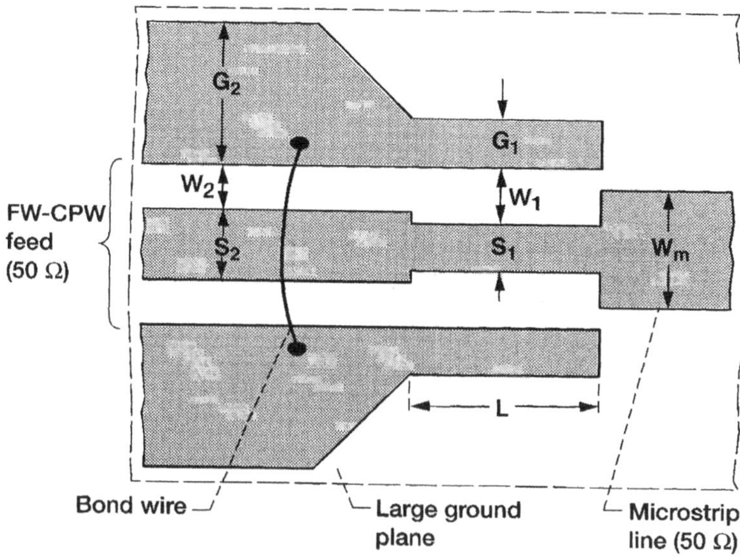

FIGURE 12.37 Schematic of coplanar waveguide with finite-width ground plane-to-microstrip transition: $S_1 = 30\ \mu m$, $W_1 = 29\ \mu m$, $G_1 = 45\ \mu m$, $S_2 = 40\ \mu m$, $W_2 = 24\ \mu m$, $G_2 = 106\ \mu m$, $W_m = 74\ \mu m$, $L = 280\ \mu m$.

In Figure 12.36*a* and *b* an aperture-coupled micromachined microstrip antenna [41] is schematically illustrated. The feed network and the antenna are fabricated on two separate silicon wafers and stacked vertically. The feed network is a coplanar waveguide with finite width ground planes (FW-CPW). The antenna is a metal patch suspended over a cavity and micromachined in the silicon wafer. Depending on the cavity depth, the quasi-static effective dielectric constant in the region of the patch is in the range of 2.8 to 3.9 [41].

The feed line transitions from a FW-CPW to a microstrip line and couples power to the patch through an *H*-shaped aperture in the common ground plane located between the two wafers. The transition is shown in Figure 12.37. The microstrip line is orthogonal and symmetrically located with respect to the aperture as shown in Figure 12.38. It is observed in [41] that a *H*-shaped aperture couples power more efficiently than a conventional rectangular aperture. The real and the imaginary parts of the antenna input impedance are dependent on the slot length L_s and the microstrip line extension L_m [41].

The measured return loss of the antenna is about −18.0 dB at 94 GHz with a 2:1 VSWR bandwidth of 1 percent. The radiation from the antenna is linearly polarized with the plane of polarization parallel to the patch side of dimension *L*. The antenna has a radiation efficiency of 58 ± 5 percent, as determined from radiometric measurements. The measured crosspolarization level is below −20.0 dB and the front-to-back ratio is below −10.0 dB in both *E*- and *H*-planes [41].

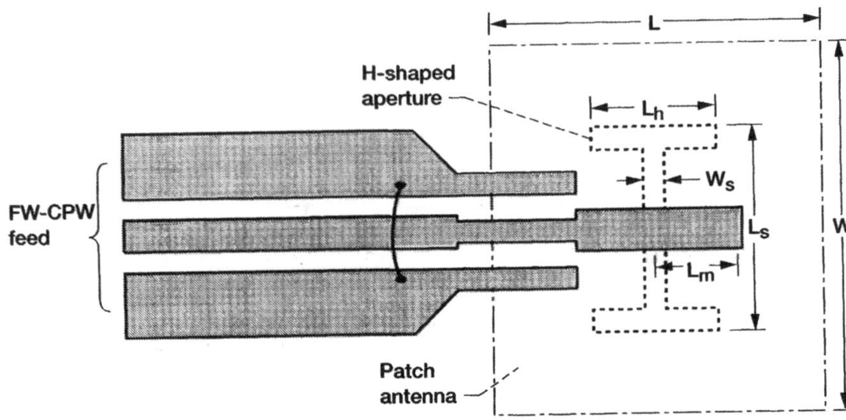

FIGURE 12.38 Schematic illustrating the integration of the microstrip line with the patch antenna: $L_h = 250\ \mu m$, $W_s = 50\ \mu m$, $L_m = 160\ \mu m$, $L_s = 500\ \mu m$, $L = 800\ \mu m$, $W = 800\ \mu m$.

REFERENCES

[1] C. L. Goldsmith, Z. Yao, S. Eshelman, and D. Denniston, "Performance of Low-loss RF MEMS Capacitive Switches," *IEEE Microwave Guided Wave Lett.*, Vol. 8, No. 8, pp. 269–271, Aug. 1998.

[2] Z. Y. Yao, S. Chen, S. Eshelman, D. Denniston, and C. Goldsmith, "Micromachined Low-loss Microwave Switches," *IEEE J. Microelectromech. Systems*, Vol. 8, No. 2, pp. 129–134, June 1999.

[3] P. Osterberg, H. Yie, X. Cai, J. White, and S. Senturia, "Self-consistent Simulation and Modeling of Electrostatically Deformed Diaphragms," *1994 IEEE Micro Electro Mechanical Sys. Conf. Proc.*, pp. 28–32, Oiso, Japan, Jan. 25–28, 1994.

[4] P. M. Zavracky, S. Majumder, and N. E. McGruer, "Micromechanical Switches Fabricated Using Nickel Surface Micromachining," *IEEE J. Microelectromechanical Systems*, Vol. 6, No. 1, pp. 3–9, March 1997.

[5] R. A. Gaspari and H. H. Yee, "Microwave GaAs FET Switching," *1978 IEEE MTT-S Int. Microwave Symp. Dig.*, pp. 58–60, Ottawa, Canada, June 27–29, 1978.

[6] D. A. Blackwell, D. E. Dawson, and D. C. Buck, "X-Band MMIC Switch with 70 dB Isolation and 0. 5 dB Insertion Loss," *1995 IEEE Microwave Millimeter-Wave Monolithic Circuits Symp. Dig.*, pp. 97–100, Orlando, FL, May 15–16, 1995.

[7] D. D. Heston, D. J. Seymour, and D. Zych, "100 MHz to 20 GHz Monolithic Single-Pole, Two-, Three-, and Four-Throw GaAs PIN Diode Switches," *1991 IEEE MTT-S Int. Microwave Symp. Dig.*, Vol. 2, pp. 429–432, Boston, MA, June 10–14, 1991.

[8] S. Pacheco, C. T. Nguyen, and L. P. B. Katehi, "Micromechanical Electrostatic K-Band Switches," *1998 IEEE MTT-S Int. Microwave Symp. Dig.*, Vol. 3, pp. 1569–1572, Baltimore, MD, June 7–12, 1998.

[9] J. J. Yao and M. F. Chang, "A Surface Micromachined Miniature Switch for Telecommunications Applications with Signal Frequencies from DC to 4 GHz," *8th Int. Conf. on Solid-State Sensors and Actuators, Eurosensors IX,* Digest of Tech. Papers, Vol. 2, pp. 384–387, Stockholm, Sweden, June 25–29, 1995.

[10] D. Hyman, A. Schmitz, B. Warneke, T. Y. Hsu, J. Lam, J. Brown, J. Schaffner, A. Walston, R. Y. Loo, G. L. Tangonan, M. Mehregany, and J. Lee, "GaAs-Compatible Surface-Micromachined RF MEMS Switches," *Electron. Lett.*, Vol. 35, No. 3, pp. 224–226, Feb. 1999.

[11] N. S. Barker and G. M. Rebeiz, "Optimization of Distributed MEMS Phase Shifters," *1999 IEEE MTT-S Int. Microwave Symp. Dig.*, Vol. 1, pp. 299–302, Anaheim, CA, June 13–19, 1999.

[12] A. Borgioli, Y. Liu, A. S. Nagra, and R. A. York, "Low-Loss Distributed MEMS Phase Shifter," *IEEE Microwave Guided Wave Lett.*, Vol. 10, No. 1, pp. 7–9, January 2000.

[13] K. A. Muller and J. G. Bednorz, "The Discovery of a Class of High-Temperature Superconductors," *Science*, Vol. 237, No. 4819, pp. 1133–1139, September 1987.

[14] P. H. Carr and B. R. McAvoy, eds., Special Issue on Microwave Applications of Superconductivity, *IEEE Trans. Microwave Theory Tech.*, Vol. 39, No. 9, September 1991.

[15] M. Nisenoff and W. J. Meyers, eds., Special Issue on Microwave and Millimeter Wave Applications of High Temperature Superconductivity, *IEEE Trans. Microwave Theory Tech.*, Vol. 44, No. 7, Part II, July 1996.

[16] S. Ramo, J. R. Whinnery, and T. V. Duzer, *Fields and Waves in Communication Electronics*, 3rd ed, New York: Wiley, 1994, secs. 2.5, 3.16, 3.17, 3.18, and 5.11.

[17] W. Rauch, E. Gornik, G. Solkner, A. A. Valenzuela, F. Fox, and H. Behner, "Microwave Properties of $YBa_2Cu_3O_{7-x}$ Thin Films Studied with Coplanar Transmission Line Resonators," *J. Appl. Phys.*, Vol. 73, No. 4, pp. 1866–1872, February 1993.

[18] A. A. Valenzuela, G. Solkner, J. Kessler, and P. Russer, "Microwave Characterization of Structured $YBa_2Cu_3O_{7-\delta}$—Thin Films in Synthesis and Characterization of High-Temperature Superconductors," *Materials Science Forum*, Vol. 130–132, J. J. Pouch, S. A. Alterovitz, and R. R. Romanofsky and A. F. Hepp, eds., Aedermannsdorf, Switzerland: Trans Tech Publications, 1993, pp. 349–372.

[19] A. Porch, M. J. Lancaster, and R. G. Humphreys, "The Coplanar Resonator Technique for Determining the Surface Impedance of $YBa_2Cu_3O_{7-\delta}$ Thin Films," *IEEE Trans. Microwave Theory Tech.*, Vol. 43, No. 2, pp. 306–314, February 1995.

[20] D. M. Pozer, Microwave Engineering, Reading, MA: Addison-Wesley, 1990, sec. 7.1.

[21] K. Watanabe, K. Yoshida, T. Aoki, and S. Kohjiro, "Kinetic Inductance of Superconducting Coplanar Waveguides," *Jpn. J. Appl. Phys.*, Vol. 33, Part 1, No. 10, pp. 5708–5712, October 1994.

[22] K. Yoshida, K. Nagasawa, T. Kiss, H. Shimakage, and Z. Wang, "Residual Surface Resistance of $YBa_2Cu_3O_{7-x}$ Thin Films—Weakly Coupled Grain Model," *IEEE Trans. Appl. Superconductivity*, Vol. 7, No. 2, pp. 1253–1256, June 1997.

[23] T. Edwards, *Foundations for Microstrip Circuit Design*, 2nd ed., Chichester, England: Wiley, 1992, secs. 4.10.1 and 8.4.7.

[24] W. Rauch, E. Gornik, A. A. Valenzuela, G. Solkner, F. Fox, H. Behner, G. Gieres, and P. Russer, "Planar Transmission Line Resonators from $YBa_2Cu_3O_{7-x}$ Thin Films and Epitaxial SIS Multilayers," *IEEE Trans. Appl. Superconductivity*, Vol. 3, No. 1, pp. 1110–1113, March 1993.

[25] C. L. Chen, H. H. Feng, Z. Zhang, A. Brazdeikis, Z. J. Huang, W. K. Chu, C. W. Chu, F. A. Miranda, F. W. Van Keuls, R. R. Romanofsky, and Y. Liou, "Epitaxial Ferroelectric $Ba_{0.5}Sr_{0.5}TiO_3$ Thin Films for Room-Temperature Tunable Element Applications," *Appl. Physics Lett.*, *Vol.* 75, No. 3, pp. 412–414, July 1999.

[26] R. F. Harrington, *Time-Harmonic Electromagnetic Fields*, New York: McGraw-Hill, 1961, pp. 23–31.

[27] H.-D. Wu and F. S. Barnes, "Doped $Ba_{0.6}Sr_{0.4}TiO_3$ Thin Films for Microwave Device Applications at Room Temperature," *Int. J. Integrated Ferroelectrics*, Vol. 22, No. 1–4, pp. 291–305, 1998.

[28] H.-C. Li, W. Si, A. D. West, and X. X. Xi, "Near Single Crystal-Level Dielectric Loss and Nonlinearity in Pulsed Laser Deposited $SrTiO_3$ Thin Films," *Appl. Physics Lett.*, Vol. 73, No. 2, pp. 190–192, July 1998.

[29] A. Kozyrev, V. Keis, V. Osadchy, A. Pavlov, O. Buslov, and L. Sengupta, "Microwave Properties of (Ba, Sr)TiO_3 Ceramic Films and Phase-Shifters on Their Base," *Abstracts of 12th Int. Symp. on Integrated Ferroelectrics*, Aachen, Germany, p. 156, March 12–15, 2000.

[30] E. Yablanovitch, T. J. Gmitter, and K. M. Leung, "Photonic Band Structure: The Face-Centered-Cubic Case Employing Nonspherical Atoms," *Phys. Rev. Lett.*, Vol. 67, No. 17, pp. 2295–2298, October 1991.

[31] *IEEE Trans. Microwave Theory Tech.*, Vol. 47, No. 11, November 1999 (Special issue on Electromagnetic Crystal Structures, Design, Synthesis, and Applications).

[32] *IEEE J. Lightwave Tech.*, Vol. 17, No. 11, November 1999 (Special section on Electromagnetic Crystal Structures, Design, Synthesis, and Applications).

[33] F.-R. Yang, K.-P. Ma, Y. Qian, and T. Itoh, "A Uniplanar Compact Photonic-Bandgap (UC-PBG) Structure and Its Applications for Microwave Circuits," *IEEE Trans. Microwave Theory Tech.*, Vol. 47, No. 8, pp. 1509–1514, August 1999.

[34] F.-R. Yang, K.-P. Ma, Y. Qian, and T. Itoh, "A Novel TEM Waveguide Using Uniplanar Compact Photonic-Bandgap (UC-PBG) Structure," *IEEE Trans. Microwave Theory Tech.*, Vol. 47, No. 11, pp. 2092–2098, November 1999.

[35] J. W. Greiser, "Coplanar Stripline Antenna," *Microwave J.*, Vol. 19, No. 10, pp. 47–49, October 1976.

[36] T. Murata and M. Fujita, "A Self-steering Planar Array Antenna for Satellite Broadcast Reception," *IEEE Trans. Broadcasting*, Vol. 40, No. 1, pp. 1–6, March 1994.

[37] W. Menzel and W. Grabherr, "A Microstrip Patch Antenna with Coplanar Feed Line," *IEEE Microwave and Guided Wave Lett.*, Vol. 1, No. 11, pp. 340–342, November 1991.

[38] L. Giauffret and J.-M. Laheurte, "Parameteric Study of the Coupling Aperture in CPW-Fed Microstrip Antennas," *IEE Proc. Microw. Antennas Propag.*, Vol. 146, No. 3, pp. 169–174, June 1999.

[39] R. Q. Lee and R. N. Simons, "Coplanar Waveguide Aperture-Coupled Microstrip Patch Antenna," *IEEE Microwave and Guided Wave Lett.*, Vol. 2, No. 4, pp. 138–139, April 1992.

[40] R. N. Simons and R. Q. Lee, "Coplanar Waveguide Aperture Coupled Patch Antennas with Ground Plane/Substrate of Finite Extent," *Electron. Lett.*, Vol. 28, No. 1, pp. 75–76, January 1992.

[41] G. P. Gauthier, J.-P. Raskin, L. P. B. Katehi, and G. M. Rebeiz, "A 94-GHz Aperture-Coupled Micromachined Microstrip Antenna," *IEEE Trans. Antennas and Prop.*, Vol. 47, No. 12, pp. 1761–1766, December 1999.

INDEX

WILEY SERIES IN MICROWAVE AND OPTICAL ENGINEERING

KAI CHANG, Editor
Texas A&M University

FIBER-OPTIC COMMUNICATION SYSTEMS, Second Edition • *Govind P. Agrawal*

COHERENT OPTICAL COMMUNICATIONS SYSTEMS • *Silvello Betti, Giancarlo De Marchis and Eugenio Iannone*

HIGH-FREQUENCY ELECTROMAGNETIC TECHNIQUES: RECENT ADVANCES AND APPLICATIONS • *Asoke K. Bhattacharyya*

COMPUTATIONAL METHODS FOR ELECTROMAGNETICS AND MICROWAVES • *Richard C. Booton, Jr.*

MICROWAVE RING CIRCUITS AND ANTENNAS • *Kai Chang*

MICROWAVE SOLID-STATE CIRCUITS AND APPLICATIONS • *Kai Chang*

RF AND MICROWAVE WIRELESS SYSTEMS • *Kai Chang*

DIODE LASERS AND PHOTONIC INTEGRATED CIRCUITS • *Larry Coldren and Scott Corzine*

RADIO FREQUENCY CIRCUIT DESIGN • *W. Alan Davis and Krishna Agarwal*

MULTICONDUCTOR TRANSMISSION-LINE STRUCTURES: MODAL ANALYSIS TECHNIQUES • *J. A. Brandão Faria*

PHASED ARRAY-BASED SYSTEMS AND APPLICATIONS • *Nick Fourikis*

FUNDAMENTALS OF MICROWAVE TRANSMISSION LINES • *Jon C. Freeman*

OPTICAL SEMICONDUCTOR DEVICES • *Mitsuo Fukuda*

MICROSTRIP CIRCUITS • *Fred Gardiol*

HIGH-SPEED VLSI INTERCONNECTIONS: MODELING, ANALYSIS, AND SIMULATION • *A. K. Goel*

FUNDAMENTALS OF WAVELETS: THEORY, ALGORITHMS, AND APPLICATIONS • *Jaideva C. Goswami and Andrew K. Chan*

ANALYSIS AND DESIGN OF INTEGRATED CIRCUIT ANTENNA MODULES • *K. C. Gupta and Peter S. Hall*

PHASED ARRAY ANTENNAS • *R. C. Hansen*

HIGH-FREQUENCY ANALOG INTEGRATED CIRCUIT DESIGN • *Ravender Goyal (ed.)*

MICROWAVE APPROACH TO HIGHLY IRREGULAR FIBER OPTICS • *Huang Hung-Chia*

NONLINEAR OPTICAL COMMUNICATION NETWORKS • *Eugenio Iannone, Francesco Matera, Antonio Mecozzi, and Marina Settembre*

FINITE ELEMENT SOFTWARE FOR MICROWAVE ENGINEERING • *Tatsuo Itoh, Giuseppe Pelosi and Peter P. Silvester (eds.)*

INFRARED TECHNOLOGY: APPLICATIONS TO ELECTROOPTICS, PHOTONIC DEVICES, AND SENSORS • *A. R. Jha*

SUPERCONDUCTOR TECHNOLOGY: APPLICATIONS TO MICROWAVE, ELECTRO-OPTICS, ELECTRICAL MACHINES, AND PROPULSION SYSTEMS • *A. R. Jha*

OPTICAL COMPUTING: AN INTRODUCTION • *M. A. Karim and A. S. S. Awwal*

INTRODUCTION TO ELECTROMAGNETIC AND MICROWAVE ENGINEERING • *Paul R. Karmel, Gabriel D. Colef, and Raymond L. Camisa*

MILLIMETER WAVE OPTICAL DIELECTRIC INTEGRATED GUIDES AND CIRCUITS • *Shiban K. Koul*

MICROWAVE DEVICES, CIRCUITS AND THEIR INTERACTION • *Charles A. Lee and G. Conrad Dalman*

ADVANCES IN MICROSTRIP AND PRINTED ANTENNAS • *Kai-Fong Lee and Wei Chen (eds.)*

OPTICAL FILTER DESIGN AND ANALYSIS: A SIGNAL PROCESSING APPROACH • *Christi K. Madsen and Jian H. Zhao*

THEORY AND PRACTICE OF INFRARED TECHNOLOGY FOR NONDESTRUCTIVE TESTING • *Xavier P. V. Maldague*

OPTOELECTRONIC PACKAGING • *A. R. Mickelson, N. R. Basavanhally, and Y. C. Lee (eds.)*

OPTICAL CHARACTER RECOGNITION • *Shunji Mori, Hirobumi Nishida, and Hiromitsu Yamada*

ANTENNAS FOR RADAR AND COMMUNICATIONS: A POLARIMETRIC APPROACH • *Harold Mott*

INTEGRATED ACTIVE ANTENNAS AND SPATIAL POWER COMBINING • *Julio A. Navarro and Kai Chang*

ANALYSIS METHODS FOR RF, MICROWAVE, AND MILLIMETER-WAVE PLANAR TRANSMISSION LINE STRUCTURES • *Cam Nguyen*

FREQUENCY CONTROL OF SEMICONDUCTOR LASERS • *Motoichi Ohtsu (ed.)*

SOLAR CELLS AND THEIR APPLICATIONS • *Larry D. Partain (ed.)*

ANALYSIS OF MULTICONDUCTOR TRANSMISSION LINES • *Clayton R. Paul*

INTRODUCTION TO ELECTROMAGNETIC COMPATIBILITY • *Clayton R. Paul*

ELECTROMAGNETIC OPTIMIZATION BY GENETIC ALGORITHMS • *Yahya Rahmat-Samii and Eric Michielssen (eds.)*

INTRODUCTION TO HIGH-SPEED ELECTRONICS AND OPTOELECTRONICS • *Leonard M. Riaziat*

NEW FRONTIERS IN MEDICAL DEVICE TECHNOLOGY • *Arye Rosen and Harel Rosen (eds.)*

ELECTROMAGNETIC PROPAGATION IN MULTI-MODE RANDOM MEDIA • *Harrison E. Rowe*

ELECTROMAGNETIC PROPAGATION IN ONE-DIMENSIONAL RANDOM MEDIA • *Harrison E. Rowe*

NONLINEAR OPTICS • *E. G. Sauter*

COPLANAR WAVEGUIDE CIRCUITS, COMPONENTS, AND SYSTEMS • *Rainee N. Simons*

ELECTROMAGNETIC FIELDS IN UNCONVENTIONAL MATERIALS AND STRUCTURES • *Onkar N. Singh and Akhlesh Lakhtakia (eds.)*

FUNDAMENTALS OF GLOBAL POSITIONING SYSTEM RECEIVERS: A SOFTWARE APPROACH • *James Bao-yen Tsui*

InP-BASED MATERIALS AND DEVICES: PHYSICS AND TECHNOLOGY • *Osamu Wada and Hideki Hasegawa (eds.)*

DESIGN OF NONPLANAR MICROSTRIP ANTENNAS AND TRANSMISSION LINES • *Kin-Lu Wong*

FREQUENCY SELECTIVE SURFACE AND GRID ARRAY • *T. K. Wu (ed.)*

ACTIVE AND QUASI-OPTICAL ARRAYS FOR SOLID-STATE POWER COMBINING • *Robert A. York and Zoya B. Popović (eds.)*

OPTICAL SIGNAL PROCESSING, COMPUTING AND NEURAL NETWORKS • *Francis T. S. Yu and Suganda Jutamulia*

SiGe, GaAs, AND InP HETEROJUNCTION BIPOLAR TRANSISTORS • *Jiann Yuan*

ELECTRODYNAMICS OF SOLIDS AND MICROWAVE SUPERCONDUCTIVITY • *Shu-Ang Zhou*

Printed and bound by CPI Group (UK) Ltd, Croydon, CR0 4YY
16/09/2026
14963441-0001